Richard Talman
Accelerator X-Ray Sources

Related Titles

Mills, D. M. (ed.)
Third-Generation Hard X-Ray Synchrotron Radiation Sources
Source Properties, Optics, and Experimental Techniques
416 pages
2002
Hardcover
ISBN 0-471-31433-1

Als-Nielsen, J., McMorrow, D.
Elements of Modern X-Ray Physics
336 pages
2001
Softcover
ISBN 0-471-49858-0

Edwards, D. A., Syphers, M. J.
An Introduction to the Physics of High Energy Accelerators
304 pages with 103 figures
1993
Hardcover
ISBN 0-471-55163-5

Richard Talman

Accelerator X-Ray Sources

WILEY-VCH Verlag GmbH & Co. KGaA

The Author

Prof. Richard Talman
Cornell University
Laboratory of Elementary Physics
Ithaca, NY, USA
talman@mail.lns.cornell.edu

Cover illustration created by the author.

All books published by **Wiley-VCH** are carefully produced. Nevertheless, authors, editors, and publisher do not warrant the information contained in these books, including this book, to be free of errors. Readers are advised to keep in mind that statements, data, illustrations, procedural details or other items may inadvertently be inaccurate.

Library of Congress Card No.: applied for.

British Library Cataloging-in-Publication Data:
A catalogue record for this book is available from the British Library.

Bibliographic information published by Die Deutsche Bibliothek
Die Deutsche Bibliothek lists this publication in the Deutsche Nationalbibliografie; detailed bibliographic data is available in the Internet at <http://dnb.ddb.de>.

© 2006 WILEY-VCH Verlag GmbH & Co. KGaA, Weinheim

All rights reserved (including those of translation into other languages). No part of this book may be reproduced in any form – nor transmitted or translated into machine language without written permission from the publishers. Registered names, trademarks, etc. used in this book, even when not specifically marked as such, are not to be considered unprotected by law.

Printed in the Federal Republic of Germany

Printed on acid-free paper

Typesetting   Uwe Krieg, Berlin
Printing   betz-Druck GmbH, Darmstadt
Binding   J. Schäffer GmbH, Grünstadt
Cover Design   aktivComm, Weinheim

ISBN-13: 978-3-527-40590-9
ISBN-10: 3-527-40590-9

Contents

Preface *XIII*

1   **Beams of Electrons or Photons** *1*
1.1   Preview *1*
1.2   Coordinate Definitions *2*
1.3   One-dimensional Transverse Propagation Equations *4*
1.4   Transfer Matrices for Simple Elements *7*
1.4.1   Drift Space *7*
1.4.2   Thin Lens *7*
1.4.3   Thick Lens *9*
1.4.4   Erect Quadrupole Lens *11*
1.5   Elliptical (in Phase Space) Beams *13*
1.6   Beam Envelope $E(s)$ *15*
1.7   Gaussian Beams: their Variances and Covariances *17*
1.8   Pseudoharmonic Trajectory Description *18*
1.9   Transfer Matrix Parametrization *21*
1.10   Reconciliation of Beam and Lattice Parameters *23*
1.10.1   Beam Evolution Through a Drift Section *23*
1.10.2   Beam Evolution Through a Thin Lens *24*
    References *25*

2   **Beams Treated as Waves** *27*
2.1   Preview *27*
2.2   Scalar Wave Equation *28*
2.3   The Short Wavelength, Geometric Optics Limit *30*
2.3.1   Determination of Rays from Wavefronts *31*
2.3.2   The Ray Equation in Geometric Optics *32*
2.3.3   Obtaining Phase Information from Intensity Measurement *34*
2.4   Wave Description of Gaussian Beams *36*
2.4.1   Gaussian Beam in a Focusing Medium *37*

Accelerator X-Ray Sources. Richard Talman
Copyright © 2006 WILEY-VCH Verlag GmbH & Co. KGaA, Weinheim
ISBN: 3-527-40590-9

2.4.2	Spatial Dependence of a Wave Near a Free Space Focus 39
2.4.3	The ABCD Law 40
2.4.4	Optics Using Mirrors 42
2.4.5	Wave Particle Duality for Electrons 44
2.5	Synchrotron Radiation: Waves or Particles? 45
2.6	X-ray Holography and Phase Contrast and Lens-free Imaging 46
	References 48

3	**Synchrotron Radiation From Accelerator Magnets** 49
3.1	Capsule History of Synchrotron Light Sources 49
3.2	Generalities 51
3.3	Potentials and Fields 53
3.4	Relations Between Observation Time and Retarded Time 54
3.5	Evaluation of Electric and Magnetic Fields 58
3.5.1	Radial Field Approximation 60
3.6	Total Power Radiated and its Angular Distribution 62
3.7	Spectral Power Density of the Radiation 65
3.7.1	Estimate of Frequency Spectrum from Pulse Duration 65
3.7.2	Radial Approximation 67
3.7.3	Accurate Formula for Spectral Power Density 68
3.8	Radiation from Multiple Charges 70
3.9	The Terminology of "Intensity" Measures 71
3.10	Photon Beam Features "Inherited from" the Electron Beam 74
3.11	Intensity Estimates for Bending Magnet beams 75
	References 81

4	**Simple Storage Rings** 83
4.1	Preview 83
4.2	The Uniform Field Ring 84
4.3	Horizontal Stability 85
4.4	Vertical Stability 87
4.5	Simultaneous Horizontal and Vertical Stability 88
4.6	Dispersion 89
4.7	Momentum Compaction 90
4.8	Chromaticity 91
4.9	Strong Focusing 92

5	**The Influence of Synchrotron Radiation on a Storage Ring** 95
5.1	Preview 95
5.2	Statistical Properties of Synchrotron Radiation 96
5.2.1	Total Energy Radiated 96
5.2.2	The Distribution of Photon Energies; "Regularized Treatment" 99

5.2.3	Randomness of the Radiation *103*
5.3	The Damping Rate Sum Rule: Robinson's Theorem *105*
5.3.1	Vertical Damping *110*
5.3.2	Longitudinal Damping *111*
5.3.3	Horizontal Damping and Partition Numbers *119*
5.4	Equilibrium between Damping and Fluctuation. *120*
5.5	Horizontal Equilibrium and Beam Width *121*
5.6	Longitudinal Bunch Distributions *128*
5.6.1	Energy Spread *128*
5.6.2	Bunch Length *131*
5.7	"Thermodynamics" of Wiggler-dominated Storage Rings *132*
5.7.1	Emittance of Pure Wiggler Lattice *132*
5.7.2	Thermodynamic Analogy *137*
	References *140*
6	**Elementary Theory Of Linacs** *141*
6.1	Acknowledgement and Preview *141*
6.2	The Nonrelativistic Linac *142*
6.2.1	Transit Time Factor *142*
6.2.2	Shunt Impedance *143*
6.2.3	Cavity Q, R/Q, and Decay Time *144*
6.2.4	Phase Stability and Adiabatic Damping *145*
6.2.5	Transverse Defocusing *151*
6.3	The Relativistic Electron Linac *153*
6.3.1	Introduction *153*
6.3.2	Particle Acceleration by a Wave *154*
6.3.3	Wave Confined by Parallel Planes *155*
6.3.4	Circular Waveguide *161*
6.3.5	Cylindrical "Pill-box" Resonator *163*
6.3.6	Lumped Constant Model for One Cavity Resonance *164*
6.3.7	Cavity Excitation *166*
6.3.8	Wave Propagation in Coupled Resonator Chain *168*
6.3.9	Periodically Loaded Structures *172*
6.3.10	Space Harmonics *174*
7	**Undulator Radiation** *179*
7.1	Preview *179*
7.2	Introduction *180*
7.3	Electron Orbit in a Wiggler or Undulator *184*
7.4	Energy Radiated From one Wiggler Pole *187*
7.5	Spectral Analysis for Arbitrary Longitudinal Field Profile *188*
7.6	Spectrum of the Radiation from a Single Pole *191*

7.6.1	Orbit Treated as Arc of Circle *191*	
7.6.2	Radiation from a Single, Short, Isolated, Magnet, $K \ll 1$ *192*	
7.7	Coherence from Multiple Deflections *196*	
7.8	Phasor Summation for $K \ll 1$ *199*	
7.9	Photon Energy Distributions *203*	
7.9.1	Energy Distribution from the $n = 1$ Undulator Fundamental *203*	
7.10	Undulator Radiation for Arbitrary K Value *204*	
7.10.1	Analytic Formulation *204*	
7.10.2	Diffraction Grating Analogy *212*	
7.10.3	Numerical/Graphical Representation of Undulator Radiation *213*	
7.10.4	Approximation of the Integrals by Special Functions *220*	
7.10.5	Practical Evaluation of the Series *222*	
7.11	Post-monochromator Profile *223*	
7.11.1	Monochromatic Annular Rings *225*	
7.11.2	Numerical Investigation of Undulator Rings *227*	
7.11.3	Is the Forward Undulator Peak Subject to Angular Narrowing? *227*	
	References *230*	

8   **Undulator Magnets**   *231*
8.1   Preview   *231*
8.2   Considerations Governing Undulator Parameters   *232*
8.3   Simplified Radiation Formulas   *234*
8.4   A Hybrid, Electo-permanent, Asymmetric Undulator   *237*
8.4.1   Electromagnet Design   *239*
8.4.2   Permanent Magnet Design—Small Gap Limit   *244*
8.4.3   Combined Electro-/Permanent- magnet Design   *246*
8.4.4   Estimated X-ray Flux   *247*
   References   *248*

9   **X-Ray Beam Line Design**   *249*
9.1   Preview   *249*
9.2   Beam Line Generalities   *250*
9.3   Accelerator Parameters   *252*
9.4   Bragg Scattering and Darwin Width   *254*
9.5   Aperture-defined Beam Line Design   *257*
9.5.1   Undulator Radiation, $n = 1$, Negligible Electron Divergence   *257*
9.5.2   Effect of Electron Beam Emittances on Flux and Brilliance   *261*
9.5.3   Brilliance with $K > 1$ and $n > 1$ *263*
9.6   X-ray Mirrors   *266*
9.6.1   Specular Reflection of X-rays   *266*
9.6.2   Elliptical Mirrors   *266*
9.6.3   Hyperbolic Mirrors   *267*

9.7	X-ray Lenses *270*	
9.7.1	Monochromatic X-ray Lens *270*	
9.7.2	Focusing a Monochromatic Undulator Radiation Ring *270*	
9.7.3	Undulator-specific X-ray Lens *272*	
9.7.4	Lens Quality *278*	
9.8	Beam cameras *279*	
9.8.1	The Pin-hole Camera *279*	
9.8.2	Imaging the Beam with Visible Light *280*	
9.8.3	Practicality of Lens-based, X-ray Beam Camera? *283*	
9.9	Aperture-free X-ray Beam Line Design *286*	
9.9.1	Aperture-free Rationale *286*	
9.9.2	Aperture-free Microbeam Line Based on Lenses *287*	
9.9.3	Effective Lens Stop Caused by Absorption *291*	
9.9.4	Choice of Undulator Parameter K *292*	
9.9.5	Estimated Flux *293*	
9.9.6	Estimated Brilliance and Qualifying Comments *295*	
	References *296*	
10	**The Energy Recovery Linac X-Ray Source** *299*	
10.1	Preview *299*	
10.2	Introduction *300*	
10.3	Emittance Evolution in a DC Electron Gun *301*	
10.4	Qualitative Description of Lattice Design Issues *306*	
10.5	Isochronous Arc Design *309*	
10.6	Evolution of Betatron Amplitudes through the Linac Sections *314*	
10.6.1	Deceleration through Linac Section to Dump *314*	
10.6.2	Triplet Design *318*	
10.6.3	Acceleration through the High Energy Linac Section *319*	
10.7	Emittance Growth Due to CSR and Space Charge *321*	
	References *325*	
11	**A Fourth Generation, Fast Cycling, Conventional Light Source** *327*	
11.1	Preview *327*	
11.2	Low Emittance Lattices *328*	
11.3	Production of High Quality X-ray Beams *333*	
11.3.1	A Formula for Brilliance $\mathcal{B}$ *333*	
11.3.2	A Strategy to Maximize $\mathcal{B}$ *334*	
11.3.3	Hypothetical Utilization of an Existing Large Ring *335*	
11.4	Acceleration Scenario *338*	
11.5	Power Considerations *342*	
11.5.1	Average Power *342*	
11.5.2	Instantaneous Power *343*	

11.6	Critical Components and Parameter Dependencies	*344*
11.7	Trbojevic-Courant Minimum Emittance Cells	*346*
11.7.1	Basic Formulas	*346*
11.7.2	Thin Lens Treatment	*348*
11.7.3	Thick Lens Treatment	*353*
11.7.4	Zero Dispersion Straight Sections	*355*
11.7.5	Nonlinearity and Dynamic Aperture	*357*
11.8	Emittance Evolution During Acceleration	*359*
11.9	Touschek Lifetime Estimate	*362*
11.10	Performance as X-ray Source	*363*
11.10.1	Brilliance from Short Undulator	*363*
11.10.2	Refinement of Brilliance Calculation	*365*
	References	*368*
12	**Compton Scattered Beams And "Laser Wire" Diagnostics**	*369*
12.1	Preview	*369*
12.2	Compton Scattering Kinematics	*370*
12.3	Some Specialized Laser, Electron Beam Configurations	*373*
12.3.1	Back-scattered Photons	*373*
12.3.2	Orthogonal Photon Incidence in the Laboratory	*374*
12.3.3	Orthogonal Electron Frame Incidence	*375*
12.4	Total Compton Cross Section	*380*
12.5	The Photon Beam Treated as an Electromagnetic Wave	*381*
12.5.1	Determination of the Electron's Velocity Modulation	*382*
12.5.2	Undulator Parametrization of Electron Motion in a Wave	*386*
12.6	Undulator Fields in Electron Rest Frame	*388*
12.6.1	Some Formulas from Special Relativity [7]	*389*
12.6.2	Treatment of an Undulator Magnet as an Electromagnetic Wave	*389*
12.7	Classical Derivation of Thomson Scattering	*391*
12.7.1	Introduction	*391*
12.7.2	Free Electron Oscillating in Electromagnetic Wave [5]	*392*
12.7.3	Electric Dipole Radiation [6]	*392*
12.7.4	Scattering Rate Expressed as a Total Cross Section	*394*
12.8	Transformation of Photon Distributions to the Laboratory	*395*
12.8.1	Solid Angle Transformation	*395*
12.8.2	Photon Energy Distribution in the Laboratory	*396*
12.8.3	A Theorem Applicable to Isotropic Distributions	*398*
12.8.4	Energy Distribution of Undulator Radiation	*399*
12.9	Rate Estimates for a Laser Wire Diagnostic Apparatus	*401*
12.9.1	The Nonrelativistic Limit	*403*
12.9.2	Laser Wire Treated as Undulator	*403*

12.9.3	Laser Wire Treated via Electron Rest Frame	*405*
12.9.4	Invariant Cross Section Applied to Laser Wire	*406*
12.9.5	Bunched Beam Rates	*408*
	References	*411*

13 Space Charge Effects and Coherent Radiation *413*
13.1 Acknowledgement and Preview *413*
13.2 Introduction to the String Space Charge Formalism *414*
13.3 Self-force of Moving Straight Charged String *418*
13.4 Self-force of Moving Straight Charged Ribbon *423*
13.5 Curve End Point Determination *430*
13.6 Field Calculation *434*
13.7 "Regularization" of the Longitudinal Force *438*
13.8 Coherent Synchrotron Radiation *440*
13.9 Evaluation of Integrals *442*
13.10 Calculational Practicalities *443*
13.11 Suppression of CSR by Wall Shielding *444*
13.12 Effects of Entering and Leaving Magnets *445*
13.13 Space Charge Calculations Using Unified Accelerator Libraries *447*
13.13.1 Numerical Procedures Used by UAL *448*
13.13.2 Program Architecture *450*
13.13.3 Numerical Procedures *452*
13.13.4 Comparison with TRAFIC4 [24] *453*
References *456*

14 The X-ray FEL *457*
14.1 Absorption and Spontaneous and Stimulated Emission *457*
14.2 Closed and Open FELs *458*
14.3 Interpretation of Undulator Radiation as Compton Scattering *459*
14.4 Applicability Condition for Semi-Classical Treatment *461*
14.5 Comparison of Storage Ring, ERL, and FEL *462*
References *465*

Index *467*

Preface

Since roughly the turn of the present century, *high coherency* has been the attribute driving the design of new x-ray sources. This property enables the extension into the x-ray regime of the interferometric methods that have been so powerful with visible light, especially as provided by visible light lasers. Already, with present day light sources, coherency is augmenting the other features that make x-rays such powerful probes of microscopic systems—namely high spatial resolution and high penetrating power.

Highly coherent x-ray sources are also referred to as *bright*, or as *brilliant*, or as *fourth generation light sources*. There are three general categories: the storage ring (SR), the energy recovery linac (ERL), and the free electron laser (FEL). Because of the close connection between coherency and stimulated emission the FEL has, by definition, high brilliance and high coherency, but storage rings and energy recovery linacs can also produce the near complete coherence needed to produce interference-sensitive phenomena.

One hates to be forced, especially in only the third paragraph, to introduce *ad hoc* terminology; but the delineation of the subject matter of this book requires a distinction to be drawn among FEL-type accelerators. Some FELs are covered and some not. In atomic lasers neither the electrons nor the photons are "free"; the electrons are bound in molecules and the photons are captured (briefly) in resonators. In the free electron laser it is only the electron that is "free". This still leaves two possibilites for the photons. The FEL can be called "closed" if the photons are restrained by a Fabry-Perot-like resonator, or "open" if the photons are free. In spite of the absence of a photon resonator there can be substantial "superradiant" photon emission in an open FEL.

In a closed FEL the electrons and the photons come into a state of equilibrium in which both distributions depend on their repetitive mutual interaction. This configuration has great physical interest and the analyses are deep and sophisticated, so there is a vast literature on the subject of closed FELs [1]. Closed FELs are uniquely appropriate tuneable souces for photons in the infrared and ultraviolet ranges. Any serious discussion of the closed FEL would add greatly to both the length and the level of difficulty of this text. Fortu-

Accelerator X-Ray Sources. Richard Talman
Copyright © 2006 WILEY-VCH Verlag GmbH & Co. KGaA, Weinheim
ISBN: 3-527-40590-9

nately none of this difficult material is needed to understand the simpler *open* FELs that are capable of producing the hard x-rays that are the subject of this book.

In an open FEL each electron bunch interacts only once with the beam of radiated photons. Further simplification of the interaction results when the electron beam energy is sufficiently high that treatment of the influence of the radiation on the particles is particularly simple.

This book is concerned only with x-rays of wavelength so short that there is negligible near-normal reflection from any realizable mirror. There can therefore be no such thing as a Fabry-Perot-like resonator and no "closed" FEL for sub-Ångstom x-rays. For x-rays this distinction between open and closed is so great as to make the use of the unqualified term "FEL" misleading. It is only the "open" FEL, which could also be referred to as "mirror-free", that is relevant to this book, and there will be little further discussion of closed FELs.

With the closed FEL having been dispensed with, most of the discussion of the open FEL becomes relatively simple and the introductory material that has to be understood before analysing them seriously is largely included in the analyses of the other fourth generation candidates. There are, however, features that distinguish the open FEL from both the energy recovery linac and the storage ring. Both the ERL and the SR can be referred to as "rings", since their electron trajectories consist of one or more closed loops, and the two types are therefore subject to very similar analyses. In neither case does stimulated emission contribute importantly to the produced beams. The open FEL differs in both of these respects. Its electron trajectories are more or less straight and stimulated emission is essential to its operation.

The approach taken is to analyse accelerator x-ray sources theoretically and idealistically. The topics emphasized are largely those for which reasonably straightforward theory is satisfactory. The three main categories are discussed individually. After analysing the ERL, a "fast cycling" (but otherwise conventional) storage ring (FCSR) is analysed. Rings that are variants of these two types are in various states of design and construction at this time. Finally the FEL is studied but, because of its immature state, the discussion is less detailed.

A series of a few chapters is devoted to each of these types, with one containing a quite detailed design. To facilitate comparisons between the FCSR and ERL approaches the numerical examples assume more or less equivalent parameters for the two cases. These parameter values are the nominal values for a light source planned to replace the electron-positron collider, CESR, at Cornell University, when that facility is decommissioned in the near future. However, all concepts are intended to be applicable to all such sources.

The ERL strategy is to start with, and preserve, high phase space particle density. The FCSR approach is to start with arbitrary phase space density and to increase the density by damping. The material in these chapters can be used to assess the relative merits of these two approaches—much of the material came into existence for just this purpose. Of course no definitive determination as to which design is superior can be reached, especially because that determination would depend on costs, which are not discussed here. Several laboratories are making these comparisons, but the jury seems still to be out as to which approach is the more cost effective.

The reader expecting to find substantial analysis of general FELs will be disappointed. For the reasons already explained, closed FELs are hardly covered at all. Issues specific to the open FEL are discussed, but only briefly, in the final chapter, along with comparisons with the other fourth generation types. The beams from open FELs are expected to enable single shot, illumination and reconstruction of small molecular structures, even in cases in which radiation damage precludes multiple exposure. This is in almost perfect contrast with the non-destructive, but weak, multiple exposures of the same sample with the highly reproducible beams produced by storage rings. The FEL and the conventional light source are expected, therefore, to remain complementary, and not really in competition, for the forseeable future.

While storage rings have been in more or less continuous evolution for half a century, FEL x-ray sources are too recent for any vintage even to have matured. Ambitious x-ray sources based on closed FELs have been funded and are progressing well. The issues that are most appropriate for treatment in this text are "self amplified spontaneous emission" (SASE) and the "seeding" that will be needed to produce x-ray beams of the high degree of reproducibility that x-ray scientists have come to demand. Both of these aspects are under active development but neither is yet of sufficient maturity for coverage at the level of this text. The chapter on FELs is therefore brief. Should this text ever have a second edition the FEL chapter will need to be much expanded if these devices come close to achieving the advance claims made for them.

The content of this book is largely derived from various courses on accelerator physics the author has given at Cornell University, at the U.S. Particle School (USPAS) and at the Canadian Light Source (CLS) at the University of Saskatchewan. Some of these courses have emphasized accelerator physics *per se*, others have emphasized the accelerator as a source of x-rays. In one case, Chapter 6 on linacs, the material originated with lectures delivered by Greg Loew at the 1982 USPAS school, which I later wrote up and augmented for publication in the USPAS book series. One course at Cornell (emphasizing x-rays especially) was shared with Sol Gruner, and one USPAS course (emphasizing computer simulation) was shared with Nikolay Malitsky.

The material in the book is therefore appropriate for courses on these subjects. Most of the courses had regular problem assignments, and some of the problems have been included in this text.

The level of difficulty of the early chapters in the book is intended to be moderate and the difficulty increases more or less monotonically from the beginning to the end of the book. The level of specificity similarly advances monotonically from the general to the specific. Each of the first six chapters is devoted to a particular component or aspect of an accelerator. The following three chapters are devoted to the insertion devices that are the sources for x-ray beams and to the design of their beam lines. The remaining chapters are considerably more specific and more advanced.

The emphasis in Chapter 12 is on beam diagnostics using Compton scattering. This material might superficially appear to be somewhat out of character with the rest of the book, until it is looked at closely enough to see the intimate connection between undulator radiation and Compton scattering. In fact much of the analysis of undulator radiation has been deferred to this chapter. At the same time, the use of a "laser wire" (another name for diagnosing a beam by shining a laser at it) has become important for measuring the ultra-small emittances of electron beams to be used for FELs or ERLs.

Chapter 13 contains a detailed analysis of coherent synchrotron radiation (CSR) and other space charge effects. This material is primarily germane to the treatment of ERLs in Chapter 10, and to microbunching in Chapter 14. Because of their long bunch lengths and their inherent damping, storage rings are largely immune from these space charge effects. They *are* however, subject to intrabeam scattering via the so-called Touschek effect. For existing light sources an entire chapter on that subject would be justified, but the whole point of the "fast cycling" in FCSR, is to ameliorate this effect to the point that it can be ignored.

Recapitulating: Chapters 1, 2, and 3 are applicable to all sources; Chapters 7, 8, and 9 apply to all x-ray beams; Chapters 4, 5, and 11 explain storage rings; Chapters 6, 10, 12, and 13 explain ERLs; and Chapters 6, 12, 13, and 14, plus Section 10.3 provide introductory material on open FELs, but without the sort of analysis needed for their detailed analysis.[1] Within each of these groupings it probably makes sense to traverse the chapters in order, but, after having completed the first grouping (Chapters 1, 2 (possibly just skimmed), and 3) the other groupings can be attacked in any order.

The figure on the cover is explained in Chapter 7. It is replicated in Figure 7.22.

The people who have helped me understand this material are too numerous to be individually acknowledged. This includes especially the many graduate students I have worked with. I will mention only people without whom large sections could not have been written. Three of them have already been men-

---

1) For a student specially interested in FELs this book can be regarded as an introduction to a detailed description of open FELs such as is contained in the TESLA design report [2].

tioned. The entire scholarly tradition in U.S. accelerator physics, modest as it is, is due more to Mel Month and his U.S. Particle Accelerator School than to anyone else. Much of this book has evolved from lectures at that school. Mike Bancroft at the Canadian Light Source first suggested developing the notes into a complete book. Lewis Kotredes made important contributions to the chapter on undulators; Ivan Bazarov to the chapter on FCSR; Ken Finkelstein to Compton diagnostics; and Mark Palmer to the discussion of damping rings. Computational results derive primarily from Karl Brown, TRANSPORT; Chris Iselin and Eberhard Keil, MAD; Lindsay Schachinger, TEAPOT; or Nikolay Malitsky, UAL. Finally, my wife Myrna has always been supportive.

References

1 Bakker, R. (1998) *The Storage Ring Free Electron Laser*, in *Synchrotron Radiation and Free Electron Lasers*, 1996 CERN Accelerator School, S. Turner, Ed., CERN 98-04, p. 337.

2 *TESLA Design Report, Part VI Appendices*, (2001), DESY 2001-011, March.

1
Beams of Electrons or Photons

1.1
Preview

This chapter deals with beams of particles, be they photons or electrons, where electrons can be taken to include also positrons. In spite of their charge difference, the kinematics of photons and electrons are essentially the same since, in the highly relativistic regime, the electron rest mass can be neglected and even bending magnets are *almost* inert as regards "optical" effects.

Optical lines are constructed from lenses, mirrors and free space. Charged particles lines are constructed from analogous magnets and field-free regions. For this reason the most fundamental features (Liouville's theorem and its generalizations, linear/nonlinear distinction, intensities, emittances, adiabatic invariance, and so on) apply equally to both sorts of beam. Unfortunately, this equivalence tends to be masked by the different terminologies that have evolved during the independent historical development of accelerator physics and optical physics. The merging of these two fields has come only comparatively recently with the development of storage rings as sources of photon beams. Though the notation in this chapter is drawn mainly from the accelerator side, in this chapter, *all formulas apply equally to photons and electrons*.[1]

This chapter analyses the paraxial (i.e. nearly parallel) propagation of particles, be they electrons or photons.

---

1) Since it is hard for an accelerator physicist to avoid using the word "lattice" to describe a beam line of lenses, an optical physicist will have to tolerate this usage. Another confusing practice drawn from accelerator tradition is to use the same symbol for seemingly unrelated quantities, knowing they will later be shown to be equal (in some sense, or under some circumstances). This comment applies especially to the symbols β and ϵ.

1.2
Coordinate Definitions

By definition, a beam is a number (usually billions and billions) of particles all traveling more or less parallel. For purposes of description it is useful to start by picking some "most-central" particle as a "reference particle" and describing its "ideal" or "reference" ray or trajectory through the system. Typically this ray passes through the centers of all the optical elements making up the beam line. Any particular point on the reference trajectory can be located by global Cartesian coordinates (X, Y, Z) relative to some fixed origin, but, to take advantage of the essentially one dimensional placement of elements it is profitable to locate elements by arc length (called s) from the origin along the central ray. Once these coordinates are known for the central particle, all the other particles in the beam can be located by relative coordinates. For these relative coordinates it is convenient to use a reference frame aligned with the central ray. "Tangential" or "longitudinal" displacements are specified by incremental arc length $z = \Delta s$ and (x, y) serve as "transverse" coordinates, with x usually being horizontal and y vertical. The longitudinal coordinates s and z increment trivially in drift[2] regions—by simple addition. For relativistic particles, s is proportional to t ($s = vt$, $v \approx c$) and s can take over from t the role of "independent variable".[3]

There is nothing "small" about the reference trajectory—in a storage ring it bends through 2π, in a light beam it may reflect through comparably large angles. On the other hand there is much that is "small" about the motion of particles relative to the reference particle and, of course, that is why local coordinates are introduced. Loosely speaking then, the coordinates (X, Y, Z, s) are "large" and the coordinates (x, y, z) are "small". The fundamental beam properties mentioned in the first paragraph apply primarily to these small displacements.

Much the same comments apply to velocity and momentum. The momentum components of the central ray are $(0, 0, P_0)$ and, in the absence of accelerating elements, P_0 is constant. The momentum components of a beam particle are customarily expressed fractionally as the ratios[4]

$$(p_x, p_y, \delta) = \left(\frac{P_x}{P_0}, \frac{P_y}{P_0}, \frac{P - P_0}{P_0} \right), \tag{1.1}$$

2) The term "drift" is accelerator jargon for field-free, empty space.
3) The practice of using a longitudinal coordinate as an independent variable perhaps dates from the early days of optics, before photons were even conceived of, when there was no reason to contemplate the rate of advance of anything along a ray, so the purely geometric character of the ray was emphasized.
4) It is more correct, when the motion can be less than fully relativistic, to define the longitudinal component in terms of energies, as $\delta = (\mathcal{E} - \mathcal{E}_0)/(cP_0)$, but, in this text, fully relativistic motion is almost always assumed.

where $\delta = \Delta P/P_0 = (P - P_0)/P_0$ is therefore the fractional momentum deviation from the central momentum. These definitions are traditional for particle description; for photon description the momentum is more traditionally expressed as "wave vector" $\mathbf{k} = \mathbf{P}/\hbar$ and the definitions have to be modified accordingly.

The name "Gauss" is often attached to beam transport transport systems and, with quite different meaning, to beams . In a Gaussian beam the above-mentioned deviations, $(x, y, z, p_x, p_y, \delta)$, are distributed according to Gaussian probability distributions. Typically it is the spreads of these variables that sufficiently characterize the distributions and the Gaussian serves primarily as a convenient "bell-shaped" probability distribution that drops off rapidly for large deviations. The standard deviation of the Gaussian is a good objective measure of beam spread that can serve as a semi-quantitative size measure in all cases and as an accurate quantitative measure in those cases where the beam distribution is truly Gaussian. We will return to Gaussian beams shortly.

Far more fundamental is "Gaussian" (also known as "paraxial") optics, which describes transport systems in which the trajectory deviations, both displacement and slope, away from the reference trajectory are small enough to be treated by "linearized" equations of motion. That is, the differential equations governing the evolution of $(x(s), y(s), \ldots)$ are linear in $(x, y, \ldots)$. Since these linearized equations are usefully expressed using matrices, such optical systems are said to satisfy "matrix optics". The fundamental characteristics listed in the first paragraph are most easily described in terms of these matrices, and the "linear" distinction listed there is synonymous with "paraxial".

The propagation of a beam of particles through a sequence of beam line elements can be viewed from three different perspectives. These *particle, beam*, and *beam line* views will be taken up, in order, in the following sections and, after that, be reconciled. A rich source of confusion is that the same symbols (especially ϵ and β) will be used in all views, so the reader will have to trust that seemingly unrelated usages will eventually be (more or less) reconciled; this is to say that the logic of assigning the same symbol to different quantities will eventually be made clear.

Also to be reconciled (in Chapter 2) are the parametrizations of light optics and particle optics. One formalism of light optics concentrates on the evolution of the parameters describing optical wave fronts. This formalism uses complex parameters and the so-called A, B, C, D parameters. Fortunately this reduces to being a rehashing of the same propagation formulas.

1.3
One-dimensional Transverse Propagation Equations

The most essential feature distinguishing material particles like electrons from photons is that electrons can be bent in magnetic fields. For the subject matter under discussion this distinction turns out to be "inessential", for much the same reason that plane mirrors, in spite of grossly redirecting light beams, have no essential optical effect. For unity of description we therefore discuss now only a beam traveling along a straight channel centered on the s-axis, planning later to incorporate the (small) effects due to magnetic bending.

To prevent the eventual departure of even slightly divergent rays, it is necessary to have focusing elements such as quadrupoles for charged particles, fiber optics or optical lenses for visible photons, curved mirrors for x-rays, and so on. The differential equation describing such focusing is[5]

$$\frac{d^2 y}{ds^2} = K(s) y. \tag{1.2}$$

This equation will be referred to as the "focusing equation" and $K(s)$ as the "vertical focusing strength". Just about everything of consequence in this chapter, and much of the next one, follows directly from this equation. The sign of K is like that of a Hooke's law force, negative for "restoring". The dependence of $K(s)$ on s permits the description of systems in which the focusing strength varies along the orbit. In particular, $K(s) = 0$ describes "drift spaces" in which case Eq. (1.2) is trivially solved, and yields the obvious result that particles in free space travel in straight lines.

It is conventional to designate dy/ds by y'. Then, from any two solutions y_1 and y_2 of Eq. (1.2) one can form the "Wronskian" $y_1 y_2' - y_1' y_2$ which has the virtue of being constant, since

$$\frac{d}{ds}(y_1 y_2' - y_1' y_2) = 0. \tag{1.3}$$

There are three candidates for describing particle directions: angle θ_y, slope y', or momentum p_y. All of these are exhibited in Figure 1.1, and one sees that

$$y' \equiv \frac{dy}{ds} = \tan\theta_y = \frac{p_y}{\cos\theta_y}. \tag{1.4}$$

This multiple ambiguity in what constitutes the coordinate conjugate to y is something of a nuisance at large amplitudes but, fortunately, all three definitions approach equality in the small-angle limit that characterizes Gaussian

---

5) The symbol y, rather than, say, x, has been chosen as the dependent variable, because storage rings are traditionally horizontal, so the vertical orbit equations will remain valid when, later, we incorporate the bending magnets that are necessarily present.

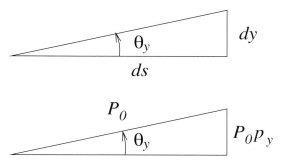

Fig. 1.1 Relations among transverse angle, momentum, and slope.

optics. One knows from Hamiltonian mechanics that p_y is the coordinate of choice but, while limiting ourselves to Gaussian optics, we will refer loosely to y' as "vertical momentum" so that we can refer to the (y, y')-plane as "vertical phase space".

Starting from any point s_0 along the beam line, one defines two special orbits, a "cosine-like" orbit $C(s, s_0)$ with unit initial amplitude and zero slope, and a "sine-like" orbit $S(s, s_0)$ with zero initial amplitude and unit slope:[6]

$$\begin{aligned} C(s_0, s_0) &= 1, & C'(s_0, s_0) &= 0, \\ S(s_0, s_0) &= 0, & S'(s_0, s_0) &= 1. \end{aligned} \tag{1.5}$$

Because Eq. (1.2) is linear and second order, *any* solution $y(s)$ and its first derivative $y'(s)$ can be expressed as that linear superposition of these two solutions that matches initial conditions $y(s_0)$ and $y'(s_0)$;

$$\begin{aligned} y(s) &= C(s, s_0) y(s_0) + S(s, s_0) y'(s_0), \\ y'(s) &= C'(s, s_0) y(s_0) + S'(s, s_0) y'(s_0). \end{aligned} \tag{1.6}$$

This can be expressed in matrix form, with $\mathbf{y} = (y, y')^\mathrm{T}$ being a "vector in phase space":

$$\mathbf{y}(s) \equiv \begin{pmatrix} y(s) \\ y'(s) \end{pmatrix} = \begin{pmatrix} C(s, s_0) & S(s, s_0) \\ C'(s, s_0) & S'(s, s_0) \end{pmatrix} \mathbf{y}(s_0) = \mathbf{M}(s_0, s) \mathbf{y}(s_0). \tag{1.7}$$

This serves to define $\mathbf{M}(s_0, s)$, the "vertical transfer matrix from s_0 to s". Since *any* solution of Eq. (1.2) can be expressed in this way, an entire beam line can be characterized by $\mathbf{M}(s_0, s)$. This matrix can be "composed" by multiplying (or "concatenating") the matrices for the successive beam line elements making up the line.

---

[6] Since unity slope is manifestly *not* a small angle, these definitions only make sense after the equations have been linearized as in Eq. (1.2).

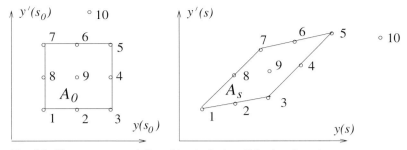

Fig. 1.2 Phase space evolution of ten typical particles in advancing from s_0 to s along the beam line.

Being interested in not just one particle, but many particles making up a beam, we define a "phase space density function" $\rho(s;y,y')$ with the property that $\rho(s;y,y')\,dy\,dy'$ is the number of particles in the range $dy\,dy'$. The evolution of a few particles in the beam in proceding from s_0 to s might resemble Figure 1.2 where corresponding particles are numbered. Particles 1 through 8 outline an area A_0 that evolves into area A_s. From calculus one knows that the factor relating these areas is the Jacobian determinant,

$$J(s_0, s) = \det \begin{vmatrix} \frac{\partial y}{\partial y_0} & \frac{\partial y}{\partial y'_0} \\ \frac{\partial y'}{\partial y_0} & \frac{\partial y'}{\partial y'_0} \end{vmatrix}. \tag{1.8}$$

From Eq. (1.7), suppressing the s_0-dependence, it can then be seen that

$$J(s) = \det |\mathbf{M}(s)| = C(s)S'(s) - S(s)C'(s) = 1. \tag{1.9}$$

which, being the Wronskian for the solutions $C(s)$ and $S(s)$, is known to be constant; that the value of the constant is 1 comes from evaluating $J(s_0)$, using Eqs. (1.5).

Particle trajectories in phase space cannot cross—this follows from the fact that instantaneous position and slope uniquely determine the subsequent motion. Hence the boundary will continue to be defined by particles 1 through 8 and a particle like 9 that is inside the box at s_0 will remain inside the box at s. Similarly, a particle like 10 will remain outside.

Putting these these things together, we obtain the result that the particle density $\rho(s)$ is, in fact, independent of s. This is Liouville's theorem, a result that can be discussed with various degrees of erudition and generality, but it is surprisingly simple in the present context. It is most succinctly stated by the requirement

$$\det |\mathbf{M}(s)| = 1, \tag{1.10}$$

which is the so-called "symplectic condition" in this special case of one dimensional motion.

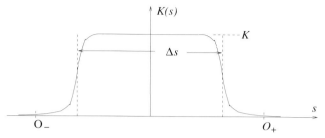

Fig. 1.3 Realistic thin lens focusing profiles are more or less constant with value K over a central region and become negligible outside points O_- and O_+. Effective length Δs is determined by matching $K\Delta s$ to the "field integral".

1.4
Transfer Matrices for Simple Elements

1.4.1
Drift Space

The most important transfer matrix is $\mathbf{M}_l$, which describes propagation through a drift space of length ℓ. Since the orbits are given by $y(s) = y_0 + y'_0 s$, $y'(s) = y'_0$, we have

$$\mathbf{M}_l = \begin{pmatrix} 1 & \ell \\ 0 & 1 \end{pmatrix} \tag{1.11}$$

In the context of this book, the basis for calling drifts "the most important element" is that the synchrotron light beam line leading from storage ring to detection apparatus is one long drift section.

1.4.2
Thin Lens

The next most important transfer matrix describes a "thin lens" where the definition of "thin" is that the thickness Δs is sufficiently small that coordinate $y(s - \Delta s/2)$, just before the lens, and $y(s + \Delta s/2)$, just after, can be taken to be equal. A typical focusing profile is shown in Figure 1.3. The lens causes a "kink" $\Delta y' = y'(s + \Delta s/2) - y'(s - \Delta s/2)$ in the orbit which, as shown in Figure 1.4, will be taken as occurring at the center of the lens. The kink can be obtained by integrating Eq. (1.2) from O_-, just before the lens to O_+, just after it:

$$\Delta y' = \int_{O_-}^{O_+} \frac{d}{ds}\left(\frac{dy}{ds}\right) ds = y \int_{O_-}^{O_+} K(s)\, ds \equiv y K \Delta s, \tag{1.12}$$

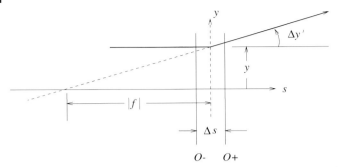

Fig. 1.4 Focusing action of a thin lens for which the focusing strength-length product is $K\Delta s$ and fields outside the range $O_- < s < O_+$ can be neglected. For the "diverging" case drawn, both K and q are positive so, algebraically, the focal length f, defined by Eq. (1.13), is negative.

which defines the product $K\Delta s$. Of course a focusing strength that changes discontinuously from 0 to K is not actually realistic, but the product $K\Delta s$, known as a "field integral", can be regarded as an abbreviation for $\int_{O_-}^{O_+} K(s)\,ds$ where O_- and O_+ are well outside the field region. If K is taken to be equal to K_0 (the value at the center of the element) then Δs is typically equal to (or, because of fringe fields, slightly greater than) the physical length of the element. The "focal length" f of the lens, defined in Figure 1.4, and the "lens strength" $q = -1/f$, are then given by[7]

$$q = -\frac{1}{f} = \frac{\Delta y'}{y} = K\Delta s. \tag{1.13}$$

Building in the (thin lens) approximation that y is constant through the lens, the transfer matrix is then given by

$$\mathbf{M}_q = \begin{pmatrix} 1 & 0 \\ q & 1 \end{pmatrix}. \tag{1.14}$$

As drawn in Figure 1.4, K and q are positive, f is negative, and the lens is "defocusing".

---

7) With the conventions of Eq. (1.13) and Figure 1.4 a focusing lens has positive focal length f. So a vertically focusing lens has positive f_y, negative K_y, and negative q. It is useful to ascribe the same parameter q to the element as regards also its effect on horizontal motion. For an optical lens with the same (negative) q-value, horizontal focusing strength K_x is negative and f_x positive, but, for a quadrupole, the focusing character is opposite in x and y planes. A quadrupole with negative q-value, according to Eq. (1.13) will be vertically focusing, with positive focal length f_y and negative K_y. With the same q-value, the same quad will have negative horizontal focal length f_x and—oh, never mind!

1.4.3
Thick Lens

The condition for the thin lens formula just given to be valid ($\Delta s \ll |f|$) is usually well satisfied for optical systems and accelerator lenses. Even if it is not, if $K(s)$ is constant (as it usually is, according to design anyway) it is easy to integrate Eq. (1.2), yielding matrix elements of **M** that are no worse than sines and cosines (or hyperbolic sines and cosines, depending on the sign of K). See Problem 1.4.1. For low energy, few element, accelerators this approach used to be considered "canonical" but, for high energy accelerators the thin lens approximation is usually adequate. In any case, making use of now readily available computer power, one can always split elements longitudinally to better validate the assumption that the elements are "thin". Quite apart from improving accuracy, it is also handy to split elements into two, in order to enable lattice function evaluations at lens centers. (The lattice functions typically go through maxima or minima near these locations; finite vacuum chamber dimensions often define limiting apertures there.) Even in the most extreme cases of intersection region quads, splitting into four thin quads is usually more than adequate, especially since the residual inaccuracy is typically less than the errors due to the neglect of other factors like fringe fields. For these reasons thick element formulas will be used only rarely in this text. When they are it will usually be to polish formulas obtained initially assuming thin lens optics.

Problem 1.4.1 *For an ideal, horizontally-focusing, thick quadupole, the focusing strength is $K_x(s) = -K$, with K positive, and trajectory equations (1.2) become*

$$\frac{d^2x}{ds^2} = -Kx, \quad and \quad \frac{d^2y}{ds^2} = Ky, \tag{1.15}$$

Show that the transfer matrices through such a quadrupole of length L are given by

$$\mathbf{M}_x = \begin{pmatrix} \cos\sqrt{K}L & \frac{1}{\sqrt{K}}\sin\sqrt{K}L \\ -\sqrt{K}\sin\sqrt{K}L & \cos\sqrt{K}L \end{pmatrix},$$

$$\mathbf{M}_y = \begin{pmatrix} \cosh\sqrt{K}L & \frac{1}{\sqrt{K}}\sinh\sqrt{K}L \\ \sqrt{K}\sinh\sqrt{K}L & \cosh\sqrt{K}L \end{pmatrix}. \tag{1.16}$$

Problem 1.4.2 *For the transfer matrices of the previous problem evaluate M_x^2 and M_y^2 and show that the results agree with Eq. (1.16) with $L \to 2L$.*

Problem 1.4.3 *A so-called "FODO cell", modeled by thin elements, is shown in Fig-*

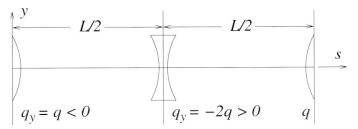

Fig. 1.5 A standard "FODO cell" of length L, beginning and ending at the centers of thin (horizontally) focusing half-quads.

ure 1.5. Show that the transfer matrices through such a cell are given by

$$\mathbf{M}_x = \begin{pmatrix} 1 - q^2 L^2/2 & L(1 + qL/2) \\ -q^2(1 - qL/2)L & 1 - q^2 L^2/2 \end{pmatrix},$$

$$\mathbf{M}_y = \begin{pmatrix} 1 - q^2 L^2/2 & L(1 - qL/2) \\ -q^2(1 + qL/2)L & 1 - q^2 L^2/2 \end{pmatrix}. \quad (1.17)$$

For $qL/2 < 1$ both $M_{x,21}$ and $M_{y,21}$ elements are negative, which corresponds to con-vergent focusing in both planes. Such a configuration is also known as a "triplet". The net convergence occurs because the relatively greater convergent deflection at large amplitude dominates the smaller divergent deflection at small amplitude.

Problem 1.4.4 *In an N-cell module of a linear accelerator there is a net focusing effect, occasionally referred to (for reasons I don't understand) as "ponderomotive" focusing, due to time-varying transverse forces at the ends of individual RF cells. This provides a sequence of opposite-sign focusing elements of strength q in successive cells of the linac. This focusing ranges from "very strong" at low energies, through "small" at intermediate energies, to negligible at high energies. A detailed model of transverse dynamics in the linac would use the structure of Figure 1.5 repeated $N/2$ times, with successive FODO parameters adjusted to compensate for the gradually increasing particle energy. It is also necessary to include an appropriate focusing quad at the entrance and defocusing quad at the exit to give the correct total entrance and exit focusing.*

Rather than retaining what is typically a rather large number of such sections in a module, one can represent their concatenated product by a single matrix in the form of Eq. (1.16). Focusing coefficients can be matched by requiring

$$K_{x,\text{eff.}} = K_{y,\text{eff.}} \equiv -K = -q^2/L, \quad (1.18)$$

Rather than retaining the accumulated length as NL, an "effective" number of cells $\bar{N}$ can be defined such that the effective length of the full sector is $\bar{N}L$. Clearly the treatment of alternating sign focusing by uniform focusing is valid only for $qL \ll 1$. Assuming this to be true, and neglecting the beam energy variation along the linac, find the value of $\bar{N}$ such that the 2,1 element of the uniform model matches the 2,1 element of the FODO model. For conventional linacs, the focusing effect can, in most cases, be ascribed predominantly to the end half-quads. Furthermore, in this approximation, the strength of the output half-quad can be derated relative to the input half-quad to account for the energy increase in the full linac module. The approach taken in this problem is a candidate for modeling transverse dynamics in ERL's, a topic that is treated in Chapter 10.

In the sequel, I will not hesitate to jump between the kinky orbits that characterize thin lenses and the smooth orbits that characterize thick elements. Usually figures will be drawn as if the orbits are smooth while explicit calculations will be performed using kinks. This may occasionally be disconcerting for the reader, but its use will be restricted to cases where the approximation is sensible.

1.4.4
Erect Quadrupole Lens

All formulas so far apply equally to optical and charged particle lenses. Since glass lenses and spherical mirrors can be focusing in both planes, Eq. (1.14) applies to both transverse planes, but a quadrupole lens that focuses in one plane (unhappily) defocuses in the other. This is because the vertical bend is due (ideally) to a horizontal magnetic field $B_x = (\partial B_x/\partial y)\, y$, where the proportionality to y is what causes the lens to focus, and what justifies calling the lens linear. For substitution into Eq. (1.2) we therefore use

$$P_0 \frac{d\mathbf{p}}{dt} = e\mathbf{v}\times\mathbf{B} = e \begin{vmatrix} \hat{x} & \hat{y} & \hat{s} \\ v_x & v_y & v_s \\ B_x & B_y & 0 \end{vmatrix}. \tag{1.19}$$

Using the result $v_s = ds/dt$, this yields, for the vertical deflection

$$P_0 \Delta p_y = e \frac{\partial B_x}{\partial y}\, \Delta s\, y, \tag{1.20}$$

which, though approximate, is exact in the small Δs limit. Using a once-integrated version of Eq. (1.4), this can be recast as the deflection caused by an ideal lens of strength[8]

$$q_y = \frac{\Delta y'}{y} = K_y \Delta s, \quad \text{where} \quad K_y = \frac{c(\partial B_x/\partial y)}{(P_0 c/e)}. \tag{1.21}$$

---

[8] The factors in Eq. (1.21) are grouped so that all quantities can be conveniently expressed in M.K.S. units; in particular $P_0 c/e$ is measured in volts when $P_0 c$ is measured in electron volts.

Because of the Maxwell equation $\partial B_x/\partial y = \partial B_y/\partial x$, this vertical focusing strength is necessarily accompanied by horizontal focusing strength of opposite sign:[9]

$$q_x = -q_y. \tag{1.22}$$

To express this constraint succinctly it is convenient to write a 4×4 transfer matrix:

$$\mathbf{M}_q = \begin{pmatrix} 1 & 0 & 0 & 0 \\ -q & 1 & 0 & 0 \\ 0 & 0 & 1 & 0 \\ 0 & 0 & q & 1 \end{pmatrix}, \tag{1.23}$$

which acts on the vector $(x, x'y, y')^T$. (Unlike this, a visible light lens can have same sign off-diagonal terms—for example both focusing. Clearly this simplifies the design of lines for visible light.)

The matrix of Eq. (1.23) has the desirable property of having no "cross-plane coupling" so vertical and horizontal motion can be treated independently. But if the quadrupole is rolled (around the s axis) by $45°$ it becomes a "skew quad". It is not hard to show that the transfer matrix is then

$$\mathbf{M}_q^S = \begin{pmatrix} 1 & 0 & 0 & 0 \\ 0 & 1 & \pm q & 0 \\ 0 & 0 & 1 & 0 \\ \pm q & 0 & 0 & 1 \end{pmatrix}, \tag{1.24}$$

where the sign ambiguity corresponds to the ambiguity in specifying the orientation of the azimuthal roll. Now the off-diagonal elements couple horizontal and vertical motion.

In the context of synchrotron light sources, the existence of such "coupling" elements limits the quality of ribbon electron beams (which ideally have finite width but negligible height) by increasing their heights and hence increasing the heights of the ribbon x-ray beams they produce. Such beams may be especially important for apparatus (such as monochromators) that rely on Bragg

---

9) Optical systems for visible light are usually constructed from elements (lenses) whose boundaries are surfaces of rotation around the beam axis. For such systems, unlike Eq. (1.22), one has $q_x = q_y$. Then the symmetry of the system makes it advantageous to work with radial coordinate r rather than Cartesian coordinates x and y. In accelerators, on the other hand, the optics always distinguishes horizontal and vertical axes. This distinction tends to obscure analogies between light optics and particle optics. For example (except for distortion) the wave fronts of visible light are spherical, while the analogous surfaces of particle optics tend to be saddle-shaped. This makes it advantageous to pretend the accelerator motion is one-dimensional, in which case the "wave fronts" are cylinders, with axes being transverse.

scattering from crystals, in which case the (angular) height of the ribbon is of pre-eminent concern. Since this is a somewhat specialized concern we will mainly simplify the discussion by assuming the absence of cross-plane coupling.

According to Eq. (1.10) every transfer matrix has unit determinant and one can confirm this to be the case for the matrices written so far. Any product of these matrices clearly has the same property, which is consistent with Eq. (1.10).[10]

Problem 1.4.5 *What Maxwell equation was used in deriving Eq. (1.22) and what does it assume concerning the currents that cause the bending magnetic fields? Would the same result hold if the magnetic fields due to the beam itself were included?*

Problem 1.4.6 *Represent, by a 4×4 matrix $\mathbf{R}_{45}$, the transformation to coordinates (X, X', Y, Y') which are related to coordinates (x, x', y, y') by a 45° "roll" around the s-axis.*

Problem 1.4.7 *For a skew quadrupole, argue that*

$$\mathbf{M}_q^S = \mathbf{R}_{-45}\mathbf{M}_q\mathbf{R}_{45} \tag{1.25}$$

and thereby derive Eq. (1.24). (As in Eq. (1.24), the signs of the off-diagonal terms are only made unambiguous by careful specification of the orientations of the rolls.)

1.5
Elliptical (in Phase Space) Beams

Because of the gigantic number of particles they contain, it is appropriate to represent entire beams by distribution functions. Beams that have elliptical shape in phase space are especially appropriate because, though the sizes, aspect ratios, and orientation vary with s, the shapes remain elliptical. The reason for this is that Eqs. (1.7) are linear. The sort of distribution envisaged is illustrated in Figure 1.6. Some example interpretations of the ellipse are: the interior of the ellipse is uniformly populated (This is approximately applicable in a proton ring.); or particles are uniformly distributed on the elliptical shell (Though not physically realizable this is useful for theoretical calculations.); or the beam particles are "Gaussian distributed", with the density function constant on ellipsoidal surfaces (This is accurate for electron storage rings). When expressed in coordinates aligned with the ellipse, the general equation

---

10) For 2×2 matrices the unit determinant condition is known as the "symplectic condition". For matrices larger than 2×2 this determinant condition is necessary but not sufficient. Still, the most general symplectic matrix can be manufactured as products of the matrices written so far, even in higher dimensions [1].

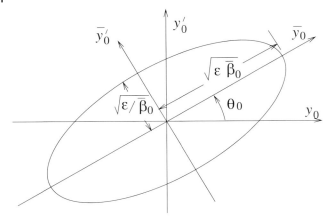

Fig. 1.6 An elliptical beam in phase space. In the $(\bar{y}, \bar{y}')$-frame, skewed by angle θ_0, the ellipse is erect.

of the beam ellipse at $s = s_0$, as shown in Figure 1.6, is

$$\frac{\bar{y}_0^2}{\bar{\beta}_0} + \bar{\beta}_0 \bar{y}_0'^2 \equiv \bar{\mathbf{y}}_0^T \begin{pmatrix} 1/\bar{\beta}_0 & 0 \\ 0 & \bar{\beta}_0 \end{pmatrix} \bar{\mathbf{y}} = \epsilon_y. \tag{1.26}$$

However the beam evolves, its outline has this equation, with appropriate parameters $\bar{\beta}(s), \theta(s), \epsilon_y$. The aspect ratio of the ellipse is governed by $\bar{\beta}(s)$ and its area is given by

$$\text{vertical phase space area} = \pi \sqrt{\epsilon_y \bar{\beta}} \sqrt{\epsilon_y / \bar{\beta}} = \pi \epsilon_y. \tag{1.27}$$

Since this area has been previously shown to be invariant, the parameter ϵ_y is, in fact, *invariant*, which is why it has been given no subscript 0, nor overhead bar, nor argument s. ϵ_y is known as the beam "emittance"—its definition is still ambiguous, however, by an overall constant factor that depends on the nature of the beam distribution, and its parametrization by ϵ_y; for example ϵ_y could be a maximum or an r.m.s. value.

The skew coordinates $\bar{\mathbf{y}}$ are related to erect coordinates by

$$\bar{\mathbf{y}}_0 = \mathbf{R}_0 \mathbf{y}_0, \quad \text{where} \quad \mathbf{R}_0 = \begin{pmatrix} \cos\theta_0 & \sin\theta_0 \\ -\sin\theta_0 & \cos\theta_0 \end{pmatrix}, \tag{1.28}$$

and the beam ellipse, expressed in erect coordinates, is

$$\gamma_0 y_0^2 + 2\alpha_0 y_0 y_0' + \beta_0 y_0'^2 \equiv \mathbf{y}_0^T \begin{pmatrix} \gamma_0 & \alpha_0 \\ \alpha_0 & \beta_0 \end{pmatrix} \mathbf{y}_0$$

$$= \bar{\mathbf{y}}_0^T \mathbf{R}_0 \begin{pmatrix} \gamma_0 & \alpha_0 \\ \alpha_0 & \beta_0 \end{pmatrix} \mathbf{R}_0^{-1} \bar{\mathbf{y}}_0 = \epsilon_y. \tag{1.29}$$

(The relation $\mathbf{R}_0^{-1} = \mathbf{R}_0^{\mathrm{T}}$ has been used.) The parameters α, β, γ will be called "*beam-based* Twiss parameters" for reasons to be explained shortly. Correlating Eq. (1.29) with Eq. (1.26) yields

$$\begin{pmatrix} \gamma_0 & \alpha_0 \\ \alpha_0 & \beta_0 \end{pmatrix} = \mathbf{R}_0^{-1} \begin{pmatrix} 1/\bar{\beta}_0 & 0 \\ 0 & \bar{\beta}_0 \end{pmatrix} \mathbf{R}_0. \tag{1.30}$$

Since all factors on the right-hand side have unit determinant, so also must the left-hand side, and hence

$$\gamma_0 = \frac{1 + \alpha_0^2}{\beta_0}. \tag{1.31}$$

Using this relation and "completing squares", ellipse equation (1.29) can be simplified to

$$\frac{y_0^2 + (\alpha_0 y_0 + \beta_0 y_0')^2}{\beta_0} = \epsilon_y. \tag{1.32}$$

When ϵ_y is expressed in terms of the particle coordinates in this way it is known as the "Courant–Snyder" invariant; $\epsilon_{CS}(y, y')$. Note that the Courant–Snyder invariant is a property of a *particle* while emittance is a property of a *beam*—an unfortunate clash of the same symbol ϵ having different meanings.

The purpose in introducing the beam ellipse was that its evolution is more significant than that of any one particle. Substituting from Eq. (1.7) into Eqs. (1.29) expressed at location s, the matrix of coefficients expressing ϵ_y at s is given by

$$\begin{pmatrix} \gamma & \alpha \\ \alpha & \beta \end{pmatrix} = (\mathbf{M}^{-1})^{\mathrm{T}} \begin{pmatrix} \gamma_0 & \alpha_0 \\ \alpha_0 & \beta_0 \end{pmatrix} \mathbf{M}^{-1}$$

$$= \begin{pmatrix} S' & -C \\ -S & C \end{pmatrix} \begin{pmatrix} \gamma_0 & \alpha_0 \\ \alpha_0 & \beta_0 \end{pmatrix} \begin{pmatrix} S' & -S \\ -C' & C \end{pmatrix}. \tag{1.33}$$

This can be expanded to be more explicit:

$$\begin{pmatrix} \beta \\ \alpha \\ \gamma \end{pmatrix} = \begin{pmatrix} C^2 & -2CS & S^2 \\ -CC' & CS' + SC' & -SS' \\ C'^2 & -2C'S' & S'^2 \end{pmatrix} \begin{pmatrix} \beta_0 \\ \alpha_0 \\ \gamma_0 \end{pmatrix}. \tag{1.34}$$

1.6
Beam Envelope $E(s)$

To interpret the meanings of the lattice functions β, α, and γ one can study the geometry of points on the beam ellipse, as shown in Figure 1.7. Of greatest

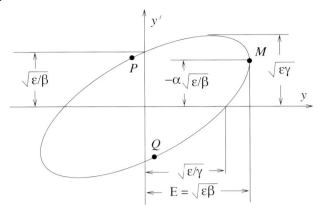

Fig. 1.7 Correlating beam ellipse parameters with ellipse geometry.

interest is the maximum (vertical in this case) excursion $E(s)$ of the ellipse, since this can be compared with a scraper position, or collimator size, to see whether the particle or ray will "wipe out". Accepting the result of doing Problem 1.6.1, the abscissa of point M in Figure 1.7 is given by

$$E_y(s) = \sqrt{\epsilon_y \beta_y(s)}, \tag{1.35}$$

The aperture "stay clear" must exceed $E_y(s)$ to avoid particle loss. It is because of the appearance of β in this formula that β is usually considered to be the most important Twiss beam parameter.

Problem 1.6.1 *Find the coordinates of point M in Figure 1.7, thereby deriving Eq. (1.35). A handy way of locating M is to require that that the gradient of ϵ_y (which is normal to a contour of constant ϵ_y) has no y' component. That is $\partial \epsilon_y / \partial y' = 0$.*

It can be seen from Figure 1.8 that the slope of the beam envelope at point M is the same as the slope of the particular ray defined by point M, which is given by $y'(M) = -\alpha_y \sqrt{\epsilon_y / \beta_y}$. Differentiating Eq. (1.35), this makes it possible to obtain the rate of change of the beam envelope;

$$E'_y(s) \equiv \frac{dE_y}{ds} = \frac{\beta'_y}{2} \sqrt{\frac{\epsilon_y}{\beta_y}}. \tag{1.36}$$

Equating y' to E'_y yields

$$\beta' = -2\alpha, \tag{1.37}$$

meaning that (except for factor -2) α is the slope of the curve of $\beta(s)$ plotted against s. The y subscripts have been left off this equation since the same equation holds for the x motion.

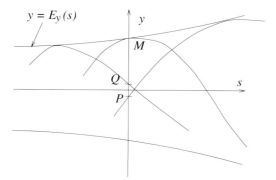

Fig. 1.8 The beam envelope $E_y(s)$ is equal to the maximum transverse excursion y at longitudinal position s. Point labels M, P, and Q correspond approximately to Figure 1.7.

1.7
Gaussian Beams: their Variances and Covariances

The joint probability distribution for a "Gaussian distributed" beam is

$$\rho(y,y') = K\exp\left(-\frac{1}{2\epsilon_{y,b}}\mathbf{y}^T\begin{pmatrix}\gamma & \alpha \\ \alpha & \beta\end{pmatrix}\mathbf{y}\right) = K\exp\left(-\frac{\epsilon_y}{2\epsilon_{y,b}}\right), \qquad (1.38)$$

where the factor K has to be chosen to fix the normalization. (Recall that the combination $\mathbf{y}^T\begin{pmatrix}\gamma & \alpha \\ \alpha & \beta\end{pmatrix}\mathbf{y}$, being the Courant–Snyder invariant, has the same value everywhere in the ring.) This probability density falls to $e^{-1/2}$ when ϵ_y (as given by Eq. (1.32)) is equal to a specified beam emittance $\epsilon_{y,b}$. This distribution is simpler when expressed in terms of erect coordinates $\bar{\mathbf{y}} = \mathbf{R}^{-1}\mathbf{y}$, where $\mathbf{R}$ is the rotation matrix of Eq. (1.28), but applicable at lattice point s. Because this transformation is a simple rotation the probability distributions are related by $\bar{\rho}(\bar{y},\bar{y}') = \rho(y,y')$, and hence

$$\bar{\rho}(\bar{y},\bar{y}') = \frac{1}{2\pi\epsilon_{y,b}}\exp\left(-\frac{\bar{y}^2}{2\epsilon_{y,b}\bar{\beta}} - \frac{\bar{y}'^2}{2\epsilon_{y,b}/\bar{\beta}}\right). \qquad (1.39)$$

The normalization is the subject of Problem 1.7.1, and has yielded $K = 1/(2\pi\epsilon_{y,b})$.

Problem 1.7.1 *By integrating over all of phase space, apply the condition that distribution (1.39) represents one particle, to confirm the normalizing factor.*

The elements of the "variance-covariance matrix" are the expectation values, or moments, $<y^2>, <yy'>, <y'^2>$:

$$\begin{pmatrix} <y^2> & <yy'> \\ <yy'> & <y'^2> \end{pmatrix} = <\mathbf{y}\mathbf{y}^T> = <\mathbf{R}^{-1}\overline{\mathbf{y}\mathbf{y}}^T\mathbf{R}>$$

$$= \mathbf{R}^{-1} \begin{pmatrix} <\overline{y}^2> & 0 \\ 0 & <\overline{y}'^2> \end{pmatrix} \mathbf{R}$$

$$= \mathbf{R}^{-1} \begin{pmatrix} \epsilon_{y,b}\overline{\beta} & 0 \\ 0 & \epsilon_{y,b}/\overline{\beta} \end{pmatrix} \mathbf{R}. \tag{1.40}$$

The expression on the right-hand side can be obtained from Eq. (1.30):

$$\begin{pmatrix} \beta & -\alpha \\ -\alpha & \gamma \end{pmatrix} = \begin{pmatrix} \gamma & \alpha \\ \alpha & \beta \end{pmatrix}^{-1} = \mathbf{R}^{-1} \begin{pmatrix} \overline{\beta} & 0 \\ 0 & 1/\overline{\beta} \end{pmatrix} \mathbf{R}. \tag{1.41}$$

Finally then, we obtain

$$\begin{pmatrix} <y^2> & <yy'> \\ <yy'> & <y'^2> \end{pmatrix} = \begin{pmatrix} \epsilon_{y,b}\beta & -\epsilon_{y,b}\alpha \\ -\epsilon_{y,b}\alpha & \epsilon_{y,b}\gamma \end{pmatrix}. \tag{1.42}$$

For the special case $\alpha = 0$ this can be easily checked.

It has been shown then that (except for factor $\epsilon_{y,b}$) the matrix of coefficients in the quadratic form for ϵ_y is the inverse of the variance-covariance matrix. Commonly (for example in the program TRANSPORT) the variance-covariance matrix is designated as σ

$$\sigma = \begin{pmatrix} <y^2> & <yy'> \\ <yy'> & <y'^2> \end{pmatrix} = \begin{pmatrix} \sigma_{11} & \sigma_{12} \\ \sigma_{12} & \sigma_{22} \end{pmatrix} = \epsilon_{y,b} \begin{pmatrix} \beta & -\alpha \\ -\alpha & \gamma \end{pmatrix}. \tag{1.43}$$

Because of this relation, the known evolution of the beam Twiss parameters (see the first of Eqs. (1.33)) can be used to obtain the evolution of σ:

$$\sigma = \mathbf{M}^T \sigma_0 \mathbf{M}. \tag{1.44}$$

Note that it is $\mathbf{M}$ appearing in this equation, whereas it was $\mathbf{M}^{-1}$ that appeared in Eq. (1.33).

1.8
Pseudoharmonic Trajectory Description

It has been seen in Eq. (1.35) that the beam envelope scales proportional to $\sqrt{\beta}$. One conjectures therefore, that individual trajectories will scale the same way and be describable in the form

$$y(s) = a\sqrt{\beta(s)}\cos(\psi - \psi_0), \tag{1.45}$$

where ψ depends on s and a is a constant amplitude. This form has to satisfy Eq. (1.2), and the $\cos(\psi - \psi_0)$ "ansatz" is based on the known solution when $K(s)$ is, in fact, constant. This is the basis for naming the trajectory description "pseudoharmonic". As β appears in this equation, it has no apparent *beam* attribute and it will be referred to as a "lattice function". Differentiating Eq. (1.45) we get

$$y'(s) = -a\sqrt{\beta(s)}\,\psi'\,\sin(\psi - \psi_0) + \frac{a\beta'}{2\sqrt{\beta}}\cos(\psi - \psi_0). \tag{1.46}$$

Substituting into Eq. (1.2) we can demand that the coefficients of sin and cos terms vanish independently, since that is the only way to maintain equality for all values of ψ_0. This leads to the equations

$$\beta\,\psi'' + \beta'\psi' = 0,$$
$$2\beta\,\beta'' - \beta'^2 - 4\beta^2\psi'^2 + 4\beta^2 K(s) = 0. \tag{1.47}$$

From the first equation it follows that $\beta\,\psi'$ is constant. To obtain the conventional description we pick this constant to be 1 and obtain

$$\psi' = \frac{1}{\beta}, \quad \text{or} \quad \psi(s) = \psi(s_0) + \int_{s_0}^{s} \frac{ds'}{\beta(s')}. \tag{1.48}$$

Since ψ is the argument of a sinusoidal function, and the argument of a harmonic wave is $2\pi s/\text{wavelength}$, this permits us to interpret $2\pi\beta(s)$ as a "local wavelength" or, equivalently, $1/\beta(s)$ is the "local wave number". Substituting into the second of Eqs. (1.47), we obtain

$$\beta'' = 2\beta\,K(s) + 2\frac{1 + \beta'^2/4}{\beta}. \tag{1.49}$$

This second order, nonlinear differential equation is usually considered to be the fundamental defining relationship for the evolution of the lattice β-function. Like any differential equation its solution is unique only if initial conditions or boundary conditions are given. The possibilities are:

- Initial conditions $\beta_0, \beta_0'(= -2\alpha_0)$ are given, for example at the beginning of a beam transfer line. A "matched transfer line" also has specified values, β_1, α_1, at its exit. (Satisfying these extra conditions requires the lens strengths in the line to be tuned appropriately.)

- Values are given at two points, β_0 and β_1. This is uncommon.

- The function $\beta(s)$ is required to be periodic with period $\mathcal{C}$ where $\mathcal{C}$ is the circumference of a storage ring (or the period of a periodic lattice); $\beta(0) = \beta(\mathcal{C}), \alpha(0) = \alpha(\mathcal{C})$. This is the most important case.

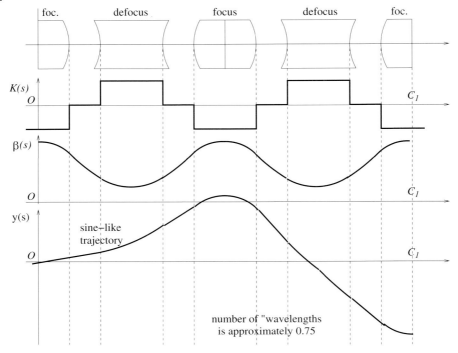

Fig. 1.9 Pictorial representation of β-function variation corresponding to a possible focusing profile $K(s)$ produced, for example, by the long quadrupoles shown. The lower figure shows the sine-like trajectory. The "tune" of this section of beam line is about 0.75 since the trajectory goes through roughly 3/4 of a full wavelength.

Because $K(s)$ depends on s, solving the equation may be quite difficult in general. An artist's conception of the β-function corresponding to a somewhat typical, stepwise-constant, focusing profile is exhibited in Figure 1.9. The figure also shows a sine-like trajectory, which executes something like three quarters of a full oscillation in this section, so the "tune" is about 0.75. (A typical storage ring would be made up of some number n_c of cells like this so the circumference would be $n_c \mathcal{C}_1$ and the tune would be about $0.75 n_c$.)

A "first integral" of Eq. (1.49) can be obtained by squaring and adding Eqs. (1.45) and (1.46) (rearranged) to give

$$\frac{y^2}{\beta} + \beta \left(y' - \frac{\beta'}{2\beta} y \right)^2 = a^2. \tag{1.50}$$

According to Eq. (1.37) we have $\beta' = -2\alpha$, so this result confirms the constancy of the Courant–Snyder invariant, (Eq. (1.32)), and a^2 can be identified with what was previously called $\epsilon_{y,b}$. There is an essential new feature however. In Eq. (1.32) the constancy referred to all points on the beam ellipse. Here it refers to all points s on a single ray or trajectory. Fortunately, as the symbols have been introduced, one can now afford to drop this distinction.

1.9
Transfer Matrix Parametrization

The transfer matrix from s_0 to s can by obtained by substituting pseudoharmonic trajectory formulas into Eq. (1.7). The sine-like and cosine-like solutions are

$$C(s,s_0) = \sqrt{\beta/\beta_0}(\cos(\psi - \psi_0) + \alpha_0 \sin(\psi - \psi_0))$$
$$S(s,s_0) = \sqrt{\beta\beta_0}\sin(\psi - \psi_0). \tag{1.51}$$

As a result the transfer matrix is

$$\mathbf{M}(s_0,s) = \begin{pmatrix} \sqrt{\frac{\beta}{\beta_0}}(\cos(\psi - \psi_0) + \alpha_0 \sin(\psi - \psi_0)) & \sqrt{\beta_0\beta}\sin(\psi - \psi_0) \\ \frac{-(1+\alpha_0\alpha)\sin(\psi-\psi_0)+(\alpha_0-\alpha)\cos(\psi-\psi_0)}{\sqrt{\beta_0\beta}} & \sqrt{\frac{\beta_0}{\beta}}(\cos(\psi - \psi_0) - \alpha \sin(\psi - \psi_0)) \end{pmatrix}. \tag{1.52}$$

Since these matrix elements have arisen from describing the "lattice" of focusing elements, one can call them "lattice-based Twiss parameters". The evolution of beams or particles with distance s along the beam line, can be represented by specifying s-dependent parameters, $\beta(s), \alpha(s), \psi(s), \ldots$, instead of in the form of Eq. (1.7).

A special (and important) case has $\beta = \beta_0$, $\alpha = \alpha_0$, for example because the transfer matrix refers to evolution once around a closed accelerator lattice, so initial and final points are "the same point". In this case Eq. (1.52) reduces to

$$\mathbf{M} = \begin{pmatrix} \cos\mu + \alpha\sin\mu & \beta\sin\mu \\ -\gamma\sin\mu & \cos\mu - \alpha\sin\mu \end{pmatrix}, \quad \text{where} \quad \gamma = \frac{1+\alpha^2}{\beta}, \tag{1.53}$$

which is known as the "once around" transfer matrix. Form (1.53) is also known as the standard Twiss parameterization of a 2×2 matrix with unit determinant. (This is why the term "Twiss parameter" occurs repeatedly in accelerator physics in spite of the fact that Twiss was not an accelerator physicist.) One should confirm that this form has the correct number of free parameters and the correct determinant.[11] This form will turn out to be most

---

[11] Since a unity determinant is the only requirement for a 2×2 matrix to be symplectic, any valid transfer matrix can be expressed as in Eq. (1.53). However, except in the special case of periodic lattices, it would be a mistake to infer that the parameters β, α, μ, are equal to, or even simply related to, the parameters β, α, ψ appearing in Eq. (1.52).

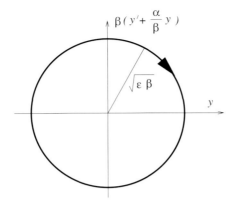

Fig. 1.10 The "normalized" or "circular" phase space trajectory is a circle. In this representation trajectory evolution is reduced to pure rotation.

useful for describing *periodic* focusing arrays. The closed storage ring mentioned above is necessarily periodic in this sense.

The pseudoharmonic formalism can be reduced to its simplest form by introducing "normalized" phase space as shown in Figure 1.10. Using the "completed squares" form of the Courant–Snyder invariant (Eq. (1.32)), the phase space trajectory becomes a circle of radius $\sqrt{\beta\epsilon}$ when the combination $\beta(y' + (\alpha/\beta)y)$ is plotted against y. In this representation trajectory evolution is reduced to pure rotation, with the angular coordinate being the "betatron phase" $\psi(s)$.

Problem 1.9.1 *Confirm that Eqs. (1.51) give the cosine-like and sine-like trajectory solutions for a beam line and that, therefore, Eq. (1.52) gives the transfer matrix for the line.*

Problem 1.9.2 *Suppose that, by some kind of operational procedure, all four elements of the "once around" transfer matrix of a storage ring have been measured. Show that "tune" $Q \equiv \mu/(2\pi)$ is given by*

$$Q = \frac{1}{2\pi} \cos^{-1}\left(\frac{\text{tr}\,\mathbf{M}}{2}\right), \tag{1.54}$$

being sure to discuss the possibility of multiple possible values for Q. Express β and α in terms of $M_{11}, M_{12}, M_{21}, M_{22}$. Show that β (which is positive by definition) is determined uniquely in spite of the multiple values of Q. Is α determined uniquely?

1.10
Reconciliation of Beam and Lattice Parameters

As yet no clean connection has been demonstrated between *beam*-based Twiss parameters and *lattice*-based Twiss parameters, even though the same symbols have been used. Furthermore, the lattice properties are fixed, once and for all, while the beam properties can vary wildly, depending upon injection tuning or mistuning. It is clear, therefore, that there can be no simple equality between beam parameters and lattice parameters. Nevertheless there is an extremely close connection that will be investigated next.

1.10.1
Beam Evolution Through a Drift Section

Suppose the region from s_0 to s is purely a drift section. According to Eq. (1.11) and the first of Eqs. (1.33), the beam evolution is given by

$$\begin{pmatrix} \gamma & \alpha \\ \alpha & \beta \end{pmatrix} = \begin{pmatrix} 1 & 0 \\ -s & 1 \end{pmatrix} \begin{pmatrix} \gamma_0 & \alpha_0 \\ \alpha_0 & \beta_0 \end{pmatrix} \begin{pmatrix} 1 & -s \\ 0 & 1 \end{pmatrix} \tag{1.55}$$

Completing the multiplication, one finds

$$\beta = \beta_0 - 2\alpha_0 s + \gamma_0 s^2,$$
$$\alpha = \alpha_0 - \gamma_0 s,$$
$$\gamma = \gamma_0. \tag{1.56}$$

One finds, upon substituting $\beta(s)$, given here, into lattice Eq. (1.49), that the beam-based $\beta(s)$ satisfies the lattice-based differential equation, at least in drift sections. If the two versions of β are equal at one point in the drift section, they will remain equal throughout the section.

Behavior of the lattice functions in the vicinity of a "beam waist" (traditionally signified by an asterisk *) will be especially important. At such a point $\alpha \equiv \alpha^* = 0$, so $\beta = \beta^*$ has an extreme value, which we take to be a minimum. At a waist $\gamma^* = 1/\beta^*$. Eqs. (1.56) describes the evolution of β away from that point. To emphasize that the formula applies only in drift regions we will replace s by z. Substituting into the first of Eqs. (1.56) we obtain

$$\beta(z) = \beta^* + \frac{z^2}{\beta^*}. \tag{1.57}$$

A "waist length" can be defined by

$$z_{\text{waist}}^{(P)} = \beta^*; \tag{1.58}$$

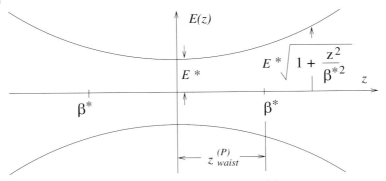

Fig. 1.11 Longitudinal variation of beam half-width $w(z)$ near a beam waist.

it is the length after which the beam transverse-dimension-squared has doubled.[12][13] Because the transverse beam size is proportional to $\sqrt{\beta}$, the variation of the beam half-width is given by

$$E(z) = E^* \sqrt{1 + (z/\beta^*)^2}, \tag{1.59}$$

as shown in Figure 1.11.

1.10.2
Beam Evolution Through a Thin Lens

Consider next the beam evolution in passing through a thin lens. According to Eq. (1.14) and the first of Eqs. (1.33), the beam evolution is given by

$$\begin{pmatrix} \gamma_+ & \alpha_+ \\ \alpha_+ & \beta_+ \end{pmatrix} = \begin{pmatrix} 1 & -q \\ 0 & 1 \end{pmatrix} \begin{pmatrix} \gamma_- & \alpha_- \\ \alpha_- & \beta_- \end{pmatrix} \begin{pmatrix} 1 & 0 \\ -q & 1 \end{pmatrix} \tag{1.60}$$

Completing the multiplication, one finds

$$\begin{aligned} \beta_+ &= \beta_-, \\ \alpha_+ &= \alpha_- - \beta q, \\ \gamma_+ &= \gamma_- - 2\alpha_- q + \beta q^2. \end{aligned} \tag{1.61}$$

Problem 1.10.1 *The discussion below Eq. (1.56) showed that evolution of beam-based (α, β, γ) parameters through a drift section is governed by the same formulas*

---

12) The superscript P on $z_{\text{waist}}^{(P)}$ refers to the implicit assumption that a particle (in contrast to the wave that will be introduced shortly) is being described.
13) Stretching the metaphor unconscionably, one notes that a person wishing to have a slim waist would have to be correspondingly short.

as is the evolution of particle-based (α, β, γ) parameters. Using Eq. (1.49), show that the same statement can be made for evolution through a thin lens. That is, use Eq. (1.49) to calculate the "kink" in the β-function caused by a thin lens, and show that the result agrees with Eqs. (1.61). (This involves a limiting process as $\Delta s \to 0$, for which it is useful to refer to Figure 1.3.)

For lines made up only of drifts and thin lenses, it has therefore been shown that beam-based and lattice-based Twiss functions, if once equal, remain equal. Since, as stated earlier, all transfer matrices can be constructed from drifts and thin lenses, the result is true for any general (uncoupled) beam line.

References

[1] V. Guillemin, S. Sternberg (1990), *Symplectic Techniques in Physics*, Cambridge University Press, p. 23.

2
Beams Treated as Waves

2.1
Preview

Treating beams as waves instead of as bunches of particles, this chapter establishes beam evolution prescriptions largely equivalent to those obtained in Chapter 1. This wave theory is substantially deeper and the mathematics somewhat more complicated than was the particle treatment. Since the material in Chapter 1 already serves as a sufficient basis for most of the subsequent chapters, detailed study of the material in the present chapter can be deferred with little loss of continuity.

The essential similarity of the propagation of light and electrons was established in Chapter 1. Much of that similarity was the result of treating both types of beams as collections of *particles*. One is familiar, however, with wave-particle duality, according to which beams also exhibit wave-like phenomena such as interference and diffraction. For visible light encountering apertures and obstacles, the wave-like behavior has been understood for centuries, largely because the light wavelength can be comparable with the dimensions of available apertures and obstacles. For x-rays, because of their far shorter wavelength, wave-like behavior was first understood only in the form of diffraction from crystals having Ångstrom-scale lattice constants, comparable to the x-ray wave lengths.

This chapter begins with a wave description of light and, without dwelling on wave-like phenomenology, makes contact with the particle-like description established previously by analysing wave propagation in the short wavelength limit.

Much of the enthusiasm in the x-ray community for progressing beyond the so-called "third generation light sources" is based on exploiting wave-like properties of x-rays. Perhaps the most important new emphasis is on *coherent* x-rays. For visible light, the coherence properties of light have become understood largely in connection with the development of lasers. Coherence is only important for accelerator-generated light when the accelerator emittance is comparable with the x-ray wavelength. Until recently the relatively large emittances available in electron storage rings have limited the exploitation of coherence properties to radiation having wavelengths far longer than x-rays.

Accelerator X-Ray Sources. Richard Talman
Copyright © 2006 WILEY-VCH Verlag GmbH & Co. KGaA, Weinheim
ISBN: 3-527-40590-9

Now there are plans to produce x-rays using free electron lasers (FELs). These have much in common with the devices described in this book, and the material in this chapter would form a natural introduction to their theory. However, for the forseeable future, even these FEL projects will be limited to relatively soft x-rays (perhaps a few kilovolts). Because of this limitation, and to limit the length of the book, as explained in the preface I have chosen to discuss only x-ray FEls. This chapter does, however, contain a brief discussion of the optical resonator, which is an important components of a closed FEL. Its relevance here is in quantifying x-ray coherence.

Though interest in the physics of coherence phenomena has provided the main motivation for the existence of this textbook, the book itself concentrates on the *production* of the x-rays, and the tailoring of their properties, rather than in exploring the physics of their *use*, but, to provide at least one motivation for the production of beams of high brilliance, this chapter does digress briefly into the topic of *phase contrast imaging*. It is efficient to discuss this material in this chapter since most of the background material will have already been covered while treating x-ray beams as waves.

2.2
Scalar Wave Equation

To study geometric optics in media with spatially-varying index of refraction $n = n(\mathbf{r})$ one should work with electric and magnetic fields but, to reduce complication (without compromising the issues to be emphasized) we will work with *scalar* waves. Any such wave must satisfy the wave equation for a wave of velocity $c/n(\mathbf{r})$;

$$\nabla^2 \Psi \equiv \nabla \cdot \nabla \Psi = \frac{n^2(\mathbf{r})}{c^2} \frac{\partial^2 \Psi}{\partial t^2}. \tag{2.1}$$

For isotropic media the function $\Psi(\mathbf{r}, t)$ can be taken as either transverse component of an electric or magnetic field of the wave. The simplest traveling wave solution is a plane wave in a medium with constant index of refraction n,

$$\Psi(\mathbf{r}, t) = a\, \Re\, e^{i(\mathbf{k}\cdot\mathbf{r} - \omega t)} = a\, \Re\, e^{ik_0(n\hat{\mathbf{k}}\cdot\mathbf{r} - ct)}. \tag{2.2}$$

Here $\Re$ stands for the real part, a is a constant amplitude, c is the speed of light in vacuum, ω is the angular frequency, and $\mathbf{k}$, the "wave vector", satisfies $\mathbf{k} = k_0 n \hat{\mathbf{k}}$, where $\hat{\mathbf{k}}$ is a unit vector pointing in the wave direction and k_0 is the "vacuum wave number". (That is, $k_0 \equiv 2\pi/\lambda_0 = \omega/c$, where λ_0 is the vacuum wavelength for the given frequency ω; linearity implies that all time variation has the same frequency everywhere.) The relation between wavefronts and wave direction of a plane wave is exhibited in Figure 2.1.

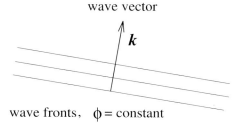

Fig. 2.1 The wave vector **k** is normal to the wavefronts of a plane wave.

The index of refraction n is a dimensionless number, typically in the range from 1 to 2 for the optics of visible light, (though slightly *less than* 1 for x-rays). Because n is the wavelength in free space divided by the local wavelength, the product $n\,dr$, or distance "weighted" by n, where dr is a path increment along $\hat{\mathbf{k}}$, is said to be the "optical path length". The "phase velocity" is given by

$$v = \frac{\omega}{|\mathbf{k}|} = \frac{c}{n}. \tag{2.3}$$

The group velocity will not be needed.

Problem 2.2.1 *Derive the result, valid for constant n, to be used shortly:*

$$\nabla(n\hat{\mathbf{k}} \cdot \mathbf{r}) = n\nabla(\hat{\mathbf{k}} \cdot \mathbf{r}) = n\hat{\mathbf{k}}. \tag{2.4}$$

Regarded as a function of position $\mathbf{r}$, the function $\phi(\mathbf{r}) = \hat{\mathbf{k}} \cdot \mathbf{r}$ has the property that surfaces of the form $\phi = $ constant form a family of parallel planes. The vector $\hat{\mathbf{k}}$ is orthogonal to these planes.

A wave somewhat more general than the one given by Eq. (2.2), but which has the same frequency, is required if $n(\mathbf{r})$ depends on position;

$$\Psi(\mathbf{r},t) = \psi(\mathbf{r})\Re e^{ik_0(\phi(\mathbf{r})-ct)}. \tag{2.5}$$

Since this wave function must satisfy the wave equation (2.1), the amplitude $\psi(\mathbf{r})$ is necessarily (though weakly) position-dependent. Since the function $\phi(\mathbf{r})$ in Eq. (2.5) takes the place of $n\hat{\mathbf{k}} \cdot \mathbf{r}$ in Eq. (2.2), it generalizes the previously mentioned optical path length; ϕ used to be known as the "eikonal", a name with no mnemonic virtue whatsoever to recommend it. Nowadays it is simply referred to as "the phase"; it advances by 2π and beyond as one moves a distance equal to one wavelength and beyond, along a ray. The relation between local wave direction and local wavefront is exhibited in Figure 2.2.

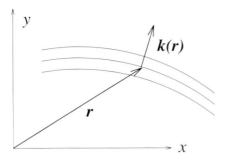

Fig. 2.2 Wavefronts of light wave in a medium with nonconstant index of refraction.

2.3
The Short Wavelength, Geometric Optics Limit

The condition characterizing "geometric optics" is for wavelength λ to be short compared to distances x over which $n(\mathbf{r})$ varies appreciably in a fractional sense. More explicitly this is $\frac{dn/n}{dx/(\lambda/2\pi)} \ll 1$, or

$$\frac{1}{n}\frac{dn}{dx} \ll k. \tag{2.6}$$

This is known as an "adiabatic" condition. (This condition is violated at boundaries, for example at the surfaces of lenses, but this can be accommodated by matching boundary conditions.) This approximation will permit dropping terms proportional to $|dn/dx|$. By matching exponents of Eq. (2.5) and Eq. (2.2) locally one can define a local wave vector $\hat{\mathbf{k}}$ such that

$$\phi(\mathbf{r}) = n(\mathbf{r})\,\hat{\mathbf{k}}(\mathbf{r}) \cdot \mathbf{r}. \tag{2.7}$$

This amounts to best-approximating the wave function locally by the plane wave solution of Eq. (2.2). Because n and $\hat{\mathbf{k}}$ are no longer constant, Eq. (2.4) becomes

$$\nabla \phi = (\nabla n(\mathbf{r}))\hat{\mathbf{k}}(\mathbf{r}) \cdot \mathbf{r} + n(\mathbf{r})\nabla(\hat{\mathbf{k}}(\mathbf{r}) \cdot \mathbf{r}) \approx n(\mathbf{r})\hat{\mathbf{k}}(\mathbf{r}), \tag{2.8}$$

where inequality Eq. (2.6) has been used to show that the first term is small compared to the second. Spatial derivatives of $\hat{\mathbf{k}}(\mathbf{r})$ have also been dropped because deviation of the local plane wave solution from the actual wave are necessarily proportional to $|dn/dx|$. A simple rule of thumb expressing the approximation is that all terms that are zero in the constant-n limit can be dropped. Equation (2.8) shows that $\nabla \phi$ varies slowly, even though ϕ varies greatly (i. e. of order 2π) on the scale of one wavelength.

One must ascertain, with ϕ given by Eq. (2.7), whether Ψ, as given by Eq. (2.5), satisfies the wave equation (2.1). Differentiating Eq. (2.5) twice, the approximation of neglecting the spatial variation of **r**-dependent factors, $n(\mathbf{r})$ and $\psi(\mathbf{r})$ relative to that of eikonal $\phi(\mathbf{r})$ can be made.

$$\nabla \Psi \approx ik_0 \nabla \phi(\mathbf{r}) \, \Psi, \quad \text{and} \quad \nabla^2 \Psi = \nabla \cdot \nabla \Psi \approx -k_0^2 |\nabla \phi|^2 \Psi. \tag{2.9}$$

With this approximation, substituting Eq. (2.1) becomes

$$|\nabla \phi(\mathbf{r})|^2 = n^2(\mathbf{r}), \quad \text{or} \quad |\nabla \phi(\mathbf{r})| = n(\mathbf{r}), \tag{2.10}$$

which is known as the "the eikonal equation". It can be seen to be equivalent to Eq. (2.8), provided $\phi(\mathbf{r})$ and $\hat{\mathbf{k}}(\mathbf{r})$ are related by Eq. (2.7). Then the eikonal equation can be written as a vector equation that fixes the direction as well as the magnitude of $\nabla \phi$,

$$\nabla \phi = n\hat{\mathbf{k}}. \tag{2.11}$$

The real content of this equation is twofold: it relates rate of phase advance $\nabla \phi$ in magnitude to the local index of refraction and in direction to the ray direction. Since this equation might have been considered obvious and been written down without apology at the start of this section, the discussion to this point can be regarded as a review of the wave equation and wave theory in the short wavelength limit.

2.3.1
Determination of Rays from Wavefronts

Referring to Figure 2.3, any displacement $\mathrm{d}\mathbf{r}_f$ lying in the equal phase surface, $\phi(\mathbf{r}) = \text{constant}$, satisfies[1]

$$0 = \frac{\partial \phi}{\partial x^i} \mathrm{d}x_f^i = \nabla \phi \cdot \mathrm{d}\mathbf{r}_f. \tag{2.12}$$

This shows that the vector $\nabla \phi$ is orthogonal to the surface of constant ϕ. (Which is why it is called the "gradient"—$\phi(\mathbf{r})$ varies most rapidly in that direction.) From Eq. (2.11) we then obtain the result that $\hat{\mathbf{k}}(\mathbf{r})$ is locally orthogonal to a surface of constant $\phi(\mathbf{r})$. "Wavefronts" are, by definition, surfaces of constant $\phi(\mathbf{r})$. Rays are directed locally along $\hat{\mathbf{k}}(\mathbf{r})$. It has been shown then that "rays" are curves that are everywhere normal to wavefronts. If the displacement $\mathrm{d}\mathbf{r}$ lies along the ray and $\mathrm{d}s$ is its length then $\mathrm{d}\mathbf{r}/\mathrm{d}s$ is a unit vector and hence

$$\hat{\mathbf{k}} = \frac{\mathrm{d}\mathbf{r}}{\mathrm{d}s}. \tag{2.13}$$

---
1) Repeated indices in Eq. (2.12), and future equations are summed over—the Einstein summation convention.

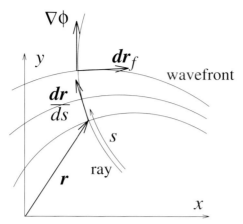

Fig. 2.3 Geometry relating a ray to the wavefronts it crosses.

Combining Eqs. (2.11) and (2.13) we obtain a *differential equation for the ray*,

$$\frac{d\mathbf{r}}{ds} = \frac{1}{n}\nabla\phi. \tag{2.14}$$

2.3.2
The Ray Equation in Geometric Optics

Equation (2.14) is a hybrid equation containing two unknown functions $\mathbf{r}(s)$ and $\phi(\mathbf{r})$ and as such is only useful if the wavefront function $\phi(\mathbf{r})$ is already known, but we can convert it into a differential equation for $\mathbf{r}(s)$ alone. Expressing Eq. (2.14) in component form, differentiating it, and then resubstituting from it yields

$$\frac{d}{ds}\left(n\frac{dx^i}{ds}\right) = \frac{d}{ds}\frac{\partial\phi}{\partial x^i} = \frac{\partial^2\phi}{\partial x^j \partial x^i}\frac{dx^j}{ds} = \frac{\partial^2\phi}{\partial x^j \partial x^i}\frac{1}{n}\frac{\partial\phi}{\partial x^j}. \tag{2.15}$$

The final expression can be re-expressed using Eq. (2.10);

$$\frac{\partial^2\phi}{\partial x^j \partial x^i}\frac{1}{n}\frac{\partial\phi}{\partial x^j} = \frac{1}{2n}\frac{\partial}{\partial x^i}\sum_j\left(\frac{\partial\phi}{\partial x^j}\right)^2 = \frac{1}{2n}\frac{\partial}{\partial x^i}|\nabla\phi|^2 = \frac{1}{2n}\frac{\partial n^2}{\partial x^i} = \frac{\partial n}{\partial x^i}. \tag{2.16}$$

Combining results yields the vector equation,

$$\frac{d}{ds}\left(n\frac{d\mathbf{r}}{ds}\right) = \nabla\mathbf{n}(\mathbf{r}). \tag{2.17}$$

This is "the ray equation". A second order, ordinary differential equation, it is the analog for "light trajectories" of the Newton equation for a point particle.

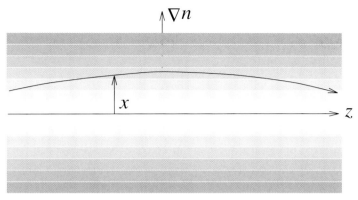

Fig. 2.4 A ray of light being guided by a graded fiber having axially symmetric index of refraction n. Darker shading corresponds to *lower* index of refraction.

In this analogy arc length s plays the role of time and the index of refraction $n(\mathbf{r})$ is somewhat analogous to the potential energy function $U(\mathbf{r})$. The analogy will be made more precise shortly.

All of the geometric optics of refraction of light in the presence of variable optical media, such as lenses, can be based on the ray equation.

Problem 2.3.1

- **Light rays in a lens-like medium.** *Consider paraxial, that is to say, almost parallel to the z-axis, rays in a medium for which the index of refraction is a quadratic function of the transverse distance from the axis;*

$$n(x,y) = n_0(1 + B\, r^2), \tag{2.18}$$

where $r^2 = x^2 + y^2$ and B is a constant. Given initial values (x_0, y_0) and initial slopes (x_0', y_0') at the plane $z = 0$, using Eq. (2.17) find the space curve followed by the light ray. See Yariv [1] or any book about fiber optics for discussion of applications of such media.

- *Next suppose the coefficient B in part (i) depends on z (arbitrarily though consistent with short wavelength approximation (2.6)), but that x and y can be approximated for small r as in part (i). In this case the "linearized" ray equation becomes*

$$\frac{d}{dz}\left(n(z)\frac{dx}{dz}\right) + k_B(z)x = 0, \quad \text{or} \quad p' + k_B x = 0, \tag{2.19}$$

where "momentum" $p(z) \equiv n(z)(dx/dz)$, prime stands for d/dz, and there is a similar equation for y. Consider any two (independent) solutions $x_1(z)$

and $x_2(z)$ of this equation. For example $x_1(z)$, can be the "cosine-like" solution with $C(0) = 1, C'(0) = 0$ and $x_2(z) \equiv S(z)$, the "sine-like" solution with $S(0) = 0, n(0)S'(0) = 1$. Show that propagation of any solution from $z = z_0$ to $z = z_1$ can be described by a matrix equation

$$\begin{pmatrix} x(z_1) \\ p(z_1) \end{pmatrix} = M \begin{pmatrix} x(z_0) \\ p(z_0) \end{pmatrix}, \tag{2.20}$$

where M is a 2×2 matrix called the "transfer matrix". Identify the matrix elements of M with the cosine-like and sine-like solutions. Show also, for sufficiently small values of r, that the expression obtained from two separate rays,

$$x_1(z)p_2(z) - x_2(z)p_1(z), \tag{2.21}$$

is conserved as z varies. Finally, use this result to show that $\det |M| = 1$. The analog of this result in mechanics is Liouville's theorem. In the context of optics it would not be difficult to make a more general proof by removing assumptions made in introducing this problem.

Problem 2.3.2 Consider an optical medium with spherical symmetry (e. g. the earth's atmosphere), such that the index of refraction $n(r)$ is a function only of distance r from the center. Let d be the perpendicular distance from the center of the sphere to any tangent to the ray. Show that the product nd is conserved along the ray. This is an analog of the conservation of angular momentum.

2.3.3
Obtaining Phase Information from Intensity Measurement

It is relatively straightforward to measure the spatial variation of the intensity of x-rays, for example those scattered from a sample. Determining the corresponding dependence of phase has been a perennial challenge of x-ray physics. A natural approach is to use interferometric methods, but there has been some progress in extracting phase information from measurements of intensity only. A method suggested by Teague [2] was to measure intensities at two or more longitudinally displaced positions and to extract phase information from the gradients of the intensity.

It is not difficult to include some physical optics in the preceeding theory of wave evolution—namely the variation of light intensity along a light ray. The resulting equation is known as the "Transport of Intensity Equation". This equation can then be used, at least in principle, to convert intensity measurements into phase measurements.

Studying the evolution of light intensity motivates paying attention not just to one ray, but rather to all nearby rays, and to the evolution of their local density. Visualize the trajectory described so far as one ray in a *steady*,

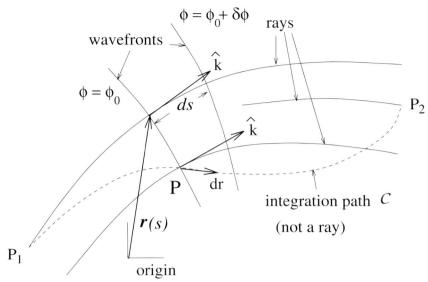

Fig. 2.5 A sample of rays and wavefronts belonging to a steady scalar "light" wave. A possible integration path not along a ray is also shown. There is one ray and one wavefront through each point.

transversely-extended beam of light. The quantity needed to describe the light intensity I is the Poynting vector $\mathbf{S} = I\hat{\mathbf{k}} = I\nabla\phi/n$ where, as before, $\mathbf{k}$ is a unit vector pointing along the ray, and the new physics needed is conservation of energy. The energy crossing directed area $d\mathbf{A}$ per unit time is $\mathbf{S} \cdot d\mathbf{A}$ and energy conservation is expressed by the *continuity equation*

$$0 = \nabla \cdot \mathbf{S} = \nabla \cdot \left(\frac{I}{n}\nabla\phi\right), \tag{2.22}$$

where Eq. (2.14) has been used, and there is no time derivative term because the beam is steady. This equation expresses the condition that there are no local sources or sinks of energy. Figure 2.5 shows a sample of rays and wavefronts belonging to the steady waves—there is one ray and one wavefront through each point. It is important to distinguish between rays and general curves in the region under study. One such general curve $\mathcal{C}$ joins the points P_1 and P_2 in the figure.

As shown in Eq. (2.14), because $\nabla\phi$ is directed along a ray, the operator $\nabla\phi \cdot \nabla = n(d/ds)$ is a directional derivative for displacements along a ray s; with s being arc length, this operator is tailor-made for use with the ray parametrization $\mathbf{r}(s)$. Performing the differentiation indicated in Eq. (2.22) and then solving for $\nabla^2\phi$ yields

$$\nabla^2\phi = -\frac{1}{I/n}\nabla\phi \cdot \nabla\left(\frac{I}{n}\right) = -\nabla\phi \cdot \nabla\left(\ln\frac{I}{n}\right) = -n\frac{d}{ds}\left(\ln\frac{I}{n}\right). \quad (2.23)$$

This is the "Transport of Intensity Equation". With n constant (for free space propagation) the equation is ascribed to Teague [2], perhaps because he first showed how it could be used to measure phases. But the present derivation has been copied from Born and Wolf [3]. With n variable the equation is applied to phase contrast imaging, which is especially appropriate for high energy x-rays where the intensity contrast between adjacent features tends to be very weak.

Equation (2.23) allows $\nabla^2\phi/n$ to be integrated, *but only along a ray*, to obtain the variation of I

$$\ln\left(\frac{I_2}{n_2}\frac{n_1}{I_1}\right) = -\int_1^2 \frac{\nabla^2\phi}{n}ds. \quad (2.24)$$

Taking ratios between two points, at longitudinal positions s_1 to s_2 yields

$$\frac{I_2}{I_1} = \frac{n_2}{n_1}e^{-\int_{s_1}^{s_2}\frac{\nabla^2\phi}{n}ds}. \quad (2.25)$$

This result is significant in that, using only ray analysis, it governs a property—intensity—that might have seemed to need a wave calculation.

Problem 2.3.3 *In the special case of a plane wave propagating in a region of constant n one has $\nabla^2\phi = 0$ and the intensity is constant along a ray. For an isotropic cylindrically-spreading wave find $\phi(\rho)$ and confirm Eq. (2.25). Do the same for an isotropic spherically-spreading wave.*

2.4
Wave Description of Gaussian Beams

Though this chapter began by studying waves, the effect of going to the short wavelength limit has been to reproduce the ray theory that was the subject of Chapter 1.

Now we want to follow the wave theory somewhat further, especially paying attention to the wavefronts of the wave. This will further develop the analogy between beams of light and beams of particles. This analysis is essential to understand the focusing and coherence properties of photon beams and of optical lasers—in particular free electron lasers [1]. [2]

Continuing to suppress polarization properties, one can treat light as a scalar wave. Let $\Psi(t;x,y,z) \sim \exp(-i\omega t)$ be such a wave, which we assume to be monochromatic and traveling more or less parallel to the z axis in a "focusing medium" such as an optical fiber. Paraxial approximations will be assumed to be valid. Suppressing the factor $\exp(-i\omega t)$, the wave equation satisfied by Ψ is

$$\nabla^2 \Psi + (k^2 - k k_{2,x} x^2 - k k_{2,y} y^2)\Psi = 0. \tag{2.26}$$

Here $k \equiv k(z)$ is the on-axis wave number and the spatial variation of the index of refraction has been specialized to quadratic dependence on the transverse coordinates x and y. Though not exibited explicitly, the coefficients are also allowed to depend on z. It is the quadratic terms $-k k_{2,x} x^2 - k k_{2,y} y^2$ that cause the wavelength to depend on transverse position and this is what causes focusing.

2.4.1
Gaussian Beam in a Focusing Medium

For consistency with the rest of this chapter we will simplify a bit by assuming that $k_{2,x} = 0$ and that Ψ is independent of x. However all the formulas generalize naturally to simultaneous x and y dependence and (especially) to azimuthally-symmetric systems for which $k_{2,x} = k_{2,y}$, in which case the motion depends only on $r = \sqrt{x^2 + y^2}$. Simplifying the notation slightly, $k_{2,y} \to k_2$, the wave equation is

$$\frac{\partial^2 \Psi}{\partial y^2} + \frac{\partial^2 \Psi}{\partial z^2} + (k^2 - k k_2 y^2)\Psi = 0. \tag{2.27}$$

Since we will study only paraxial beams centered on the z-axis we represent Ψ as

$$\Psi(y,z) = \psi(y,z)\, e^{ikz}, \tag{2.28}$$

where $\psi(y,z)$ is assumed to be a slowly varying "modulating" factor, satisfying the inequalities $\partial^2 \psi/\partial z^2 \ll k^2 \psi$ and $\partial^2 \psi/\partial z^2 \ll k \partial \psi/\partial z$. Substituting into Eq. (2.27), we find that $\psi(y,z)$ must satisfy

$$\frac{\partial^2 \psi}{\partial y^2} + 2ik\frac{\partial \psi}{\partial z} - k k_2 y^2 \psi = 0. \tag{2.29}$$

---

2) There are "factor of two" differences between the present treatment of cylindrical waves, dependent only on the transverse coordinate y and the azimuthally symmetric (i. e. dependent only on the radial component $r = \sqrt{x^2 + y^2}$) treated in Ref. [1].

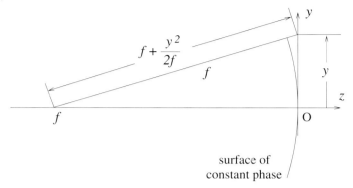

Fig. 2.6 For a focus to have existed at the longitudinal position $-ff$ relative to point O, the wavefront must be a sphere of radius f. The "extra path length" at transverse displacement y is therefore approximately $f + y^2/(2f) - f$.

With a view towards obtaining two conditions, one for motion for which y is negligible, and another where y motion is important, we seek a solution of this equation in the form

$$\psi = \exp\left(iP(z) + ik\frac{y^2}{2}\frac{1}{f(z)}\right). \tag{2.30}$$

The motivation for this parametrization can be inferred from Figure 2.6. Expressing wave phases on the transverse plane through O, for a plane wave parallel to the axis, the phase would be zero everywhere on the plane. Suppose the phase is *actually* given by $ky^2/(2f)$. (This is the first nonvanishing term in a power series for which the constant term is set to zero, the second vanishes by symmetry, and the coefficient is, for convenience, espressed as $k/(2f)$.) The beam therefore appears to be diverging from a focus at longitudinal position $-f$ relative to O. The preceeding rays need not actually have emerged from a point, but the phase variation on the plane is consistent with this history. The z-dependent parameter $f(z)$ is therefore the effective local radius of curvature of the wavefront passing point O at position z. An overall phase kz has already been "factored out" by definition (2.28), and the term $P(z)$ allows for a phase shift that depends on z but not on y. As well as being dependent on z, for reasons to be explained, $f(z)$ will later be allowed to be complex and be called the "complex radius of curvature". When substituting from Eq. (2.30) into Eq. (2.29) one can separate terms proportional to y^2 from terms having coefficients independent of y. To be valid for all y both coefficients must vanish. This leads to

$$\frac{1}{f^2} + \frac{d}{dz}\left(\frac{1}{f}\right) + \frac{k_2}{k} = 0, \quad \text{and} \quad \frac{dP}{dz} = \frac{i}{2f}. \tag{2.31}$$

2.4.2
Spatial Dependence of a Wave Near a Free Space Focus

In free space Eqs. (2.31) simplify to

$$\frac{1}{f^2} + \frac{d}{dz}\left(\frac{1}{f}\right) = 0, \quad \text{and} \quad \frac{dP}{dz} = \frac{i}{2f}. \tag{2.32}$$

From the first of these $f(z) = z + b$, where b is an arbitrary constant. Defining

$$\frac{1}{f} = \frac{1}{z - iz_0} = \frac{z + iz_0}{z^2 + z_0^2} = \frac{1}{z(1 + z_0^2/z^2)} + \frac{i}{z_0(1 + z^2/z_0^2)}, \tag{2.33}$$

the second equation yields

$$P(z) = \frac{i}{2}\ln\left(1 + \frac{iz}{z_0}\right), \tag{2.34}$$

where the constant of integration has been chosen as $-(i/2)\ln(-iz_0)$. Substituting these expressions into Eq. (2.30) yields

$$\psi = \exp\left(-\ln\sqrt{1 + i\frac{z}{z_0}} + k\frac{y^2}{2}\left(\frac{i}{z(1 + z_0^2/z^2)} - \frac{1}{z_0(1 + z^2/z_0^2)}\right)\right). \tag{2.35}$$

Using the relation $\ln(a + ib) = \ln\sqrt{a^2 + b^2} + i\tan^{-1}(b/a)$ the first term in the exponent becomes

$$\exp\left(-\ln\sqrt{1 + i\frac{z}{z_0}}\right) = \frac{1}{\sqrt{1 + (z/z_0)^2}}\exp\left(-i\tan^{-1}\frac{z}{z_0}\right). \tag{2.36}$$

To cast Eq. (2.35) into a form appropriate for comparison with Figure 1.11, we define

$$w_0^2 = \frac{z_0}{k}, \quad w(z) = w_0\sqrt{1 + \frac{z^2}{k^2 w_0^4}}, \quad \text{and} \quad R(z) = z\left(1 + \frac{z_0^2}{z^2}\right). \tag{2.37}$$

Finally, by Eq. (2.28), the wave field is given

$$\Psi(y, z) = \Psi_0 \frac{w_0}{w(z)} \exp\left(-i\tan^{-1}\frac{z}{z_0}\right)\exp\left(ikz + \frac{y^2}{2}\left(\frac{ik}{R(z)} - \frac{1}{w^2(z)}\right)\right). \tag{2.38}$$

This is rather complicated, but the magnitude of Ψ depends only on the real part of the exponent, which is $-y^2/(2w^2(z))$. So the distribution is Gaussian transversely, with $w(z)$ being the r.m.s. beam width. The first exponential factor, being purely imaginary, has no effect on the absolute value and, furthermore is independent of y. This factor (known as the "Gouy phase factor") will be unimportant in the sequel.

We now apply this functional form to describe the wave near a "point focus". In geometric optics all rays pass precisely through a point focus. But this is inconsistent with wave behavior, where at most a "necking-down" of the wave is possible. Such a region is analogous to the beam waist, exhibited in Figure 1.11, which applied to a beam made up of a bunch of particles having beam width $E(z)$. By the second of Eqs. (2.37) the functional form of beam width $w(z)$ is the same as that of $E(z)$ as given by Eq. (1.59).

For emphasis, let us repeat the last point. Both the Courant–Snyder formalism and the wave description of this section contain more "physics" than is contained in geometric optics. The *beam waist* is a characteristic feature of a particle beam in free space, and a *focus* is a characteristic feature of a wave in free space. Curiously, even though the Courant–Snyder contains no wave-like ingredient, it gives the same behavior near a waist that the wave theory gives near a focus. This means there is a kind of wave-particle duality even in classical physics.

In laser physics terminology, mode (2.38) is known as the "fundamental mode", though that terminology refers to an azimuthally symmetric, three dimensional wave, while we are dealing with a 2D wave with only one transverse coordinate. This "lowest mode" has resulted because only terms of order y^2 have been retained. Producing the standing wave corresponding to this mode, by reflecting the beam back and forth between spherical (or rather, in our case, cylindrical) mirrors, will be discussed later.

2.4.3
The ABCD Law

One wishes also to apply the formalism that has been developed to the evolution of waves in a "focusing medium" for which the index of refraction term k_2/k in Eq. (2.31) is nonvanishing. The reason for having chosen the symbol $R(z)$ in Eq. (2.38), is that in a pure cylindrical wave, the spatial dependence near the z-axis is as

$$\exp ikR = \exp ik\sqrt{z^2 + y^2} \approx \exp\left(ikz + \frac{y^2}{2}\frac{ik}{R}\right). \tag{2.39}$$

The exponent here contains two of the three terms in the second exponential of Eq. (2.38). The third term can also be included by allowing the previously-introduced radius of curvature f to be complex. Separating $1/f$ into real and imaginary parts,

$$\frac{1}{f(z)} = \frac{1}{R(z)} + \frac{i}{kw^2(z)}. \tag{2.40}$$

The real part represents the actual curvature of the wave and the imaginary part represents the phase evolution corresponding to the spatial spreading of

2.4 Wave Description of Gaussian Beams

the wave. If the wave were allowed to expand indefinitely, its representation by phase dependence on a transverse plane through the point O would not be sensible. But, in practice, the same optical components that keep the rays paraxial, also keep the wavefronts more or less coincident with transverse planes.

Though Eq. (2.38) was derived to describe free space evolution of the wave, we need not demand that $\Psi(y, z)$ as given by Eq. (2.38) came from that source. Rather we can regard $R(z)$ and $w(z)$ as a parametrization of an arbitrary wave, whose properties may have been formed by propagation through previous optical elements. $R(z)$ is the local radius of curvature of the wavefront and $w(z)$ is a measure of the local beam width (actually height since y is a vertical coordinate.) For this representation to be useful we must be able to calculate the evolution of $R(z)$ and $w(z)$ through arbitrary optical elements.

In order for Eq. (2.38) to describe field evolution in a focusing medium, the differential equation satisfied by f is Eq. (2.31). This can be converted into a linear equation by changing the dependent variable from f to $\mathcal{F}$ according to

$$\frac{1}{f} = \frac{\mathcal{F}'}{\mathcal{F}}, \quad \text{so} \quad \mathcal{F}'' + \frac{k_2}{k}\mathcal{F} = 0. \tag{2.41}$$

where, as usual, d/dz is indicated by a prime. The general solution of this equation is

$$\mathcal{F}(z) = a \sin\sqrt{\frac{k_2}{k}} z + b \cos\sqrt{\frac{k_2}{k}} z,$$

$$\mathcal{F}'(z) = a\sqrt{\frac{k_2}{k}} \cos\sqrt{\frac{k_2}{k}} z - b\sqrt{\frac{k_2}{k}} \sin\sqrt{\frac{k_2}{k}} z. \tag{2.42}$$

Using the first of Eqs. (2.41) to transform the dependent variable back from $\mathcal{F}$ to f yields

$$f(z) = \frac{f_0 \cos\sqrt{\frac{k_2}{k}} z + \sqrt{\frac{k}{k_2}} \sin\sqrt{\frac{k_2}{k}} z}{-f_0 \sqrt{\frac{k_2}{k}} \sin\sqrt{\frac{k_2}{k}} z + \cos\sqrt{\frac{k_2}{k}} z}. \tag{2.43}$$

For the evolution $z_0 \to z_1$ this transformation is conventionally written as

$$f_1 = \frac{A_{10} f_0 + B_{10}}{C_{10} f_0 + D_{10}}. \tag{2.44}$$

The A, B, C, D coefficients can be read off by evaluating, at $z = z_0$, the sines and cosines in the ratio given by Eq. (2.43). Subsequent evolution $z_1 \to z_2$ is given by

$$f_2 = \frac{A_{21} f_1 + B_{21}}{C_{21} f_1 + D_{21}}. \tag{2.45}$$

The concatenated transformation takes the same form,

$$f_2 = \frac{A_{20} f_0 + B_{20}}{C_{20} f_0 + D_{20}}. \tag{2.46}$$

Direct substitution shows that this concatenation satisfies

$$\begin{pmatrix} A_{20} & B_{20} \\ C_{20} & D_{20} \end{pmatrix} = \begin{pmatrix} A_{21} & B_{21} \\ C_{21} & D_{21} \end{pmatrix} \begin{pmatrix} A_{10} & B_{10} \\ C_{10} & D_{10} \end{pmatrix}. \tag{2.47}$$

This evolution rule is known as the "ABCD law".

By a miracle of the sort that makes physics so satisfactory, the ABCD matrices introduced here are none other than the transfer matrices introduced in Section 1.2. The reason for this is that Eq. (1.2) and the second of Eqs. (2.41) are the same. The matrix elements are the sine-like and cosine-like solutions of this equation. As a result the coefficients in Eq. (2.43) are the same as the transfer matrix elements for Eq. (1.2). Since all linear optical elements are special (or limiting) cases of a uniform focusing medium, and since the same matrix concatenation holds, the result is true for arbitrary beam lines.

The ABCD rules govern the evolution of $f(z)$ (which is to say, of $R(z)$ and $w(z)$). They apply equally to photon and electron beams though, as far as I know, the formalism has never been applied, except implicitly, to electron beams. In the world of lasers there is a formalism that treats $1/f$ as a complex impedance and develops analogs between optical lines and lumped constant electrical circuits [4, 5].

2.4.4
Optics Using Mirrors

Designing optical beamlines for x-rays is greatly hampered by the absence of lossless refractive media at short wavelengths.[3] The formalism just derived can be used for the analysis of optics based on mirrors. The laser "spherical resonator" shown in Figure 2.7 is an important example. Since the transfer matrix for a mirror is

$$\begin{pmatrix} 1 & 0 \\ -2/R & 1 \end{pmatrix}, \tag{2.48}$$

Equation (2.44) produces

$$\frac{1}{f_1} = -\frac{2}{R} + \frac{1}{f_0}. \tag{2.49}$$

---

3) There is however an appreciable range of wavelengths for which x-rays can be reflected and this makes glancing incidence spherical or cylindrical mirrors practical.

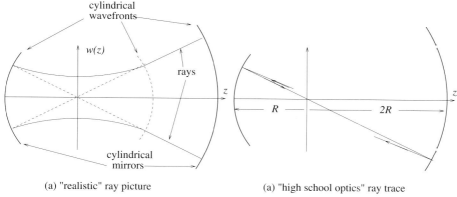

Fig. 2.7 Optical resonator defined by two cylindrical mirrors. (a) More or less realistic rays and wavefronts. (b) Elementary ray tracing using focal length equals radius/2.

Substituting from Eq. (2.40), and ignoring imaginary parts,

$$\frac{1}{R_1} = -\frac{2}{R} + \frac{1}{R_0}. \tag{2.50}$$

When this formula is applied to the optical system shown in Figure 2.7 one sees that the image of a point source at the center of one mirror is at the center of the other. The formula also partially validates the "high school optics" formula

$$\frac{1}{\text{object-dist.}} + \frac{1}{\text{image-dist.}} = \frac{1}{\text{focal-length}}, \tag{2.51}$$

where the focal length of a mirror is half its radius. One sees though that the actual rays are not at all what one draws with classical ray tracing. The actual beam envelope shown is the same as would be obtained using the β-function formalism of accelerator physics. Of course the beam divergences shown in the figure are unrealistically great.

Concentric mirror pairs like those shown in Figure 2.7 can form a resonator for light at IR, UV and visible wavelengths. By inserting such a resonator in an electron storage ring (with concentric holes to let the beam through) it is possible to achieve stimulated, free electron laser (FEL) radiation, that is synchronized with the bunch structure of the accelerator. This configuration is impractical for x-rays, however, because mirror reflectivities become negligible for normal incidence.

The subject matter of this text is restricted to x-rays harder than, say, one kilovolt. For this reason closed FELs like those mentioned in the previous paragraph will not be discussed further.

Even though mirrors are no good for x-rays at normal incidence there can be high reflectivity at glancing incidence. Mirrors are therefore practical even for hard x-rays, though the geometries are far less favorable, and give greater distortion, than those shown in Fig. (2.7). Examples are given in Chapter 9.

2.4.5
Wave Particle Duality for Electrons

Because the quantum mechanical properties of electrons play no further role in this book, the material in this section is less essential than the material in the rest of this chapter. It is included only to complete the parallel treatment of beams or waves consisting of photons or electrons.

A mechanical system analogous to a ray of light, has a single particle of kinetic energy $T = (p_x^2 + p_y^2 + p_z^2)/(2m)$ traveling more or less parallel to the z axis in a "potential well" such that its potential energy is $V = k'_{2,x} x^2/2 + k'_{2,y} y^2/2$, with Hamiltonian $H(x,y;p_x,p_y,p_z) = T + V$ and total energy $\mathcal{E}$. The Schrödinger equation for such a system is

$$i\hbar \frac{\partial \Psi}{\partial t} = -\frac{\hbar^2}{2m} \nabla^2 \Psi + V(\mathbf{r}) \Psi. \tag{2.52}$$

In the classical limit one neglects $\hbar$, for example by substituting

$$\Psi(\mathbf{r},t) = A e^{iS(\mathbf{r},t)/\hbar}, \tag{2.53}$$

and then setting $\hbar = 0$. The result is

$$\frac{1}{2m}(\nabla S)^2 + V(\mathbf{r}) = -\frac{\partial S}{\partial t}. \tag{2.54}$$

This is the Hamilton-Jacobi equation for the analogous mechanical system. In this formalism $\mathbf{p} = \nabla S$ and $\mathcal{E} = -\partial S/\partial t$, so Eq. (2.54) amounts to conservation of energy. Because the gradient operator picks out the normal to a surface of constant S, the equation $\mathbf{p} = \nabla S$ relates the particle direction to surfaces of constant S in the same way that rays are related to wavefronts in optics—which is that rays are everywhere normal to wavefronts. The trajectories derived from this equation are necessarily the same as the solutions of our original propagation equation (1.2).

The connections between particle quantities and wave quantities are the deBroglie and Planck relations,

$$\mathbf{k}(\mathbf{r}) = \frac{\mathbf{p}(\mathbf{r})}{\hbar}, \quad \text{and} \quad \omega = \frac{p_0^2}{2m\hbar} = \frac{\hbar k_0^2}{2m}. \tag{2.55}$$

Let us then introduce a new dependent variable $\mathbf{k}(\mathbf{r})$, related to $S(\mathbf{r})$ by

$$\frac{S(\mathbf{r})}{\hbar} = \mathbf{k}(\mathbf{r}) \cdot \mathbf{r} - \frac{p_0^2}{2m\hbar} t, \quad \text{so} \quad \Psi(\mathbf{r},t) = A e^{i\mathbf{k}(\mathbf{r})\cdot \mathbf{r}} e^{-i\frac{\hbar k_0^2}{2m} t}. \tag{2.56}$$

The frequency ω introduced here is rather artificial in classical mechanics and has no significance other than being $\mathcal{E}$ is different units. In any case, $\hbar k_0$ is a constant equal to the momentum a particle of this frequency has when, because it is traveling precisely along the z-axis, it has neither potential energy nor transverse kinetic energy.

With Ψ having the time dependence shown in Eq. (2.56), and after relating the parameters $k'_{2,x}$ and $k'_{2,y}$ to $k_{2,x}$ and $k_{2,y}$, one sees that Eq. (2.26) and Eq. (2.54) are the same. In the short wavelength limit then, the rays of a solution to the wave equation are trajectories of a material particle propagating in the corresponding potential.

2.5
Synchrotron Radiation: Waves or Particles?

A question to be answered at least qualitatively is: When is it valid to consider synchrotron radiation as being made up of discrete photons following paths called rays? When we consider an intense undulator we will find that wave features are essential, but we will now see that treating synchrotron radiation from the regular arc bending magnets as particles is justified for sufficiently hard x-rays.

For light of frequency ν, wavelength $\lambda \stackrel{\text{e.g.}}{=} 10^3 \text{Å}$; (being near ultraviolet, this is beyond the long wavelength end of the range of wavelengths we are likely to be interested in) the wave number k is given by

$$k = \frac{2\pi\nu}{c} = \frac{2\pi}{\lambda} \quad (\stackrel{\text{e.g.}}{=} 2\pi \times 10^7 \text{ m}^{-1}). \tag{2.57}$$

It is possible to form, say at $z = 0$, a "minimal circular point image" of radius w_0 using plane waves of this wavelength, all traveling more or less parallel to the z axis. But the width w of such a wave necessarily increases for $z \neq 0$. By symmetry the spreading is $\pm z$-symmetric, so the leading deviation from w_0 is quadratic in z. According to the second of Eqs. (2.37), the half-width of a cylindrically spreading beam is given by

$$w^2(z) = w_0^2 + \frac{z^2}{k^2 w_0^2}. \tag{2.58}$$

Essentially the same result is given by Mandel and Wolf [6]. This spreading is closely related to the Heisenberg uncertainty principle, since the transversely localized beam spot has to have a matching angular spread.

In the pure particle picture discussed in Chapter 1, the beam spreading is given by Eq. (1.57), as illustrated in Figure 1.11. With w being proportional to $\sqrt{\beta}$ these dependences are strikingly similar. We can now define a "wave-based waist half-length" $z_{\text{waist}}^{(W)}$

$$z_{\text{waist}}^{(W)} = kw_0^2, \qquad (2.59)$$

as the half-length over which the (squared) width doubles according to wave theory. This can be compared to $z_{\text{waist}}^{(P)}$ as given by Eq. (1.58). Since the x-ray beam waist is determined by the electron beam, we set $w_0^2 = \epsilon_{y,b}\beta^*$ and obtain

$$\frac{z_{\text{waist}}^{(W)}}{z_{\text{waist}}^{(P)}} = k\epsilon_{y,b} \quad (\stackrel{\text{e.g.}}{=} 2\pi \times 10^7 \times 10^{-8}). \qquad (2.60)$$

For the particle picture to be justified this ratio should be larger than 1. For the example numbers that have been used (ultraviolet and shorter) the condition is satisfied for the value $\epsilon_{y,b} = 10^{-8}$ m assumed here. On the other hand emittances smaller than this can certainly be achieved, which makes the condition harder to meet.

In the jargon of synchrotron light sources, the radiated beam is said to be "diffraction limited" when the ratio in Eq. (2.60) is *small* compared to 1. When trying to achieve x-ray beams of high brilliance (and hence high coherence) (as described in Chapter 11) the challenge will be to meet this condition for the values of λ appropriate for *hard* x-rays.

2.6
X-ray Holography and Phase Contrast and Lens-free Imaging

As in Eq. (2.38) the amplitude of any *diverging* single mode, cylindrical, electromagnetic wave has the form

$$\Psi(y,z) = \Psi_0 \frac{w_0}{w(z)} \exp\left(ikz + \frac{y^2}{2}\left(\frac{ik}{z\left(1+\frac{z_0^2}{z^2}\right)} - \frac{1}{w_0^2\left(1+\frac{z^2}{k^2w_0^4}\right)}\right)\right), \qquad (2.61)$$

for $z > 0$. Such a formula would apply even to a beam of x-rays, provided their wavelength were long enough for the beam to be coherent. Until recently, because x-ray beams were incoherent sums of many such waves, such a formula had no relevance for x-ray beams, but, with modern high brightness sources, especially for soft x-rays, there is the possibility of "diffraction limited" operation for which Eq. (2.61) is applicable.

When a sample is placed in such a coherent beam and, further downstream, the illumination is photographed, the result is a known as a Gabor in-line hologram. An image can then be formed by coherent illumination of the hologram. In a refinement of this approach known as "phase contrast imaging", there is no need to record the hologram. Rather, part of the beam is stopped and the rest sent through a focusing lens that forms a phase contrast image of the sample. These, and many similar variants, depend on the interference between rays following different paths.

Even with coherence high enough to make Eq. (2.61) applicable, these techniques are still not necessarily practical for x-rays. Without a converging lens no image can be reconstructed from the hologram, nor can the phase contrast imaging be done. The possibility of performing these sorts of imaging therefore requires the existence of x-ray lenses. There has been substantial recent progress toward achieving this goal. Further discussion of configurations like these is contained in Chapter 9. A phase-contrast imaging configuration is shown in Figure 9.11.

Even with the transverse coherence implied by Eq. (2.61) and the availability of x-ray lenses these sorts of imaging can also be foiled by insufficient longitudinal coherence. The longitudinal coherence length of a beam from an undulator with some large number $N_w \stackrel{\text{e.g.}}{=} 1000$ poles has a coherence length of only 1000 wavelengths. For 1 Å photons this is only 0.1 μm. For two paths to interfere, their path length difference cannot exceed this length. For path lengths being tens of meters this would be all but impossible to achieve. Fortunately, by monochromatization, as the energy spread is reduced, the coherence length increases proportionally (at the cost of proportionally decreasing flux.) This can make it possible to match path lengths closely enough to enable the interference needed to make these types of imaging possible.

I have stated that a focusing lens is necessary for x-ray imaging, but this is not entirely true. A kind of lens-free imaging is possible using a coherent x-ray beam. Snigerev et al. [7] describe a scheme for phase contrast imaging by coherent x-rays in the 10–50 keV range. Holograms photographed some tens of centimeters past a sample (that causes phase shift but no absorption) provide a kind of image of an object that would otherwise be invisible to x-rays. Their beam is described by the wave function given above in Eq. (2.61). The presence of the refractive sample introduces a further y dependent phase shift factor $\exp(\phi(y))$. The illumination in the hologram is obtained as a standard Huygens-Kirchoff integral over the wave function. Unmodified this would give uniform illumination, but the $\exp(\phi(y))$ factor causes a pattern agreeing with their observation.

There are also lens-free methods of image reconstruction under development that are based on the numerical processing of digitized hologram patterns.

References

1 A. Yariv (1997), *Optical Electronics in Modern Communications*, 5th Edn., Oxford University Press, Oxford.

2 M. Teague (1993), *Deterministic phase retrieval: a Green's function solution*, J. Opt. Soc. Am. **73** (11), 1434.

3 M. Born, E. Wolf (1970), *Principles of Optics*, 4th Edn., Pergamon Press, Oxford.

4 H. Kogelnik (1965), *On the propagation of Gaussian beams through lenslike media*, Appl. Opt., **4**, 1562.

5 A. Gerrard, J. Burch (1975), *Introduction to Matrix Methods in Optics*, Dover, New York.

6 L. Mandel, E. Wolf (1995), *Optical Coherence and Quantum Optics*, Cambridge University Press, Cambridge.

7 A. Snigerev et al. (1995), *On the possibilities of x-ray phase contrast microimaging by coherent high-energy synchrotron radiation*, Rev. Sci. Instrum., **66**, (12), 5486.

3
Synchrotron Radiation From Accelerator Magnets

3.1
Capsule History of Synchrotron Light Sources

A kind of shorthand for the historical development of the storage ring as a radiation source has grown up, in which synchrotron light sources have been assigned to "generations", one through four. The terminology seems especially appropriate since the time between generations is not very different from the time between new human generations, with the fourth generation beginning near the turn of the century. But, like many qualitative classifications, this one is understood differently by different individuals. The generations defined here are more tied to the particle physics origins of accelerators than most x-ray scientists would favor. The four generations are:

1. The physics of synchrotron radiation was well established on electron accelerators long before storage rings were even conceived of. Understanding accelerator performance depended on understanding the influence of synchrotron radiation on the circulating beams. Early accelerators were designed with only elementary particle physics in mind. The synchrotron light was welcome only to the extent it indicated there was beam in the machine. Otherwise it was just a nuisance. Eventually, though, it was appreciated that the radiation was "brighter" than that available with conventional sources, and could be used as incident beams for low energy physics. This usage, usually just parasitic, defined the first generation.

2. It was *storage rings* that first provided x-ray beams unambiguously superior to traditional UV and x-ray sources for doing low energy physics. Because of the strong dependence of synchrotron radiation on magnetic field it was soon realized that radiation intensity along one (or a few) beamlines could be greatly increased by introducing wiggler magnets into the ring. These so-called "insertion devices" had minimal influence on the performance of the storage ring, and the lattice designs continued to be predicated largely on their use for elementary particle physics. Some usage for low energy physics continued to be parasitic, but ded-

Accelerator X-Ray Sources. Richard Talman
Copyright © 2006 WILEY-VCH Verlag GmbH & Co. KGaA, Weinheim
ISBN: 3-527-40590-9

icated usage also became common. Colliding beam physics and light source physics both sought high beam current and, at least initially, both sought small electron emittances. Common design, therefore continued to be sensible. The boundary between second and third generations is especially vague, but it soon became obvious that the synchrotron light and colliding beam applications were incompatible, especially because of the large number of beam lines needed for low energy energy physics. These could only be provided in dedicated light sources.

3. Once divorced from elementary particle physics, accelerator design advanced rapidly in optimizing storage rings for their dedicated use as light sources. Great progress was made in low emittance lattice design. Modern facilities dedicated entirely to producing photon beams from insertion devices are known as third generation light sources. Some high energy (greater than 5 GeV) examples are ESRF in France, APS at Argonne, and SPRING8 in Japan. There have been many intermediate energy (2 − 3 GeV) examples, with NSLS at Brookhaven the first with optimized-for-light-source lattice design and CLS at Saskatoon the most recently commissioned (at the time this is being written). A low energy (less than 2 GeV) example is ALS, at Berkeley. The single most important measure of quality of such rings is the "brilliance", a property that will be central to the continuing discussion in this text. At third generation facilities it also became clear that "undulators" (wigglers with relatively weak magnetic fields) produce especially useful beams for many purposes.

4. One seeks designs that go beyond existing light sources by having extraordinary properties such as ultrashort bunches or ultrahigh brightness. What constitutes the most promising route to provide these features is controversial. Two candidates are free electron lasers (FEL) and energy recovery linacs (ERL). Since these are based on novel principles they certainly deserve to be designated as "post-third" generation. There are other candidates for greatly improved improved light source performance based on the de- and re-commissioning of large storage ring facilities built initially for colliding beam physics. Their large circumferences and expansive (and previously paid-for) infrastructure, make it economical for them to be re-built using largely conventional design, with performance comparable with the above-mentioned post-third-generation sources. One example is PETRA which is currently being so reconfigured as PETRA III at DESY, in Germany. Another example, analysed in detail in this text, is a proposed reconfiguration of the CESR facility at Cornell. Its minimum-emittance, rapid-cycling, design makes its potential brilliance greater than that of the ERL mentioned

above. In this text, because these generations succeed an earlier, expiring, generation, these proposals are also referred to as "fourth generation".

3.2 Generalities

An electron traveling in a uniform magnetic field emits synchrotron radiation. In this chapter we will assume the orbit is a continuous circle, but the formulas can be easily adapted to shorter segments (as long as they are not *too* short.) The so-called "arc magnets" of an ordinary electron storage ring amply satisfy this requirement, but some insertion devices (and proton accelerators) may require special treatment.

Though the theoretical formulas describing synchrotron radiation from a charged particle following a circular orbit are well within the difficulty level of texts like Jackson [2], their accurate evaluation is rather difficult. The brief discussion of technological development given above showed that, in many ways, it is undulator radiation that is of the greatest importance for exploiting light sources to produce external photon beams. Formally, the theory of bending magnet and undulator radiation is the same, but it is our good fortune that the formulas are more easily evaluated in the more important case, namely undulator radiation. This will be described in a later chapter. However, because synchrotron radiation from the bending magnets continues to be the dominant influence on the circulating beam in storage rings, it remains necessary to calculate also the bending magnet radiation. In this chapter the most important ideas and formulas for bending magnet radiation will be introduced, and plausible approximation schemes contemplated, but exact evaluation will not be stressed.

The text I recommend (and follow) for everything up to, but not yet including, synchrotron radiation, is Griffiths [1] For detailed synchrotron radiation calculations there is, of course, Jackson [2].[1] Other good references are Schwinger [3], Hofmann [4,5] and Wiedemann [6].

The most subtle physics associated with synchrotron radiation is the concept of retarded time. Once that has been grasped all that is left is mathematical manipulation. Unfortunately these manipulations bring in unwelcome complication such as Macdonald functions and Airy integrals. This mathematical complexity tends to obscure the physics. A major purpose of this chapter is to attempt approximations for which only elementary mathematical

---

1) Though Jackson has a very clear section on undulator radiation, his formulation is insufficiently general to cover undulator resonances higher than the lowest energy one. This is unfortunate since it has turned out, in practice, that undulator resonances of third, fifth, and even higher are the resonances mainly used.

functions are adequate. Most of these approximations are well justified, but two are rather poorly controlled. The first of these is a "physics" assumption. It will only retain the dominant (horizontal) polarization, and over-estimate it substantially at that.. For the purpose of understanding accelerator dynamics an error of this magnitude might be tolerable but it is still uncomfortably large. We will later be able to repair this approximation. The second poorly justified approximation, is of a purely computational nature. It will leave an uncontrolled uncertainty, at roughly the 10% level, that can only be eliminated by more careful evaluation of the integrals.

The basis for the error estimates just given is that an essentially exact formulation (due originally to Schwinger) is known. This being the case, one might well ask, "why not just study the exact theory?" The reason, as already stated, is that the exact theory is mathematically difficult and may not provide much useful intuition. A consequence of this is that one must later approach new devices, such as undulators or free electron lasers, without much intuitive guidance. It is hoped that the approximate formulation presented here will improve intuition concerning where the dominant contributions and polarizations come from and into such subtleties as coherence and diffraction. More precise numerical determinations can be found in the references already given.

These justifications for approximate treatment may be feeble, but the treatment in this chapter can serve as warm-up preparation for following an exact treatment. The needed physical formulas are presented and all dimensional factors are correctly included. Only the numerical factors multiplying the results will be less accurate than needed for some purposes. The more important case of undulator radiation will be handled accurately in a later chapter.

The most important parameter in the sequel will be $\gamma = \mathcal{E}_e/(m_e c^2)$. The following equations, and especially the final approximation, will be assumed to be "second nature" to the reader:

$$\begin{aligned}\gamma &= \sqrt{1 - \frac{v^2}{c^2}}, \\ \frac{1}{\gamma^2} &= 1 - \frac{v^2}{c^2}, \\ \frac{v}{c} &= \sqrt{1 - \frac{1}{\gamma^2}} \approx 1 - \frac{1}{2\gamma^2}.\end{aligned} \quad (3.1)$$

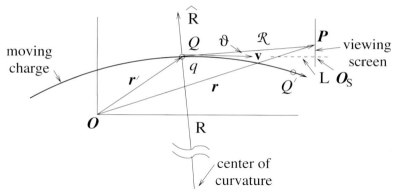

Fig. 3.1 Symbol definitions: **r** = radius vector of "field point" P; **r**′ = radius vector of "source" point Q; $\mathcal{R}$ = distance from source point to field point. R is the local bending radius and $\hat{R}$ is the outward unit normal. When light from Q gets to P the particle is already at Q', very nearly in the plane of the viewing screen.

3.3
Potentials and Fields

The basic formulas on which synchrotron radiation is based are the so-called scalar and vector "retarded potentials":

$$V(\mathbf{r},t) = \frac{1}{4\pi\epsilon_0} \iiint \frac{\rho(\mathbf{r}',t_r)}{\mathcal{R}} d^3\mathbf{r}',$$

$$\mathbf{A}(\mathbf{r},t) = \frac{\mu_0}{4\pi} \iiint \frac{\rho(\mathbf{r}',t_r)\mathbf{v}(\mathbf{r}',t_r)}{\mathcal{R}} d^3\mathbf{r}', \qquad (3.2)$$

where **r**′ and **r** are know as "source point", Q and "field point", P respectively, as shown in Figure 3.1. The integrals run over all charge-containing volumes. Formulas essentially like these should be familiar from elementary E.&M. The present formulas differ from electro- and magnetostatics only in the seemingly inconsequential respect of allowing for the time of signal propagation from **r**′ to **r**. That is, to obtain the fields at P at time t the integrands are to be evaluated at a "retarded time" t_r such that

$$t = t_r + \frac{\mathcal{R}}{c}. \qquad (3.3)$$

If the charge density ρ is time-independent this modification has no effect, and Eqs. (3.2) reduce to the Coulomb and Biot-Savart laws.

This retardation modification makes an important difference only when charges are moving at close to the speed of light, which is of course the case in a synchrotron light source. There is a kind of "sonic boom" phenomenon in which, if a source particle approaches the field point P at high speed, signals from an appreciable path interval "pile up" at P.[2]

Once the integrals in Eqs. (3.2) have been evaluated, the electric and magnetic fields are obtained from the potentials by[3]

$$\mathbf{E} = -\nabla V - \frac{\partial \mathbf{A}}{\partial t}, \quad \mathbf{B} = \nabla \times \mathbf{A}. \tag{3.4}$$

Working out these derivatives is rather difficult, however.

3.4
Relations Between Observation Time and Retarded Time

The radiation pattern is often compared to the headlight of a locomotive rounding a curve. This is illustrated in Figure 3.2 which shows a coordinate system on the transverse viewing screen. The radiation peak sweeps along the x-axis in this plane. The y axis passes through that tangent to the circular orbit that is normal to the viewing screen. When a reference electron touched this tangent at point Q at an appropriately earlier time it was aimed at the origin O_s. The radiation falls off rapidly for large positive and large negative values of y and, at any point near the x-axis, the fields are large only during a brief interval appropriately delayed from the time when the "headlight" aimed at the point O_s, the origin of the screen coordinates.

The observation point P is taken to be at $x = 0$, but vertically displaced to y, at a vertical angle ψ from the origin.

It is convenient to define the zero of retarded time ($t_r = 0$) and the zero of observer time ($t = 0$) to mark the passage of the radiating particle through the point of tangency Q. If we regard the arrival of a photon at point P as a "signal" sent from the circulating particle as it passed through point Q then that signal arrives at P at observer time $t = L/(c \cos \psi)$.

During time t_r the moving particle traverses an arc of length vt_r. Because the particle moves on a circle, the longitudinal range over which there are appreciable contributions to the field at P is very short; in particular the range is much less than L. To calculate t in terms of t_r the strategy will be to identify and calculate effects that delay the arrival at "P" of the "signal" emitted at time t_r compared to the time for photon transit direct from Q to P. Because

2) Because the particle speed v cannot exceed c, light from two points on the particle orbit cannot arrive at P simultaneously, so synchrotron light is not *exactly* analogous to a sonic boom. But with the speed of light being six orders of magnitude higher than the speed of sound, a gigantic "magnification" is nevertheless possible. A more nearly exact analog of a sonic boom is Cherenkov radiation.
3) To correlate formulas in this chapter with formulas in papers by Albert Hofmann, one must appreciate that his $\mu_0 \mathbf{A}$ is our $\mathbf{A}$.

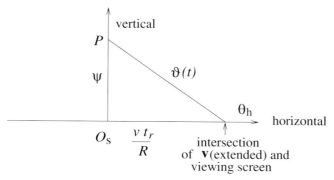

Fig. 3.2 Passage of the radiation maximum across the viewing screen through observation point P. The distance along the tangent line from tangent point Q to viewing screen is L. The coordinates are shown as angles at tangent point Q.

of the high electron speed, and the nearly tangential geometry both signals arrive at P at very nearly the same time, but there are three delaying effects.

Consider a "signal" sent (i.e. a photon emitted) at time t_r from a point slightly past Q. Because the electron is a bit slow, a delay

$$t_r\left(1 - \frac{v}{c}\right) \approx \frac{t_r}{2\gamma^2}, \qquad (3.5)$$

is introduced. There is also a delay due to the slightly non-optimal path taken by the electron—it is not "aimed" toward P. This delay is

$$t_r(1 - \cos\psi) \approx t_r \frac{\psi^2}{2}. \qquad (3.6)$$

Furthermore, because the electron follows a circular rather than a straight line path, it has not advanced quite as far toward P as it "should have". The effect is a delay

$$\frac{c^2 t_r^3}{6R^2}. \qquad (3.7)$$

This delay is illustrated in Figure 3.3. Combining these deviations we get

$$\begin{aligned} t &= \frac{L/c}{\cos\psi} + t_r\left(\frac{1}{2\gamma^2} + \frac{\psi^2}{2}\right) + \frac{c^2 t_r^3}{6R^2} \\ &\stackrel{?}{\approx} \frac{L}{c\cos\psi} + \frac{1 + \gamma^2\psi^2}{2\gamma^2} t_r. \end{aligned} \qquad (3.8)$$

The $L/(c\cos\psi)$ term is unimportant—it will be suppressed shortly by simply redefining the origin of observer time, and it does not affect dt/dt_r. The term

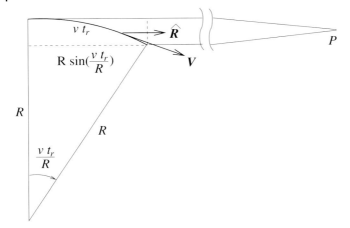

Fig. 3.3 Top view of geometry for calculation of retarded time. Following the curved path introduces delay $t_r - (R/c)\sin(vt_r/R) \approx c^2 t_r^3/(6R^2)$. The figure is also intended to serve as a reminder that a Fraunhofer limit is valid—that is, the observation point is so far away as to be effectively at infinity. In this approximation the path length from a transverse plane to point P is independent of position in the plane. This simplifies the retarded time calculation.

$c^2 t_r^3/(6R^2)$ has been singled out for special discussion in the final version of Eq. (3.8). This term causes the relation between t and t_r to be cubic, which greatly complicates the integrals that need to be evaluated, especially since the inverse relation is needed. Since the radiation is dominated by the region near $t = 0$ the term proportional to the t_r^3 term will, in fact, turn out to be relatively unimportant. This may be somewhat surprising though, since the presence of this term is the only evidence that the electron is actually traveling in a circle rather than a straight line.

The numerator of retarded vector potential **A**, in Eq. (3.2), depends relatively weakly on **v**, which varies along the circular orbit. The highly peaked nature of the integral is caused by the retarded time dependence of the denominator. As $|t_r|$ increases the cubic term in Eq. (3.8) eventually overwhelms the linear term, effectively "cutting off" the integral.[4] This is how the circle parameters influence the formulas for the observed radiation. The approximation in the last form of Eq. (3.8) is not, therefore, legitimate for evaluating the retarded potential integrals, but, for relating t and t_r in the region of dominant intensity near the center of the radiation pattern, the approximation may be applicable. In a later chapter, when undulator radiation is being discussed, there is no corresponding cut-off of the integrals, and this approximation will greatly simplify the calculations.

4) Terms of order t_r^5 have been dropped. This is justified by the fact already mentioned that the t_r^3 term cuts off the integral before higher order terms can have appreciable effect.

3.4 Relations Between Observation Time and Retarded Time

In any case, the determination of scalar potential V and vector potential $\mathbf{A}$ has been reduced to the evaluation of the integrals in Eq. (3.2) after substituting for t_r from Eq. (3.8) (inverted). Following Hofmann, the inversion can be performed explicitly, using standard formulas for a cubic equation. Introduce symbols

$$t_P = t - \frac{L/c}{\cos\psi}, \quad \omega_0 = \frac{c}{R}, \quad \omega_c = \frac{3\gamma^3\omega_0}{2}, \quad \text{and} \quad w = \frac{2\omega_c t_P}{(1+\gamma^2\psi^2)^{3/2}}, \tag{3.9}$$

the first of which amounts to translating the time origin to the nominal signal arrival time at P, the second introduces the revolution frequency ω_0, the third introduces the "critical frequency" ω_c, and w is just a convenient abbreviation. Equation (3.8) becomes

$$(\gamma\omega_0 t_r)^3 + 3(1+\gamma^2\psi^2)\gamma\omega_0 t_r - 4\omega_c t_P = 0. \tag{3.10}$$

Since the "discriminant" D of this equation,

$$D = (1+\gamma^2\psi^2)^3 + (2\omega_c t_P)^2, \tag{3.11}$$

is clearly positive, there is one real root of Eq. (3.10). This root is given by

$$t_r = \frac{\sqrt{1+\gamma^2\psi^2}}{\gamma\omega_0}\left((w+\sqrt{1+w^2})^{1/3} + (w-\sqrt{1+w^2})^{1/3}\right). \tag{3.12}$$

This formula is too complicated for heuristic discussion at the level of this text, but it is very practical for numerical calculations.

For performing the derivatives in Eq. (3.4) it is sufficient to relate the retarded time differential dt_r to the observation time differential dt. Rather than using Eq. (3.8), a more compact expression as a scalar product can be obtained by first differentiating Eq. (3.3) with respect to t_r and then using the geometry of Figure 3.1 to obtain $d\mathcal{R}/dt_r$. The result is

$$dt = dt_r\left(1 + \frac{1}{c}\frac{d\mathcal{R}}{dt_r}\right) = dt_r\left(1 - \hat{\mathcal{R}}\cdot\frac{\mathbf{v}(t_r)}{c}\right). \tag{3.13}$$

It is because the factor in parenthesis on the right-hand side can be extremely small that dt_r can greatly exceed dt. This situation occurs only when $v \approx c$ and $\mathbf{v}$ is directed approximately toward P.

Problem 3.4.1 *Confirm Eqs. (3.5) through (3.7).*

Problem 3.4.2 *A typical value for t_r in Figure 3.3 is $R/(\gamma c)$ where $R \approx 50$ m and $\gamma \approx 10^4$. Assume that the lower ray in the figure refers to this time. If the distance to point P in the figure is $L = 50$ m, estimate the range of path length variations,* and hence the range of retarded time variations, for the band of rays shown in the figure. (Take the bottom and top rays to have the same (maximum) length, so the ray half way between them will have the minimum length.) Compare this time range from minimum to maximum with the observation time interval between the arrivals of the top and bottom rays. This calculation is supposed to confirm the validity of the "Fraunhofer" approximation mentioned in the caption of Figure 3.3.

Problem 3.4.3 In Figure 3.3 the unit vector $\widehat{\mathcal{R}}$ points toward P (angle ψ above the plane of the orbit) from the point on the orbit at the same "typical" retarded time $t_r = R/(\gamma c)$ as in the previous problem. Using the same numerical values as in the previous problem, approximate the parenthesized factor on the right-hand side of Eq. (3.13). Use Taylor series to obtain the leading terms in ψ and $1/\gamma$ before plugging in numbers.

Problem 3.4.4 From Eq. (3.8) determine $dt/dt_r|_{t_r=0}$ and confirm that the result agrees with Eq. (3.13). Use the fact that $\vartheta = \psi$ at $t_r = 0$.

3.5
Evaluation of Electric and Magnetic Fields

The electric and magnetic fields are obtained by substituting from Eq. (3.13) into Eqs. (3.4). The calculation is spelled out in Section 10.3.2 of Griffiths. Because it is lengthy the calculation will not be repeated here (except for identifying, below, where the dominant contribution comes from.) The result of doing the calculation, and dropping "near-field" terms, is that electric and magnetic fields at P are given by

$$\mathbf{E}(\mathbf{r},t) = \frac{q}{4\pi\epsilon_0 c} \frac{1}{\mathcal{R}} \left(\frac{\widehat{\mathcal{R}} \times \left((\widehat{\mathcal{R}} - \mathbf{v}/c) \times \dot{\mathbf{v}}/c \right)}{(1 - \widehat{\mathcal{R}} \cdot \mathbf{v}/c)^3} \right)_{\text{ret.}}$$

$$\mathbf{B}(\mathbf{r},t) = \frac{1}{c} \widehat{\mathcal{R}} \times \mathbf{E}(\mathbf{r},t). \tag{3.14}$$

The term that has been dropped has spatial dependence $1/\mathcal{R}^2$ instead of $1/\mathcal{R}$. Roughly speaking, it describes the Coulomb field "attached" to the charge and is unimportant for the observer at P since it is appreciable only near the point Q' where q has arrived at time t. It is a sufficient approximation to take the factor $1/\mathcal{R}$ outside the retarded time calculation since $\mathcal{R}$ varies little over the relevant range. All formulas so far have been known for about a century.

All quantities in Eq. (3.14) are to be evaluated at the retarded time. It is important to specify the meaning of $\dot{\mathbf{v}}$ unambiguously. For circular (radius R) motion at constant speed v, its magnitude is $|d\mathbf{v}/dt| = v^2/R$, which is constant, and it points toward the instantaneous center of curvature. Using this, a result that will be needed later is

$$\mathbf{E}(\mathbf{r}, t) = \frac{q/\mathcal{R}}{4\pi\epsilon_0 c} \left[\frac{1}{1 - \widehat{\mathcal{R}} \cdot \mathbf{v}/c} \frac{d}{dt} \left(\frac{\widehat{\mathcal{R}} \times (\widehat{\mathcal{R}} \times \mathbf{v}/c)}{1 - \widehat{\mathcal{R}} \cdot \mathbf{v}/c} \right) \right]_{\text{ret.}}$$

$$= \frac{q/\mathcal{R}}{4\pi\epsilon_0 c} \left[\frac{1}{1 - \widehat{\mathcal{R}} \cdot \mathbf{v}/c} \frac{d}{dt} \left(\frac{-\mathbf{v}_\perp/c}{1 - \widehat{\mathcal{R}} \cdot \mathbf{v}/c} \right) \right]_{\text{ret.}}, \quad (3.15)$$

where $\mathbf{v}_\perp$ is the component of $\mathbf{v}$ normal to $\widehat{\mathcal{R}}$. This surprisingly simple result can be proved by straightforward differentiation, treating $\widehat{\mathcal{R}}$ as constant (which is the same "Fraunhofer" approximation as has been used before.) The derivative d/dt acts on a function in which $\mathbf{v}$ is the only variable and, when d/dt does act on $\mathbf{v}$ it generates $\dot{\mathbf{v}}$, whose meaning has just been specified.

The denominator factor in Eq. (3.14) is

$$1 - \widehat{\mathcal{R}} \cdot \mathbf{v}/c = 1 - \cos\vartheta \sqrt{1 - \frac{1}{\gamma^2}} \approx \frac{1}{2\gamma^2}(1 + \gamma^2 \vartheta^2). \quad (3.16)$$

where ϑ is the angle between $\mathbf{v}$ and $\widehat{\mathcal{R}}$.

Following Hofmann, to evaluate the numerator of Eq. (3.14), it is useful to introduce (x, y, z) coordinates with origin at point of tangency Q, z axis tangential and x axis radially outward, and to express the needed vectors in terms of t_r:

$$\begin{aligned}
\mathbf{v} &= -v\sin(vt_r/R)\,\hat{\mathbf{x}} + v\cos(vt_r/R)\,\hat{\mathbf{z}} \\
\dot{\mathbf{v}} &= -(v^2/R)\cos(vt_r/R)\,\hat{\mathbf{x}} - (v^2/R)\sin(vt_r/R)\,\hat{\mathbf{z}} \\
\widehat{\mathcal{R}} &= \sin\psi\,\hat{\mathbf{y}} + \cos\psi\,\hat{\mathbf{z}} \\
\widehat{\mathcal{R}} - \mathbf{v}/c &= (v/c)\sin(vt_r/R)\,\hat{\mathbf{x}} + \sin\psi\,\hat{\mathbf{y}} + (\cos\psi - (v/c)\cos(vt_r/R))\,\hat{\mathbf{z}}
\end{aligned}$$
(3.17)

Restricting the observation point to be vertically above the origin O_s of the observation plane, as the formula for $\widehat{\mathcal{R}}$ assumes, is not a restriction since the symmetry of the situation causes the radiation pattern (except for arrival time) to be independent of horizontal angle. Using the triple cross product formula and simplifying yields

$$\frac{cR}{v^2}\widehat{\mathcal{R}} \times \left((\widehat{\mathcal{R}} - \mathbf{v}/c) \times \dot{\mathbf{v}}/c \right) = \left(\cos(vt_r/R) - (v/c)\cos\psi \right) \hat{\mathbf{x}} \\ - \cos\psi\,\sin\psi\,\sin(vt_r/R)\,\hat{\mathbf{y}} \\ + \sin^2\psi\,\sin(vt_r/R)\,\hat{\mathbf{z}}. \quad (3.18)$$

Using $v/c \approx 1 - 1/(2\gamma^2)$ and small angle approximations, and replacing v by c where applicable, this becomes

$$\widehat{\mathcal{R}} \times \left(\left(\widehat{\mathcal{R}} - \frac{\mathbf{v}}{c}\right) \times \frac{\dot{\mathbf{v}}}{c}\right) = \frac{c}{2\gamma^2 R}\left(1 + \gamma^2\psi^2 - \frac{(c\gamma t_r)^2}{R^2}\right)\hat{\mathbf{x}} - \frac{c^2\psi t_r}{R^2}\hat{\mathbf{y}}. \quad (3.19)$$

Substituting from Eqs. (3.16) and (3.19) into Eq. (3.14), the electric field is given by

$$\mathbf{E}(\mathbf{r}, t) = \frac{q}{4\pi\epsilon_0 c}\frac{8\gamma^6}{\mathcal{R}}\frac{\frac{c}{2\gamma^2 R}\left(1 + \gamma^2\psi^2 - (c\gamma t_r/R)^2\right)\hat{\mathbf{x}} - \frac{c^2\psi t_r}{R^2}\hat{\mathbf{y}}}{(1 + \gamma^2\psi^2 + (c\gamma t_r/R)^2)^3}, \quad (3.20)$$

where $\vartheta^2 = \psi^2 + (ct_r/R)^2$ has been used. Magnetic field $\mathbf{B}(\mathbf{r}, t)$ is then readily determined using the second of Eqs. (3.14).

Problem 3.5.1 *Check the calculations leading to Eq. (3.19), especially because the y-component differs by a factor of two from the value given by Hofmann. (His corresponding formula on page 18 has y-component less by a factor of 2, and the same factor is missing from E_y in his Eq. (24) on the same page. I don't know if the error propagates to later formulas. It would not affect his Fig. (14) and most of his subsequent results are based on the Fourier-transformed field which is differently derived. The signs here are opposite to Hofmann's because our opposite choice of positive x direction.) Apart from the importance of checking these comments, the nature of the approximations can only be appreciated by working through them in detail in problems such as this.*

Before continuing the development with this (essentially) exact formula, we will derive a formula which, though not a terribly good approximation, gives a good intuitive picture of the dominant radation pattern.

3.5.1
Radial Field Approximation

A more reckless approach to evaluating expression (3.18) would have been to perform the simplification

$$\widehat{\mathcal{R}} \times \left(\left(\widehat{\mathcal{R}} - \frac{\mathbf{v}}{c}\right) \times \frac{\dot{\mathbf{v}}}{c}\right) \approx -\frac{\dot{\mathbf{v}}}{c}\left(1 - \widehat{\mathcal{R}} \cdot \frac{\mathbf{v}}{c}\right). \quad (3.21)$$

The rationale supporting this simplification is that $\dot{\mathbf{v}}$ is approximately orthogonal to $\widehat{\mathcal{R}}$, which makes the first term in the triple cross product expansion vanish. What makes it reckless is that the retained term has a factor that is also small precisely at $t_r = 0$, and that $\dot{\mathbf{v}}$ is orthogonal to $\widehat{\mathcal{R}}$ only precisely at the center of the radiation pattern. These reservations notwithstanding, using the approximation in Eq. (3.14) yields

$$\mathbf{E}(\mathbf{r},t) \approx -\frac{\mu_0}{4\pi} \frac{q}{\mathcal{R}} \left[\frac{\dot{\mathbf{v}}}{(1-\widehat{\mathcal{R}}\cdot\mathbf{v}/c)^2} \right]_{\text{ret.}}. \qquad (3.22)$$

Comparison with Eqs. (3.2) and (3.4) shows that this contribution to **E** comes from the term $-\partial \mathbf{A}/\partial t$—one of the denominator factors $(1-\widehat{\mathcal{R}}\cdot\mathbf{v}/c)$ comes from the previously mentioned piling up[5], and the other comes from the d/dt operation in Eq. (3.4).

In the approximation of Eq. (3.22), the radiation polarization is such that the electric field, being parallel to $\dot{\mathbf{v}}$, is purely horizontal (assuming the bend plane is horizontal). This is the rationale for referring to Eq. (3.22) as a "radial field approximation", but it should be recognized that a term has been left out that contributes appreciably to the horizontally polarized radiation, at least on the tails of the pattern. (These terms actually reduce the total radiation from that given by the approximate formula.) In the approximation that the electric field is horizontal, the magnetic field is vertical.

With the charged particle q traveling in a circle of radius R we have $\dot{\mathbf{v}} = -\widehat{\mathbf{R}} c^2/R$ and

$$\mathbf{E}(\mathbf{r},t) \approx \frac{q}{4\pi\epsilon_0} \frac{1}{\mathcal{R} R} \left[\frac{\widehat{\mathbf{R}}}{(1-\widehat{\mathcal{R}}\cdot\mathbf{v}/c)^2} \right]_{\text{ret.}}. \qquad (3.23)$$

The range of validity of approximation (3.21) remains to be investigated,[6] but this formula is certainly easier to use and remember than the exact formula (3.14). (For undulators the same approximation going into its derivation will be far more valid.)

Some typical values are $R = 100\,\text{m}$, $\mathcal{R} \approx 20\,\text{m}$. Except for the factor $(1-\widehat{\mathcal{R}}\cdot\mathbf{v}/c)^{-2}$, the magnitude of the field given by Eq. (3.23) would be that of a point charge q at (the very great) distance $\sqrt{\mathcal{R} R}$. The denominator factor was calculated in Eq. (3.16) to be $(1+\gamma^2\vartheta^2)/(2\gamma^2)$. The maximum enhancement factor is therefore $4\gamma^4 \stackrel{\text{typ.}}{=} 4\times 10^{16}$. This *huge* factor can convert what would

---

5) Griffiths (for example) shows that the integrals in Eq. (3.2) acquire a factor $(1-\widehat{\mathcal{R}}\cdot\mathbf{v}/c)^{-1}$ because of an effective elongation of charged volumes by this factor with no change in charge density ρ.

6) Remember that the range of validity of the approximation has not been investigated, so the reader is advised to be skeptical of seemingly over-simple results, and to refer to exact calculations for more accurate formulas when actually designing apparatus or analysing data. The approximation made in Eq. (3.21) retains only the dominant contribution to horizontal polarization, and keeps no vertical polarization component. Hofmann refers to horizontal polarization as "σ-mode" and vertical polarization as "π-mode".

otherwise seem like a *small* electric field given by Eq. (3.23) into an extremely *large* electric field. The same formula shows that the electric field has fallen to one quarter of its maximum value already when $\vartheta = \vartheta_{\text{typ.}} = 1/\gamma \stackrel{\text{typ.}}{=} 10^{-4}$. This fall-off occurs both vertically and (at the instant the tangent ray passes point of tangency P) horizontally.

To make the formula for $\mathbf{E}(\mathbf{r}, t)$ explicitly dependent on the observation time[7], t it is necessary to evaluate the denominator factor Eq. (3.16), which depends on the angle ϑ, as shown in Figure 3.2. Since the horizontal angle is given as a function of t_r, it is necessary to employ relation (3.8). Here I will use the approximated version;

$$\vartheta^2 = \psi^2 + \left(\frac{vt_r}{R}\right)^2 \approx \psi^2 + \left(\frac{2c\gamma^2/R}{1+\gamma^2\psi^2}\right)^2 t^2 \qquad (3.24)$$

Here the time t origin has been shifted to be zero when the tangential photon arrives at point P; this suppressed the $L/(c\cos\psi)$ term of Eq. (3.8). Substituting into Eq. (3.23) yields

$$\mathbf{E}(\mathbf{r}, t) \approx \frac{q}{4\pi\epsilon_0} \frac{4\gamma^4}{\mathcal{R}R} \left(1 + \gamma^2\psi^2 + \left(\frac{2c\gamma^3/R}{1+\gamma^2\psi^2}\right)^2 t^2\right)^{-2} \hat{\mathbf{R}}. \qquad (3.25)$$

Clearly, even apart from approximations already mentioned, this formula is valid only if the magnet can be treated as "long", in the sense

$$L > \frac{2R}{\gamma}. \qquad (3.26)$$

If this condition is satisfied, the full radiation cone of angle $\pm 1/\gamma$ will sweep past an observer at P. Otherwise, the radiation pattern will be limited to a shorter time interval. Inequality (3.26) is *not* satisfied in an undulator.

3.6
Total Power Radiated and its Angular Distribution

More physically significant and measurable than the electric and magnetic fields is the power carried by the radiation. Eventually, when the beam is interpreted as a bundle of photons, the number of photons will be obtained from the beam power by using the Planck formula for the energy of individual photons. Radiation power density is described by the Poynting vector. The

---

7) From here on the observation time $t = 0$ origin will be taken to be the nominal signal arrival time at point P; that is, the symbol t_P will be replaced by t.

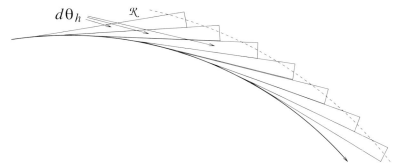

Fig. 3.4 Angular ranges segmented for purposes of calculating total energy carried away from the ring by a single passage of the moving charge q. To avoid worrying about "double counting" and "leakage" through tangential edges, flux can be calculated through the dashed circle (actually a cylinder.) The area and obliquity factors cancel in Eq. (3.29).

Poynting vector observed at point P is

$$\mathbf{S} = \frac{1}{\mu_0} \mathbf{E} \times \mathbf{B} = \frac{1}{\mu_0 c} \mathbf{E} \times (\widehat{\mathcal{R}} \times \mathbf{E}) = \frac{E^2}{\mu_0 c} \widehat{\mathcal{R}}, \qquad (3.27)$$

where $\mathbf{B}$ has been obtained from Eq. (3.14) and $\mathbf{E}$ is known to be perpendicular to $\widehat{\mathcal{R}}$ (in the radiation field.) The power passing through transverse area $(\mathcal{R}\,d\psi)(\mathcal{R}\,d\theta_h)$ in the observation plane is given by $|\mathbf{S}|\mathcal{R}^2\,d\Omega$, where θ_h is a horizontal angle away from the local center (where the maximum impinges at $t = 0$) and $d\Omega = d\psi\,d\theta_h$. From the single charge q this flux is strongly peaked, both in t and in vertical angle ψ. The total energy dU passing through a band of width $\mathcal{R}\,d\theta_h$ is given by

$$d = (\mathcal{R}\,d\theta_h) \int_{-\infty}^{\infty} dt \int_{-\pi/2}^{\pi/2} (\mathcal{R}d\psi)\,\frac{E^2}{\mu_0 c}. \qquad (3.28)$$

The total energy carried away from the ring can be calculated as suggested in Figure 3.4. By taking the limit $d\theta_h \to 0$ any dependence of the integral on horizontal angle is suppressed, yet the radiation emitted while completing one complete revolution can be obtained by setting $d\theta_h = 2\pi$. Also, the fall-off with increasing ψ is so fast that the vertical limits can be taken to be $\pm\infty$. The total energy carried away after one revolution of charge q is therefore given by

$$U_0 \approx \frac{2\pi}{\mu_0 c} \int_{-\infty}^{\infty} dt \int_{-\infty}^{\infty} d\psi\, \mathcal{R}^2 E^2$$

$$= \frac{2q^2 c \gamma^8}{\pi \epsilon_0 \mathcal{R}^2} \int_{-\infty}^{\infty} d\psi \int_{-\infty}^{\infty} \frac{dt}{\left(1+\gamma^2 \psi^2 + \left(\frac{2c\gamma^3/\mathcal{R}}{1+\gamma^2\psi^2}\right)^2 t^2\right)^4}. \quad (3.29)$$

Radial approximation (3.25) has been used. It is reassuring that this result is independent of $\mathcal{R}$ which was chosen arbitrarily. To complete the evaluation requires the integrals

$$\int_{-\infty}^{\infty} \frac{dx}{(a^2+b^2 x^2)^4} = \frac{5\pi}{16 a^7 b}, \quad \text{and} \quad \int_{-\infty}^{\infty} \frac{dx}{(1+a^2 x^2)^{5/2}} = \frac{4}{3a}, \quad (3.30)$$

with the result

$$U_0 \approx \frac{5}{16} \frac{q^2 \gamma^5}{\epsilon_0 R} \int_{-\infty}^{\infty} \frac{d\psi}{(1+\gamma^2 \psi^2)^{5/2}} = \frac{5}{12} \frac{q^2 \gamma^4}{\epsilon_0 R}. \quad (3.31)$$

According to Sands [8], the exact value (in GeV) of U_0 is

$$U_0 = C_\gamma \frac{E^4}{R}, \quad \text{where} \quad C_\gamma = \frac{4\pi}{3} \frac{r_e}{(m_e c^2)^3} = 0.885 \times 10^{-4}\,\text{m (GeV)}^{-3},$$

$$r_e = \frac{e^2}{4\pi \epsilon_0 m_e c^2} = 2.81784 \times 10^{-15}\,\text{m}. \quad (3.32)$$

Our estimate (3.31) differs from the exact value in the ratio $\frac{5/12}{1/3} = 1.25$, i.e. our estimate is 25% on the high side. The overestimate of horizontally-polarized power more than makes up for the neglect of vertically-polarized power. The estimate is improved in the following problem.

Problem 3.6.1 *Formula (3.20) can be used for a more accurate determination of the total power radiated. Note that the numerator factor for E_x is, except for the sign of the last term, the same as the denominator factor. After adding twice this term, the numerator cancels against one of the denominator factors, and reduces to our approximate formula Eq. (3.29). Evaluate the integrals that result when Eq. (3.20) is substituted into Eq. (3.29) and the cubic term is dropped from the $t_r \to t$ relation. Compare the result with Eq. (3.31). Using* Maple *the integrals are very manageable; I get $5/12 \to 13/48 + 7/240 = 3/10$, closer to the accurate value $1/3$. The horizontally polarized energy is roughly ten times the vertically polarized energy.*

Continuing with the approximate, radial component only, formulation, the (vertical) angular distribution of the radius can be read off from Eq. (3.31):

$$\frac{dU_0}{d\psi} \approx \frac{5\pi}{4} \frac{q^2}{4\pi\epsilon_0 R} \frac{\gamma^5}{(1+\gamma^2 \psi^2)^{5/2}}. \quad (3.33)$$

Since the horizontal distribution is uniform the power radiated per solid angle is given by

$$\frac{dU_0}{d\Omega} \approx \frac{5}{8} \frac{q^2}{4\pi\epsilon_0 R} \frac{\gamma^5}{(1+\gamma^2\psi^2)^{5/2}}. \tag{3.34}$$

3.7
Spectral Power Density of the Radiation

The Fourier transform of **E** is defined by

$$\tilde{\mathbf{E}}(\omega) = \frac{1}{\sqrt{2\pi}} \int_{-\infty}^{\infty} e^{i\omega t}\, \mathbf{E}(t)\, dt. \tag{3.35}$$

Eventually negative values of ω will be interpreted as corresponding to positive physical frequencies having the same magnitude.

3.7.1
Estimate of Frequency Spectrum from Pulse Duration

Since we have a closed form expression for the electric field, it is straightforward to find its Fourier transform, and from that the energy spectrum of the radiation. Before doing this we can obtain a rough estimate by referring to Figure 3.5 and Eq. (3.9). The figure illustrates the retarded time calculation for a typical pulse duration, where "typical" refers to the fact that the radiation is largely contained in a cone of angular radius $1/\gamma$ about the electron direction. Setting to zero the radiation outside this cone, the start of the radiation pulse comes from point A and the end comes from point A'. The difference in arrival times of these signals is equal to the time difference Δt between electrons taking a curved path and photons taking a straight path;

$$\Delta t = \frac{2R}{\gamma c}\left(\frac{c}{v} - \gamma \sin \frac{1}{\gamma}\right)$$
$$\approx \frac{2R}{\gamma c}\left(1 + \frac{1}{2\gamma^2} - 1 + \frac{1}{6\gamma^2}\right) = \frac{4}{3}\frac{R}{\gamma^3 c}. \tag{3.36}$$

One sees that this time is less than the revolution time $2\pi R/c$, by a factor of order $\gamma^3 \approx 10^{12}$. Expressed in terms of frequencies, there will be frequency components at the observation point that are greater than the electron's revolution frequency by a factor of order γ^3. The radiation pulse arriving at the phototube can be regarded as a roughly square pulse of length

$$\Delta t = \frac{4}{3}\frac{R}{\gamma^3 c}. \tag{3.37}$$

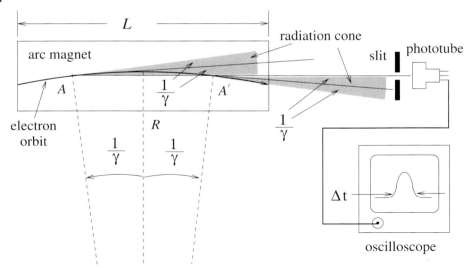

Fig. 3.5 Figure used to estimate spectral content of synchrotron radiation from arc magnets. The angle $1/\gamma$ is greatly exaggerated to make the figure legible. The shaded regions indicate the cones in which the electric field amplitude is appreciable. (This figure is adapted from Hofmann [4].)

(Evaluating at a typical vertical angle $\psi_{\text{typ.}} = 1/\gamma$ would change the coefficient $4/3 \to 7/3$.) A typical frequency is therefore approximately $1/(2\Delta t)$, which motivates the definition of a typical frequency,

$$\omega_c = \frac{3}{2} \frac{c\gamma^3}{R}, \tag{3.38}$$

This sort of argument does not fix the numerical factor, but the numerical factors in Eqs. (3.37) and (3.38) have been chosen so that ω_c would match the conventionally defined "critical frequency" introduced previously in Eq. (3.9). The product $\omega_c t$ is a convenient combination to be used as the independent variable for plotting the time dependence of the radiation pulse.

The corresponding "critical photon energy" u_c, expressed as a photon energy, is

$$\begin{aligned} u_c = \hbar \omega_c &= \frac{3}{2} \frac{hc}{2\pi R/\gamma^3} \\ &= \frac{3}{2} \frac{1.2406 \times 10^{-6} \text{ eV m}}{2\pi R/\gamma^3}. \end{aligned} \tag{3.39}$$

For $\gamma = 10^4$, $R \approx 100$ m, this works out to about 3 keV. Approximately half of the radiated energy comes in the form of the photons whose energies lie above u_c.

3.7.2
Radial Approximation

A frequency ω_ψ typical at vertical angle ψ can be defined by

$$\omega_\psi = \frac{3}{2}\frac{c\gamma^3}{R}\frac{1}{(1+\gamma^2\psi^2)^{3/2}}. \tag{3.40}$$

ω_ψ is a kind of "ψ-dependent critical frequency" such that $\omega_{\psi=0} = \omega_c$. The functional dependence on ψ will be justified straightaway in simplifying the Fourier integral giving the spectral amplitude. In the radial approximation the only electric field component is horizontal. It can be calculated using Eq. (3.25). A simplifying feature of this expression is that the integrand is an even function of t. Using this fact to restrict the range of integration, one obtains

$$\frac{\tilde{E}_x(\omega,\psi)}{\frac{q}{4\pi\epsilon_0}\frac{4\gamma^4}{\mathcal{R}R}} \approx \sqrt{\frac{2}{\pi}}\frac{1}{(1+\gamma^2\psi^2)^2}\left(\frac{(1+\gamma^2\psi^2)^{3/2}}{2c\gamma^3/R}\right)^4$$

$$\times \int_0^\infty dt\cos\omega t\left(\left(\frac{(1+\gamma^2\psi^2)^{3/2}}{2c\gamma^3/R}\right)^2 + t^2\right)^{-2}$$

$$= \sqrt{\frac{2}{\pi}}\frac{(3\omega_\psi^{-1}/4)^4}{(1+\gamma^2\psi^2)^2}\int_0^\infty dt\frac{\cos\omega t}{\left((3\omega_\psi^{-1}/4)^2+t^2\right)^2}$$

$$= \sqrt{\frac{9\pi}{128}}\frac{1}{(1+\gamma^2\psi^2)^2}\frac{1}{\omega_\psi}\left(1+(3/4)|\omega|/\omega_\psi\right)e^{-(3/4)|\omega|/\omega_\psi}$$

$$= \sqrt{\frac{9\pi}{128}}\frac{1}{\omega_c}\left(\frac{\omega_\psi}{\omega_c}\right)^{1/3}\left(1+\frac{3|\omega|}{4\omega_\psi}\right)e^{-\frac{3|\omega|}{4\omega_\psi}}. \tag{3.41}$$

Being a Fourier transform, $\tilde{E}_x$ could have been complex, but it is, in fact, real. We will return later to the physical implication of this.

To obtain the photon energy spectrum we must calculate the power spectrum of the radiation. This can be obtained by expressing $E_x(t)$ in terms of $E_x(\omega)$ by Fourier inversion:

$$E_x(t) = \frac{1}{\sqrt{2\pi}}\int_{-\infty}^\infty \tilde{E}_x(\omega)e^{-i\omega t}d\omega, \tag{3.42}$$

and then substituting this into the first of Eqs. (3.29) to give[8]

$$U_0 \approx \frac{1}{2\pi\mu_0 c} \int (\mathcal{R}^2 d\Omega) \int_{-\infty}^{\infty} \int_{-\infty}^{\infty} \tilde{E}_x(\omega) \tilde{E}_x(\omega') \, d\omega d\omega' \int_{-\infty}^{\infty} e^{-i(\omega+\omega')t} \, dt$$

$$= \frac{2}{\mu_0 c} \int (\mathcal{R}^2 d\Omega) \int_{0}^{\infty} |\tilde{E}_x(\omega)|^2 \, d\omega. \tag{3.43}$$

In the last line the identity

$$\int_{-\infty}^{\infty} e^{iat} \, dt = 2\pi \delta(a), \tag{3.44}$$

has been used, as well as $\tilde{E}_x(\omega)\tilde{E}_x(-\omega) = |\tilde{E}_x(\omega)|^2$, and negative frequencies have been folded to positive values. The integrand of Eq. (3.43) can be interpreted as the energy per unit frequency range and per unit solid angle radiated during one revolution;

$$\frac{dU_0}{d\Omega d\omega} \approx \frac{q^2 \gamma^2}{4\pi\epsilon_0 c} \frac{1}{4} \left(\frac{\omega_\psi}{\omega_c}\right)^{2/3} \left(1 + \frac{3\omega}{4\omega_\psi}\right)^2 e^{-(3/2)\omega/\omega_\psi}, \quad \omega \geq 0. \tag{3.45}$$

The (most readily measurable) quantity $dU_0/d\omega$, is obtained by integrating over $d\Omega = 2\pi d\psi$ from $\psi = -\pi/2$ to $\psi = \pi/2$ or, to good accuracy, from $-\infty$ to ∞.

3.7.3
Accurate Formula for Spectral Power Density

It turns out to be not very difficult to obtain a more nearly exact formula for the spectral power density, even including both polarization components. Returning to Eq. (3.35), that was derived earlier for use at this point, we obtain

$$\frac{\tilde{\mathbf{E}}(\omega,\psi)}{\frac{q}{4\pi\epsilon_0 c \mathcal{R}}} = \frac{1}{\sqrt{2\pi}} \int_{-\infty}^{\infty} e^{i\omega t} \left[\frac{1}{1 - \hat{\mathcal{R}} \cdot \mathbf{v}/c} \frac{d}{dt}\left(\frac{-\mathbf{v}_\perp/c}{1 - \hat{\mathcal{R}} \cdot \mathbf{v}/c}\right)\right]_{\text{ret.}} dt. \tag{3.46}$$

Changing the variable of integration from t to t_r (using Eqs. (3.13) and (3.8)) yields

$$\frac{\tilde{\mathbf{E}}(\omega,\psi)}{\frac{q}{4\pi\epsilon_0 c \mathcal{R}}} = \frac{1}{\sqrt{2\pi}} \int_{-\infty}^{\infty} e^{i\omega t_r \left(\frac{1}{2\gamma^2} + \frac{\psi^2}{2} + \frac{c^2 t_r^2}{6R^2}\right)} \frac{d}{dt_r}\left(\frac{-\mathbf{v}_\perp/c}{1 - \hat{\mathcal{R}} \cdot \mathbf{v}/c}\right) dt_r. \tag{3.47}$$

Since we are now treating t_r as the parameter describing the electron's orbit (as in Eqs. (3.17) the retarded time designation has been removed and d/dt has been replaced by d/dt_r.

---
[8] The manipulations in Eq. (3.42) amount to being a formal derivation of Parseval's formula for Fourier transforms.

We can now take advantage of the form of the integrand to integrate by parts

$$\frac{\tilde{\mathbf{E}}(\omega,\psi)}{\frac{q}{4\pi\epsilon_0 cR}} = \frac{\omega}{\sqrt{2\pi}} \int_{-\infty}^{\infty} e^{i\omega t_r \left(\frac{1}{2\gamma^2} + \frac{\psi^2}{2} + \frac{c^2 t_r^2}{6R^2}\right)} \frac{-\mathbf{v}_\perp}{c} dt_r$$

$$= \frac{\omega}{\sqrt{2\pi}} \int_{-\infty}^{\infty} e^{i\omega t_r \left(\frac{1}{2\gamma^2} + \frac{\psi^2}{2} + \frac{c^2 t_r^2}{6R^2}\right)} \left(-\sin\frac{vt_r}{R}\hat{\mathbf{x}} + \psi\cos\frac{vt_r}{R}\hat{\mathbf{y}}\right) dt_r$$

$$\approx \frac{\omega}{\sqrt{2\pi}} \int_{-\infty}^{\infty} e^{i\omega t_r \left(\frac{1}{2\gamma^2} + \frac{\psi^2}{2} + \frac{c^2 t_r^2}{6R^2}\right)} \left(-\frac{ct_r}{R}\hat{\mathbf{x}} + \psi\hat{\mathbf{y}}\right) dt_r. \quad (3.48)$$

(Compare Eq. (14.79) of Jackson.) Here the approximations $\sin\psi \approx \psi$ and $\sin^2\psi = 0$ require no comment, but we have also used $\sin(vt_r/R) \to ct_r/R$ and $\cos(vt_r/R) \to 1$. What makes this valid is that the exponential factor oscillates so rapidly that contributions to the integral are negligible outside the region of validity of these approximations.

Since the x-integrand is an odd function of t_r, $\tilde{E}_x(\omega)$ is pure real. But the y-integrand is an even function of t_r, so $\tilde{E}_y(\omega)$ is pure imaginary. As a result x and y components are 90° out of phase. Should their amplitudes be equal, this would be circular polarization. In general it is elliptical polarization, with the major and minor axes erect.

The exact evaluation of the integrals in Eq. (3.48) brings in the nasty Airy functions.

One can calculate the energy content of x and y fields independently. Having again calculated the Fourier transform $\tilde{\mathbf{E}}(\omega)$ (in the present more accurate treatment) we substitute it into a generalized form of Eq. (3.43)

$$U_0 = U_{0,x} + U_{0,y}$$
$$= \frac{R^2}{\mu_0 c} \int d\Omega \int_{-\infty}^{\infty} \left(\tilde{E}_x(\omega)\tilde{E}_x(-\omega) + \tilde{E}_y(\omega)\tilde{E}_y(-\omega)\right) d\omega. \quad (3.49)$$

We know that the radiation comes in the form of photons of energy $u = \hbar\omega$. Designating the number of photons in the energy range u to $u + du$ as $n_u(u) du$ we have

$$n_u(u) = \frac{1}{u}\frac{dU_0}{du} = \frac{1}{u\hbar}\frac{dU_0}{d\omega}. \quad (3.50)$$

Problem 3.7.1 *It has been argued previously that simply dropping the retardation term proportional to t_r^3 introduces only a small error. Investigate this by evaluating the integrals in Eq. (3.48) with the term $c^2 t_r^2/(6R^2)$ dropped. Perform numerical comparisons with exact formulas, given, for example, by Hofmann.*

Problem 3.7.2 *Especially for designing a "beam camera" (which measures the properties of an electron beam by viewing the emitted synchrotron radiation) it is useful* to estimate σ_ψ which is the r.m.s. value of the vertical angle of the radiation pattern. In particular (since detectors commonly are sensitive only over a limited band of wavelengths) it is important to know how $\sigma_\psi^2 = <\psi^2>$ depends on ω. Starting from Eq. (3.45), evaluate

$$\sigma_\psi^2(\omega) = \frac{1}{U_0/(2\pi)} \int_0^\infty \psi^2 \frac{dU_0}{d\Omega\, d\omega} d\psi. \tag{3.51}$$

Problem 3.7.3 *Integrate $dU_0/(d\Omega\, d\omega)$ given by Eq. (3.45) over directions to find an approximate formula for $dU_0/d\omega$.*

3.8
Radiation from Multiple Charges

Actual electron rings contain (a very large number) N of electrons and the observed radiation is some superposition of the radiation from each electron. One can idealize the N electrons as being uniformly arrayed around the ring or as forming one "superelectron", a point charge of charge Ne. In the former case one obtains essentially a steady ring of current which, from magnetostatics, we know emits no radiation. In the latter case the formulas up to this point need only be modified by the replacement $e \to Ne$ with the consequence that $U_0 \to N^2 U_0$. With N being a number of order 10^{12}, we are therefore uncertain, so far, by 24 orders of magnitude as to how much radiation actually occurs.

It was Schwinger [3] who first gave what is now considered to be a correct treatment of this aspect of the radiation. To a first approximation the electrons radiate incoherently and the total radiation is obtained by multiplying all formulas to this point by N.[9] During the 1950s this question was somewhat controversial and, if I am not mistaken, Schwinger's calculation was first confirmed experimentally by Corson at Cornell's first electron synchrotron. Corson's experiment consisted of measuring the energy loss of electrons by measuring the rate at which they "spiral in" after the RF accelerating field has been turned off suddenly.

Suppose that the position of the j'th (of N) particles is positioned in the ring such that the center of its radiation pulse arrives at P at time Δt_j. The total energy can be calculated using Eq. (3.43), but with $\widetilde{E}_x$ summed over all particles. (We work only with the x-component, but the y-component calculation is the same.) Since the Fourier transform of the j field acquires a factor $e^{i\omega \Delta t_j}$, the energy radiated by all particles is given by

$$U_N = \frac{\mathcal{R}^2}{\mu_0 c} \int d\Omega \int_{-\infty}^\infty \widetilde{E}_x(\omega)\widetilde{E}_x(-\omega)\, d\omega \sum_{j,k=1}^N e^{i\omega(\Delta t_j - \Delta t_k)}. \tag{3.52}$$

[9] At the same time Schwinger calculated the suppression of extremely long wavelengths due to the conductive vacuum chamber walls.

Since there are N^2 terms in the final factor, its value can, in principle, range over the vast (previously mentioned) range from 0 to N^2. Since this factor can be written

$$\sum_{j,k=1}^{N} e^{i\omega(\Delta t_j - \Delta t_k)} = N + \sum_{j \neq k} e^{i\omega(\Delta t_j - \Delta t_k)}, \tag{3.53}$$

we have

$$U_N = N U_0 + \frac{\mathcal{R}^2}{\mu_0 c} \int d\Omega \int_{-\infty}^{\infty} \widetilde{E}_x(\omega) \widetilde{E}_x(-\omega) d\omega \sum_{j \neq k}^{N} e^{i\omega(\Delta t_j - \Delta t_k)}$$

$$= N U_0 + \frac{\mathcal{R}^2}{\mu_0 c} \int d\Omega \int_{0}^{\infty} \widetilde{E}_x^2(\omega) d\omega \, 4 \sum_{j > k}^{N} \cos \omega(\Delta t_j - \Delta t_k). \tag{3.54}$$

If the arrival times Δt_j are randomly distributed then the final factor averages to zero. The situation is not quite this simple though because (a) the particles are, in fact, bunched in time and (b) the factor $4 \sum_{j>k}^{N} \cos \omega(\Delta t_j - \Delta t_k)$ has r.m.s. fluctuation $4\sqrt{N(N-1)/2} \sqrt{\overline{\cos^2}} \approx 2N$. For large N the fluctuations are fractionally unimportant, but the beam bunching cannot be neglected at low frequency. "Low" here means photon wavelengths comparable with the electron bunch length. For purposes of high energy synchrotron light sources (with bunch lengths of order 1 cm) such photons are in the far infrared and can be regarded as negligible. For rings with ultrashort bunches, such as ERLs this does not necessarily remain true.

Coherent radiation and other forces depending on N are analysed at great length in Chapter 13 and some implications for light sources are discussed in the next section.

3.9
The Terminology of "Intensity" Measures

A synchrotron light source can be used to produce beams of infrared, visible, ultraviolet, soft or hard x-ray, or even multi-Mev gamma beams. But, since the lion's share of the energy is usually in the x-ray region, for brevity in what follows, the beam will usually be referred to as an x-ray beam.

To discuss the storage ring as a source of synchrotron radiation, although we initially emphasized electrons, we are ultimately more interested in the photons. It is the high "intensity" available from the storage ring that justifies using such large and expensive apparatus and justifies our continuing effort to quantify and increase intensity. The word "intensity" has been placed in quotation marks to warn that it will be used loosely to stand for any one of the numerous measures going by such names as brightness, brilliance, current, illuminance, intensity, inverse emittance, luminance, (spectral) radiance, and so on. To the extent these terms are not equivalent, they refine different essentials of the radiation such as angular, spatial, or frequency spread, or power content.

As well as its loose usage just mentioned, the word *intensity* (without quotation marks) will be used with its unambiguous, quantitative definition in terms of power density and the Poynting vector. Of the other terms listed above, only "brightness" and "brilliance" are to be used to any extent in this text. *Brightness* is most useful as the measure of beam quality for x-ray measurements, such as fluorescence, that depend only weakly on x-ray directionality. *Brilliance* is the useful measure of intensity for directionally sensitive measurements such as crystal diffraction. Since this text emphasizes directionally sensitive applications, it is *brilliance* that will be analysed almost exclusively.

Regrettably there is no unanimity concerning the meaning of these two terms. In the world of optics the term "brightness" is defined to be an emission rate of photons per unit area and per unit time. This is the usage to be followed in this text. In the past, in the synchrotron light world [6] a more specialized term "spectral brightness" was introduced that, for one thing, included energy quantification—hence the prefix "spectral". The term "spectral brightness" also included angular quantification, but without any prefix added to register this fact. In other words, the prefix "spectral" was intended to cover the inclusion of *two* new concepts, one spectral, one angular. Perhaps because of the confusion caused by this, the term "spectral brightness" was later superceded by the term "brilliance" [7]. Lately there has been an effort, motivated by the desire for uniformity, to go back to the term "spectral brightness".

In this section I give a detailed explanation why adopting "brilliance" was a good idea and why, therefore, going back to the earlier term is a bad idea (and one that is not adopted in this text). Though standardization efforts are entirely commendable, to be successful they must also be logical.

As well as explaining *my* meanings of the terms "brightness" and "brilliance" the following few paragraphs will explain why I expect the term "spectral brightness", like other of its antecedents, to continue its lapse into obsolescence. Until this occurs, the reader who favors the term "spectral brightness" can repair the present text by systematically replacing "brilliance", wherever it occurs, by "spectral brightness".

Consider three sentences: "The sun is bright.", "The snow is bright.", and "The sodium headlight is bright.". These are all good and meaningful sentences. Words other than "bright" could be used in these sentences with no change of meaning; e.g. "brilliant", "radiant", "lustrous", "luminous". All of these sentences have both a superficial sense (one is forced to squint in all cases) and a more specialized sense (conveyed by the other words in the sentences). For technical usage in science, words conveying quantities having physical dimensions have to convey the same sort of specialization, but without the benefit of the accompanying verbiage.

In this text, and in this context, the only specializations that matter have to do with color and with directionality. Color is easier and can be dispensed with first. "Sodium headlight" instantly conveyed the meaning "yellow" to the literate reader. In physics, instead of "color" one usually uses more technical jargon such as wavelength or frequency or energy. In adjectival form "color" is replaced here, as in most of science, by the tonier word "spectral".

All x-ray experiments are color-sensitive, and most are hypersensitive. For any experiment, then, a "nominal" or "central" color can be specified. In most cases it is only photons in a narrow "bandwidth" (BW) about the nominal, that are useful to the experiment. The important measures of "intensity" include a specification (conventionally taken to be 1/1000) of the fractional bandwidth. Only photons within this bandwidth are included in any particular "intensity" measure being quoted. One part in a thousand is broader than appropriate for some experiments and narrower for others. But, because the convention is uniformly understood, it is easy for experimenters to multiply by that factor appropriate to their bandwidth sensitivity to estimate the rate of useful photons impinging on their particular apparatus.

Pedantry would require, therefore, that *all* significant measures of x-ray "intensity" should contain the prefix "spectral-". But, for the reasons just given, this prefix would really be superfluous, and, with little fear of ambiguity, need never be included. This being the English language, the "spectral-" would get dropped anyway, sooner rather than later. Here it is dropped from the start.

The issue of directionality is far more subtle. Having dispensed with sodium light, we are left with distinguishing between the way the snow and the sun are differently bright. Experimentally, with a lens, one can start a fire using sunlight but not using snow light. A more technical distinction to be made is that snow light is diffuse and comes from a spread-out source while sunlight comes from what the viewer perceives to be a point-like source. At a qualitative level we will say that the snow is "bright" and the sun is "brilliant". Consistent with this usage, the terms "brightness" and "brilliance" will be defined quantitatively in the following sections. Another term "flux density", closely connected (actually synonymous with) "brightness", will also be introduced.

To repeat, for emphasis, a point made above, these terms have the implicit meaning of "spectral-brightness" and "spectral-brilliance" but, with no possible ambiguity, the "spectral-" prefixes have been dropped.

Unfortunately the previous paragraph was too quick in using the phrase "no possible ambiguity", and did not conceive of anyone returning to the use of the word "spectral" to help in distinguishing sunlight from snow light. Color is simply not relevant in this distinction. It is for this reason the term "spectral brightness" appears, in this sentence, for the last time in this text.

3.10
Photon Beam Features "Inherited from" the Electron Beam

To a large extent a photon beam "inherits" its measures of "intensity" from the electron beam producing it. The main measures of electron intensities are the total number N of circulating particles and I, the total beam current. They are related by

$$I = eNf = eN\frac{c}{\mathcal{C}} \quad \left(\stackrel{\text{e.g.}}{=} 0.1\,\text{A}\right), \tag{3.55}$$

where f is the revolution frequency and $\mathcal{C}$ is the circumference. A prototypical differential measure of electron "intensity" is $I/(\epsilon_{x,b}\epsilon_{y,b})$ where "emittance" $\epsilon_{y,b}$ was introduced in Section 1.7. Even if the electrons form a perfect pencil beam (no spatial or angular spread) and are perfectly monochromatic (no momentum spread) the radiated photons have a continuous frequency spread (usually scaled to the critical frequency ω_c) and a finite angular spread. The characteristic radiation angle is

$$\theta_{\gamma,\text{typ.}} = \frac{1}{\gamma} = \frac{mc^2}{E} \quad \left(\stackrel{\text{e.g.}}{=} \frac{0.5\,\text{MeV}}{5\,\text{GeV}} = 10^{-4}.\right) \tag{3.56}$$

This is a very small angle. Is it negligible? That is, is this spread small compared to the angular spread of electrons in the storage ring? In one sense it is certainly small, since the electron's horizontal angle ranges through 2π.[10] On the other hand, the typical vertical electron angle is

$$\theta_{y,\text{typ.}} = \sqrt{\frac{\epsilon_{b,y}}{\beta_y}} \quad \left(\stackrel{\text{e.g.}}{=} \sqrt{\frac{10^{-8}\,\text{m rad}}{1\,\text{m}}} = 10^{-4}.\right) \tag{3.57}$$

According to these estimates, $\theta_{\gamma,\text{typ.}}$ and $\theta_{e,\text{typ.}}$ are comparable in magnitude, so neither can be neglected. Of course the factors entering Eq. (3.57) are, to some extent, subject to adjustment. For example, β_y can be easily increased by 10 or 100, so the vertical electron angle can be made negligible. But one knows that a proportional increase in spatial spread is inevitable.

---

10) The estimates in this section will have to be reconsidered when insertion devices (wigglers or undulators) are introduced. For now we just consider the pure synchrotron radiation from the ring magnets.

The ultimate photon distribution is a convolution of the electron distribution with the radiation pattern. Based on the above estimates, the horizontal angle of the photon relative to its radiating electron can sometimes be neglected, but vertical distributions are influenced both by the electron distribution and by the distribution of vertical photon angles relative to the electron.

3.11
Intensity Estimates for Bending Magnet beams

In this section, to provide a feel for the orders of magnitude and dimensionality of the various measures of beam intensity, estimates will be given for an x-ray beam emitted from a regular arc magnet. This is somewhat academic since, by this time, the important beams from light sources are produced by wigglers and undulators. More accurate formulas will be derived in the chapter describing those beam lines. Beams from arc magnets *do* continue to be used for diagnostic purposes however. The formulas in this section are intended to give only "ball park" estimates preparatory to more accurate determination later on.

An arc magnet x-ray beam consists of the synchrotron radiation several meters or more from an arc magnet in the storage ring. The beam is always restricted, at least horizontally, either intentionally or unintentionally, by a collimator or other obstacle.

Various oversimplifications will be made. The photon cone angle will be ignored initially, as will the correlation between photon energy and photon emission angle. Because photons are radiated around the full periphery of the ring it is obvious that, to produce a "beam", at a minimum, a vertical collimating slit is required. The width ΔD of this slit will be assumed to be "small" in a sense to be explained below. Assumptions like these cause the various measures of photon intensity to be rather uncertain. Some measures are given in Table 3.1.

Tab. 3.1 Measures of x-ray intensity. "BW" stands for "bandwidth".

Parameter	Unit	Invariant?
power	W	yes
flux	photons $(0.1\%BW)^{-1} s^{-1}$	yes
brightness (flux density)	photons $(0.1\%BW)^{-1} mm^{-2} s^{-1}$	no
brilliance	photons $(0.1\%BW)^{-1} mrad^{-2} mm^{-2} s^{-1}$	yes
cleanliness	photons $(0.1\%BW)^{-1} mrad^{-2} mm^{-2} J^{-1}$	yes

The least ambiguous property of an x-ray beam is its *power*. Almost equivalently, since the radiation comes in the form of photons, one can (in principle) count the photons individually (and accumulate their energies) as they come down the line. With few exceptions, any apparatus making use of the beam has a "nominal" or "central" wavelength λ_{nom}; it is only photons very close to this wavelength that are useful. For purposes of comparing the relative usefulness of different x-ray sources, this bandwidth is *conventionally* taken to be $\lambda_{\text{nom}}(1 \pm 1/2000)$. "Flux" is defined to be the number of photons per second within this band. A *nominal* transverse area of one millimeter-squared and a *nominal* solid angle of one milliradian-squared are also introduced for use shortly.

The more concentrated these photons are transversely, the more useful they are, because more photons can impinge on the same (small) target. One defines "brightness" or "flux density" to be "number of photons within $\pm 0.05\%$ of nominal energy per square-millimeter of transverse area.

The reason brightness is not very useful is that it is not an *invariant* measure; i. e. it is altered by optical elements (lenses and drift spaces) along the beamline. To obtain an invariant measure, one defines the "brilliance" to be number of photons within $\pm 0.05\%$ of nominal energy per nominal area per nominal solid angle per second. By Liouville's theorem, this quantity *is* preserved by arbitrary paraxial optical systems.[11]

Brilliance is therefore much more useful than brightness. But no single parameter can distill into a single number all relevant spectral features. Another relevant measure is the brilliance divided by the beam power. Being the ratio of two invariants this ratio is also invariant. In the absence of a conventional name I will refer to this ratio as "cleanliness". Even high quality beams are "dirty" in the sense that only a small fraction of the photons in the beam have any chance whatsoever of being counted or otherwise registered in the experimental apparatus at the end of the beamline. Beyond this, these excess photons can have undesirable effects, often having to do with the the backgounds they cause or the heat they generate when they are absorbed. Generally speaking, undulator beams are cleaner than wiggler beams which are cleaner than bending magnet beams.

For an apparatus that happens to make use of precisely those photons within the nominal bandwidth (0.1%) and the nominal spot-size × solid-angle product (which is 1 mrad^{2}mm^2), the useful-power/beam-power ratio is given by the cleanliness parameter. For low flux beams this measure may be inessential because monochromators can filter out the useless photons effectively, but, for high flux beams, power expense may be appreciable and power handling difficult.

---

11) The distinction between brightness and brilliance is much like the distinction, in beam physics, between r.m.s. transverse beam "spot size" σ_x and transverse emittance ϵ_x. They are related by $\sigma_x(s) = \sqrt{\epsilon_x \beta_x(s)}$. Because the lattice function $\beta_x(s)$ depends on s, "spot size" $\sigma_x(s)$ also depends on s. But the ratio $\epsilon_x = \sigma_x^2(s)/\beta_x(s)$ is invariant; i. e. independent of s.

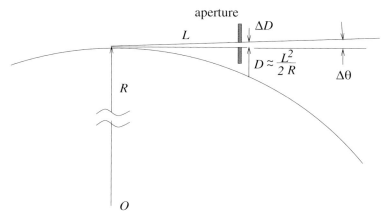

Fig. 3.6 Prototypical (first generation) synchrotron light beamline.

Some estimates that illustrate the considerations involved follow. A prototypical synchrotron light beamline is shown in Figure 3.6. For various practical reasons it is difficult to make the length L of the beamline really short. It is convenient for the separation distance from the accelerator, $D \approx L^2/(2R)$, to be, say, 2 m or more, so

$$L \approx \sqrt{2RD} \quad \left(\stackrel{\text{e.g.}}{=} \sqrt{2 \times 100 \times 2} \approx 20\,\text{m.} \right) \tag{3.58}$$

According to Eq. (3.32) the total energy radiated, per electron, per turn, is given by

$$U_0^{\text{rad}} = C_\gamma \frac{\mathcal{E}_e^4}{R} \quad \left(\stackrel{\text{e.g.}}{=} 1\,\text{MeV} \right), \tag{3.59}$$

where $C_\gamma = 0.885 \times 10^{-4}\,\text{m/GeV}^3$ and the electron energy $\mathcal{E}_e$ has been taken to be 5 GeV, here and in the following numerical examples. This formula gives U_0^{rad} in GeV units. A rather small fraction, $\Delta\theta/(2\pi)$, of all the radiated energy passes through the aperture ΔD. Combining these results with Eqs. (3.55), the total power through the aperture is

$$\frac{\text{power}}{\Delta D} = \frac{I}{e} C_\gamma \frac{\mathcal{E}_e^4}{R} \frac{1}{L} \frac{1}{2\pi} \quad \left(\stackrel{\text{e.g.}}{=} I \times 10^6\,\text{eV/e} \times \frac{10^{-2}}{20} \frac{1}{2\pi} = 80\,I\,\frac{\text{W}}{\text{cm}}, \right) \tag{3.60}$$

where the "per centimeter" refers to the slit width D. A result to be derived later is that the total number of synchrotron radiation photons emitted, per revolution, by a single electron in a uniform ring, is given by

$$\mathcal{N}_0 = \frac{ecC_\gamma \mathcal{E}_e^3 B}{0.3079\gamma u_c} = 0.0662\,\gamma, \tag{3.61}$$

where the "critical energy" u_c ($\stackrel{\text{e.g.}}{=}$ 3 keV) was defined in Eq. (3.39). One should note the remarkable facts that this number is independent of magnetic field B and that the dependence on $\mathcal{E}_e$ is simple proportionality to γ.[12] Rearranging Eqs. (3.60) and (3.61), the total number of photons passing through the aperture per second is

$$\mathcal{N} = \frac{I}{e} 0.0662\, \gamma\, \frac{\Delta D}{2\pi L} \quad \left(\stackrel{\text{e.g.}}{=} \frac{1}{1.6\text{e-}19\,\text{C}}\, 662\, I\, \frac{\Delta D}{L}\, \frac{1}{2\pi} = 0.66\text{e}21\, I\, \frac{\Delta D}{L}\, \text{s}^{-1}. \right) \tag{3.62}$$

Here the factor I/e gives the number of electrons passing a given point in the ring per second.

By convention the "flux" $\mathcal{F}$, counts only photons in an energy range $\Delta u = 10^{-3} u_{\text{nom}}$ where u_{nom} is the central photon energy of the experiment served by the beamline. This is an appropriate figure of merit for apparatus that responds only to a narrow band of energies. The energy spectrum is described by a (normalized to 1) probability distribution $n_u(u)$ such that $n_u(u)\,\mathrm{d}u$ is the probability a particular photon energy u lies in the range u to $u + \mathrm{d}u$. The "number spectrum" is then given by

$$\frac{\mathrm{d}\mathcal{N}}{\mathrm{d}u} = \mathcal{N}\, n_u(u)\,. \tag{3.63}$$

For purposes of estimation, let us take $u_{\text{nom}} = u_c$. It is common to measure the photon energy u in units of the "critical energy" u_c. (i.e. units such that $u_c = 1$.) With u measured in these units it will be shown in the next chapter that $n_u(1) \approx 0.5/3.2$.[13] Then $\mathcal{F}$ acquires the factor $n_u(1)\, 10^{-3}$. Compared to Eq. (3.62):

$$\mathcal{F}|_{u=u_c} = 10^{17}\, I\, \frac{\Delta D}{L}\, \frac{\text{photons (s)}^{-1}}{0.1\% \text{BW}}\,. \tag{3.64}$$

The spectral density factor is explained more fully in the next chapter.

As defined earlier, the beam *brightness* is equal to the density (per transverse area) of $\mathcal{F}$. If L is, say, doubled, with ΔD left constant, then the flux $\mathcal{F}$ is cut in half, the beam height is doubled, and the brightness is cut to one quarter. The brightness can therefore be said to have "inverse square law" dependence.

The brilliance is an intensity that is differential in both source area and in solid angle at the source. A complication in determining it in the present case is the correlation between the longitudinal location of the source and the transverse position at the slit. Photons at the outer edge of the slit tend to have

---

12) The fact that the number $\mathcal{N}_0$ can be defined *at all* is because the spectrum is *not* divergent at long wavelengths—there is no infrared divergence.

13) The value $n_u(1) \approx 0.5/3.2$ has been extracted from figures in the article by Sands. The factor can also be obtained from Figure 5.1.

come from more upstream than do photons at the inner edge. This correlation is especially significant when $\Delta D/L \gg 1/\gamma$, which is not uncommon. Let us therefore assume, at least initially, that $\Delta D/L \ll 1/\gamma$. This means, essentially, that the depth of the source region is small compared to the distance from source to slit. We then pretend that all photons have come from the same plane, transverse to the central electron trajectory and upstream from the slit by a distance L. It may, later, be important to account for "depth of focus" effects neglected in this approximation.

At this point, let us interpret "the beam" to consist of precisely those photons passing through the slit (per second, if you wish.) When traced back to the source plane these particles were distributed, both horizontally and vertically, and both in position and angle, over the electron beam distributions at the same position. In fact the distribution of photons in transverse phase space at the source is identical to the distribution of electrons at the same point. Let us further assume that the transverse phase coordinates at the slit are uncorrelated with the transverse phase space coordinates at the slit. According to Eq. (1.27), the product of populated transverse phase space areas at the source is

$$\Delta x \Delta x' \Delta y \Delta y' = \pi^2 \epsilon_{x,b} \epsilon_{y,b}. \tag{3.65}$$

(Recall that this assumes a somewhat unrealistic, uniform-inside, vanishing-outside, phase space distribution.) Using a result from Chapter 1 we know that the phase space density of an x-ray beam is constant as the beam evolves from source to slit. For the beam brilliance at the slit we therefore obtain

$$\mathcal{B} = \frac{\mathcal{F}}{\Delta x \Delta x' \Delta y \Delta y'} = \frac{\mathcal{F}}{\pi^2 \epsilon_{x,b} \epsilon_{y,b}}. \tag{3.66}$$

Substituting numerical values as in the expression for $\mathcal{F}$ in Eq. (3.64),

$$\mathcal{B}|_{u=u_c} = \frac{I}{e} 0.0662 \, \gamma \, \frac{\Delta D}{2\pi L} \frac{10^{-3} 0.5/3.2}{\pi^2 \epsilon_{x,b} \epsilon_{y,b}}. \tag{3.67}$$

In this case, if the distance L is doubled with ΔD held constant then, unlike the brightness, the brilliance $\mathcal{B}$ is cut only in half.

Suppose that, with L held fixed, slit width ΔD is increased. By Eq. (3.64), the brightness is constant and the flux $\mathcal{F}$ increases proportional to ΔD. Only when ΔD becomes appreciable on the scale of the orbit radius does this proportionality break down. Curiously enough, by Eq. (3.66), $\mathcal{B}$ increases proportionally to ΔD, at least initially. But, when ΔD becomes comparable with L/γ, (which is typically a rather small, millimeter scale, length) the assumptions that have been made begin to break down. Beyond that point $\mathcal{B}$ becomes independent of ΔD.

For CESR, a typical value is $\epsilon_{x,b} \approx 2 \times 10^{-7}$ m rad. The factor $\epsilon_{y,b} \approx \epsilon_{x,b}/100$, is somewhat less well known (and, when running x-ray beams parasitically during colliding beam operation, varies with time), but, if necessary, $\epsilon_{y,b}$ can be known to several percent accuracy. It must be remembered though, that there are different conventions governing the definition of emittance. For example, the CESR values just mentioned are r.m.s. values, not necessarily quite right for immediate substitution into Eq. (3.67).

The brilliance parameter $\mathcal{B}$ is used primarily as a figure of merit specifying wiggler and undulator beams in "later generation" synchrotron light sources that have ultra-small emittances. The conventional units of the phase space factor are mm^2 mr^2, which reduces the numerical value of brilliance by 10^{12} compared with M.K.S. units. Even so, since beam dimensions in dedicated synchrotron light sources are typically small compared to a mm, the numerical values of $\mathcal{B}$ can exceed the numerical value of $\mathcal{F}$ by many orders of magnitude. Since the photon spreads for these beams contribute importantly to the brightness, one must be very careful in using the brightness as a figure of merit. Technically, the brightness is the maximum value of a differential cross section. Since the distribution is highly peaked, Eq. (3.67) cannot be applied to high brilliance beams.

The factor $\Delta D/L$ can be regarded as fixing the *effective length* of the source region. This factor is well known but certain idealizations have gone into Eq. (3.64), that can invalidate the formula, especially at long x-ray wavelengths. Recall also that the formula is only valid in the limit of small $\Delta D/L$. Finally, the brilliance given by Eq. (3.67) assumes detection apparatus centered on $u = u_c$. For other energies an appropriate value of $n_u(u)$ has to be obtained, for example from Figure 5.2.

It is also important to remember that, since only radiation from regular arc bends has been considered, the formulas in this section apply only to first generation light sources in which the radiation is taken from arc magnets. Obviously wiggler and undulator radiation has to be considered separately and will, in general, give much greater brightness.

References

1 Griffiths, D. (1999), *Introduction to Electrodynamics*, 3rd edn., Prentice Hall, Upper Saddle River, NJ.

2 Jackson, D. (1999), *Classical Electrodynamics*, 3rd edn., John Wiley, New York, 1999

3 Schwinger, J. (1945) *On Radiation by Electrons in a Betatron*, transcribed by Miguel Furman, LBNL-39088/CBP Note-1996.

4 Hofmann, A. (1998), *Characteristics of Synchrotron Radiation*, in *Synchrotron Radiation and Free Electron Lasers*, CERN 98-04.

5 Hofmann, A. (2004), *The Physics of Synchrotron Radiation*, Cambridge Monographs on Particle Physics, **20**.

6 Wiedemann H. (1999), 2nd edn., *Particle Accelerator Physics I and II*, Springer-Verlag, Berlin

7 APS Report ANL-88-9.

8 Sands, M. (1971), *The Physics of Electron Storage Rings*, in *International School of Physics "Enrico Fermi"*, Academic Press, New York.

9 Lawson, Lapostalle and Gluckstern, (1973) *Emittance, Entropy, and Information, Particle Accelerators*, **5**, 61.

4
Simple Storage Rings

4.1
Preview

An accelerator physicist has trouble finishing a sentence that does not include words like "tune", "dispersion", "chromaticity", etc. Though terms like these have elementary and unambiguous meaning within the field, those meanings differ from conventional usage in English or even in other fields of physics. It seems useful to introduce terms like these in the most elementary context possible. That is one purpose of this chapter.

Another purpose is to further develop the beam size and emittance quantities introduced in Chapter 1 and, in particular, to emphasize the importance of making them as small as possible (consistent with high beam current I). It will be seen later that the so-called "brilliance" $\mathcal{B}$ is the best measure, for many applications, of the quality of an x-ray beam.[1] It will also be shown that $\mathcal{B}$ depends on electron beam properties mainly through the proportionality[2]

$$\mathcal{B} \sim \frac{I}{\epsilon_x \epsilon_y}. \tag{4.1}$$

where ϵ_x and ϵ_y are horizontal and vertical beam emittances, respectively. These emittance factors depend on the damping and excitation effects caused by synchrotron radiation. Before these calculations can be described it is necessary to understand the essentials of storage ring design—another purpose of the chapter.

One could credibly "rewrite history" by pretending that all developments in storage rings were motivated by the desire to improve their performance as synchrotron radiation sources. From Eq. (4.1) one sees that this amounts to reducing the emittance factors and increasing beam current. As it happens,

---

1) In some quarters the *brilliance* parameter is called *spectral brightness*. The controversy concerning terminology is discussed at length in Section 3.9.
2) Implicit in Eq. (4.1) is the assumption that the spread of photon emission angles is negligible. For early accelerators this assumption was abundantly true. For later generation synchrotron light sources, designed for ultra-low emittances, it is necessary to take better account of the radiation pattern on the brightness.

Accelerator X-Ray Sources. Richard Talman
Copyright © 2006 WILEY-VCH Verlag GmbH & Co. KGaA, Weinheim
ISBN: 3-527-40590-9

accelerator development for elementary particle physics has, in fact, actually been driven largely by the desire to accomplish these same things. One push has been to reduce magnet costs by reducing the volume of magnetic field required. This has motivated the reduction of transverse beam sizes. Another reason for reducing emittances has been to reduce beam spot sizes at collision points in order to increase luminosity in colliding beam operation. Beam currents are increased for the same reason. For these reasons, the evolution in storage ring design to be described here is more or less consistent with the actual chronology, which was driven more by elementary particle physics considerations.

The sequence of accelerator developments motivated by the desire to reduce emittances includes uniform field (cyclotron geometry); weak focusing; strong focusing (combined function or separated function); and special purpose, low emittance, dedicated light sources. In each case one must define and determine parameters such as "tunes" Q_x and Q_y and beta-functions β_x and β_y. The latter have already been introduced in Chapter 1. This chapter concentrates on the weak focusing synchrotron, but then shows how the transverse beam size can be reduced by strong focusing.

It will be left to subsequent chapters to determine the actual magnitudes of emittances ϵ_x and ϵ_y. On the one hand, according to Eq. (4.1), ϵ_x and ϵ_y determine the synchrotron radiation brightness but, on the other hand, it turns out that they are themselves determined by the properties of the synchrotron radiation. To make a long story short, all emittances in an electron storage ring are determined by a competition between damping and stochastic excitation, both of which are due to the synchrotron radiation.

4.2
The Uniform Field Ring

The leading requirement of a (more or less) circular accelerator is that the particles be bent through $360°$, say in a circle of radius R ($\stackrel{\text{e.g.}}{=}$ 100 m.) Assume a uniform magnetic field B_0, directed along the (vertical) y-axis. The non-relativistic equation governing the motion is

$$\frac{mv^2}{R} = evB_0, \quad \text{or} \quad \frac{pc}{R} = ecB_0. \tag{4.2}$$

The latter form (which, unlike the former, is relativistically valid) will be all that is used in what follows. It can be expressed in alternative forms:

$$\frac{d\mathbf{p}}{dt} = e\mathbf{v} \times \mathbf{B}, \quad \text{or}$$

$$B_0 = \frac{pc/e}{cR} \quad \left(\stackrel{\text{e.g.}}{=} \frac{5 \times 10^9 \text{ V}}{3 \times 10^8 \text{ m s}^{-1} \times 10^2 \text{ m}} = 0.16 \text{ Tesla} \equiv 1.6 \text{ KG.} \right) \quad (4.3)$$

A particle with pc/e given by exactly $cB_0 R$ and traveling in a perfect circle, will be called the "central" or "reference" particle, and will be said to follow the "reference trajectory".

4.3
Horizontal Stability

In practice, the particles in an accelerator follow trajectories that deviate from the reference trajectory, as shown, for example in Figure 4.1. Simple geometry shows that a particle deviating by angle $\Delta x'$ at point O will again pass through O after one revolution. Letting the radial offset $x(s)$ be defined by $r(s) = R + x(s)$, one determines (to "eye-ball" accuracy only) from the lower figure that the transverse displacement appears to be given by

$$x(s) = R\Delta x' \sin \frac{s}{R}. \quad (4.4)$$

Since the point O could have been taken anywhere in the ring this dependence could have been any linear combination of $\sin(s/R)$ and $\cos(s/R)$, and these are the only forms satisfying the linearized, paraxial focusing equation (1.2). Working out the second derivative of Eq. (4.4), one finds the second order differential equation satisfied by $x(s)$ to be

$$\frac{d^2 x}{ds^2} = -\frac{1}{R^2} x. \quad (4.5)$$

This equation has the form of Eq. (1.2). By construction it is satisfied by $x(s)$ as given in Eq. (4.4). This shows there is an effective horizontal focusing force caused by the geometry of circular motion. The strength of this focusing can be quantified by a focusing coefficient

$$K_R = -\frac{1}{R^2}. \quad (4.6)$$

We will soon find that this is a *very weak* focusing. This confirms the statement, made in Chapter 1, that the focusing associated with the bending of a charged particle in a magnet is often negligible. Here is where we are repairing this neglect to improve the treatment by accounting for this weak focusing effect. One defines the "tune" Q_x to be the number of betatron oscillations per

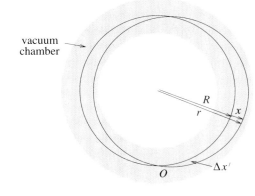

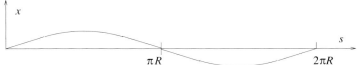

Fig. 4.1 Orbit of particle with non-vanishing horizontal slope offset $\Delta x'$ at the origin. The vacuum chamber is shown shaded. The orbit returns exactly to the starting point after completing a circle. Therefore tune $Q_x = 1$.

particle revolution. Since, the motion exhibited in Figure 4.1 has exactly one complete betatron oscillation per revolution, $Q_x = 1$. Another way of expressing the motion is the so-called pseudoharmonic description of chapter 1;

$$x(s) = a\sqrt{\beta_x(s)}\,\cos\psi_x(s). \tag{4.7}$$

Clearly, in the present situation, $\beta_x(s)$ has to be constant, independent of s. But, recalling formula (1.48),

$$\frac{d\psi_x}{ds} = \frac{1}{\beta_x}. \tag{4.8}$$

Ignoring a constant of integration, this requires

$$\psi_x(s) = \frac{s}{\beta_x}, \tag{4.9}$$

and hence, to be consistent with Eq. (4.4), that

$$\beta_x = R. \tag{4.10}$$

Finally, the tune is given by

$$Q_x = \frac{\psi_x(s = 2\pi R)}{2\pi} = 1. \tag{4.11}$$

which confirms what was already noted from the figure.

The rigorous definition of what constitutes *stability* is somewhat subtle mathematically. But, taking *stability* to mean that the amplitude remains bounded forever, this motion is certainly stable.

Surprisingly enough, we will find immediately that the vertical situation is *unstable* in the same sense. We will also find that $Q_x = 1$ is an undesirable tune, both because it is an integer and because it is *small*. and that $\beta_x = R$ is an undesirably large β-function. However, these are the values we have to work with for the time being.

4.4
Vertical Stability

Unfortunately, the particle motion just described is unstable vertically, because any nonvanishing component of vertical velocity will cause a particle to drift inexorably in the vertical (y) direction until it strikes a magnet pole and is lost. To overcome this problem one can shape the magnet poles as shown in Figure 4.2. The field nonuniformity is conventionally parametrized by introducing "field index" n according to which

$$B_x = -\frac{B_0 n}{R} y. \tag{4.12}$$

In free space, according to Maxwell's equations, one has $\nabla \times \mathbf{B} = 0$. We will ignore the gentle curvature of the magnet as s varies, in order to be able to treat the magnetic field as if it depends only on x and y. It then follows that $\partial B_y/\partial x = \partial B_x/\partial y$, which enables us to approximate the spatial dependence of the vertical field component by

$$B_y = B_0 \left(1 - \frac{n}{R} x\right). \tag{4.13}$$

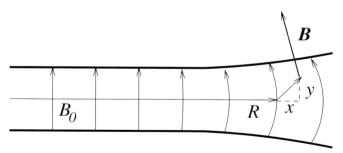

Fig. 4.2 Magnet poles shaped to produce a horizontal component of magnetic field B_x to prevent particles from drifting vertically. This is a "weak focusing" magnet. Since there are never any particles in the left two thirds of the field region shown, the magnetic field need not, and in practice would not, be present there.

Now the force equation (4.3) becomes

$$\frac{d\mathbf{p}}{dt} = e B_0 \begin{vmatrix} \hat{\mathbf{x}} & \hat{\mathbf{y}} & \hat{\mathbf{s}} \\ 0 & 0 & c \\ -\frac{ny}{R} & 1 - \frac{nx}{R} & 0 \end{vmatrix} \qquad (4.14)$$

Using $ds = c dt$ and $p_y = y' p$, this yields, for the vertical equation of motion,

$$y'' = -\frac{c B_0 n}{(pc/e) R} y = -\frac{n}{R^2} y. \qquad (4.15)$$

This equation also has the form of Eq. (1.2). Provided the field index satisfies $n > 0$, the equation exhibits the vertical restoring force that the magnet was designed to provide.

4.5
Simultaneous Horizontal and Vertical Stability

Providing vertical stability has not come for free. With the field modified as shown in Figure 4.2, it is clear that a particle for which $x > 0$ finds itself in a somewhat reduced magnetic field, which is the opposite of what would bend it back toward the reference orbit. In other words, some horizontal defocusing has been introduced. The effect of this force can described by modifying Eq. (4.5), using the x-component of Eq. (4.14);

$$x'' = -\frac{1-n}{R^2} x. \qquad (4.16)$$

Provided the field index satisfies $1 - n > 0$, there is net horizontal focusing. Since it is necessary for both horizontal and vertical motion to be stable, we combine their two conditions into

$$0 < n < 1. \qquad (4.17)$$

We can now extract tunes and β-functions for both planes, starting from the betatron phases required to satisfy Eqs. (4.15) and (4.16);

$$\psi_x = \sqrt{\frac{1-n}{R^2}} s, \quad \psi_y = \sqrt{\frac{n}{R^2}} s. \qquad (4.18)$$

Again employing Eq. (1.48), we therefore have

$$\beta_x = \frac{R}{\sqrt{1-n}}, \quad \beta_y = \frac{R}{\sqrt{n}}. \qquad (4.19)$$

Since both $|n|$ and $|1 - n|$ are less than 1, both β-functions are greater that R. Finally, proceeding as in Eq. (4.11),

$$Q_x = \sqrt{1-n}, \quad Q_y = \sqrt{n}, \quad \text{and hence} \quad Q_x^2 + Q_y^2 = 1. \qquad (4.20)$$

This final condition makes it clear that the focusing in both planes must be relatively weak.

4.6
Dispersion

Just as no particle lies precisely on the reference orbit, no particle has precisely the nominal momentum. The fractional momentum offset $\delta = \Delta p/p$ was defined in Eq. (1.1). A possible orbit for such a particle is a circle uniformly outside the reference circle of radius R, with radius given by $R + D\delta$. This is the defining relation for the "dispersion function" $D(s)$, which, in general, depends on s. But in all cases treated so far, D is independent of s.

For a uniform magnetic field, applying Eq. (4.2) to this "off-momentum" orbit yields

$$R + D\delta = \frac{pc\,(1+\delta)}{ecB_0}, \quad \text{or} \quad D = R. \tag{4.21}$$

For a weak focusing magnet, the same equation can be applied, but it is necessary to take account of the reduced magnet field strength for $\Delta x > 0$;

$$R\left(1 + \frac{\Delta x}{R}\right) = \frac{pc\,(1+\delta)}{ecB_0\left(1 - n\frac{\Delta x}{R}\right)}. \tag{4.22}$$

Cross-multiplying the factor $(1 - n\Delta x/R)$, Taylor expanding both sides of this equation, setting $\Delta x = D\delta$, and equating first order terms yields, for the dispersion of a weak focusing ring,

$$D = \frac{R}{1-n} \quad \left(\stackrel{\text{also}}{=} \frac{R}{Q_x^2}.\right) \tag{4.23}$$

(The parenthesized final relation notes, in passing, a standard "rule of thumb" approximate relation between dispersion and tune, which happens to be exact for the weak focusing lattice.) By modern standards the dispersion given by Eq. (4.23) is *gigantic*, but it was what earlier accelerator users had to deal with. For CESR, with $R \approx 100$ m, if one had $n = 0.5$ the dispersion would be $D = 200$ m. Then, for an off-momentum value of 10^{-3}, (which is typical) the off-momentum orbit would be displaced by $200 \times 10^{-3} = 20$ cm. Being roughly twice the full width of the vacuum chamber, this would clearly be unacceptable—both in the sense that the beam would not survive and that the transverse extent of the beam would be unacceptably large as a source of synchrotron radiation.

4.7
Momentum Compaction

Because of dispersion, the orbit circumference $C(p)$ depends on p. A parameter that enters discussion of off-momentum orbits, and can be quite confusing, is the so-called "momentum compaction", which is obtained from $C(p)$ using the relation,

$$\alpha = \frac{p}{C(p)} \frac{dC}{dp}. \tag{4.24}$$

α is equal to the fractional increase in circumference $\Delta C/C$ divided by the fractional increase in momentum $\delta = \Delta p/p$. This quantity is important for analysing longitudinal stability, which depends on the time of arrival of the electron at the RF accelerating cavity. For an off-momentum particle, this time depends on its excess time of flight relative to the reference particle. For non-relativistic particles the velocity deficit, relative to the reference velocity, contributes to this deficit, but for electrons (which are relativistic), the entire deficit is due to the excess circumference, which is what α quantifies; in *linearized* form

$$\frac{\Delta C}{C} = \alpha \delta. \tag{4.25}$$

For a uniform field magnet, with $R \sim p$, one obtains immediately from Eq. (4.24) that $\alpha = 1$. The name "momentum compaction" derives from this. This is because momentum compaction less than 1 implies that the off-momentum circumference increases less rapidly than linearly with p, so the transverse density of trajectories is "compactified" relative to what would be true in a uniform magnetic field. For a weak focusing magnet, copying D from Eq. (4.23),

$$\alpha = \frac{p_0}{R} \frac{d}{dp} \left(R + D \frac{p - p_0}{p_0} \right) = \frac{1}{1 - n} \quad \left(\stackrel{also}{=} \frac{1}{Q_x^2}. \right) \tag{4.26}$$

It might be thought that the value of α would always be of order 1, but modern lattices are designed to have $\alpha \ll 1$. In fact, the final equation of Eq. (4.26), which has been listed mainly to enable this comment, continues to be true for high tune lattices having $Q_x \gg 1$. One can even contemplate lattices having $\alpha = 0$ in order to obtain very short bunch lengths. An example appropriate for an ERL will be described in Chapter 10. For fully relativistic particles, such a lattice is said to be "isochronous".

4.8 Chromaticity

Since the focusing strengths of typical magnetic elements vary inversely with momentum, the tunes tend to become smaller as the momentum increases. This is known as "chromaticity". One defines chromaticity parameters by,

$$Q'_x = \frac{dQ_x}{d\delta}, \quad Q'_x = \frac{dQ_y}{d\delta}. \tag{4.27}$$

If nothing is done both chromaticities are negative, but usually sextupoles are included in the lattice to compensate the chromaticities to zero or slightly positive values. With all sextupoles turned off the chromaticities given by Eq. (4.27) are said to be the "natural chromaticities". The presence of sextupoles causes opposite-sign terms not shown. In the formulas given so far for a weak focusing magnet, no dependence of δ has been exhibited, but one sees from the inverse R dependence indicated by Eq. (4.12), that $n(\delta) \approx n(1 - \delta)$. Hence we have

$$Q'_x = \frac{-dn/d\delta}{2\sqrt{1-n}} = \frac{n}{2\sqrt{1-n}}, \quad Q'_y = -\frac{\sqrt{n}}{2}. \tag{4.28}$$

Nothing has been said about "resonances". Because the damping of betatron oscillations is so weak, any effect that occurs on every turn has the possibility of accumulating and, eventually, destroying the beam. By appropriate choice of tunes most resonances can be avoided. But, if the chromaticity is large, or the energy spread is large, or both, the spread of tunes coming from the energy spread may make it impossible to avoid damaging resonances. This is the reason it has proved necessary to "compensate" the chromaticities of light sources, and it is the purpose of the sextupoles mentioned above. Understanding the problems caused by sextupoles brings in nonlinear mechanics, chaos, and other things too fierce to mention. Fortunately, in synchrotron light sources, the the equilibrium betatron amplitudes are sufficiently small that these complications are unimportant. But overly strong focusing can seriously compromise the injection of electrons into the ring, including "top-off" injection, e. g. to replace particles lost to Touschek effect.

The following problems are intended more as preparation for material to come than as practice on what has gone before.

Problem 4.8.1 *Find the eigenvalues of* **M**, *the "transfer matrix in Twiss form"*

$$\mathbf{M} = \begin{pmatrix} \cos\mu + \alpha\sin\mu & \beta\sin\mu \\ -\frac{1+\alpha^2}{\beta}\sin\mu & \cos\mu - \alpha\sin\mu \end{pmatrix}. \tag{4.29}$$

Problem 4.8.2 *The transfer matrices for the FODO section shown in Figure 1.5 are given in Eq. (1.17). Matching the vertical matrix to the form given in the previous* problem gives the equation

$$\begin{pmatrix} \cos\mu + \alpha\sin\mu & \beta\sin\mu \\ -\frac{1+\alpha^2}{\beta}\sin\mu & \cos\mu - \alpha\sin\mu \end{pmatrix}$$
$$= \begin{pmatrix} 1 - q^2L^2/2 & L(1-qL/2) \\ -q^2(1+qL/2)L & 1 - q^2L^2/2 \end{pmatrix}. \quad (4.30)$$

Show that $\alpha = 0$ and, using that result, show that

$$\beta = \frac{1}{q}\sqrt{\frac{1-qL/2}{1+qL/2}}, \quad (4.31)$$

where q is assumed to be positive. This gives the value of the beta function value both at the beginning and the end of the cell and, in a periodic lattice, the value at every cell boundary. Also find the betatron phase advance $\Delta\psi$ through the cell.

Problem 4.8.3 *The following relativistic relations among mass m, momentum p, velocity v, energy E, and relativistic factor γ are assumed to be well-known:*

$$\gamma = \frac{1}{\sqrt{1-\frac{v^2}{c^2}}}, \quad E = mc^2\gamma, \quad p = mv\gamma, \quad E^2 = p^2c^2 + m^2c^4. \quad (4.32)$$

Derive the relations

$$\frac{pc}{E} = \frac{v}{c}, \quad \text{and} \quad \frac{dp}{dE} = \frac{1}{v}. \quad (4.33)$$

Problem 4.8.4 *For an off-momentum particle, fractional momentum offset δ from a nominal, central momentum p_0 is defined by $p = p_0(1+\delta)$. Show that*

$$\left.\frac{dv}{d\delta}\right|_{\delta=0} = \frac{v_0}{\gamma_0^2}. \quad (4.34)$$

Problem 4.8.5 *This problem is concerned with longitudinal motion of less than fully relativistic particles; $v < c$. A particle of momentum p_0 in a uniform magnetic field B travels in a circle of radius $R_0 = (p_0c/e)/(cB)$. First find the particle's revolution period T_0. When the particle's momentum is increased to $p_0(1+\delta)$ the revolution period changes to T. Using the result of the previous problem, find $dT/d\delta|_{\delta=0}$. Use first order differentials only, and don't forget to include two effects—altered circle radius and altered speed.*

4.9
Strong Focusing

The problem with the lattice described so far is that the focusing is weak—why else would the lattice be called "weak focusing". This is guaranteed by

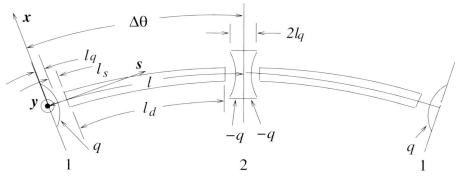

Fig. 4.3 An idealized, thin lens, FODO-arc cell, showing dimensioning and element strength parameters.

stability conditions (4.17) which requires the magnitude of field index n to be less than 1. This restriction is *not* brought about by any difficulty in making magnets with higher n-value. Values higher by factors of 10, or even 100, are practical. Such magnets are said to be "strong focusing".

To see what can be done some results are needed from Chapter 1. According to Eq. (1.42) the beam height is given by

$$<y^2> = \epsilon_{y,b}\beta_y. \tag{4.35}$$

Assuming for the time being that $\epsilon_{y,b}$ cannot be changed, $<y^2>$ can be reduced only by reducing β_y. This can be accomplished by using strong focusing.

"Strong focusing" and "alternating gradient focusing" are approximate synonyms. Both terms imply the focusing coefficient $K(s)$ cycles between (large) positive and negative values. The value of $K(s)$ can be made large within a bending magnet, in which case the lattice is said to be "combined function". Alternatively, the bending magnets can have more or less uniform magnetic fields and all focusing concentrated in quadrupoles, in which case the lattice is said to be "separated function". As far as betatron oscillations are concerned, there is little difference between the the combined function and the separated function approaches. For important reasons to be discussed in the next few chapters, combined function lattices are inappropriate for light sources. Our few remaining comments concerning simple lattice properties will therefore assume separated function lattices.

The prototypical strong focusing lattice cell is the FODO cell shown in Figure 1.5 and analysed in Problem 4.8.2. Since that cell had no bending magnet it would be inappropriate for the arc of a circular machine. The same, length $L = 2\ell$ cell, but with bending magnets included, is shown in Figure 4.3.

It has been stated previously, and can now be confirmed, that bending magnets are (nearly) inert optically. By Eq. (4.6) the effective inverse focal length of one bending magnet in the lattice shown, is ℓ_d/R^2 (where the dipole length ℓ_d is commonly in the range $0.5\ell < \ell_d < \ell$.) This value can be compared to q, for which the next paragraph will suggest a value of, perhaps, $q = 0.5/\ell$. The focusing strength of bend relative to quadrupole is therefore of order $(\ell/R)^2$. Except for very small rings, having only a few bending magnets, this ratio is much smaller than one.

We can therefore ignore the presence of the bending magnets and accept the result of Problem 4.8.2, which gives the rough proportionality $\beta \sim 1/q$. It is clear that β can be made small by making q large. Recalling that q is the inverse focal length of the FODO lenses, this indeed corresponds to making the focusing strong.

If q is made too large the square root in the expression for β becomes imaginary which, not surprisingly, is unacceptable, because it implies the lattice is unstable. For given L the value of β cannot, therefore, be made arbitrarily small. But L itself can be made small so, in principle, β *can* be made arbitrarily small. In practice there are all kinds of factors which set limits on this. Many of these factors are discussed in Chapter 11.

It has been shown conclusively that strong focusing can be used to reduce the transverse beam size in a storage ring. However, refering back to Eq. (4.1), one sees that it is the emittances ϵ_x and ϵ_y that need to be made small to improve the brilliance of the ring as an x-ray source. In the next few chapters it will be seen how this can be achieved using strong focusing, along with other lattice refinements.

5
The Influence of Synchrotron Radiation on a Storage Ring

5.1
Preview

It is synchrotron radiation that distinguishes electron storage rings from proton storage rings. The large difference in mass of these particles causes radiation to be dominant for electrons and negligible for protons. The leading effect of the radiation is the simple frittering away of energy. The strong dependence on electron energy of this energy loss sets a practical limit to the achievable energy of an electron ring. The e^+e^- ring LEP at CERN in Geneva, Switzerland has achieved an energy of about 100 GeV, but nothing beyond that is currently judged to be economically feasible.[1]

To lend concreteness to the formulas, numerical values are frequently assigned to the various parameters; they are mainly taken from CESR (Cornell Electron Storage Ring). Phenomena will be scrutinized for possible modifications that could lead to reduced emittance, shortened bunch length, etc.

For existing electron accelerators, once paying the power bill needed to replenish radiated energy has been accepted and discounted, the remaining effects of radiation are a mixed blessing. The steady loss of energy has the desirable concomitant effect of the betatron damping that accompanies its replenishment. This leads to a beneficial emittance reduction. (For protons at the highest energies presently contemplated, at the LHC, this benefit will be just beginning to be useful.) Unfortunately this radiation is emitted in the form of photons, at a random rate, and with random energies. This randomness, also known as "quantum fluctuation", causes the stochastic excitation of both transverse and longitudinal oscillations. There is a competition between this excitation mechanism and the previously mentioned damping. This leads to an eventual equilibrium in which all beam dimensions are determined, independent of prehistory. Beams smaller/larger than this are excited/damped to these dimensions in a time that is typically of the order of milliseconds. Furthermore, except for nonlinear effects, all beam distributions

---

1) There have been design studies at Fermilab for a VLLC (Very Large Lepton Collider) with energy of perhaps 200 GeV, which would only make economic sense in conjunction with the huge tunnel needed for the VLHC (Very Large Hadron Collider).

Accelerator X-Ray Sources. Richard Talman
Copyright © 2006 WILEY-VCH Verlag GmbH & Co. KGaA, Weinheim
ISBN: 3-527-40590-9

become Gaussian. Analysis of these effects is an interesting application of stochastic physics, and makes up the main content of this chapter.

Many of the results to be discussed were first derived by Sands [1]. In writing this chapter, for the sake of continuity, all results have been derived from first principles but, if questions arise, returning to Sands's article could be useful. As far as I know, there are no essential differences between Sands' formulas and those presented here.

5.2
Statistical Properties of Synchrotron Radiation

5.2.1
Total Energy Radiated

For the highly relativistic situation characteristic of light sources, it is not necessary to distinguish between energy $\mathcal{E}$ and momentum p (or pc if one insists).

There is a central energy $\mathcal{E}_0$ which, in a large machine like LEP, varies appreciably around the ring because of radiation. For typical synchrotron light sources this effect is negligible, because the energy an electron loses per turn is such a tiny fraction of its own energy. At the CLS (Canadian Light Source) $\Delta \mathcal{E}/\mathcal{E} \simeq 0.0002$. Nevertheless, any particular electron energy $\mathcal{E} \simeq pc$ will deviate from the nominal energy $\mathcal{E}_0$. The energy of a particular emitted photon will be denoted by u, the total energy radiated in one turn by U_0 and the revolution period by $T_0 = 1/f_0$. Stochasticity results because of the randomness of u, and U_0 is also, in principle, a random variable, but the number of photons emitted per revolution is so great that it is legitimate to treat U_0 as non-probabilistic.

The radius of curvature R and magnetic field B are related by

$$R = \frac{pc/e}{cB} \tag{5.1}$$

In what follows, B and R will be treated as necessarily positive quantities. (Reverse bends are rare except in wigglers, where they are essential and where radiation effects are intentionally important. In those cases (straightforward) alterations must be applied to all formulas to follow.) In the same spirit it is common to make an "isomagnetic" assumption, probably introduced by Sands, in which the magnetic field, wherever it is nonvanishing, has the same value B. This causes the gross orbit to be a perfect circle except for interpolated straight line segments. Since there is no radiation in straight sections, the presence of these straights causes only trivial modification, such as circumference C being larger than $2\pi R$, and f_0 correspondingly smaller. Since the isomagnetic assumptions is never entirely valid, it will be necessary to correct the formulas later to analyse practical accelerators. Remedying oversimplifications such as this may test our understanding of the statistics of the phenomena under study.

The steps to be taken in analyzing emittance growth due to photon radiation in a certain sector of the accelerator are: first, calculate the total energy radiated in the sector; second, knowing the mean energy of radiated photons, calculate the number of photons radiated; third, calculate the effect of a single photon emission on the betatron amplitude of a particle in the accelerator; and fourth, using the fact that emission times and energies are random, find the accumulated effect of these excitations. Finally, to find the final equilibrium, it will still be necessary to analyze the damping effect accompanying the replenishment of the radiated energy.

The total energy radiated at all photon energies from a single electron during time T is given [1,3] by[2]

$$U_T = P_\gamma T = \frac{e^2 c^3}{2\pi} C_\gamma \mathcal{E}^2 B^2 T = \sum_{j=1}^{\mathcal{N}_T} u_j, \qquad (5.2)$$

where $C_\gamma = 0.885 \times 10^{-4}\,\mathrm{m/GeV}^3$, the total number of photons emitted is $\mathcal{N}_T$, and their energies are $u_j, j = 1, \ldots, \mathcal{N}_T$. The instantaneous radiated power P_γ is made up of forward-traveling photons; for present purposes they will be taken as emitted exactly in the forward direction.

In a practical accelerator the field B varies with arc length coordinate s and the nominal energy radiated in one turn U_0 can be expressed as an integral along the central orbit

$$U_0 = \frac{e^2 c^2}{2\pi} C_\gamma \mathcal{E}^2 \oint B^2(s)\,ds. \qquad (5.3)$$

If B has the same value everywhere (except where it vanishes) the total energy U_0 radiated in a complete revolution is

$$U_0 = C_\gamma \frac{\mathcal{E}^4}{R}. \qquad (5.4)$$

---

2) In this chapter, following Sands, the attempt is made to express most absolute radiative intensities, fluxes, etc. in terms of U_0, the energy radiated by one electron each turn. For the same reason (to help in keeping magnitudes straight) Hofmann refers most quantities to a single absolute quantity; in his case radiated power P_0. This is the same as our P_γ. Of course U_0, a per-turn quantity, and P_0, an instantaneous quantity, are related, at least approximately, by $(2\pi R/c) P_\gamma = U_0$. For real rings, since the presence of bend-free regions cause circumference $C > 2\pi R$, the average power is less than P_γ evaluated within a bending magnet. But the energy loss per turn is still U_0. So P_γ is the more fundamental quantity, but it is U_0 which is handier for quick approximations because it is the quantity that makes its presence known in the control room of the storage ring.

This formula is as simple as it is because it is actually the defining equation for constant C_γ. It is necessary to be careful when using this equation since, conventionally, C_γ is given, as above, in other than M.K.S. units. As defined earlier in Eq. (3.39), a "typical" or "critical" photon energy parameter u_c is defined by

$$u_c = \frac{3}{2}\hbar c \gamma^3 \frac{cB}{\mathcal{E}/e}. \tag{5.5}$$

Beware that there may not be universal agreement on the definition of u_c. In chapter 4 it was shown that half the radiated energy falls below u_c, as defined by Eq. (5.5), and half above, but this assumed isomagnetic magnets. To serve as an example, a consistent set of numerical values (drawn from CESR) for the physical parameters and for u_c and other derived quantities are listed in Table 5.1.

Tab. 5.1 Parameters describing synchrotron radiation in a hypothetical pure FODO lattices resembling the CESR storage ring.

Quantity	Symbol	Formula or reference	Value	Units
rest energy	mc^2		0.511	MeV
relativistic factor	γ_0		10^4	
central energy	$\mathcal{E}_0$	$\gamma_0 mc^2$	5.11	GeV
magnetic field	B		0.1939	T
radius of curvature	R	(5.1)	87.9	m
magnet length	l_B		6.57	m
one magnet bend angle	$\Delta\theta_1$	l_B/R	0.0747	
time in one magnet	T_1	l_B/c	21.9	ns
energy radiated per magnet	U_{T_1}	(5.2)	8.166	keV
critical energy	u_c	(5.5)	3.367	keV
mean energy	$<u>_\gamma$	(5.13)	1.037	keV
mean squared energy	$<u^2>_\gamma$	(5.13)	4.619	keV^2
energy radiated per turn	U_0	$2\pi U_{T_1}/\Delta\theta_1$	0.687	MeV
photons per turn		$U_0/<u>_\gamma$	662	
number of bend half-cells	n		84	
half-cell length	ℓ	(4.3)	8.44	m
phase advance per cell	2ϕ		70	deg.
"curly-H"	$\mathcal{H}$	(5.65)	0.250	m
emittance	ϵ_x	(5.76)	0.1089	μm
maximum β_x in arcs	$\hat{\beta}_x$		45	m
σ_x at $\hat{\beta}_x$	$\hat{\sigma}_x$		2.22	mm
fractional energy spread	σ_δ	(5.90)	0.47	$\times 10^{-3}$

5.2.2
The Distribution of Photon Energies; "Regularized Treatment"

When the energy u is expressed in units of u_c by $\xi = u/u_c$ the probability distribution of photon energies $n_\xi(\xi)$ is a universal function. Because the number of photons in the energy range du at u is $n_u(u)\, du = n_\xi(\xi)\, d\xi$, one has $n_u(u) = n_\xi(\xi)\, d\xi/du = n_\xi(u/u_c)/u_c$. When $\mathcal{N}$ photons are emitted altogether, the number in the range du at u is $\mathcal{N} n_u(u)\, du$. These definitions require that n_u and n_ξ are normalized[3] according to

$$\int_0^\infty n_u(u')\, du' = \int_0^\infty n_\xi(\xi')\, d\xi' = 1. \tag{5.6}$$

An analytic expression for $n_\xi(\xi)$ is given in Sands and in Landau and Lifshitz, in terms of a function $S(\xi)$ which is related to the "MacDonald" function which can itself be expressed in terms of Bessel functions,

$$\xi\, \tilde{n}_\xi(\xi) \equiv S(\xi) = \frac{9\sqrt{3}}{8\pi} \xi \int_\xi^\infty K(5/3, \xi')\, d\xi', \tag{5.7}$$

where $K(5/3, \xi)$ is a modified Bessel function with fractional index $5/3$. The tilde on number density $\tilde{n}_\xi$ is to indicate that the distribution may not be (and in fact, *is not*) normalized. Hofmann expresses the frequency spectrum of the radiation in terms of the same function $S(\xi)$;

$$\frac{dU_0}{d\omega} = \frac{U_0}{\omega_c} S\left(\frac{\omega}{\omega_c}\right), \tag{5.8}$$

where $\omega_c = u_c/\hbar$.

The number density is divergent near the origin—in that limit $S(\xi) \sim \xi^{1/3}$ so $\tilde{n}_\xi \sim \xi^{-2/3}$. This is a very weak divergence though, in the sense that the integral giving the total number of photons converges. As it happens, the weak divergence at low energy is unphysical anyway, since the derivation of Eq. (5.7) breaks down there. It will be seen though, that the "unphysical photons" given by that formula have no significant effect on the measurable quantities of interest. We therefore proceed to use Eq. (5.7), for example to check the normalization of $\tilde{n}_\xi(\xi)$, using Maple. Unfortunately, perhaps due to the complicated behavior near the origin, or because the integrand is itself expressed as an integral whose limit is the variable of integration, Maple refuses to do the integral. "Working around" complications like this will be one of the objects of the following treatment.

3) It is obviously unphysical for the upper limit to be set to ∞, but for present purposes all integrands will be sufficiently convergent that it does not matter.

When calculating radiative effects the following combination of circumstances is common:

- We do not know how to predict emission at long wavelengths.
- We do not need to know about radiation at long wavelengths.
- When formulas valid at short wavelengths are extrapolated (analytically continued) to long wavelenths, mathematical complications ruin the physical predictions of the theory, for example in the form of divergent integrals.

In quantum electrodynamics this conundrum goes by the name "the infrared problem". The number of photons emitted is infinite but the number of "important" photons is finite. Should one *need* to understand the situation at long wavelengths one may have to drop the quantum mechanical formulation and revert to classical electromagnetism. Otherwise, with long wavelengths being of no interest, reformulation of the short wavelength analysis to suppress the divergence they cause is known as "regularization".

We face a similar situation; probability density $\tilde{n}_\xi$ becomes infinite at long wavelengths ($\xi \to 0$). The problem is not a true infrared divergence since the total number of photons (proportional to the integral of $\tilde{n}_\xi$) is finite. Nevertheless the divergence of $\tilde{n}_\xi$ leads to mathematical complications that we prefer to avoid. This has led me to the following, rather idiosynchratic, "regularization" procedure and discussion.

For the purpose of this formulation, I have developed the empirical or "regularized" formula

$$S_{\text{reg}}(\xi) = \begin{cases} 0 & \text{if } 0 < \xi < 0.000057, \\ 1.0599\, \xi^{0.275} e^{-0.965\xi} & \text{if } 0.000057 < \xi < \infty, \end{cases} \quad (5.9)$$

exhibited in Figure 5.1 along with $S(\xi)$ as defined by Eq. (5.7). If two curves are not visible, it is because the approximation is so good. Here "regularized" means the function is theoretically valid in all important regions and mathematically handy in physically insignificant regions. In this case the spectrum "cuts off" at a tiny value $\xi = 0.000057$, the exact numerical value of which will be justified shortly. The function $S(\xi)/\xi$ is plotted in Figure 5.2. Both of these plots agree well with plots in Hofmann's article [4].

Both Sands and Hofmann assume normalization,

$$\int_0^\infty S(\xi)\, d\xi = 1. \quad (5.10)$$

As already mentioned, Maple refuses to perform the integral. But with $S(\xi)$ approximated as in Eq. (5.9) Maple performs the integral satisfactorily and finds $\int_0^\infty S_{\text{reg}}(\xi)\, d\xi = 1.00001$, in good agreement with Sands and Hofmann.

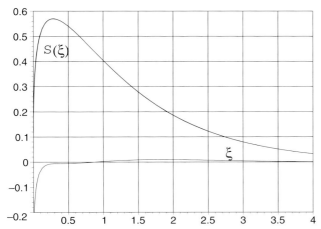

Fig. 5.1 Using Maple, the function $S(\xi)$ defined in Eq. (5.7) is plotted as a function of ξ. Also plotted is the "regularized" approximation $S_{\text{reg}}(\xi)$ defined by Eq. (5.9). Because the approximation is fairly good the fact that there are two curves is almost undetectable. Also plotted, therefore, is the ten-times-multiplied difference of the two expressions. It can be seen that the approximation breaks down only at the long wavelengths.

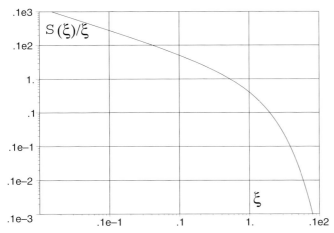

Fig. 5.2 The function $S(\xi)/\xi$ as calculated by Eq. (5.7). This function is proportional to the photon number density $n_\xi(\xi)$ where $\xi = u/u_c$. It is shown in the text that the area under this graph is $<u>_\gamma/u_c \simeq 3.264$. Therefore, to obtain $n_\xi(\xi)$, the value read from the curve has to be multiplied by 0.3080. This curve agrees with Hofmann's Figure 20 [4], except that his caption calls it a "normalized photon spectrum"; by this I think he must mean that the area under the curve corresponds to integrated power equal to 1 unit.

Let us now check the normalization of $\tilde{n}_{\xi}(\xi)$ as defined in Eq. (5.7). Maple finds

$$\int_0^\infty \frac{1}{\xi} S_{\text{reg}}(\xi) \, d\xi = 3.2468. \tag{5.11}$$

To our possible surprise, $\tilde{n}_{\xi}(\xi)$ is *not* normalized. That is why the tilde was placed on $\tilde{n}_{\xi}$. We can see, however, why this will not affect future results. Only *moments* enter the subsequent theory. Because the "probability distribution" is not normalized, the expressions for these moments have to be expressed as ratios of integrals;

$$<\xi>_\gamma = \frac{\int_0^\infty S_{\text{reg}}(\xi) \, d\xi}{\int_0^\infty \xi^{-1} S_{\text{reg}}(\xi) \, d\xi} = 0.30799$$

$$<\xi^2>_\gamma = \frac{\int_0^\infty \xi \, S_{\text{reg}}(\xi) \, d\xi}{\int_0^\infty \xi^{-1} S_{\text{reg}}(\xi) \, d\xi} = 0.40694. \tag{5.12}$$

All subsequent results will be expressed in term of moments of $n_{\xi}(\xi)$; both Sands and Hofmann state these average energy and average energy-squared quantities to be

$$<u>_\gamma = \int_0^\infty u' n_u(u') \, du' = \frac{8\sqrt{3}}{45} u_c = 0.3079 u_c,$$

$$<u^2>_\gamma = \int_0^\infty u'^2 n_u(u') \, du' = \frac{11}{27} u_c^2 = 0.4074 u_c^2, \tag{5.13}$$

where, in these expressions, $n_{\xi}(\xi)$ *is* assumed to be a true, normalized-to-1 probability distribution. Here the symbol $<>_\gamma$ implies averaging over photon energies, (as we have been doing) and as contrasted with averaging over lattice elements, or over particle turns, or betatron phase. (The γ subscript is unconnected with the relativistic factor γ.)

It is no coincidence that the right-hand sides of Eqs. (5.12) agree with the right-hand sides of Eqs. (5.13)—the mysterious cut-off, $\xi = 0.000057$ was chosen to make the first moments match. The fact that the second moments also match is a consistency check that shows the procedure to have been consistent, at least concerning the ratios of the moments. The contribution to the second moment coming from the unphysical, low energy end of the spectrum, is negligible. Since it is desirable to have a simple, explicit, normalized, formula for number density, we revise Eq. (5.9), dividing by ξ and also by normalizing factor 3.2468,

$$n_{\xi}(\xi) = 0.3080 \frac{S_{\text{reg}}(\xi)}{\xi}$$

$$= \begin{cases} 0 & \text{if } 0 < \xi < 0.000057, \\ 0.3264 \, \xi^{-0.725} e^{-0.965 \xi} & \text{if } 0.000057 < \xi < \infty. \end{cases} \tag{5.14}$$

The purpose behind insisting that the probability distribution be normalized is for the mental comfort lent by concreteness. It is convenient to have formulas that describe the radiation as an unambiguous, finite number of photons for which standard probabilistic measures are applicable.

We had better be sure, therefore, to calculate this concrete number of photons correctly. Let us denote the number radiated per turn, per electron, by $\mathcal{N}_0$, and (as in Eq. (5.13)) the average energy by $<u>_\gamma$. By definition of the term "average energy" these quantities are related to the total energy U_0 by[4]

$$\mathcal{N}_0 = \frac{U_0}{<u>_\gamma} = \frac{ecC_\gamma \mathcal{E}^3 B}{0.3079 \gamma u_c} = 0.0662 \gamma. \tag{5.15}$$

Accepting that $\mathcal{N}_0$ is the number of photons, since the radiation comes in the form of photons of energy $u = \hbar \omega$, using Eq. (5.8) and subsidiary variable $\xi = u/u_c$ we also have to insist on the chain of equalities

$$\mathcal{N}_0 n_\xi(\xi) = \mathcal{N}_0 n_u(u) \frac{du}{d\xi} = \mathcal{N}_0 n_u(u) u_c = \frac{u_c}{u} \frac{dU_0}{du}$$

$$= \frac{\omega_c}{\omega} \frac{dU_0}{d\omega} \frac{1}{\hbar} = \frac{\omega_c}{\omega} \frac{U_0}{\omega_c} S\left(\frac{\omega}{\omega_c}\right) \frac{1}{\hbar} = \frac{U_0}{u_c} \frac{S(\xi)}{\xi}, \tag{5.16}$$

where the distributions $n_u(u)$ and $n_\xi(\xi)$ have to be normalized probability distributions. Simplifying the outermost equation and substituting from Eqs. (5.15), (5.13) and (5.15) yields

$$n_\xi(\xi) = \frac{<u>_\gamma}{u_c} \frac{S(\xi)}{\xi} = 0.3079 \frac{S(\xi)}{\xi}. \tag{5.17}$$

This confirms the correctness of $n_\xi(\xi)$ as given by Eq. (5.14) as the valid, normalized probability distribution of the photon energies.

5.2.3
Randomness of the Radiation

The number $\mathcal{N}_0$ of radiated photons just introduced is subject to statistical fluctuations causing the actual number radiated in one turn to be, say,

---

4) Equation (5.15) can be the basis for a curious bit of *trivia*: every electron in every present or planned symmetric B-factory— just a cute way of saying that $\gamma = 10^4$— radiates 662 photons ($\pm \sqrt{662}$ statistical fluctuation) every turn. Half, 331, are radiated in each half, and the same would be true if the magnets in one half were replaced by bend fields one tenth as long, ten times as strong. Redistributing the bends to make the whole ring homogeneous would not make a difference. It seems then that for any guide field whatsoever (with no reverse bends, as for example in a wiggler) there will be 662 photons. Having "established" this remarkable result, one must acknowledge that it has little value—it is the energy radiated, not the number of photons, by which the radiation does most of its dirty work.

$662 \pm \sqrt{662}$, which amounts to several percent variation from turn to turn. This variability rapidly averages out over tens of turns.

But the radiation from a single passage through one magnet is subject to substantial fluctuation. Applying formulas derived above, except for fluctuations, the total number of photons radiated in a magnet of length Tc is $\mathcal{N}_T$, which is related to U_T, the total energy radiated in the magnet, by

$$\mathcal{N}_T = \frac{U_T}{<u>_\gamma}. \tag{5.18}$$

The fluctuation in this number is $\pm\sqrt{\mathcal{N}_T}$. The radiated power $P_{\gamma,u}$, expressed as a distribution of energy such that $P_{\gamma,u}(u)\,du$ is the time rate of emission of energy in the range du at u, is given by

$$P_{\gamma,u}(u) = \frac{\mathcal{N}_T}{T} u\, n_u(u). \tag{5.19}$$

which is subject to corresponding fluctuation. Averaged over this fluctuation, this is the integrand of the integral yielding $<u>_\gamma$ in Eq. (5.13). $P_{\gamma,u}(u)$ is therefore proportional to the curve plotted in Figure 5.1.

Fluctuations in photon emission are central to the dynamics of the electrons in a storage ring. The emittance growth rate due to radiation turns out to be proportional to the average sum of squares of photon energies for the $\mathcal{N}_T$ photons with total energy U_T radiated during time T. From formulas derived so far this is given by

$$<\sum_{j=1}^{\mathcal{N}_T} u_j^2>_\gamma = \mathcal{N}_T <u^2>_\gamma = \frac{U_T}{<u>_\gamma} <u^2>_\gamma = 1.323 u_c U_T. \tag{5.20}$$

The important dependences of this simple result are that (by Eq. (5.2)) U_T is proportional to $\mathcal{E}^2 B^2$ and that (by Eq. (5.5)) u_c is proportional to $\gamma^3 B/\mathcal{E}$. The dependence on B forces regions of different magnetic fields to be treated independently, i.e. integrated over separately. The dependence on $\mathcal{E}$ is important in determining the final beam equilibrium.

The main source of fluctuation is the randomness in time of the photon emissions, but the randomness in the energies of the emitted photons also plays a role—one must not simply pretend that all radiated photons have the average energy. The betatron amplitude induced by any one emission of energy u depends on the betatron phase the electron happens to have at the time. But, averaged over betatron phase, the betatron amplitude induced is proportional to u^2. If all photons are (unrealistically) supposed to have the same energy, $<u>_\gamma$, the strength of the stochastic growth would be proportional to $<u>_\gamma^2$. In fact, the growth mechanism is stronger than this by a factor equal to the ratio $<u^2>_\gamma / <u>_\gamma^2 = 4.297$. This reflects the fact that the emittance growth due to a single emission is proportional to u^2 and that a relatively small number of higher energy photons can make a relatively large contribution to the average square. It is this consideration which validates our having dropped low energy photons, for example in Eq. (5.14).[5]

The total energy U_T radiated during time T can be obtained from Eq. (5.2) and from that an almost unambiguous total number of photons $\mathcal{N}_T = U_T / <u>_\gamma$ can be obtained. To obtain this formula it has been assumed that individual emissions are uncorrelated and that $\mathcal{N}_T$ and U_T are not subject to fluctuations. For times too short (say comparable to the mean time between emissions) this would not be valid, but the formula will be applied only for times long enough to ensure small fractional deviation of $\mathcal{N}_T$; i.e. $\sqrt{\mathcal{N}_T} \gg 1$.

Problem 5.2.1 *Use a mathematical language, such as Maple, Mathematica, Matlab, or a programmable hand calculator, to find the minus-first, zeroth and first moments of the function $S(\xi)$ defined in Eq. (5.7). Since I failed to goad Maple into doing this exercise, it is not simple. Perhaps a more skillful user, or use of a less recalcitrant language can accomplish the task. A person strong in applied math and willing to wade through the literature can do the problem analytically. Failing (or instead of) all these approaches, work out the zeroth, first and second moments of the function $n_{\tilde{\xi}}(\xi)$ defined in Eq. (5.14).*

Problem 5.2.2 *Using formulas derived in what was called the radial approximation in chapter 4, derive an analytic expression for $n_{\tilde{\xi}}(\xi)$.*

5.3
The Damping Rate Sum Rule: Robinson's Theorem

The phase space coordinates of an individual particle at location s in the ring are given by (x, f, y, g, z, h), where $f \equiv x'$, $g \equiv y'$, and $h \equiv \delta$ have been introduced to suppress the primes, and to permit the matching of upper case letters for matrix elements with their corresponding, lower case coordinates. In passing through a differential longitudinal path element ds, evolution of

---

5) Approximation Eq. (5.14) is intended purely for studying the effect of radiation on beam dynamics. It clearly becomes invalid for radiation of ultraviolet and longer wavelengths, even though it is nonzero in part of that range.

these components is described by a transfer matrix:

$$\mathbf{M}(\mathrm{d}s) = \mathbf{1} + \mathrm{d}\mathbf{M}$$

$$= \begin{pmatrix} 1+\mathrm{d}X_1 & \mathrm{d}X_2 & \mathrm{d}X_3 & \mathrm{d}X_4 & \mathrm{d}X_5 & \mathrm{d}X_6 \\ \mathrm{d}F_1 & 1+\mathrm{d}F_2 & \mathrm{d}F_3 & \mathrm{d}F_4 & \mathrm{d}F_5 & \mathrm{d}F_6 \\ \mathrm{d}Y_1 & \mathrm{d}Y_2 & 1+\mathrm{d}Y_3 & \mathrm{d}Y_4 & \mathrm{d}Y_5 & \mathrm{d}Y_6 \\ \mathrm{d}G_1 & \mathrm{d}G_2 & \mathrm{d}G_3 & 1+\mathrm{d}G_4 & \mathrm{d}G_5 & \mathrm{d}G_6 \\ \mathrm{d}L_1 & \mathrm{d}L_2 & \mathrm{d}L_3 & \mathrm{d}L_4 & 1+\mathrm{d}L_5 & \mathrm{d}L_6 \\ \mathrm{d}H_1 & \mathrm{d}H_2 & \mathrm{d}H_3 & \mathrm{d}H_4 & \mathrm{d}H_5 & 1+\mathrm{d}H_6 \end{pmatrix}. \quad (5.21)$$

The differential "d" symbols indicate elements that are first order in $\mathrm{d}s$. For example, in pure drift regions $\mathrm{d}X_2 = \mathrm{d}Y_4 = \mathrm{d}s$. Eventually $\mathrm{d}s$ will be made arbitrarily small. The determinant $|\mathbf{M}(\mathrm{d}s)|$ is almost, but not necessarily exactly, equal to 1 since the matrix represents radiation effects as well as the effects of bending and accelerating elements. Transfer through a sequence of intervals $\mathrm{d}s_1, \mathrm{d}s_2, \ldots$, is represented by the matrix

$$\mathbf{M} = \prod_i \mathbf{M}(\mathrm{d}s_i). \quad (5.22)$$

First calculating the determinant of each of its factors (keeping only first order differentials) the determinant of $\mathbf{M}$ can be approximated to first order as

$$|\mathbf{M}| = 1 + \sum_i (\mathrm{d}X_{1,i} + \mathrm{d}F_{2,i} + \mathrm{d}Y_{3,i} + \mathrm{d}G_{4,i} + \mathrm{d}L_{5,i} + \mathrm{d}H_{6,i}) + \ldots. \quad (5.23)$$

Note that the off-diagonal matrix elements of Eq. (5.21) do not enter to the order exhibited. They contribute terms of quadratic and higher order.

For "Hamiltonian" elements such as magnets, drifts, and RF cavities, it is known that the determinant $\det|\mathbf{M}(\mathrm{d}s)| = 1$ and that the six eigenvalues of $\mathbf{M}(\mathrm{d}s)$ come in three complex conjugate pairs lying on the unit circle. Supposing these eigenvalues to have been determined for the interval $\mathrm{d}s$, and generalizing slightly by permitting their magnitudes to deviate slightly from 1, we define the eigenvalues as[6]

$$\lambda_x(\mathrm{d}s) = \mathrm{e}^{-\mathrm{d}\alpha_x \pm \mathrm{i}\mathrm{d}\mu_x}, \quad \lambda_y(\mathrm{d}s) = \mathrm{e}^{-\mathrm{d}\alpha_y \pm \mathrm{i}\mathrm{d}\mu_y}, \quad \lambda_s(\mathrm{d}s) = \mathrm{e}^{-\mathrm{d}\alpha_s \pm \mathrm{i}\mathrm{d}\mu_s}.$$

(5.24)

Since the determinant of a matrix is equal to the product of the eigenvalues, the determinant of the matrix appropriate for evolution through $\mathrm{d}s$ can be

---

6) Labeling the eigenvalues by x, y, and s suggests the theory is valid only if coupling is absent, but that is not the case. x, y, and s, are simply labels for the three local eigenvectors. Of course the αs introduced here have nothing to do with the lattice Twiss functions, but the $\mathrm{d}\mu$'s integrate up to the standard tunes.

expressed in terms of $d\alpha_x$, $d\alpha_y$ and $d\alpha_s$;

$$|\mathbf{M}(ds)|$$
$$= (e^{-d\alpha_x+id\mu_x}e^{-d\alpha_x-id\mu_x})(e^{-d\alpha_y+id\mu_y}e^{-d\alpha_y-id\mu_y})(e^{-d\alpha_s-id\mu_s}e^{-d\alpha_s-id\mu_s})$$
$$\approx 1 - 2(d\alpha_x + d\alpha_y + d\alpha_z), \tag{5.25}$$

where higher order terms have again been dropped. Letting $\alpha_x = \int d\alpha_x$, $\alpha_y = \int d\alpha_y$, and $\alpha_s = \int d\alpha_s$, all assumed to be small compared to 1, and proceeding to the $ds = 0$ limit, we obtain

$$|\mathbf{M}| = 1 - 2(\alpha_x + \alpha_y + \alpha_s). \tag{5.26}$$

The determinant $\mathbf{M}$ just given in Eq. (5.26) can also be approximated directly from Eq. (5.21), keeping only terms of first order in ds, and that is how we will evaluate the damping decrements α_i. If a term in the expansion of the determinant contains one off-diagonal element then it must contain at least one other, making it at least second order in ds. This limits our task to calculating the on-diagonal perturbations. In evaluating the on-diagonal elements $dX_1, dF_2, \ldots$, knowing that magnetic elements and drifts leave the value of the determinant unaltered (at 1), we need only include energy altering effects.

Within the path length ds under discussion there is the possibility of energy change due to radiation or due to RF cavities, or other longitudinal electric fields,

$$d\mathcal{E} = -dU^{\text{rad}} - dU^{\text{rf}}. \tag{5.27}$$

The rationale for the sign of dU^{rad} (previously defined as a positive quantity) is that it should lead to a reduction in $\mathcal{E}$. The rationale for the sign of dU^{rf} is that this quantity is to be regarded as a potential energy, such that moving to lower potential increases $\mathcal{E}$.

We now evaluate the six on-diagonal perturbations one-by-one. It will turn out that dX_1, dY_3 and dL_5 vanish (to first order in ds) but to see this it is useful to first understand what contributes to dF_2, dG_4, and dH_6.

If the particle energy increases from $\mathcal{E}$ to $\mathcal{E} - dU^{\text{rf}}$ by virtue of potential energy change dU^{rf} (a negative quantity), its momentum will change as shown in Figure 5.3. Though the particle's transverse momentum is unchanged, its longitudinal momentum *increases* by factor $1 - dU^{\text{rf}}/\mathcal{E}$, and this has the effect $x' \to x'(1 + dU^{\text{rf}}/\mathcal{E})$; i.e. a slight reduction in slope. The vertical slope y' is reduced by the same factor. This means that

$$F_2 = G_4 = 1 + \frac{1}{\mathcal{E}_0}\frac{\partial U^{\text{rf}}}{\partial s}ds, \tag{5.28}$$

where it has been adequate to replace the slightly variable quantity $\mathcal{E}$ by the central value $\mathcal{E}_0$. These changes of slope also alter x and y, but the changes are proportional to ds^2. For this reason $dX_1 = dY_3 = 0$.

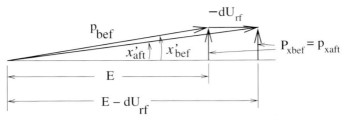

Fig. 5.3 Damping of transverse amplitude accompanying longitudinal acceleration. (As well as assuming $pc = p_s c = \mathcal{E}$, for dimensional consistency the figure assumes $c = 1$.)

The remaining term to be evaluated, dH_6, accounts for change in longitudinal momentum (which at high energy is, except for units, equivalent to energy). Superficially it might seem one should account for the change $\mathcal{E} \to \mathcal{E} - dU^{\text{rf}}$ due to RF acceleration, but since the reference orbit is also accelerated, $\mathcal{E}_0 \to \mathcal{E}_0 - dU^{\text{rf}}$, the ratio $\mathcal{E}/\mathcal{E}_0$ is unaffected by RF acceleration. This can be repeated for emphasis: though the energy of the reference orbit suffers a discontinuous change at an RF cavity, any particular particle suffers the same change. As a consequence, $\mathcal{E} - \mathcal{E}_0$ is continuous through the cavity. This may seem unnatural, and one could imagine a consistent perturbative formalism that would reckon the RF acceleration as perturbative, but that is not the case for *this* formalism. One also expects a term $(dU^{\text{rf}}/dt)(\ell/c)$ giving the change in particle energy for a particle whose arrival time at the RF cavity is ℓ/c. But, because this term contributes only to off-diagonal term dH_5, in the present context it can also be left out.

There is, however, a significant contribution to dH_6 from dU^{rad}. Referring to Eq. (5.2) one notes that dU^{rad} is proportional to $\mathcal{E}^2$, causing the mean energy radiated by a particular particle to deviate from the mean energy radiated by the reference particle. We have $dU^{\text{rad}} = dU_0^{\text{rad}}(\mathcal{E}/\mathcal{E}_0)^2 \simeq (1 + 2\delta)\, dU_0^{\text{rad}}$ and

$$\delta = \frac{\mathcal{E} - \mathcal{E}_0}{\mathcal{E}_0} \longrightarrow \frac{\mathcal{E}_0(1+\delta) - dU_0^{\text{rad}}(1+2\delta) - \mathcal{E}_0 + dU_0^{\text{rad}}}{\mathcal{E}_0}$$

$$= \left(1 - 2\frac{dU_0^{\text{rad}}}{\mathcal{E}_0}\right)\delta, \qquad (5.29)$$

which implies

$$H_6 = 1 - 2\frac{dU_0^{\text{rad}}}{\mathcal{E}_0} \qquad (5.30)$$

(The earlier result that there is no contribution to dH_6 from accelerating fields could have been obtained in this same calculation using the fact that dU^{rf} is independent of δ.)

Combining results the transfer matrix is given to adequate accuracy by

$$\mathbf{M}(\mathrm{d}s) \simeq \begin{pmatrix} 1 & 0 & 0 & 0 & 0 & 0 \\ 0 & 1+\mathrm{d}U^{\mathrm{rf}}/\mathcal{E}_0 & 0 & 0 & 0 & 0 \\ 0 & 0 & 1 & 0 & 0 & 0 \\ 0 & 0 & 0 & 1+\mathrm{d}U^{\mathrm{rf}}/\mathcal{E}_0 & 0 & 0 \\ 0 & 0 & 0 & 0 & 1 & 0 \\ 0 & 0 & 0 & 0 & 0 & 1-2\mathrm{d}U_0^{\mathrm{rad}}/\mathcal{E}_0 \end{pmatrix}. \quad (5.31)$$

Evaluating the determinant yields

$$\det |\mathbf{M}(\mathrm{d}s)| = 1 + 2\frac{\mathrm{d}U_0^{\mathrm{rf}}}{\mathcal{E}_0} - 2\frac{\mathrm{d}U_0^{\mathrm{rad}}}{\mathcal{E}_0}. \quad (5.32)$$

(From here on it will be unnecessary to distinguish between $\mathrm{d}U_0$ and $\mathrm{d}U$ since their difference is of higher order than $\mathrm{d}s$.) Provided the terms on the right-hand side remain small compared to 1, concatenating over successive intervals $\mathrm{d}s$ is equivalent to integrating over $\mathrm{d}U^{\mathrm{rf}}$ and $\mathrm{d}U_0^{\mathrm{rad}}$. Furthermore, when integrated over one complete revolution, one can assume that the RF has been adjusted to just replace the radiated energy. As a result

$$U_0^{\mathrm{rf}} + U_0^{\mathrm{rad}} = \oint \mathrm{d}U_0^{\mathrm{rf}} + \oint \mathrm{d}U_0^{\mathrm{rad}} = 0. \quad (5.33)$$

In equilibrium this adjustment is automatic. Combining results we obtain for the once-around transfer matrix

$$\det |\mathbf{M}| = 1 - 4\frac{U_0^{\mathrm{rad}}}{\mathcal{E}_0}. \quad (5.34)$$

We are now in a position to complete the step described below Eq. (5.26). Combining that equation with (5.34) yields

$$\alpha_1 + \alpha_2 + \alpha_3 = 2\frac{U_0^{\mathrm{rad}}}{\mathcal{E}_0}. \quad (5.35)$$

This analysis was first performed by Robinson [5], and, independently and at about the same time (though in somewhat less generality) by Orlov [6]. It is known as "Robinson's theorem"—in words: for any lattice the sum of the damping decrements is twice the fractional synchrotron radiation energy emitted per turn. The theorem does not depend on the statistical properties of the radiated photons—only the fractional accumulated energy loss matters.

Historically Robinson's theorem played a very significant role in the early days of electron positron storage rings. The lattices of early electron rings were constructed from combined function magnets. As we shall see, such lattices exhibit extremely strong longitudinal damping and hence, by Robinson's theorem, horizontal anti-damping. This problem was overcome by separated function lattices in which the bending and focusing is performed by separate elements.

5.3.1
Vertical Damping

By Robinson's theorem the sum of damping rates can be calculated easily, even for a complicated lattice, but the individual damping rates depend sensitively on lattice details. It is, however, usually valid to assume (by design anyway) that there is no "cross-plane coupling" between x and y motion. In such an ideally uncoupled accelerator the once-around transfer matrix takes the form

$$\mathbf{M} = \begin{pmatrix} X_1 & X_2 & 0 & 0 & X_5 & X_6 \\ F_1 & F_2 & 0 & 0 & F_5 & F_6 \\ 0 & 0 & Y_3 & Y_4 & 0 & 0 \\ 0 & 0 & G_3 & G_4 & 0 & 0 \\ L_1 & L_2 & 0 & 0 & L_5 & L_6 \\ H_1 & H_2 & 0 & 0 & H_5 & H_6 \end{pmatrix}, \qquad (5.36)$$

and one of the three eigenvalues is unambiguously identifiable with vertical coordinate y. In this case the preceding analysis can be applied to the 2×2 vertical transfer matrix to determine α_y, independent of horizontal and longitudinal motion. (See Problem 5.3.1.)

Almost invariably, in practice, the numerical values of the matrix elements are such that the smallest eigenvalue corresponds to longitudinal oscillation and the other two eigenvalues to transverse oscillation, but this feature cannot be inferred from the algebraic structure of Eq. (5.36).

Problem 5.3.1 *For the case of an uncoupled lattice, just discussed, complete the calculation and show that the vertical damping rate is given by*

$$\alpha_y = \frac{1}{2} \frac{U_0^{\text{rad}}}{\mathcal{E}_0}. \qquad (5.37)$$

In this case vertical damping accounts for 1/4 of the damping assured by the Robinson sum rule.

Problem 5.3.2 *It can be argued that the discussion of damping of pure vertical oscillations in the text and in the previous problem is internally inconsistent because the magnetic field has been assumed to be independent of y and yet stable motion with non-zero y implies non-zero focusing fields that depend on y. For the case that all focusing is performed by independent quadrupoles (for which the on-axis field vanishes), show that the damping due to radiation in the quadrupoles is negligible for sufficiently small y-amplitudes.*

5.3.2
Longitudinal Damping

Oscillations of fractional momentum variable δ, along with corresponding variation of longitudinal variable ℓ, are known as "synchrotron oscillations". Since the frequency of longitudinal oscillation is typically much less than the frequency of betatron oscillation, the longitudinal phase advance per revolution is much less than the betatron phase advance. It is perhaps for this reason that it is more customary to analyse longitudinal motion by differential equations than by transfer matrices, and this is the way we begin.

The reader is assumed to be familiar with the equations of damped harmonic oscillation; for example of a mass and spring oscillator with a very weak shock absorber, or very weak friction. All the equations in this section can be obtained from those equations by setting the mass to 1 and interpreting the restoring and friction forces appropriately. In pursuing this analogy the word "force" will be used loosely, even for quantities whose dimensions are not really those of force. Similarly "kinetic" and "potential" energies will be introduced for which the dimensions are other than energy.

As regards damping, horizontal and longitudinal motion of electrons in a storage ring are more intertwined with each other than either is with vertical motion. Specific lattice details complicate this coupling. There is an especially important distinction between combined function lattices having on-axis fields where focusing occurs, and separated function lattices for which all focusing occurs in quadrupoles having zero on-axis magnetic fields. Problem 5.3.2 was intended to provide a hint, in advance, of the importance of this distinction. Here we concentrate on isolating the longitudinal damping, beginning with a digression into simple harmonic motion.

One knows that δ does not retain its same non-zero value, but rather executes synchrotron (longitudinal/momentum) oscillation, which satisfies[7]

$$\ddot{\delta} + \mu_s^2 \delta = F_d = -2\alpha_s \dot{\delta}. \tag{5.38}$$

By using the previously-defined symbol μ_s for the "natural frequency" coefficient in this equation, we are implicitly taking the storage ring revolution time to be the time unit in this equation. A particular approximate solution,

---

[7] Since the quantity oscillating is δ it would be consistent with our other notation to use δ as the subscript on parameters μ and α. But since "s" (for synchrotron) is a standard notation, and the variable conjugate to δ is s which shares the same parameters, we use "s". Also, as will be explained later, α_s may vary as a function of phase in the longitudinal oscillation cycle. This will make it necessary to replace α_s by an appropriately averaged value later on. We will continue to use the same symbol however, and the significance of the quantity α_s will correspond to the damping rates α_x, α_y, etc. introduced previously.

starting from rest with amplitude δ_0, is

$$\delta \simeq (\delta_0 e^{-\alpha_s t}) \cos \mu_s t,$$
$$\dot{\delta} \simeq -\mu_s (\delta_0 e^{-\alpha_s t}) \sin \mu_s t - \alpha_s \delta,$$
$$\ddot{\delta} \simeq -\mu_s^2 \delta - 2\alpha_s \dot{\delta}, \tag{5.39}$$

where $\alpha_s \ll 1$, which has allowed α_s^2 terms to be dropped. (This should be checked.) For many purposes, even the terms proportional to α_s can also be neglected and the damping factors $e^{-\alpha_s t}$ set to 1. This then supports a "longitudinal phase space" graph with δ as abscissa and $\dot{\delta}/\mu_s$ as ordinate, in which the phase space point stays on a circle of radius δ_0.

We have also anticipated another answer by using the symbol α_s here—the damping will correspond to the damping previously defined in Eq. (5.24) provided the unit of time is taken to be the revolution period. For this choice of time, for a slowly varying quantity such as δ, the change in one revolution $\Delta\delta$ and the time derivative $\dot{\delta}$ are equal (to the intended accuracy.)

$$\Delta\delta \equiv \dot{\delta}_0. \tag{5.40}$$

Where the symbol Δ appears in the immediate sequel, it will usually have this same significance.

An unfortunate feature of accelerator physics is that different practitioners introduce symbols that are especially useful for their own purposes, but which, because they are redundant, are confusing for others. One such parameter is the so-called "synchrotron frequency". It is defined by

$$\Omega_s = \mu_s f_0, \tag{5.41}$$

where f_0 is the electron revolution frequency. In other words, being the product of oscillations-per-turn times turns-per-second, this is the frequency of longitudinal oscillations in ordinary time units (seconds). Since this frequency is readily visible on a spectrum analyser in the control room of a storage ring, it is operators who find Ω_s more convenient than the theoretician's "tune" parameter μ_s. As mentioned before, the effect of non-zero α_s is to cause the phase space "circle" to spiral slowly in, its fractional loss of radius each turn being α_s. So α_s could also be, but won't be here, translated to regular time units.

For the calculations to be performed now, δ is to be held constant during one complete turn—this is a sensible approximation only if $\mu_s \ll 2\pi$, as we assume. It is only the approximate validity of this assumption and "multiple time scale approximation", $Q_x \gg Q_s$, that makes it meaningful to separate out an explicit longitudinal damping rate. While its momentum is essentially constant, the motion of an off-momentum particle can be plotted as in Fig-

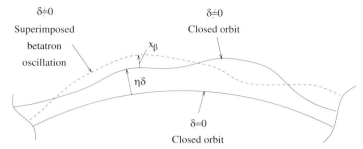

Fig. 5.4 The betatron amplitude x_β of an off-momentum particle oscillates relative to the off-momentum closed orbit.

ure 5.4 and regarded as the superposition of a betatron amplitude x_β and a synchrotron amplitude $\eta\delta$.[8]

In Eq. (5.38) the "dissipative force" $F_d = -2\alpha_s \dot{\delta}$ has been written on the right-hand side of the equation in order to encourage treating it perturbatively. To calculate the damping caused by this perturbation, one can calculate the "work" it does during a complete cycle of lossless motion; such motion is described by Eq. (5.38) with $\alpha_s = 0$, and the work is given by

$$\oint F_d \dot{\delta}\, dt = -2\alpha_s \oint \dot{\delta}^2\, dt = -\alpha_s \delta_0^2 \mu_s^2. \tag{5.42}$$

Putting completely out of mind the fact that δ is itself derived from the particle energy, we define an entirely independent "oscillator energy".

$$\mathcal{E}_{\rm osc} = \frac{1}{2}\mu_s^2 \delta_0^2 = \frac{1}{2}\dot{\delta}^2 + \frac{1}{2}\mu_s^2 \delta^2. \tag{5.43}$$

(It would not be wrong to work instead with the "action", or "longitudinal Courant–Snyder invariant" but the elementary treatment of simple harmonic motion, first understood in one's youth, may make it more familiar to relate work and energy.) In the same spirit in which F_d has been called "force" the terms in Eq. (5.43) can be called "total", "kinetic" and "potential" energy respectively. The force F_d causes a systematic reduction of $\mathcal{E}_{\rm osc}$ according to work done

$$\overline{\Delta \mathcal{E}_{\rm osc}} = \oint F_d \dot{\delta}\, dt = -\alpha_s \delta_0^2 \mu_s^2. \tag{5.44}$$

The overline indicates averaging over a complete synchrotron oscillation period, and $\dot{\mathcal{E}}_{\rm osc}$ has been replaced by change per turn $\Delta \mathcal{E}_{\rm osc}$ which, as explained

---

8) For our purposes the separation of longitudinal and horizontal oscillation based on their different time scales is a satisfactory approximation. This formalism is so firmly fixed in the minds of most accelerator physicists as to make it difficult to suppress it for phenomena such as synchrobetatron oscillations, where the approximation, though serviceable, is rather clumsy.

previously, means the same thing. The integration in Eq. (5.44) has been performed using Eq. (5.39). Combining these formulas and using $\mathcal{E}_{\text{osc}} \sim \delta_0^2$,

$$\alpha_s = -\frac{1}{2}\overline{\frac{\Delta \mathcal{E}_{\text{osc}}}{\mathcal{E}_{\text{osc}}}} = -\overline{\frac{\Delta \delta}{\delta_0}}. \tag{5.45}$$

This may seem to have been an unnecessarily roundabout procedure until one appreciates that the dissipative force F_d could have had any dependence whatsoever on δ and $\dot\delta$ (provided it and its accumulated effects are small) and the resulting Eq. (5.45) would still be valid. We will use Eq. (5.45) below to obtain α_s from a calculation of $\Delta\delta$.

Problem 5.3.3 *Look up in an introductory physics text, or rederive, the "work/energy theorem" of mechanics and confirm that its use here has been justified.*

For a one dimensional eigenmotion having $\lambda_s = e^{-\alpha_s \pm i\mu_s}$ as eigenvalue, the phase space radius acquires a factor $e^{-\alpha_s}$ each time t increases by 1, which agrees with Eq. (5.45), and is consistent with our interpretation of α_s. Note though that the attenuation in Eq. (5.39) is uniform—the fractional shrinking per turn of the phase space radius is independent of oscillation phase. For synchrotron radiation damping this will not be true, but that will be overcome by averaging—the rate α_s will be strictly applicable only when interpreted as attenuation averaged over a complete period of synchrotron oscillation. This completes the digression concerning simple harmonic motion.

We have seen in the previous section that emittance changes can result when the radiated energy depends on the particle energy and path. Starting again from Eq. (5.2), the possible dependences are through $\mathcal{E}^2$, B^2 and C_B, where C_B is the path length in regions of nonzero magnetic field B. Also it will not be possible to avoid the effect of transverse spatial variation of B needed for transverse focusing since, especially in quadrupoles; the off-momentum closed orbit passes through regions of nonvanishing field even if the central orbit does not.

We wish to calculate the average damping of purely longitudinal motion. In principle this could be done by concatenating the differential transfer matrices of Eq. (5.31), but dispersion mixes x and δ motions, which makes that difficult. Instead the loss of "oscillator energy" of an off-momentum particle will be calculated using the formula (5.45) just derived.

Consider the motion of an off-momentum particle having no betatron component and traveling along the solid off-momentum closed orbit shown in Figure 5.4. In one turn its energy will be sapped by the excess energy radiated (over and above that radiated on the central orbit),

$$\frac{\Delta U^{\text{rad}}}{\frac{e^2c^2}{2\pi}C_\gamma} \simeq \int_0^{C_B+\Delta C_B} \mathcal{E}_0^2(1+2\delta)B^2\left(1+2\frac{\Delta B(s)}{B}\right)ds \qquad (5.46)$$
$$+ \int_0^{C_Q} \mathcal{E}_0^2(\Delta B)^2\,ds - \frac{U_0^{\text{rad}}}{\frac{e^2c^2}{2\pi}C_\gamma}$$

where all quantities with subscript 0 are evaluated on the central orbit and $\Delta B(s)$ is the inescapable, off-axis, extra magnetic field needed for focusing—it is first order in δ and can be written

$$\Delta B \simeq \left.\frac{\partial B_y}{\partial x}\right|_0 \frac{dx}{d\delta}\,\delta = K_1(s)\eta_x(s)\,\delta, \qquad (5.47)$$

with η_x being the usual horizontal closed-orbit dispersion function, and the field gradient can be expressed in terms of the quadrupole gradient coefficient $K_1 = \frac{1}{pc/e}\frac{\partial B_y}{\partial x}|_0$. Understanding the long term stability of electron storage rings comes down to understanding all contributions to Eq. (5.46).

An allowance has been made for radiation in quadrupoles by including an integral over the partial circumference C_Q that contains quadrupoles. (Actually, for consistency, terms for higher multipole elements should be included as well, but they would also be negligible.) Since only terms of first order in δ have been retained and the integrand of the quadrupole term is proportional to δ^2, this term will not, in fact, contribute importantly to damping. Radiation in quadrupoles has been included here to permit making this point plus three others. (i) If, because of errors, the closed orbit does not pass through the center of a quadrupole there will be corresponding radiation and some resultant damping—in a large ring this can require serious attention to preserve the intended damping. (ii) The denominator factor B in the $(B + \Delta B)^2$ Taylor expansion cannot cause trouble by vanishing—by definition of C_B. (iii) Finally, and most important, in a strong focusing lattice, the excess magnetic field needed to bend an excess momentum particle through 2π can be, and is, found partially in quadrupoles. *Even so* quadrupoles do not contribute importantly to damping. This simple fact has a profound effect on electron storage rings. It might even be said to be what makes them possible, as it overcomes the anti-damping, to be recognized straightaway, of horizontal oscillations.

Having dispensed with quadrupoles, according to Eq. (5.46), there are three contributors to extra radiation: if the particle's energy deviates from nominal, as well as radiating excessively, it finds itself in a higher than nominal magnetic field, or it has excess path length in the bend field. The first of these effects is dealt with first; it comes from the $\mathcal{E}^2$ factor in the master radiation formula (5.2), which has expanded to $\mathcal{E}_0^2(1+2\delta)$ in the integrand of Eq. (5.46). If this were the only source of excess radiation accompanying momentum offset δ, the fractional excess energy loss per unit momentum offset would be given by $(1/U_0^{\text{rad}})(dU^{\text{rad}}/d\delta) = 2$. Following Sands, this part is subtracted, and the fractional excess energy is represented by a lattice dependent quantity $\mathcal{D}$, known as "curly-$\mathcal{D}$", that is defined by

$$\mathcal{D} = \frac{1}{U_0^{\text{rad}}} \frac{dU^{\text{rad}}}{d\delta} - 2. \tag{5.48}$$

Recombining all contributions, the damping decrement of δ can then be expressed in terms of $\mathcal{D}$;

$$\Delta\delta \simeq -\frac{\Delta U^{\text{rad}}}{\mathcal{E}_0} = -\frac{dU^{\text{rad}}}{d\delta}\frac{1}{\mathcal{E}_0}\delta = -(2+\mathcal{D})\frac{U_0^{\text{rad}}}{\mathcal{E}_0}\delta. \tag{5.49}$$

There are important cases in which one has $|\mathcal{D}| \ll 1$ so the value of $\Delta\delta$ is immediately known. (This was Sands's motive for separating terms in this way.) In general, for any practical lattice, separated or combined-function, isomagnetic or not, $\mathcal{D}$ can be calculated by numerical integration, and that is what is mainly done in practice. Here we will make some simplifications and approximations, mainly rather good ones, to get simple analytic approximations and develop intuition.

To simplify the calculation somewhat, without essential error in most cases, and again following Sands, we make the isomagnetic approximation according to which the magnetic field on the design orbit is either zero or has the same value B_0. Manipulation of Eq. (5.46) yields

$$\frac{\Delta U^{\text{rad}}}{U_0^{\text{rad}}} = 2\delta + \frac{2}{B_0 C_B}\int_0^{C_B} \Delta B\, ds + \frac{\Delta C_B}{C_B}. \tag{5.50}$$

Segments of the perturbed orbit that newly pass through the nominal field B_0 because $\delta \neq 0$ are included in the $\Delta C_B/C_B$ term.[9]

There is one global requirement that $\Delta B(s)$ must meet—the closed orbit of even an off-momentum particle must close smoothly after one turn. The total bend after a full turn must, therefore, be 2π for all δ. Because bending in a magnetic field is inversely proportional to $1+\delta$, for $\delta > 0$ there must be *positive* contributions from $\Delta C_B > 0$ and/or $\Delta B(s) > 0$. Expressed quantitatively, the closed orbit condition is a "sum rule",

$$\delta = \frac{\Delta C_B}{C_B} + \frac{1}{B_0 C_B}\int_0^{C_B} \Delta B\, ds + \frac{1}{B_0 C_B}\int_0^{C_Q} \Delta B\, ds. \tag{5.51}$$

[9] In a strong focusing lattice with pure, uniform-field bending magnets, the term $\Delta C_B/C_B$ is typically very small, because of strong "momentum compaction"; this will be made more quantitative shortly. If some focusing is provided by pole face rotation of the bending magnets, passage through the wedge-shaped region at the magnet end will give a contribution to the $\Delta C_B/C_B$ term.

The terms are as simple as they are because of the isomagnetic assumption. This formula shows how quadrupoles can *help* with the bending even though they *don't hurt* by causing radiation. This relation can be used to re-express Eq. (5.50) by eliminating the integral in favor of a different integral;

$$\frac{\Delta U^{\text{rad}}}{U_0^{\text{rad}}} = 4\delta - \frac{\Delta C_B}{C_B} - \frac{2}{B_0 C_B} \int_0^{C_Q} \Delta B \, ds. \tag{5.52}$$

In the isomagnetic approximation then, $\mathcal{D}$ is given by either of two equivalent forms

$$\begin{aligned}\mathcal{D} &= \frac{1}{C_B} \frac{dC_B}{d\delta} + \frac{2}{B_0 C_B} \int_0^{C_B} \frac{dB}{d\delta} \, ds, \\ &\stackrel{\text{or}}{=} 2 - \frac{1}{C_B} \frac{dC_B}{d\delta} - \frac{2}{B_0 C_B} \int_0^{C_Q} \frac{dB}{d\delta} \, ds.\end{aligned} \tag{5.53}$$

The ranges of integration are along the on-momentum, central orbit, and the integrands are evaluated on that orbit.

The second expression is handy if there are no quadrupoles, as in a combined function lattice, because the integral is simply absent. For a lattice that is both combined-function and strong-focusing the other term is also small so $\mathcal{D} \approx 2$. It then follows from Robinson's theorem that the lattice is horizontally unstable. (See Eq. (5.58).) This is why a strong focusing, combined function lattice cannot be used as an electron storage ring.

Useful further reduction of $\mathcal{D}$ is different for different styles of accelerator. For example, a common, not unnatural, but certainly not universal design, uses only uniform field "sector magnets". In these magnets the closed orbit enters and leaves all dipoles normal to the poles, $dB/d\delta = 0$ (because the field is uniform), and the path length in drift regions is independent of δ. This permits elimination of $C_B(\delta)$ in favor of the (more easily known) total circumference $C(\delta)$, which is greater by the accumulated lengths of all drift sections. One then defines "transition gamma", γ_t, (its inverse square γ_t^{-2} is called the "momentum compaction factor") by

$$\frac{1}{\gamma_t^2} = \frac{1}{C_0} \frac{dC(\delta)}{d\delta} \bigg|_{\delta=0}, \tag{5.54}$$

where $dC/d\delta = dC_B/d\delta$ because there is no linear dependence of path length on δ in drift regions. Then we have

$$\frac{1}{C_B} \frac{dC_B}{d\delta} = \frac{C_0}{C_B} \frac{1}{C_0} \frac{\Delta C}{\delta} = \frac{C_0}{C_B} \frac{1}{\gamma_t^2}, \tag{5.55}$$

and

$$\mathcal{D} = \frac{C_0}{C_B} \frac{1}{\gamma_t^2}. \tag{5.56}$$

In this expression, though the factor C_0/C_B may be somewhat larger than 1, it is of *order* 1. Since the other factor $1/\gamma_t^2$ is typically far less than 1, $\mathcal{D}$ is negligible, or at least small.

With quadrupole elements are present, the circumference increases far less than proportional to $1+\delta$, —which is the source of the name "momentum compaction". In high tune lattices one commonly has $\gamma_t \simeq Q_x$, which leads to $\gamma_t^{-2} \ll 1$. This is why longitudinal damping in a separated function accelerator is much less than in a combined function accelerator. Superficially undesirable, this is in fact good since it leaves room for horizontal damping under the Robinson sum rule ceiling.

We have seen the sort of calculation needed to calculate $\mathcal{D}$ and seen that it depends on lattice details such as whether the lattice is combined function (focusing built into bending magnets) or separated function (separate quadrupole magnets). All this detail is condensed into the value of $\mathcal{D}$.

As a reminder, $(\mathcal{D}+2)\delta$ gives the linearized dependence on δ of the energy damping decrement. Our job is not yet finished however, since we actually need the damping decrement of the longitudinal Courant–Snyder invariant ϵ_s^P. (Recall that the superscript P designates this as a *particle*, rather than as a *bunch* property.) $\mathcal{D}$ provides the damping at a given value of δ, but δ itself oscillates through a longitudinal oscillation cycle.

Figure 5.5 is supposed to be helpful in this averaging. From Eq. (5.39), with time measured in units of the revolution time, the frequency of simple harmonic motion is μ_s and the coordinate conjugate to δ that causes the phase space orbit to be circular is $\dot{\delta}/\mu_s$, which is the ordinate in the figure. The radius of the phase space circle is equal to δ_0, the maximum value that δ achieves, and the instantaneous synchrotron oscillation phase angle is given by $\mu_s t$, where $Q_s = \mu_s/(2\pi)$ is the "synchrotron tune". Typically $Q_s \ll 1$ so the synchrotron phase advance, as the electron completes a whole turn of the storage ring, is much less than 2π. The deviation $\Delta\delta$ reduces the radius δ_0, but the effect is proportional to $\cos^2 \mu_s t$, with one factor $\cos \mu_s t$ being due to proportionality to δ and the other due to the shift being parallel to the δ-axis in phase space rather than radial. Averaging over a full synchrotron period and using Eq. (5.45), one obtains

$$\frac{\overline{\Delta \epsilon_s^P}}{\epsilon_s^P} = \frac{\overline{\Delta\delta}}{\delta} = -\alpha_s. \tag{5.57}$$

This shows that the same parameter α_s serves as the damping decrement for both instantaneous momentum deviation δ and longitudinal emittance ϵ_s^P, even though the former is a one-dimensional coordinate and the latter is an area. As the signs have been introduced α_s is a "decay rate" rather than a "growth rate."

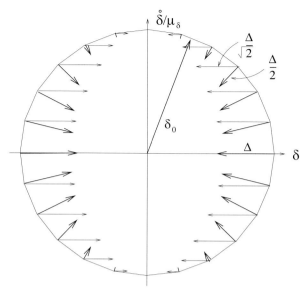

Fig. 5.5 Longitudinal phase space plot with δ as horizontal axis and $\dot\delta/\mu_\delta$ as vertical axis, for $Q_s = 1/24$, small compared to 1. Faint horizontal arrows (exaggerated in length) are the decrement of δ during that particlular revolution. Δ is proportional to the *excess* energy lost to radiation during one turn at the instantaneous value of δ. Heavy radial arrows represent the corresponding decrement of the radius of the phase space circle.

5.3.3
Horizontal Damping and Partition Numbers

We have obtained relations Eq. (5.35), for the sum of damping rates $\alpha_x + \alpha_y + \alpha_s$, Eq. (5.37) for the (uncoupled) vertical damping rate α_y, and Eq. (5.57) for the longitudinal damping rate α_s. Evidently these can be combined to obtain α_x.

$$\alpha_x = (1-\mathcal{D})\frac{U_0^{\rm rad}}{2\mathcal{E}_0}, \quad \alpha_y = (1)\frac{U_0^{\rm rad}}{2\mathcal{E}_0}, \quad \alpha_s = (2+\mathcal{D})\frac{U_0^{\rm rad}}{2\mathcal{E}_0}. \tag{5.58}$$

Sands felt (correctly as it turned out) that it would be easier to remember these rates if they were measured in units of $U_0^{\rm rad}/(2\mathcal{E}_0)$,[10] and quoted as three "partition numbers" J_x, J_y, forced by Robinson's theorem to add up to 4. Then

---

10) Actually, since the quantities α_i as Sands defines them are rates per unit time rather than rates per revolution, his rates are measured in units of $U_0^{\rm rad}/(2T_0\mathcal{E}_0)$ where T_0 is the revolution period.

Eq. (5.58) becomes

$$\alpha_x = J_x \frac{U_0^{rad}}{2\mathcal{E}_0}, \quad \alpha_y = J_y \frac{U_0^{rad}}{2\mathcal{E}_0}, \quad \alpha_s = J_s \frac{U_0^{rad}}{2\mathcal{E}_0},$$
$$J_x = 1 - \mathcal{D}, \quad J_y = 1, \quad J_s = 2 + \mathcal{D}. \tag{5.59}$$

For stable operation, all partition numbers must be positive. We can now understand a comment made earlier, that too much energy damping is bad, since it can cause J_x to be negative.

If there were no "heating effect" and all the Js were positive, all beam emittances would decay to zero. But there *is* heating, which we investigate next.

Problem 5.3.4 *Calculate α_s for an (impractical) lattice for which the entire central orbit is in a constant uniform magnetic field. Also calculate α_s for the (weak focusing) lattice, for which (according to Eq. (5.12)) the magnetic field is longitudinally constant on the entire central orbit, but has radial dependence $B_y(r) = B_{y0}(1 - \frac{n}{R}r)$.*

Problem 5.3.5 *In the text an analogy has been established between synchrotron oscillations and simple harmonic motion of a viscous-damped, point-mass and spring combination. In this analogy δ corresponds to point coordinate x and $\dot{\delta}$ to $\dot{x}$. What is there different about the damping in the two cases that violates this analogy, why does this not invalidate the result Eq. (5.57), and how could the analogy be improved?*

Problem 5.3.6 *Some modern light source lattices, such as ALS (Advanced Light Source) and CLS (Canadian Light Source) have "hybrid lattices", neither fully separated-function nor fully combined-function. The horizontal focusing in provided by separate quadrupoles, the vertical focusing by combined-function bending magnets. Find the partition numbers for this lattice in terms of the field index n in the dipoles (see e.g. Eq. (5.12)) and $\bar{\eta}$, the average value of the horizontal dispersion function in the bending magnets. Also give a reason (having to do with horizontal emittance) why the opposite choice, namely defocusing quads, focusing bending magnets, would be an unfavorable choice for a light source.*

Problem 5.3.7 *For a separated function lattice (no gradient in the dipoles) with rectangular magnets (negligible circumference dependence on momentum) give an approximate formula for $\mathcal{D}$ as a sum, over the quadrupoles in the ring, of each quad strength weighted by an appropriate lattice function evaluated at its center.*

5.4
Equilibrium between Damping and Fluctuation.

We return to the subject of radiation fluctuations, discussing vertical, horizontal, and longitudinal motion separately. Actually vertical motion can be dispensed with in two sentences. With photons radiated in the exact forward direction (an excellent approximation in this context) there is no mechanism for exciting vertical oscillations. Since there is, however, damping, all vertical coordinates damp down to negligible values.

For horizontal betatron motion the competition between damping and growth is analyzed next, After that, longitudinal motion is studied. The treatments of horizontal and longitudinal equilibration are entirely parallel. Calculation of the longitudinal heating is somewhat simpler than for the horizontal heating, but figuring out the ultimate phase space aspect ratio is easier in the transverse than in the longitudinal case.

There is a well-established stochastic formalism using the Fokker-Planck equation for analyzing problems like this. I prefer, instead, just to use "physical arguments", though it might be more accurate to say "arguments sound enough to be be persuasive to physicists". Only one result is assumed from probability theory and that is the central limit theorem. In physicist's terms, this theorem states that repeatedly convoluted distributions become Gaussian and r.m.s. deviations combine "quadratically".

5.5
Horizontal Equilibrium and Beam Width

The calculation of beam equilibrium to be performed is complicated by the need for *three* separate averagings: over betatron phase, over photon energies, and over lattice elements. (Over lattice elements it is actually a sum rather than an average, and over photons it is both summing and averaging, but that is getting ahead of the story.)

Consider the off-momentum particle illustrated in Figure 5.6. It has been specialized by assuming it to be initially free of betatron motion and traveling on its appropriate off-momentum closed orbit. At some instant the particle emits a photon of energy u exactly in the forward direction. The emission changes the particle energy but not its direction;

$$\delta \to \delta - \frac{u}{\mathcal{E}_0}. \tag{5.60}$$

The coordinates of the local closed orbit corresponding to a particle with fractional energy offset $\delta = -u/\mathcal{E}_0$ are $(-\eta_x u/\mathcal{E}_0, -\eta'_x u/\mathcal{E}_0)$. But the physical location of the particle is exactly the same as before the emission, which is to say on the closed orbit appropriate to the previous momentum. This means that horizontal betatron phase space components $(\eta_x u/\mathcal{E}_0, \eta'_x u/\mathcal{E}_0)$ have been impulsively imparted to the particle. In general, a beam particle already has non-zero betatron components corresponding to some Courant–Snyder

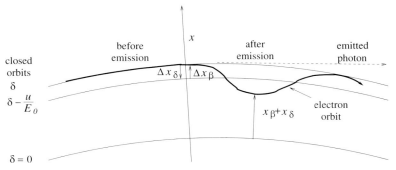

Fig. 5.6 Betatron oscillation induced by forward emission of a photon of energy u for an off-momentum particle initially having zero betatron amplitude.

invariant

$$\epsilon_{x0} = \frac{x^2 + (\beta_x x' + \alpha_x x)^2}{\beta_x} \stackrel{\text{also}}{=} \gamma_x x^2 + 2\alpha_x x x' + \beta_x x'^2, \tag{5.61}$$

before the emission takes place. Its coordinates at location s are

$$x_0 = \sqrt{\beta_x(s)\epsilon_{x0}} \cos \psi_x, \quad x_0' = -\sqrt{\frac{\epsilon_{x0}}{\beta_x(s)}} \left(\sin \psi_x + \frac{\alpha_x}{\beta_x} \cos \psi_x \right), \tag{5.62}$$

where $\alpha_x = -\beta_x'/2$ and $\psi_x(s)$ is its horizontal betatron phase. After the photon emission the phase space coordinates have become

$$\begin{aligned} x_1 &= \sqrt{\beta_x \epsilon_{x0}} \cos \psi_x + \eta_x u/\mathcal{E}_0, \\ x_1' &= -\sqrt{\frac{\epsilon_{x0}}{\beta_x}} \left(\sin \psi_x + \frac{\alpha_x}{\beta_x} \cos \psi_x \right) + \eta_x' u/\mathcal{E}_0. \end{aligned} \tag{5.63}$$

It can be seen from Figure 5.7 that a photon emission can increase or decrease the Courant–Snyder invariant, depending upon the betatron phase. Averaging over phase angle ψ proceeds as

$$\overline{(a \cos \psi + \Delta)^2 + a^2 \sin^2 \psi - a^2} = \overline{2a\Delta \cos \psi + \Delta^2} = \Delta^2. \tag{5.64}$$

Because the value of ψ_x where radiation occurs, is random, terms proportional to $\sin \psi_x$ or $\cos \psi_x$ average to zero when averaging over ψ_x. (Averaging over the betatron phase is indicated by an overhead line.) The phase-averaged change in the Courant–Snyder invariant, $\overline{\Delta \epsilon_x^u} = \overline{\epsilon_{x,1} - \epsilon_{x,0}}$, due to the emission of a photon of energy u at location s is defined to be $\mathcal{H}(s)(u/\mathcal{E}_0)^2$, where

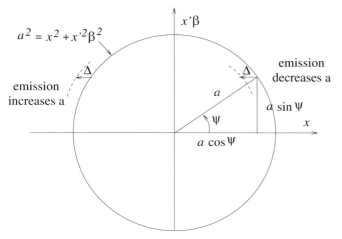

Fig. 5.7 An impulsive increment Δ to the betatron displacement can increase or decrease the Courant–Snyder invariant. But, averaged over phase, there is a net increase that is proportional to Δ^2. α_x has been taken to be zero.

$\mathcal{H}(s)$ goes by the name "curly H" and is given by

$$\mathcal{H}(s) = \frac{\overline{\Delta \epsilon_x^u}}{(u/\mathcal{E}_0)^2} = \frac{\gamma_x \overline{x_1^2 - x_0^2} + 2\alpha_x \overline{x_1 x_1' - x_0 x_0'} + \beta_x \overline{x_1'^2 - x_0'^2}}{(u/\mathcal{E}_0)^2}$$
$$= \gamma_x \eta_x^2 + 2\alpha_x \eta_x \eta_x' + \beta_x \eta'^2 \equiv \frac{\eta_x^2 + (\beta_x \eta_x' + \alpha_x \eta_x)^2}{\beta_x}. \tag{5.65}$$

It is important to appreciate that $\mathcal{H}(s)$ depends only on lattice functions, not on individual particle coordinates. In words, curly-$\mathcal{H}$ is the "Courant–Snyder invariant" calculated for the dispersion function $\eta(s)$ as if it were a particle orbit. The motivation behind the introduction of curly-$\mathcal{H}$ was to "factor out" the dependence on u.

The averaging so far has been over the betatron phase, permitting the expectation value of growth due to the emission of a single photon of energy u to be calculated. The growth due to the uncorrelated emission of many such photons is obtained by multiplying by the number of such photons. The total number of photons with energy in interval du at u must therefore be calculated next. The final result is obtained by summing over all photon energies.

It is also necessary to fold in the damping effect calculated earlier. In equilibrium the growth and damping cancel. In practical situations the time constant governing approach to that equilibrium is *long* compared to the revolution time $T_0 = 1/f_0$ around the accelerator. This *a priori* unguaranteed feature will be used to simplify the calculation now, and will be justified later by the result.

Because the lattice functions and the spectral functions depend on s the instantaneous rates should be calculated first and then integrated over lattice elements. Unfortunately our earlier discussion of fluctuations made assumptions that break down for too-short path intervals. But the long time constant mentioned in the previous paragraph makes it legitimate to treat ϵ_x as essentially constant for the duration T_0 of a single turn, say the ith turn, and to accumulate changes over a full revolution. We can use a clumsy, but temporary, notation, $\overline{\Delta\epsilon_x(u)}$, to stand for the betatron-phase-averaged, cumulative for energies less than u, increment of the Courant–Snyder invariant during one full turn. (The symbol u has been moved from being a superscript to being an argument as a lame indication that u is now the upper limit of a range rather than a parameter.) From calculations performed so far, the increment $d\overline{\Delta\epsilon_x(u)}$ coming from photon emissions in the range u to $u + du$ can be given in terms of $\mathcal{H}$ and radiation formulas;

$$d\overline{\Delta\epsilon_x(u)} = du \frac{u^2}{\mathcal{E}_0^2} \oint \mathcal{H}(s) \frac{P_\gamma(s)\,ds/c}{<u(s)>_\gamma} n_u(u). \tag{5.66}$$

Here $P_\gamma ds/c$ has given the energy emitted in path ds as in Eq. (5.2); this energy has then been converted to the total number of photons by Eq. (5.18), and their distribution in energy expressed as $n_u(u)$. The energy-accumulated, Courant–Snyder one turn increment will be denoted

$$\Delta\epsilon_x^P = \int_0^\infty \frac{d}{du'}\overline{\Delta\epsilon_x(u')}du' = \overline{\Delta\epsilon_x(\infty)}. \tag{5.67}$$

On the symbol $\Delta\epsilon_x^P$ the superscript P for "particle" has been introduced to make the point that this is an individual particle parameter, and to distinguish it from a parameter ϵ_x^B, with B for "beam", to be introduced shortly. I have also taken the opportunity to limit the proliferation of averaging and overline symbols. The averaging over the betatron phase and integration over photon energies just described will be implicitly assumed to have occurred in what follows.

A few sentences of justification may be in order. In a varying magnetic field the length over which B can be treated as constant is so short as to make the fluctuation in the number of photons emitted fractionally significant, contrary to assumption. But if one visualizes the integal of Eq. (5.66) as broken into a "Riemann" sum of terms that are re-arranged and gathered together into groups of terms for intervals on which the field has (essentially) the same value, then the fractional fluctuation in the number of photons emitted within each such group is much reduced—let us say made negligible. Since rearranging the terms in the Riemann sum does not change the integral, and our model treats all emissions as uncorrelated, Eq. (5.66) has been legitimized. A further pedantic point, made only to encourage contemplation of this formula, is that the factor $n(u)$ cannot be taken outside the integral, even though u is being held fixed, because the distribution function $n(u)$ depends on the magnetic field, which depends on s. In an isomagnetic lattice $n(u)$ could be taken outside, making moot all the concerns of this paragraph.

It remains to integrate over all values of u; this yields

$$\Delta \epsilon_x^P = \frac{1}{\mathcal{E}_0^2} \oint \mathcal{H}(s) \frac{P_\gamma(s)\, ds/c}{<u(s)>_\gamma} \int_0^\infty u'^2 n_u(u')\, du'$$

$$= \frac{1}{\mathcal{E}_0^2} \oint \mathcal{H}(s)\, P_\gamma(s) \frac{ds}{c} <u^2(s)>_\gamma <u(s)>_\gamma. \tag{5.68}$$

Note that this quantity is independent of the pre-existing betatron amplitude.

This result was anticipated in Eq. (5.20). In particular the integrand is proportional to $u_c\, dU_T/ds$. However the present result is more general in that the lattice dependent factor $\mathcal{H}(s)$ has been correctly folded in under the integral. Nevertheless an approximate growth rate can be based on the estimate of the sum of squares of radiated energies by $1.323\, u_c\, U_0^{\text{rad}}$—roughly speaking, the sum of squared energies is the product of the critical energy and the energy loss per turn. If all radiation in one turn occurred at a single location where $\eta_x' = \alpha_x = 0$ (so that $\mathcal{H} \approx \eta_x^2/\beta_x$) the expected horizontal emittance growth after one turn would be $\Delta \epsilon_x^P \approx 1.3 u_c U_0^{\text{rad}} (\eta_x^2/\beta_x)/\mathcal{E}_0^2$.

At this point it is important to distinguish between the Courant–Snyder invariant ϵ_{sc}^P of an individual particle, defined in Eq. (5.61), with superscript P having been added to emphasize the point, and the "beam emittance" ϵ_x^B, encountered now for the first time (in this context);

$$\epsilon_x^B = \frac{\sigma_x^2(s)}{\beta_x(s)}. \tag{5.69}$$

Here $\sigma_x(s)$ is the r.m.s. beam width at location s or, in other terminology, $\sigma_x^2(s)$ is the "variance" of x. Recall that ϵ_x^B is independent of s. (Choosing the same symbol ϵ for these two different quantities, as is universally done, is certainly confusing, and at this point it is misleading in a way that makes a factor-of-two error likely. The symbol σ_x is equally deserving of being decorated with the medal "B", but this is unnecessary since there is no conceivable misinterpretation of σ_x as an individual particle parameter.)

Because the transverse distribution is the result of a random walk, by the central limit theorem, the distribution is Gaussian, at least in the linear regime we are assuming. There is an arbitrary, conventional, numerical factor in the definition of ϵ_x^B having to do with the bunch distribution. For electron bunches, since all distributions are accurately Gaussian, it is natural to choose this factor to be 1. (Since proton beams are not necessarily Gaussian, the conventional "proton world" definitions of ϵ_x^B define phase space areas containing "most" of the beam particles.)

A variable $v = \beta_x x' + \alpha_x x$ has the convenient feature that phase space trajectories are circles when the abscissa is x and the ordinate is chosen to be v. The probability distributions of x and v in this representation are identical, normalized, independent, Gaussians;

$$P_x(x) = \frac{1}{\sqrt{2\pi}\sigma_x} e^{-\frac{x^2}{2\sigma_x^2}}, \quad P_v(v) = \frac{1}{\sqrt{2\pi}\sigma_x} e^{-\frac{v^2}{2\sigma_x^2}}. \tag{5.70}$$

As written here the distributions apply after equilibrium has been established. In a dynamic situation (just after injection for example) the same distributions might be more or less accurate, but the beam sigma $\sigma_{x,i}$ would depend on turn index i.

Here, and in the rest of the section, we are observing the beam at a fixed reference location in the ring, where the lattice functions α_x and β_x are fixed and known. The two-dimensional joint probability distribution of x and v is $P_{x,v}(x,v) = P_x(x)P_v(v)$. From this one can calculate the expectation value (i.e. average over all particles in the beam) on the ith turn

$$< x_i^2 + v_i^2 >_{\text{beam}} = 2\sigma_{x,i}^2, \tag{5.71}$$

where $\sigma_{x,i}^2$ depends on i only in dynamic situations.

Calling the effect of radiation fluctuations "excitations", the trick now is to treat the excitations $(\Delta x, \Delta v)$ accumulated during one turn as Gaussian distributed with shortly-to-be-determined variance κ^2,

$$P_{\Delta x}(\Delta x) = \frac{1}{\sqrt{2\pi}\kappa} e^{-\frac{\Delta x^2}{2\kappa^2}}, \quad P_{\Delta v}(\Delta v) = \frac{1}{\sqrt{2\pi}\kappa} e^{-\frac{\Delta v^2}{2\kappa^2}}. \tag{5.72}$$

These are the effective Gaussian distributions of the one-turn accumulations of radiation fluctuations. Superposition of these deviations on a pre-existing (also Gaussian) distribution will be handled by convolution.

(Even though individual emissions are far from Gaussian, by the central limit theorem, for not-too-bizarre distributions, the sum of many emissions is Gaussian distributed with variance equal to the sum over individual variances. Like the distributions of x and v, that are identical because of isotropy in phase space, the distributions of Δx and Δv must be equal in order to preserve isotropy. It would scarcely be quibbling at this point to ask "Who says isotropy must be preserved?" A pleasing answer would be based on solving the stochastic equations and proving it to be true. A less satisfying answer is that decoherence causes tangential "shear" motion in phase space and that shear causes "mixing" or "filamentation" which maintains isotropy in phase space for practical tune spreads. Though this argument is somewhat of a swindle, coming as it does "from left field", it nevertheless is descriptive of actual beam behavior. Without pursuing the issue of which argument is correct, and conscious of the great simplification that will result, we shamelessly claim that our approach correctly describes the "physics", and proceed.)

The value κ^2 needed to match the expectation value $\Delta\epsilon_x^P$ (a quantity previously calculated in Eq. (5.68)) satisfies

$$\Delta\epsilon_x^P = \frac{<\Delta x^2 + \Delta v^2>_{\text{excitation}}}{\beta_x} = \frac{2\kappa^2}{\beta_x} \quad \text{or} \quad \kappa^2 = \frac{\beta_x \Delta\epsilon_x^P}{2}. \tag{5.73}$$

There is no point in proceeding without having assimilated the purpose and meaning of this equation as it is the linchpin of the entire development. It links the beam-related quantity κ^2 to the particle-related quantity $\Delta\epsilon_x^P$. It also contains the answer to a question that is sometimes of practical importance: "What are the beam dimensions of an initially-zero emittance beam after one turn?"

The probability distributions of x_{i+1}, v_{i+1} after one turn are convolutions of (5.70) and (5.72). Gaussian variances being additive, the new variances are both equal to $\sigma_x^2 + \kappa^2 = \sigma_x^2 + \beta_x \Delta\epsilon_x^P / 2$. This is a timely point to include also the effect of damping, which has the effect $\sigma_x^2 \to \sigma_x^2(1 - 2\alpha_x)$, with α_x obtained from Eq. (5.59). Combining the effects of excitation and damping,

$$<x_{i+1}^2 + v_{i+1}^2>_{\text{beam}} = 2\sigma_x^2(1 - 2\alpha_x) + \beta_x \Delta\epsilon_x^P. \tag{5.74}$$

In equilibrium $\sigma_{x,i+1} = \sigma_{x,i}$, which implies

$$\epsilon_x^B = \frac{\sigma_x^2}{\beta_x} = \frac{\Delta\epsilon_x^P}{4\alpha_x}. \tag{5.75}$$

Incorporating Eq. (5.68), the equilibrium beam emittance is given by

$$\epsilon_x^B = \frac{\sigma_x^2}{\beta_x} = \frac{\Delta\epsilon_x^P}{4\alpha_x} = \frac{\oint \mathcal{H}(s) \frac{P_\gamma(s)\,ds/c}{<u(s)>_\gamma} <u^2(s)>_\gamma}{2\mathcal{E}_0 J_x \oint P_\gamma(s)\,ds/c}$$

$$= \frac{u_c}{B} \frac{1.323 \oint \mathcal{H}(s) B^3(s)\,ds}{2\mathcal{E}_0 J_x \oint B^2(s)\,ds} \tag{5.76}$$

where α_x has been obtained from Eq. (5.59), U_0^{rad} from Eq. (5.3), and the ratio of radiation moments from Eq. (5.13). In the isomagnetic case this reduces to

$$\epsilon_x^B \approx \frac{1.323 u_c}{2\mathcal{E}_0 J_x} <\mathcal{H}>. \tag{5.77}$$

The upper equation is valid in general, the lower makes the isomagnetic assumption. Knowing from Eq. (5.5) that the factor $\frac{u_c(s)}{B(s)} = \frac{3}{2}\frac{e\hbar}{m}\gamma^2$ is independent of B, that factor has been moved outside the integral, and this leaves the emittance given by an integral weighted by B^3 divided by an integral weighted by B^2. This implies that increased damping due to excess radiation in local regions of specially high magnetic field (increasing the denominator) cannot make up for the increased quantum fluctuations in the radiation that increase the numerator. This shows that the emittance is minimum in the isomagnetic case.

By tailoring the magnetic field profile, and the beta functions and dispersion function, one has great control over the achievable beam emittance. Occasionally, for colliding beams, the emittance is intentionally increased. More typically the emittance is minimized. Chapter 11 describes lattice designs optimized for the minimum emittances appropriate for synchrotron light sources. Another approach to reducing emittance is by including wigglers at low dispersion points. This approach has been pursued recently for the damping rings needed for linear colliders. Such "wiggler dominated" rings provide excellent examples for the formulas derived in this chapter. This is the subject of Section 5.7.

Problem 5.5.1 *As emphasized in the text, photons are emitted so nearly forward that the corresponding transverse electron recoil can be neglected. But, since there is otherwise no fundamental source of beam height, the corresponding neglect of vertical recoil is not justified. The purpose of this problem is to estimate the beam height from this source. Show that the increase in vertical C-S invariant after one turn, due to photons (vertical emission angle ψ) is*

$$\Delta \epsilon_y^P = \frac{1}{\mathcal{E}_0^2} \oint \beta_y(s) \frac{P_\gamma(s)\,\mathrm{d}s/c}{<u(s)>_\gamma} <u^2 \psi^2(s)>_\gamma . \tag{5.78}$$

From this, estimating the r.m.s. vertical emission angle as $0.6/\gamma$, show that the vertical over horizontal ratio of single-turn emittance growths can be estimated by

$$\frac{\Delta \epsilon_y^P}{\Delta \epsilon_x^P} \approx \frac{\beta_y \beta_x}{\eta_x^2} \left(\frac{0.6}{\gamma}\right)^2 . \tag{5.79}$$

If the horizontal and vertical damping partition numbers are equal, the equilibrium beam emittances will be in this same ratio.

5.6
Longitudinal Bunch Distributions

5.6.1
Energy Spread

Excitation of energy oscillations can be analyzed in much the same terms as horizontal betatron oscillations, but there are two simpler features in this case.

One simplification has to do with the longitudinal analog to $\beta_x(s)$. An analogous function, $\beta_s(s)$, can be defined to describe the modulation of energy oscillations as a function of circumferential coordinate s. But it is valid to assume that longitudinal quantities vary so slowly as to be essentially independent of s. This is equivalent to taking $\beta_s = $ constant, and for present purposes we can take the constant to be 1. It is sufficient, therefore, to define an analog to ϵ_x^P by

$$\epsilon_s^P = \delta^2 + \omega^2, \tag{5.80}$$

where ω is the phase-space coordinate complementary to δ, scaled to make the phase space trajectory circular. For linearized longitudinal oscillations ω is proportional to the longitudinal displacement from the reference particle, but discussion of that connection can be deferred. The superscript P stands for "particle" as this is a parameter of an individual particle. As before, the motivation behind analyzing ϵ_s^P, rather than δ itself, is that its constancy in the unperturbed situation permits averaging over oscillation phases while calculating the effect of the emission of a photon.

The second simplifying feature is that, though a photon emission changes δ by the amount $\Delta\delta = -u_j/\mathcal{E}_0$, it leaves ω unchanged. There is no need to introduce an s-dependent factor analogous to $\mathcal{H}(s)$. The synchrotron-phase averaged increase in single particle invariant is

$$\overline{\Delta\epsilon_s^u} = \frac{u^2}{\mathcal{E}_0^2}. \tag{5.81}$$

This formula, the analog of Eq. (5.65), can be described simply by saying that, if an $\mathcal{H}$ had been introduced, its value would have been 1. Equation (5.81) must be summed over photons emitted during one full turn to obtain $\Delta\epsilon_s^P$, the corresponding increase in single particle invariant. Analogous to Eq. (5.68) the result is

$$\Delta\epsilon_s^P = \frac{1}{\mathcal{E}_0^2} \oint \frac{P_\gamma(s)\,ds/c}{<u(s)>_\gamma} \int_0^\infty u'^2 n_u(u')\,du' \tag{5.82}$$
$$= \frac{1}{\mathcal{E}_0^2} \oint \frac{<u^2(s)>_\gamma}{<u(s)>_\gamma} P_\gamma(s)\,ds/c.$$

In the isomagnetic case this becomes

$$\Delta\epsilon_s^P = \frac{U_0^{\text{rad}}}{\mathcal{E}_0^2} \frac{<u^2(s)>_\gamma}{<u(s)>_\gamma} = \frac{1.323 u_c U_0^{\text{rad}}}{\mathcal{E}_0^2}. \tag{5.83}$$

This result is closely connected to Eq. (5.20) which can be expressed as

$$< \sum_{j=1}^{\mathcal{N}_0} \Delta\delta_j^2 >_\gamma = \frac{1.323 u_c U_0^{\text{rad}}}{\mathcal{E}_0^2}. \tag{5.84}$$

(At this point one might inquire why it is not $<u^2>_\gamma - <u>_\gamma^2$, the variance of the probability distribution of photon energies, rather than $<u^2>_\gamma$, that enters. This replacement might be appropriate if the electron were subject to no longitudinal restoring force and were simply dribbling away its energy in fluctuating lumps u_j. But, as was true for transverse excitations, it is only the term $<u^2>_\gamma$ that survives averaging over synchrotron oscillation phases.)

Corresponding to Eqs. (5.70), δ and ϖ distribution functions for the beam are

$$P_\delta(\delta) = \frac{1}{\sqrt{2\pi}\sigma_s} e^{-\frac{\delta^2}{2\sigma_s^2}}, \quad P_\varpi(\varpi) = \frac{1}{\sqrt{2\pi}\sigma_s} e^{-\frac{\varpi^2}{2\sigma_s^2}}, \tag{5.85}$$

Fluctuations in longitudinal amplitudes for one complete turn are distributed as in Eqs. (5.72)

$$P_{\Delta\delta}(\Delta\delta) = \frac{1}{\sqrt{2\pi}\kappa} e^{-\frac{\Delta\delta^2}{2\kappa^2}}, \quad P_{\Delta\varpi}(\Delta\varpi) = \frac{1}{\sqrt{2\pi}\kappa} e^{-\frac{\Delta\varpi^2}{2\kappa^2}}. \tag{5.86}$$

There is no harm in using the same symbol κ here as it will be eliminated immediately in this argument, as it was above. For that matter, the variable ϖ will also disappear shortly.

As in Eq. (5.73), the value of κ^2 needed to match $\Delta\epsilon_s^P$ is

$$\kappa^2 = \frac{\Delta\epsilon_s^P}{2}, \tag{5.87}$$

and, analogous to Eq. (5.74), the beam evolves during the ith turn according to

$$<\delta_{i+1}^2 + \varpi_{i+1}^2>_{\text{beam}} = 2\sigma_{\delta,i}^2(1-2a_s) + \Delta\epsilon_s^P. \tag{5.88}$$

Problem 5.6.1 Defining $\epsilon_s^B = \sigma_\delta^2$, Complete the preceeding argument to show that, in equilibrium, in the isomagnetic case, the longitudinal beam invariant is given by

$$\epsilon_s^B = \frac{\Delta\epsilon_s^P}{4\alpha_s}. \tag{5.89}$$

where $\Delta\epsilon_s^P$ comes from Eq. (5.82) and α_s comes from Eq. (5.59).

Problem 5.6.2 Continuing from the previous problem, derive the formula

$$\sigma_\delta = 0.813\sqrt{\frac{u_c}{(2+\mathcal{D})\mathcal{E}_0}}, \tag{5.90}$$

for the r.m.s. beam energy spread in the isomagnetic case, where the partition number J_s, defined in Eq. (5.59), has been introduced, and replaced by $2+\mathcal{D}$. This formula is especially useful in the common case, e.g. separated function lattices, that $\mathcal{D} \ll 2$.

In general, calculating the energy spread requires integrating over the lattice as in Eq. (5.76),

$$\sigma_\delta^2 = \frac{\Delta\epsilon_s^P}{4\alpha_s} = \frac{\oint \frac{P_\gamma(s)\,ds/c}{<u(s)>_\gamma} <u^2(s)>_\gamma}{2\mathcal{E}_0 J_s \oint P_\gamma(s)\,ds/c} = \frac{1.323}{2\mathcal{E}_0 J_s} \frac{u_c}{B} \frac{<B^3>_{\rm orbit}}{<B^2>_{\rm orbit}}. \tag{5.91}$$

Here, as was true when calculating the horizontal emittance, the factor $u_c(s)/B(s)$, being independent of B, can be taken outside the integral, leaving σ_s^2 proportional to $<B^3>/<B^2>$. Comparing with Eq. (5.76), this means there is a strong tendency for energy spread to be proportional to horizontal beam width. This tendency can however be overcome, for example to obtain small horizontal emittance, by designing the lattice in such a way that $\mathcal{H}$ is small at locations where B is large.

5.6.2
Bunch Length

Some experiments for which synchrotron light beams are used study ultrafast time dependences, and for these the duration of individual photon bursts is important. This duration is given by the photon bunch length (divided by c.) Though the burst observed at detection screen P from any one electron is much shorter than the electron bunch, because the electrons are traveling at the speed of light, and emissions from individual electrons are incoherent, the photon and electron bunch lengths are the same. Having calculated the electron energy spread, we are now able to calculate the electron bunch length.

The very understanding of longitudinal stability was judged sufficiently important to earn McMillen and Veksler Nobel Prizes for their invention of the synchrotron. For this reason longitudinal oscillations are often referred to as "synchrotron oscillation". They are also referred to as "phase oscillations" to emphasize the phase of the RF cavity at the time a beam bunch passes. Phase stability is the basis of all high energy accelerators. The essential element needed for this stability (and to make up for the energy taken off in synchrotron radiation) is the RF cavity. When passing through an RF cavity a particle acquires extra energy ΔE that can be expressed in terms of ct which, (scaled by a factor c), is the arrival time t at the cavity relative to that of the reference particle,

$$\Delta\mathcal{E} = e\hat{V}\sin\left(\frac{\omega_{\rm rf}}{c}ct + \phi_0\right) - U_0^{\rm rad}. \tag{5.92}$$

In this equation allowance has also been made for a possible lumped energy loss each turn given by $U_0^{\rm rad}$. In practice this would be due to synchrotron radiation or wall-impedance loss, distributed continuously around the ring, but we assume that it can be adequately represented by a single loss occurring at the RF cavity. This same formula is applicable to the "ramping up" of energy during the acceleration process, but we will assume that a steady beam is being described. The assumption that lattice parameters are being held constant requires that the nominal $ct = 0$ particle receives no net energy so $e\hat{V}\sin\phi_0 = U_0$ in Eq. (5.92).

Transverse dynamics, as described in previous chapters, has been greatly simplified by the paraxial or *linearized* approximation. Using a similar linearization, analogies can be set up between longitudinal parameters and the Twiss parameters of betatron motion. Unfortunately, for the linearization to be a good approximation would require the RF power to be larger than really needed. Since this is never done in practice, a valid treatment of longitudinal parameters necessarily brings in nonlinear analysis. A very similar analysis is applicable to both linear accelerators and circular accelerators. In this text, for linear accelerators, this theory is explained in Section 6.2.4. Even though the analysis of bunch length in a storage ring differs somewhat from the linac analysis, it seems too repetitive to justify its inclusion in this text. An analysis that is explicitly applicable to storage rings and uses assumptions and terminology consistent with usage in this text is contained in the UAL Physics User's Guide [7].

One essential result is a formula for the synchrotron tune $Q_s = \mu_s$ where

$$\mu_s^2 = \frac{T_0 \eta_{\text{syn.}} \hat{V} \omega_{\text{rf}} \cos\phi_0}{cp_0/e}. \tag{5.93}$$

Here $\eta_{\text{syn.}}$ is the so-called "slip factor"; for rough estimates one can use a value that is approximately valid for a FODO lattice of circumference C_0;

$$\eta_{\text{syn.}} \approx \frac{C_0}{2\pi Q_x^2}. \tag{5.94}$$

The r.m.s. bunch length σ_{ct} can be obtained from the energy spread by

$$\sigma_{ct} \approx \frac{C_0 \eta_{\text{syn.}}}{\mu_s} \sigma_\delta. \tag{5.95}$$

5.7
"Thermodynamics" of Wiggler-dominated Storage Rings

5.7.1
Emittance of Pure Wiggler Lattice

This section is largely the result of, and has profited from, discussions with Mark Palmer concerning damping rings.

5.7 "Thermodynamics" of Wiggler-dominated Storage Rings

High brightness x-ray rings and linear collider damping/cooling rings are very similar devices. In both cases the goal is to produce high current, low emittance beams. Another shared feature is that both have wiggler/undulator magnets built into the lattice. For damping rings the purpose of the wiggler magnets is to increase the damping rate. For x-ray rings the wiggler magnets are present as "insertion devices" from which the x-ray beams are produced. Though x-ray physicists make important distinctions between them, undulator magnets are really just weak wiggler magnets. In this terminology the magnets in damping rings should certainly be referred to as "wigglers" while both wigglers and undulators are present in x-ray rings. These distinctions are unimportant for the present discussion and all such magnets will be referred to as "wigglers".

The magnetic fields in wigglers, because their very purpose is to intentionally increase synchrotron radiated energy, tend to be far stronger than the bend fields in storage ring arcs. For x-ray sources this radiation is the end product of the facility; for damping rings the loss of particle energy is beneficial (because of the damping that accompanies its replenishment). All this is beneficial. The randomness of individual photon emissions gives a countervailing, deleterious, beam "heating effect". By judicious lattice design (mainly placing the wigglers at zero dispersion locations) the cooling effect can be made to dominate the heating effect, yielding a net benefit. The high magnetic fields in the wiggler also cause the radiated x-rays to be "harder". This tends to increase the electron beam energy spread—a deleterious effect which is acceptable in many cases, and one that will be ignored here.

The important parameters of a wiggler are its period λ_w, its number of periods N_w, and its wiggler parameter $K = \Theta\gamma$, where Θ is the maximum angular excursion in the device and γ is the usual relativistic factor. Generally speaking, λ_w has substantially larger values for damping rings (perhaps as large as a meter) than for an x-ray source (perhaps several centimeters). It will be seen below, for wigglers located in "dispersion-free" regions, that one measure of the deleterious effect of a wiggler is proportional to λ_w^2. By this measure one therefore expects the effects of an undulator in a light source to be benign, even mainly beneficial. The discussion will therefore be biased toward the analysis of wigglers in damping rings, where they are more central to the functioning of the device.

The lattice design considerations just mentioned are too complicated for detailed analysis here. Existing computer codes can be used to calculate the equilibrium emittance of arbitrarily complicated lattices. As usual in accelerator design, starting from tentative lattices, one compares, selects, and iterates, with optimal emittance being just one criterion.

Rather than going into this detail, by analyzing an idealized ring, we will attempt to determine a lower limit on the emittance—or, stated differently, the theoretical maximum benefit that can result from introducing wigglers into a storage ring. Some results can be based on Eq. (5.77)

$$\epsilon_x = \frac{C_q \gamma^2}{J_x \rho} <\mathcal{H}>, \tag{5.96}$$

where $C_q = 3.84 \times 10^{-13}$ m, J_x, the partition number, can, for present purposes, be taken to be 1, and ρ is the bend radius in bending magnets. The angle brackets indicate averaging over longitudinal coordinate s over the full ring and, at any value of s, $1/\rho$ is assumed either to vanish or to have the same "isomagnetic" non-zero magnitude. This assumption is essential to the validity Eq. (5.96). The factor $\mathcal{H}$, is given in terms of the usual Twiss functions and horizontal dispersion function η by

$$\mathcal{H} = \gamma_x \eta^2 + 2\alpha_x \eta \eta' + \beta_x \eta'^2. \tag{5.97}$$

When the isomagnetic assumption is invalid it is necessary to use a more general formula;

$$\epsilon_x = \frac{C_q \gamma^2}{J_x} \frac{<\mathcal{H}/\rho^3>}{<1/\rho^2>}. \tag{5.98}$$

The two formulas for ϵ_x can be seen by inspection to be equivalent for isomagnetic lattices.

Let us consider a lattice that is *totally dominated* by one ideally located wiggler. This assumes that the wiggler is located at a position in the lattice where (in its absence) the dispersion η is zero, and β_x is much larger than the wiggler period λ_w. To close the ring there obviously have to be magnetic fields other than in the wiggler, but they are assumed to be so weak that they cause no energy to be radiated and no emittance growth due to quantum fluctuations. Any emittance growth caused by the wiggler has to be associated with dispersion generated by the wiggler itself since, except for this self-generated dispersion, curly-H would vanish everywhere in the wiggler.

Consider the following two approximations for the magnetic field of a wiggler:

$$B_y^{(NG)} = B_0\, U\!\left(\cos\left(\frac{2\pi s}{\lambda_w}\right)\right) - B_0\, U\!\left(-\cos\left(\frac{2\pi s}{\lambda_w}\right)\right). \tag{5.99}$$

where U is a step function having value 1 for positive argument and 0 for negative argument. This formula, which models the field as a "square wave" is valid only in the "narrow gap" limit. It has been tailored to satisfy the isomagnetic assumption. Alternatively

$$B_y^{(WG)} = B_0 \cos k_x x \, \cosh k_y y \, \cos\left(\frac{2\pi s}{\lambda_w}\right) \tag{5.100}$$

provides an accurate formula for the dominant component of the magnetic field in a "wide gap" wiggler in which the gap height of the wiggler is *large* compared to the wiggler period. Neglecting the scalloping of the orbit, relative to $x = y = 0$, the first two factors are both equal to 1. They are exhibited only for generality. From the formula given, using Maxwell's equations, it is possible to reconstruct the dependence on x, y, and s of all three field components.

The earlier discussion suggests that the WG formulas will tend to be valid for light sources and the NG formulas will tend to be valid for damping rings. But, since this will not necessarily be true, this association has not been built into the notation.

Proceding as in Section 4.6, the differential equation satisfied by dispersion $\eta(s)$ can be shown to be;

$$\frac{d^2\eta}{ds^2} = \frac{1}{\rho}. \tag{5.101}$$

Letting ρ_0 be the (constant) radius of curvature, and requiring $\eta(0) = 0$, we obtain, in the narrow gap case,

$$\eta^{(NG)} = -\frac{\lambda_w s}{4\rho_0} + \frac{s^2}{2\rho_0}, \quad \frac{d\eta^{(NG)}}{ds} = -\frac{\lambda_w}{4\rho_0} + \frac{s}{\rho_0}, \quad \text{for} \quad 0 < s < \frac{\lambda_w}{2}. \tag{5.102}$$

Because the isomagnetic assumption is valid in the narrow gap case, Eq. (5.96) can be used;

$$\epsilon_x^{(NG)}\Big|_{\text{wiggler only}} = \left(\frac{C_q \gamma^2}{J_x \rho_0}\right) \frac{1}{\lambda_w} \int_0^{\lambda_w} \frac{\eta^2 + (\eta \alpha_0 + \eta' \beta_0)^2}{\beta_0} ds. \tag{5.103}$$

The factor following the parenthesized coefficent, is $< \mathcal{H} >$, where the average has been taken over one period of the wiggler. In practice the Twiss functions α_0 and β_0 are likely to vary along the wiggler. The 0 subscripts are intended to indicate that any such variation is being neglected so that the emittance calculated from one period will be applicable to all periods of all wigglers. This formula is the basis for the comment made earlier that wigglers should be placed in $\eta \approx 0$ regions. Since only a crude estimate of best possible performance is being sought, let us suppose that the condition $\eta = 0$ has been exactly achieved. In that case, Eq. (5.103) reduces to

$$\epsilon_x^{(NG)} > \left(\frac{C_q \gamma^2}{J_x \rho}\right) \frac{\beta_0}{\lambda_w/2} \int_0^{\lambda_w/2} \left(\frac{d\eta^{(NG)}}{ds}\right)^2 ds = \frac{1}{48} \frac{C_q \gamma^2}{J_x} \frac{\lambda_w^2 \beta_0}{\rho_0^3}. \tag{5.104}$$

The $=$ sign has been replaced by a $>$ sign on the assumption that the oscillatory term $\alpha_0 \eta$ is insignificant (because β_0 is large) and because the η^2 term is positive. Also the inequality has been applied to the whole ring, not just the emittance ascribable to the wiggler. In designing a damping ring it is sometimes valid to assume that the total energy radiated in the rest of the ring is negligible relative to the energy radiated in the wigglers and that quantum fluctuations are also negligible in the rest of the ring. If the "rest of the ring" is made up of (notoriously bad emittance-causing) FODO cells then the radiated power in the wigglers may have to exceed the power radiated in the rest of the ring by an order of magnitude or more for the emittance to be fully dominated by the wigglers. That is what is being assumed to be the case.

Next let us calculate the equilibrium emittance in the wide gap case. The radius of curvature ρ of a particle in the wiggler is

$$\frac{1}{\rho(s)} = \frac{1}{\rho_0} \cos\left(\frac{2\pi s}{\lambda_w}\right). \tag{5.105}$$

Substituting this and Eq. (5.100) into Eq. (5.101) yields

$$\frac{d^2 \eta^{(WG)}}{ds^2} = \frac{1}{\rho_0} \cos\left(\frac{2\pi s}{\lambda_w}\right). \tag{5.106}$$

After integration and setting $\eta(0) = 0$, this yields

$$\eta^{(WG)} = \frac{\lambda_w^2}{4\pi^2 \rho_0}\left(1 - \cos\left(\frac{2\pi s}{\lambda_w}\right)\right), \quad \frac{d\eta^{(WG)}}{ds} = \frac{\lambda_w}{2\pi \rho_0} \sin\left(\frac{2\pi s}{\lambda_w}\right). \tag{5.107}$$

Problem 5.7.1 Complete the determination of $\epsilon_x^{(WG)}$ by substituting from Eqs. (5.107) into Eq. (5.98) and making the same approximations as in the narrow gap case. Show that

$$\epsilon_x^{(WG)} > \frac{2}{15\pi^3} \frac{C_q \gamma^2}{J_x} \frac{\lambda_w^2 \beta_0}{\rho_0^3}. \tag{5.108}$$

For reference, this same calculation is performed by Wiedemann [8] (whose result is smaller by a factor of 4, perhaps because his bend angle per pole seems to be off by a factor of 2) and by Minty and Zimmerman [9] (whose result agrees with Eq. (5.108).) Some possible numerical values are: $B_0 = 1.9\,T$, $\mathcal{E}_e = 2\,GeV$, $\beta_0 = 13\,m$, $\lambda_w = 0.4\,m$, $J_x = 1$. With these values one obtains $\epsilon_x^{(WG)} > 1.4\,nm$.

There is a distressingly large difference between $\epsilon_x^{(WG)}$ and $\epsilon_x^{(NG)}$. The factor is $(1/48)/(2/(15\pi^3)) = 4.8$, assuming, as we are, that the peak magnetic field is the same in both cases. In practice the value of B_0 might be made as big as possible in both cases, and hence probably bigger in the narrow gap case. But we assume that B_0 is the same in both cases. One effect of this is that the self-induced dispersion is greater in the narrow gap case.

How are we to understand such a strong dependence on field profile from a calculational perspective? One effect is the difference in self-induced dispersion just mentioned. Another important consideration is that $d\eta/ds$ is maximum at the location where the magnetic field changes sign. In the wide gap case, the field continues to be small for a substantial interval about this point. This strongly suppresses $\mathcal{H}$ in this region. On the other hand, in the narrow gap case, the $\epsilon_x^{(NG)}$ integral is dominated by the same interval because, though the field switches sign, it does not change in magnitude.

Can we understand the difference from a physics perspective? Since a discontinuous field violates Maxwell's equations we might be dubious about the validity of the $\epsilon_x^{(NG)}$ calculation for any physically useful gap height. But let us ignore this and assume that Eq. (5.98) is truly valid for both cases. The presence of the square edges causes the frequency spectrum to be very different in the two cases—the x-rays are far *harder* in the square edge case. This can be seen from the frequency domain, since it takes very short wavelength Fourier components to represent a square edge. Furthermore it is the r.m.s. photon energy that is the main factor in establishing the strength of quantum fluctuations—a few hard photons can do far more harm than far more soft photons. From this perspective it is therefore not surprising that the equilibrium emittance is far greater in the narrow gap case.

Another potentially important consideration is that, by using the same peak field in both cases, the total energy radiated is twice as great in the narrow gap as in the wide gap case. Among other things, this means that the neglect of "the rest of the ring" is less valid in the wide gap than in the narrow gap case. Curiously enough, in spite of the fact that the energy radiated is greater in the narrow gap case, its effectiveness in reducing the emittance is much less.

5.7.2
Thermodynamic Analogy

The pure wiggler lattice just analysed is not really practical. In an actual ring both heating and cooling effects are distributed in the rest of the ring as well as in the wigglers. It is relatively straightforward to apply the formalism developed earlier in this chapter to this compound configuration. Explicit formulas for combining the separate effects of wiggler and rest-of-ring are given, for example, by Wiedemann [8]. But, since those formulas are fairly complicated, a simpler formulation, even if only approximate, would seem to be useful. That is the purpose of the "pseudo-thermodynamic" analogy to be developed next. One can hope also to use the result to avoid violating the "second law of pseudo-thermodynamics".

An emittance such as ϵ_x has much the same character as a "temperature". In the ribbon-shaped beam typical of electron storage rings, the vertical emittance ϵ_y is far less than horizontal emittance ϵ_x—vertical can be characterized as cold, horizontal hot. Their different termperatures are due to the relatively greater horizontal heating. As well as their dependence on cooling (equal in both cases) and heating (far greater in the horizontal plane) the relative size of ϵ_x and ϵ_y depends on the coupling (the analog of thermal contact) between the two planes. If the coupling is increased adiabatically until it dominates the heating differential, both emittances will have moved to their previous average which, for an initially ribbon-shaped beam, is equal to half the previous horizontal value. Their sum is preserved.

Emittance also has other aspects that are reminiscent of entropy. Under "normal" circumstances emittance *increases*. Even where emittance decreases, as in the damping associated with synchrotron radiation or stochastic cooling, the beam whose emittance decreases is only a subsytem of a larger system. If the "emittance" of the rest of the system were definable, the total value would no doubt increase. But beam emittance is an *intensive* quantity while entropy is an *extensive* quantity. To complete an analogy with thermodynamics we must therefore introduce some kind of "pseudo-capacity" which multiplies the emittance to produce a "pseudo-entropy". This refinement was not explicitly necessary to apply "pseudo-thermodynamics" to the case of x, y coupling in the previous paragraph because, in that case, the pseudo-capacities of the two subsystems were equal.

Hoping to justify the assignment later, let us conjecture that the "pseudo-capacity" of a subsystem can be defined to be the total power radiated from the subsystem. For convenience, to the pre-existing, wiggler-free, ring let us assign the numerical value of the pseudo-capacity to be 1. The dimension of pseudo-capacity should therefore be watts, but we will take it to be dimensionless. Since this power is exactly replenished by accelerating cavities, one can also specify the RF power as being 1 in the same units. For a numerical example we assume the emittance (now to be called pseudo-temperature) of a given ring to be 200 nm at electron energy of 5 GeV.

Suppose one wishes to reduce the emittance of the ring just mentioned by inserting wiggler magnets in its (assumed to be available) free spaces. Using pseudo-thermodynamics, let us try to estimate the emittance improvement that will result.

To get a big improvement one wants the pseudo-capacity of the wiggler system to be large compared to the pseudo-capacity of the pre-existing ring. But let us further suppose we are unwilling to make major changes in the RF system. Since we have no excess power, we fail before we start.

But suppose the application under development[11] allows the ring energy to be reduced from, say, 5 GeV to, say 2 GeV. Since the power radiated is proportional to the fourth power of energy, this gives us a gigantic power reduction factor $(5/2)^4 = 40$. One is then free to keep inserting wigglers, and increasing their strengths, until the full power available from the RF is radiated in the wiggler. Keeping the RF power constant, the pseudo-capacities will then be: ring, 0.025; wiggler, 0.975.

11) The calculation in this section was suggested in connection with the design of a damping ring for a linear collider.

Let us use formulas from the previous sub-section to calculate the pseudo-entropy of the wiggler, assuming its design is halfway between the wide gap (WG) and narrow gap (NG) cases. Its pseudo-temperature is therefore $1.4 \times 4.8/2 = 3.36$ nm; the pseudo-capacity being 0.975, the pseudo-entropy is 3.28 nm.

To calculate the pseudo-entropy of the rest of the ring we can start with its given 5 GeV emittance, which is 200 nm. We can use its known, γ^2 proportionality (see Eq. (5.98)) to find its 2 GeV pseudo-temperature to be $200 \times (2/5)^2 = 32$ nm; the pseudo-capacity being 0.025, the pseudo-entropy is 0.8 nm.

Bringing wiggler and rest-of-ring subsystems into thermal contact is like bringing a small hot element into contact with a large cold one. In that case the final temperature is the weighted mean (with capacities as weights) of the two temperatures. In our case, with the pseudo-capacities having been adjusted to sum to 1, the numerical value of the final pseudo-temperature is the sum of the two pseudo-entropies, which is $3.28 + 0.8 = 4.1$ nm. Using wigglers we therefore expect to obtain an emittance of 4.1 nm at 2 GeV.

Since the final emittance is already dominated by the wigglers there would be little benefit in increasing the RF power further. There would, however, be a major improvement obtained by using wide gap wigglers, whose pseudo-temperature is much less than has been used in the calculation. This would, however, require much longer free space availability in the ring since the wiggler would need to be much longer. In a thermodynamic comparison between WG and NG cases, instead of holding the peak field constant, one should hold the radiated power constant. But, as just explained, this would have little effect on the comparison.

According to our definitions, Eq. (5.98) has an attractive interpretation. Ignoring constants of proportionality, the denominator factor $1/\rho^2$, which we know to be proportional to radiated power per unit length, is interpreted as pseudo-capacity density, and the numerator factor $\mathcal{H}/\rho^3$ is pseudo-entropy density. After accumulation their ratio is pseudo-temperature, which seems appropriate. This was the basis for choosing radiated power to stand in for capacity in this analogy.

References

1 Sands, M. (1971), *The Physics of Electron Storage Rings*, in *International School of Physics, "Enrico Fermi"*, Academic Press.

2 Jackson, J. (1999), *Classical Electrodynamics*, 3rd edn., John Wiley, New York.

3 Landau, L. and Lifshitz, E. (1975), *The Classical Theory of Fields*, 4th edn., Pergamon Press.

4 Hofmann, A. (1998), *Characteristics of Synchrotron Radiation*, in *Synchrotron Radiation and Free Electron Lasers*, CERN 98-04.

5 Robinson, K. (1958), *Phys. Rev.* **111**, 373.

6 Orlov, Y. (1958), JETP, **35**, 525.

7 Malitsky, N., Talman, R. (2005), *Text for UAL Accelerator Simulation Course*, U.S. Particle Accelerator School, Ithaca, N.Y., Available at `http://www.ual.bnl.gov`.

8 Wiedemann, H. (1999), *Particle Accelerator Physics*, 2nd edn., Section 10.4.2, Springer-Verlag, Berlin.

9 Minty, M., Zimmerman, F. (2003), *Measurement and Control of Charged Particle Beams*, Section 4.3.3, Springer-Verlag, Berlin.

6
Elementary Theory Of Linacs

6.1
Acknowledgement and Preview

The majority of the material in this chapter is drawn from an article by G. Loew and R. Talman, *Elementary Principles of Principles of Linear Accelerators*, in *Physics of High Energy Accelerators*, U.S. Particle Accelerator School, AIP No. 105, M. Month, ed., 1982. This article was based on lectures at the school by Loew that were later written up and augmented by Talman.

The emphasis in this chapter will be almost entirely on electrons; they spend most of their life in an accelerator traveling at nearly the speed of light. Nevertheless one cannot avoid discussing the brief interval during which electrons are accelerated from, say, a few hundred kV to, say, 10 MeV. At the latter point the electron velocity is approximately

$$\frac{v}{c} \approx 1 - \frac{1}{2}\left(\frac{m_e c^2}{\mathcal{E}_e e}\right)^2 \approx 0.999. \tag{6.1}$$

As the electrons are accelerated from this point their speed can, for all practical purposes, be taken to be c.

In the earliest days of accelerator physics technological contraints restricted RF frequencies to relatively low frequencies, say, 50 kHz. One consequence of this was that the very first particles that could be accelerated were ions considerably more massive than protons. Roughly speaking, as feasible frequencies became higher and higher, the linear acceleration of progressively lighter particles became practical. For protons a typical frequency is 200 MHz, for electrons 3 GHz. The usual formulation of a proton's progress through a linac is discrete—the proton proceeds through a finite sequence of accelerator gaps and one calculates the energy gain in each. In an electron accelerator, consistent with its higher frequency, there are far more accelerating regions, and it is traditional to treat the electron as if it is "riding on a continuous wave". The treatment of acceleration through discrete gaps is perhaps more elementary than the continuum wave theory. This makes it reasonable to briefly study (non-relativistic) linacs before advancing to electron accelerators. In an electron linac the electrons are nonrelativistic for only a few accelerating gaps.

Accelerator X-Ray Sources. Richard Talman
Copyright © 2006 WILEY-VCH Verlag GmbH & Co. KGaA, Weinheim
ISBN: 3-527-40590-9

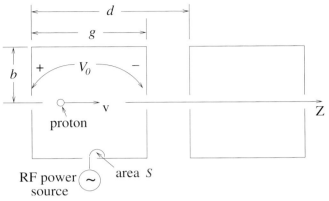

Fig. 6.1 Two cells of a pill box linac. The cavities' outer wall forms a circular cylinder centered on the beam line.

This initial theory will apply to, at most, these gaps and, in fact, it will not apply accurately even there because the fractional energy gain per gap is so high. An accurate treatment requires detailed numerical tracking through each gap; this is too technical for treatment here.

6.2
The Nonrelativistic Linac

6.2.1
Transit Time Factor

The free space wavelength λ_0 corresponding to RF frequency $f = 200$ MHz is given by

$$\lambda_0 = \frac{c}{f} = 1.5 \,\text{m}. \tag{6.2}$$

Taking the injection (kinetic) energy to be $\mathcal{E}_K = 1.17$ MeV, the injection velocity is $v/c = 0.05$, and the distance traveled in one cycle is

$$d = \frac{v}{c} \lambda_0 = 7.5 \,\text{cm}. \tag{6.3}$$

Two cells of a rudimentary linac made of a sequence of "pillbox cavities" are shown in Figure 6.1. The geometry of this cavity is sufficiently simple that all its properties can easily be calculated analytically. To make the accerating gap g shorter (and the longitudinal field proportionally higher) one can use "nose cones" which are small-radius conducting collars attached to the cavity end walls and centered on the beam. One form of distortion like this leads

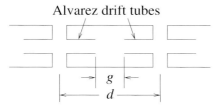

Fig. 6.2 A short length of Alvarez linac.

to the "Alvarez" linac illustrated in Figure 6.2. In this, also known as a "drift tube linac" particles move successively from within drift tubes (where they are shielded from the decelerating electric field) to gaps, say of length g, where the electric field is accelerating. Clearly the gap length should not exceed $d/2$ and, unless $g \ll \lambda_0$, the energy gained by the particle crossing the gap is less than qV_0, where V_0 is the maximum instantaneous voltage drop across the gap; the fractional reduction from this value is known as the "transit time factor" T_{tr}, which can be approximated as follows.

Treating the longitudinal electric field E_z as spatially uniform, its time dependence is given by

$$E_z = \frac{V_0}{g} \cos \omega t. \tag{6.4}$$

Approximating the particle's speed by its value v at the gap center, its longitudinal position is given by

$$z_p = vt, \tag{6.5}$$

and its kinetic energy gain through the whole gap is

$$\begin{aligned} \Delta \mathcal{E}_K &= \int_{-g/2}^{g/2} \frac{qV_0}{g} \cos \frac{\omega z_p}{v} \, dz_p \\ &= qV_0 \frac{\sin(\omega g / 2v)}{\omega g / 2v} = qV_0 T_{tr}. \end{aligned} \tag{6.6}$$

Assuming that the RF frequency is constant along the linac (which is practically always the case), as the particle's speed increases the drift tube lengths have to get longer. Normally there are many drift cells in each cavity (all mutually resonant) and many cavities.

6.2.2
Shunt Impedance

The single most problematical feature of linear accelerators is their hunger for power. How much power do we need to obtain a given energy gain? Power consumption occurs in cavity wall heating, in power transmission from RF source to cavity, in reflection due to mismatch at the cavity and, (the only beneficial consumption) in power imparted to the beam. For room temperature cavities (usually made of copper) wall heating is typically the dominant loss mechanism, and this loss usually prevents CW operation. Instead, pulses with durations measured in microseconds and repetition rates of perhaps 100 Hz are typical. Only with the advent, in recent years, of superconducting RF cavities, has CW operation become practical. For now we concentrate on power loss in the walls of room temperature cavities.

The power lost in one cell, P_{lost}, is proportional to V_0^2, and the constant of proportionality, known as "shunt impedance" R_{lin} of the cell, is defined by

$$R_{lin} = \frac{V_0^2}{P_{\text{lost}}}. \tag{6.7}$$

(Note that, by convention, there is no 1/2 in this formula, even though the fields vary sinusoidally. The subscript "lin" for "linac" is intended as a reminder. When essentially the same quantity is introduced into a lumped circuit cavity model it is represented by $R_{cir} = R_{lin}/2$, with "cir" for "circuit".) The shunt impedance per unit length r is therefore given by

$$r = \frac{R_{lin}}{d} = \frac{(V_0/d)^2}{P_{\text{lost}}/d}, \tag{6.8}$$

which is expressed this way so that both numerator and denominator have practical, and hence memorable, significance. With no beam loading, and losses other than wall heating neglected, P_{lost} can be replaced by P_{source} and the achievable value of V_0 becomes

$$V_0 = \sqrt{R_{lin} P_{\text{source}}}. \tag{6.9}$$

Combining this with Eq. (6.6), the peak energy gain is

$$\Delta \mathcal{E}_K = q \sqrt{R_{lin} T_{tr}^2 P_{\text{source}}}. \tag{6.10}$$

6.2.3
Cavity Q, R/Q, and Decay Time

By definition, at RF frequency ω, the power loss P_{lost}, the cavity Q (which is the inverse fractional energy loss per radian of RF phase advance), and the cavity stored energy W_{st} are related by

$$Q = \frac{\omega W_{\text{st}}}{P_{\text{lost}}}. \tag{6.11}$$

Expressing P_{lost} two ways yields

$$-\frac{dW_{\text{st}}}{dt} = \frac{\omega}{Q} W_{\text{st}}, \tag{6.12}$$

which shows that the time for the electric field to fall to $1/e$ of its original value after excitation is turned off is

$$t_F = \frac{2Q}{\omega}. \tag{6.13}$$

This is also known as the "filling time", which accounts for the subscript. Combining Eq. (6.11) with Eq. (6.7) yields the important ratio

$$\frac{R_{lin}}{Q} = \frac{1}{\omega} \frac{V_0^2}{W_{\text{st}}}. \tag{6.14}$$

Both factors on the r.h.s. depend only on cavity geometry and (for typically-high Q values) are essentially independent of wall material and condition, and quality of other constructional details like joints and welds. On the other hand R_{lin} and Q individually depend strongly on these details.

For fixed cavity shape the resonant frequency ω_r varies inversely with the linear size and the stored energy is proportional to $(V_0/\text{length})^2$ times volume. Presuming that the excitation frequency is close to the natural frequency, R_{lin}/Q is independent of frequency, or rather, though R_{lin}/Q, depends on the cavity shape, it is unaffected by overall change of linear scale.

6.2.4
Phase Stability and Adiabatic Damping

The longitudinal dynamics of particles in linear and circular accelerators is surprisingly similar, but the fractional energy gain per turn in a circular accelerator is small. By contrast the particle energy in a single passage through a linear accelerator can increase by a few orders of magnitude. For this reason the longitudinal dynamics in a circular ring can be treated as an elementary special case of linear acceleration.

When a bunch of particles is injected into a linac there is an inevitable spread of energies. Superficially this suggests that, over a sufficiently long distance, all particles will eventually drop out of phase with the accelerating field. In fact, "phase stability" prevents this from happening.

The important idea is that a particle whose velocity is somewhat too low gradually drops back in the bunch. If the accelerating field is properly configured this shift causes the particle to encounter a higher than average accelerating force. This permits the particle to catch up and even overtake the center of the bunch. The result can be stable longitudinal oscillation of every particle. For electrons, once they are relativistic, this restoration process is relatively insignificant, but for protons the resulting longitudinal oscillations remain significant along an entire linear accelerator.

In a circular ring these oscillations are known as "synchrotron oscillations"; They are important even for electrons because, though their velocities are sensibly constant, their times of flight through accelerator arcs depend on momentum. So the synchrotron oscillations of electrons and protons are qualitatively fairly similar. One essential difference is that protons pass through a "transition" at which the time of flight reduction with increased velocity is exactly cancelled by the longer flight path taken by their above-average-momentum. For normal electron rings the particles are already above transition when they are injected.

In a linac the drift tubes, gaps and fields are so arranged that one can speak of a "reference" or "synchronous" particle which is accelerated through the structure without any longitudinal oscillation; this particle defines the bunch center. In practice, this reference trajectory is achieved by empirically adjusting the phases of the drive to successive sections of the linac.

It turns out to be simpler, rather than time t, to use the distance Z along the accelerator as the independent variable. In this section upper case letters will be used to represent the dependent variables of an individual particle. This will free up lower case symbols to represent deviations from the corresponding variable of the reference trajectory. For example, consider a particle that passes the point Z at time T, at which time the phase of the RF accelerating field is $\Phi = \omega T$. The reference particle passes the same point at time T_s and when the phase is $\Phi = \omega T_s$. The *relative* arrival time, phase and longitudinal position are therefore

$$\begin{aligned} t &= T - T_s, \\ \phi &= \Phi - \Phi_s = \omega(T - T_s) = \omega t, \\ z &= -V\phi, \end{aligned} \tag{6.15}$$

where, f "V"or the moment, stands for velocity. Referring back to Eq. (6.4), the rate of increase of energy of the synchronous particle is given by

$$\frac{d\mathcal{E}_{Ks}}{dZ} = q\bar{E}_{z0} \cos \Phi_s, \tag{6.16}$$

where $\bar{E}_{z0}$ is the appropriately average electric field near Z, including the transit time factor. For now we are assuming positive acceleration (when discussing the ERL it will be necessary to switch this) so $q\bar{E}_{z0} \cos \Phi_s > 0$. For maximum acceleration of the synchronous particle Φ_s would be adjusted to zero, but we will see shortly that this would not yield phase stability. Let us however require $|\Phi_s| < \pi/2$—with the inequality typically meaning "much less". For the general particle the rate of energy gain is

$$\frac{d\mathcal{E}_K}{dZ} = q\bar{E}_{z0} \cos \Phi. \tag{6.17}$$

The energy offset $e = \mathcal{E}_K - \mathcal{E}_{Ks}$ therefore satisfies

$$\frac{de}{dZ} = q\bar{E}_{z0}\big(\cos(\Phi_s + \phi) - \cos\Phi_s\big). \tag{6.18}$$

Our intention here (as elsewhere in the book) is not to give the most general formulation, but rather to explain the ideas using formulas that are as simple as possible (while retaining the essence.) For the time being this will include using nonrelativistic mechanics. It will be necessary (and not difficult) to rectify this when analysing relativistic electrons.

For small ϕ, Eq. (6.18) shows that there is an energy "correction" proportional to the phase offset ϕ, but that is only half the story. There is also an accumulating phase shift due to the energy offset e. T and T_s are given in terms of velocities V and V_s by

$$T = \int_0^Z \frac{dZ'}{V}, \quad T_s = \int_0^Z \frac{dZ'}{V_s}. \tag{6.19}$$

(Any excess flight time accompanying transverse oscillation is negligible in practical linacs.) From these and Eq. (6.15) the phase offset satisfies

$$\frac{d\phi}{dZ} = \omega\left(\frac{1}{V} - \frac{1}{V_s}\right). \tag{6.20}$$

(It was the simplicity of this relation and the natural occurrence of Z as the variable in Eq. (6.17) that motivated the use of Z as the independent variable.) Velocities and energies are related by

$$\mathcal{E}_K = \frac{1}{2}mV^2, \quad \text{and} \quad \mathcal{E}_{Ks} = \frac{1}{2}mV_{Ks}^2, \tag{6.21}$$

which lead to

$$e = \frac{1}{2}m(V^2 - V_s^2) \approx mV_s(V - V_s). \tag{6.22}$$

Combining this equation with Eq. (6.20), and copying from Eq. (6.18) to form a pair of coupled differential equations yields

$$\begin{aligned}\frac{de}{dZ} &= q\bar{E}_{z0}\big(\cos(\Phi_s + \phi) - \cos\Phi_s\big) \\ &\approx -q\bar{E}_{z0}\sin\Phi_s\,\phi, \\ \frac{d\phi}{dZ} &= -\frac{\omega}{mV_s^3}e.\end{aligned} \tag{6.23}$$

Such coupled equations describe a possibly oscillatory system having "phase space" coordinates (ϕ, e). The approximate form of the upper equation is not always valid, but when it is, the resultant longitudinal analysis is said to be "linearized".

Mentally interpreting Z as "time" we obtain "Newton's law" for this system by combining these equations;

$$"F(\phi)" = \frac{d^2\phi}{dZ^2} = -\frac{\omega}{mV_s^3}q\bar{E}_{z0}\left(\cos(\Phi_s + \phi) - \cos\Phi_s\right)$$
$$\approx \frac{\omega}{mV_s^3}q\bar{E}_{z0}\sin\Phi_s\,\phi, \tag{6.24}$$

where "$F(\phi)$" can play the role of "force" (even though its dimensions do not match that role.) The condition for stability is that the force be "restoring". That is, the coefficient of ϕ should be negative, which implies $\Phi_s < 0$. (A particle for which $\phi < 0$, being near the front of the bunch, gets a lesser longitudinal kick according to the second of Eqs. (6.23), which slows it down as is appropriately "restoring". On the other hand, a particle with $e > 0$ necessarily advances toward the front of the bunch, i. e. to more negative ϕ, because of its excess velocity; that is the content of the lower of Eqs. (6.23). If particles pass through "transition" in the accelerator, the sign of this relation reverses.)

Since the r.h.s. of Eq. (6.24) is a function only of ϕ, it is natural to introduce an "effective potential energy $\mathcal{V}(\phi)$" whose (negative) derivative is the "force";

$$\mathcal{V}(\phi) = -\int F(\phi)\,d\phi$$
$$= \frac{\omega}{mV_s^3}q\bar{E}_{z0}\left(\sin(\Phi_s + \phi) - \phi\cos\Phi_s\right). \tag{6.25}$$

This time $\mathcal{V}$ (not the same as V) is potential energy and, since they have already become tiresome, the quotation marks have been dropped from the symbols in this equation. From here on, when terms like force, kinetic and potential energy, Hamiltonian, and so on, are used they are to be interpreted as "generalized" quantities whose precise meanings have to be inferred from the context, and whose dimensions will usually not match the normal meanings of the terms. The accelerating wave and the generalized potential function are plotted in Figure 6.3 (a) and (b) with the horizontal line in (a) indicating two phases at which the energy gain is the same and such that $\phi = 0$ in Eq. (6.25). These are candidate phases for the synchronous particle. We have already seen that the negative Φ_s intersection is the appropriate one for stability.

The "frequency" of small amplitude oscillations can be read off from the linearized form of Eq. (6.24);

$$\frac{2\pi}{\lambda_z} = \sqrt{\frac{\omega q\bar{E}_{z0}|\sin\Phi_s|}{mV_s^3}}. \tag{6.26}$$

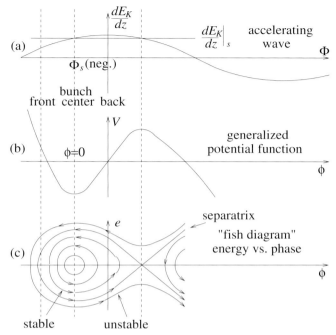

Fig. 6.3 Phase stability graphs.

Here, since the independent variable is Z, the result is the wave number $2\pi/\lambda_z$; λ_z is the distance along the accelerator over which one cycle of longitudinal oscillation occurs. Note that λ_z increases with increasing V_s. As the particles become more relativistic this effect becomes even more pronounced. The correct relativistic expression is obtained by the replacement $V_s \to \gamma_s V_s$ where γ_s is the usual relativistic factor.

As usual with oscillators, it is enlightening to write the equations of motion in Hamiltonian form and to discuss the motion in "phase space" for which the axes are ϕ and e as in Figure 6.3(c). The generalized Hamiltonian

$$H = q\bar{E}_{z0}\left(\sin(\Phi_s + \phi) - \phi\cos\Phi_s\right) + \frac{\omega}{2mV_s^3}e^2. \tag{6.27}$$

along with the Hamilton equations

$$\frac{de}{dZ} = \frac{\partial H}{\partial \phi}, \quad \text{and} \quad \frac{d\phi}{dZ} = -\frac{\partial H}{\partial e}, \tag{6.28}$$

recover Eqs. (6.23). (Our choice of the pair (ϕ, e) as phase space coordinates has the following perversity: It seems natural to think of ϕ as a canonical position and e as a canonical momentum, but the signs in Eq. (6.23) correspond to the opposite identification. This could be rectified by switching the sign of H, but negative "kinetic energy" would seem unattractive. The problem can be said to have arisen because of the negative sign in the third of Eqs. (6.15).)

If a Hamiltonian has no explicit dependence on the independent variable it follows that H is a constant of the motion. In our treatment V_s is treated as a constant parameter. This is valid for "adiabatic" motion, in which V_s has small fractional change per oscillation wavelength λ_z. Accepting this as valid, the value of H is constant, with value, say, H_1;

$$H(\phi, e) = H_1 \tag{6.29}$$

This provides the equations of motion of possible phase space trajectories including all those illustrated in Fig.. 6.3(c). For small amplitude oscillations this equation reduces to

$$H_1 = \frac{1}{2} q \bar{E}_{z0} |\sin \Phi_s| \phi^2 + \frac{\omega}{2mV_s^3} e^2, \tag{6.30}$$

which is the equation of an ellipse. As with any simple harmonic motion, the maximum momentum can be calculated from the maximum displacement, or vice versa. In other words, the aspect ratio of the ellipse is fixed by the form of the Hamiltonian

$$\frac{e_{max}}{\phi_{max}} = \sqrt{\frac{q\bar{E}_{z0}|\sin \Phi_s| mV_s^3}{\omega}} \tag{6.31}$$

But H is not, in fact, independent of Z; the whole point of the accelerator is to increase the value of V_s, and the value of $H(\phi, e)$ is not even approximately constant. But, as long as the motion is adiabatic, as defined above, the motion in phase space will follow an almost elliptic path close to one of those shown in Figure 6.3(c). It is not difficult to show (e. g. Landau and Lifshitz, *Mechanics*, that the area of the ellipse is an "adiabatic invariant". Then, since the area of an ellipse is proportional to the product of major and minor axes,

$$e_{max}\phi_{max} = \text{constant}. \tag{6.32}$$

Taken together, Eqs. (6.31) and (6.32), by fixing both their product and and their ratio, fix e_{max} and ϕ_{max} individually;

$$e_{max} \propto \left(\bar{E}_{z0}|\sin \Phi_s|V_s^3\right)^{1/4},$$
$$\phi_{max} \propto \left(\bar{E}_{z0}|\sin \Phi_s|V_s^3\right)^{-1/4}. \tag{6.33}$$

The latter relation shows that the bunch damps longitudinally and becomes short compared to the phase acceptance of the linac sections. Even the energy spread is, in a sense, damped. It is normally the *fractional* energy spread

$e_{max}/\mathcal{E}_{Ks}$ that is significant. For this quantity the exponent is $-3/4$ in the relationship like (6.33).

Little more will be said concerning details of longitudinal motion. For large amplitude motion the unapproximated Hamiltonian of Eq. (6.27) has to be used. Its main features are illustrated in Figure 6.3(c). Phase space is separated into stable and unstable regions by the so-called "separatrix". Particles have to be injected into the stable region; i. e. their energies and longitudinal positions have to be close to the "elliptic fixed point" at the center of the stable region. Some kind of impulsive, non-Hamiltonian process is needed to transform the injection line optics (also Hamiltonian) to the accelerator Hamiltonian. For protons one such non-Hamiltonian process involves starting with negatively ionized protons that are "stripped" into protons by a foil at the entrance to the linac.

For a linac that is to serve as the injector to a storage ring all or most of the electrons initially inside the linac separatrix will contribute to the storage ring circulating beam. This is because of the damping mechanism that is a characteristic feature of electron storage rings. The characteristic time for this equilibration is typically thousands of revolution periods. For an ERL the situation is entirely different. By limiting the propagation to a single passage, both the beam growth due to quantum fluctuation and the beam damping due to synchroton radiation are made negligible, and the output phase space density is essentially equal to the input phase density. This makes the maximization of phase space density injected into the linac a critical item for ERLs.

6.2.5
Transverse Defocusing

There is a transverse focusing (or rather defocusing) that occurs in the drift sections of a drift tube linac. It is illustrated in Figure 6.4. The force is inward (focusing) as the particle enters the gap and outward as it leaves. As explained before and exhibited in Figure 6.3(a), for stability, the magnitude of the electric field has to be increasing as the particle passes the center of the gap, so the transverse force on exit is stronger than on entrance. As a result there is a net defocusing through the gap. The inevitability of this defocusing can be seen by considering the particle's motion in its own rest frame. In this frame there is only a time-independent electric field because the phase velocity of the field is equal to the particle velocity. From the result known as "Earnshaw's theorem" there can only be unstable equilibrium in a static electric field. Since the equilibrium has been adjusted to be stable longitudinally it has to be unstable transversely.

To quantify the defocusing effect it is necessary to model the z and t dependence within the gap. (We are reverting to lower case symbols for z and t for

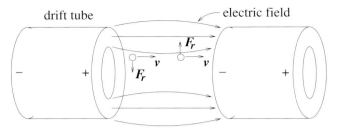

Fig. 6.4 Transverse focusing in an accelerating gap.

the reference particle.) The longitudinal electric field can be represented by a traveling wave;

$$E_z(z,t) = \bar{E}_{z0} \cos\left(\omega t - \omega \int_0^z \frac{dz'}{V_s}\right). \tag{6.34}$$

Here the argument has been arranged so that the particle stays on the peak of the traveling wave—a possible small constant phase displacement from the crest is not shown.

In a later section, using Fourier transformation to work in the frequency domain, such traveling waves will be derived for general accelerating structures, but for now only Eq. (6.34) is required. Taking r as radial displacement and θ as azimuthal angle the near-axis values of the other field components can be inferred by using Maxwell's equations. From the symmetry of the problem $H_z = H_r = E_\theta = 0$. Then, by the Gauss law, in the absence of free charge, $\nabla \cdot \mathbf{E} = 0$, which gives

$$\frac{1}{r}\frac{\partial}{\partial r}(rE_r) = -\frac{\partial E_z}{\partial z}. \tag{6.35}$$

For small r, E_r is proportional to r, so Eqs. (6.34) and (6.35) together yield

$$E_r = -\frac{r}{2}\bar{E}_{z0}\frac{\omega}{V_s}\sin\left(\omega t - \omega \int_0^z \frac{dz'}{V_s}\right). \tag{6.36}$$

Similarly, using the Ampère law,

$$\frac{1}{r}\frac{\partial}{\partial r}(rH_\theta) = \epsilon_0 \frac{\partial E_z}{\partial t}, \tag{6.37}$$

which leads to

$$H_\theta = -\frac{r\omega}{2}\epsilon_0 \bar{E}_{z0}\sin\left(\omega t - \omega \int_0^z \frac{dz'}{V_s}\right). \tag{6.38}$$

Using the Lorentz force in Newton's equation, the transverse equation of motion is

$$\frac{dp_r}{dt} = q(E_r - V_s\mu_0 H_\theta) = -\frac{\omega q \bar{E}_{z0}}{2V_s}(1-\beta_s^2)\, r \sin \Phi_s. \tag{6.39}$$

In the final equation the phase factor has been replaced by the explicit phase Φ_s of the synchronous particle. Equations (6.24) and (6.39) have opposite signs since, other than positive factors, they share the common factor $q\bar{E}_{z0}\sin\Phi_s$. If one motion is stable the other is unstable. Since we have to require stable longitudinal motion it follows that there is transverse defocusing.

Space charge effects also tend to enlarge the transverse beam size, but the effect just described limits operation even at low beam current. Early corrective schemes used electrostatic grids to produce transverse focusing. These grids were rather unsatisfactory since they reduced the beam intensities by factors of two or more due to absorption. It was the discovery of alternating gradient focusing by Blewett, Courant, Livingston, and Snyder that really overcame the problem. By introducing quadrupoles inside the drift tubes (yet outside the beam, and powered by wires running out through the mounting posts of the drift tubes) net transverse focusing is produced by alternating-sign quadrupoles. As the particles become relativistic the factor $1 - \beta_s^2$ strongly suppresses the defocusing force, thereby reducing the need for compensation.

6.3
The Relativistic Electron Linac

6.3.1
Introduction

Until recently most electron accelerators have been based on room temperature copper cavities. At the electric field values needed for interestingly high electron energy the resistive wall heating has precluded running such accelerators "CW". Rather they have been pulsed, typically for times of a few microseconds with repetition rates of order 100 Hz. A section of such a linac may have of the order of 100 cells. This makes it natural to develop a theory of periodic structures, with the ends terminated by impedance matching devices. A frequency of 2850 MHz has come to be something of a standard frequency for these devices.

With the advent of superconducting RF cavities, CW operation has become practical. Somewhat lower frequencies, such as 1300 MHz seem to be more nearly optimal for superconducting operation. As a consequence the cell sizes are larger than for room temperature linacs. Also, optimization of cell shape has led to more nearly spherical cavity shapes. A concomitant of larger cavity size, for fixed electron bunch length, is the increased seriousness of beam-excited "higher modes". Because of the high cost of refrigeration, it is essential that these higher modes are not absorbed at the low temperature required for superconducting RF. This has necessitated the development of higher mode damping structures close to the ends of linac sections. For a 100-cell linac section such damping sections are too far away to be useful. This has made it appropriate to design sections with far fewer cells, for example 7 or 9. These are small enough numbers that it is practical to solve for the entire field pattern of the section without exploiting the periodicity. However, even 7 is a large enough number that the theory of waveguides with infinite periodicity is largely applicable even to superconducting linacs.

6.3.2
Particle Acceleration by a Wave

As mentioned before, electrons become fully relativistic very near the front end of a linear accelerator. Here we assume they are fully relativistic when injected into the linac. Up to this point we have assumed acceleration occurs in a sequence of appropriately-phased accelerating gaps. From here on we will treat the accelerating field as a longitudinally-continuous traveling wave with phase velocity just right (namely the speed of light) so that the electron maintains a constant phase relationship (presumably a phase at which the field accelerates) with the wave.

Since an electromagnetic wave in free space travels at the speed of light it is natural to contemplate using the electric field of such a wave as the accelerating field. Unfortunately such a wave has the property that its **E** and **H** fields are transverse, i.e. at right angles to its wave vector **k**. In fact the unit vectors **Ê**, **Ĥ**, and **k̂** form an orthonormal triad. If, to take advantage of a nonvanishing acceleration, the electron is moving in a direction not quite parallel to **k**, the electron will soon diverge from the wave since the wave undoubtedly has restricted transverse extent

There are two ways around this difficulty: either one can keep bending the beam back into the wave or one can reflect the wave back onto the beam. The first route leads to the so-called "inverse free electron laser"—whether or not this route is promising, it is not the topic of this chapter and will not be pursued further. The second possibility is the principle of all linacs using RF waveguides. An appropriate waveguide can channel the electromagnetic wave to keep it superimposed on the particle trajectory. It can also be arranged that the electric field at the electron's position always points in the forward (accelerating) direction. To have a cumulative effect, however, the particles and the wave must be synchronous. As we will now see, it is the need to achieve these two conditions simultaneously which is the main challenge for the accelerator structure designer.

6.3.3
Wave Confined by Parallel Planes

Figure 6.5 shows two plane waves propagating on the same side of an infinite conducting plane lying in the (y, z) plane. The wave vectors $\mathbf{k}_1$ and $\mathbf{k}_2$ are parallel to the (x, z) plane at angles $\pm\theta$ relative to the z-axis—this is consistent with one wave being incident on and the other having reflected from the conducting plane. The continuous lines represent wavefronts of these waves, separated from each other by $\lambda_0/2$ where the free space wavelength λ_0 is given in terms of the wave frequency ω by

$$\lambda_0 = c \frac{2\pi}{\omega}. \tag{6.40}$$

All arrows shown in the figure are electric field vectors. The superposition of these two waves is known as a TM (transverse magnetic) wave. In fact the

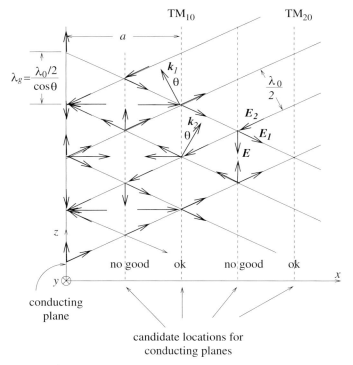

Fig. 6.5 Graphical construction showing the TM_{10} wave as the superposition of two free space waves reflecting back and forth off the sides of a rectangular waveguide. Only electric field vectors are shown. On the boundary they are normal to the conducting walls, as required. The only nonvanishing magnetic field component is H_y. The same wave propagating in a guide twice as wide would be called TM_{20}.

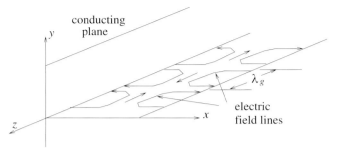

Fig. 6.6 Perspective view of the same TE$_{10}$ wave as in Figure 6.5. Note the presence of a longitudinal (accelerating) component of electric field **E**.

only nonvanishing magnetic field component is H_y. At a few of the intersection points the summed electric field $\mathbf{E} = \mathbf{E}_1 + \mathbf{E}_2$ is exhibited. One notes that **E** is perpendicular to the conducting plane at $x = 0$, which satisfies one of the boundary condition requirements. Furthermore, there is no magnetic field component normal to the surface, which is the other boundary requirement. A perspective view of the same field pattern is shown in Figure 6.6. As well as exhibiting the longitudinal electric field component desired for acceleration, the figure shows that this component reverses sign periodically (as it must to avoid build up of unphysically large end-to-end potential difference).

The broken lines in Figure 6.5 represent "candidate" locations at which it may be possible to place a conducting plane at which the same two boundary conditions would again be satisfied, so the two waves can be regarded as bouncing back and forth between the two planes. One sees that the plane at $x = a/2$ is no good, since the electric fields are parallel to it. But the $x = a$ plane is satisfactory. With a conducting plane there, the wave is known as TM$_{10}$. A rectangular waveguide with width a can propagate such a wave. The next satisfactory position would describe propagation of essentially the same pair of waves, in a waveguide twice as wide, forming a TM$_{20}$ wave, and so on.

Changing our perspective and terminology somewhat we refer to the superposition of two waves, each tilted a bit from the central axis, as a *single* wave traveling parallel to the z-axis. All waveguide fields can be visualized as superpositions of two or more, or even an infinite number of reflecting waves. From now on, all such superpositions will be referred to as a single wave. (Even the TM$_{10}$ wave under discussion would not be a valid wave in a rectangular waveguide of finite height b, since the magnetic boundary condition at the top and bottom surfaces would not be satisfied. But such a wave, still called TM$_{10}$ could be represented as the superposition of four reflecting free space waves.)

The repetition period λ_g along the z-axis is known as the "guide wavelength". By construction its value can be seen to be

$$\lambda_g = \frac{\lambda_0}{\cos\theta}. \tag{6.41}$$

We are now confronted by one of the obstacles tending to foil our efforts. After waiting a time $2\pi/\omega$ the entire pattern will have advanced by distance λ_g along the guide. The speed with which the pattern moves, known as the "phase velocity" is given by

$$v_p = \lambda_g \frac{\omega}{2\pi} = \lambda_g \frac{c}{\lambda_0} = \frac{c}{\cos\theta}. \tag{6.42}$$

Since $\cos\theta < 1$ *the phase velocity is greater than c*. Since the particle to be accelerated is necessarily traveling at essentially speed c our wave cannot yet be used for cumulative acceleration.

Before addressing the phase velocity problem, we should establish a more general analytic formulation of waveguide modes and waveguide propagation. This formalism is based directly on Maxwell's equations. The term "wave" will refer to any propagating solution of these equations that satisfies the appropriate boundary conditions and no attempt will be made to represent such waves as superpositions of simpler reflecting free-space waves. For simplicity we will still consider only waves in the region between parallel planes separated by distance a. The fields will therefore be independent of y.

Since we want to have a longitudinal electric field, the so-called "transverse electric" TE modes are inappropriate. So the discussion can be restricted to TM modes and a standard formalism exists in which one seeks the longitudinal field component in the form

$$E_z(x,z,t) = E_z(x)\,e^{-\gamma z}\,e^{j\omega t}. \tag{6.43}$$

(The remaining field components will later be obtained from E_z.) Substituting this form into the wave equation, one finds that $E_z(x)$ satisfies the so-called Helmholtz equation

$$\left(\frac{\partial^2}{\partial x^2} + \gamma^2 + \frac{\omega^2}{c^2}\right) E_z = 0. \tag{6.44}$$

Along with the requirement that $E_z(0) = E_z(a) = 0$, this is an eigenvalue equation for which there are solutions only for a discrete set of values of the combination $\gamma^2 + \omega^2/c^2$. For each of these eigenvalues, with ω being fixed, there is a corresponding value for γ. Since γ^2 may be negative, γ is, in general, complex

$$\gamma = \alpha + j\beta. \tag{6.45}$$

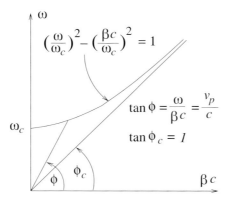

Fig. 6.7 Brillouin diagram for a guided wave in an unloaded guide.

One solution can be written down by inspection

$$E_z = C e^{-\gamma z} \sin \frac{\pi x}{a}, \qquad (6.46)$$

where C is a constant amplitude to be determined, and the condition

$$-\frac{\pi^2}{a^2} + \gamma^2 = -\frac{\omega^2}{c^2}, \qquad (6.47)$$

has to be satisfied.

Higher modes can also be obtained by choosing the x dependence to be $\sin(n\pi x/a)$ where n is an odd integer. If there is no attenuation α vanishes and, for any particular mode, relation (6.47) takes the form

$$\beta^2 = \frac{\omega^2 - \omega_c^2}{c^2}, \qquad (6.48)$$

where ω_c is known as the "cut-off" frequency, since $\omega < \omega_c$ causes β^2 to be negative. Restricting discussion to the lowest mode Eq. (6.47) becomes

$$\beta = \sqrt{\left(\frac{\omega}{c}\right)^2 - \left(\frac{\pi}{a}\right)^2}, \quad \text{and} \quad \omega_c = c\frac{\pi}{a}. \qquad (6.49)$$

When hyperbolic relation (6.48) is plotted as in Figure 6.7 it is known as a "Brillouin diagram". It plots the required frequency ω as a function of the desired wave number β. It is useful to think about the situations near the extremes of the graph. At high frequency the curve is asymptotic to the free space relationship between frequency and wave number. In Figure 6.5 this corresponds to wavefronts that are very close together ($\lambda_g \to \lambda_0$, $\beta \to \omega/c$) and the two components of the wave propagate almost parallel to the axis. This situation is also known as "overmoded" since there are many nearby modes and a wave packet with transverse dimension small compared to a can be produced. As discussed earlier, such a wave is not useful for acceleration.

At the low frequency extreme of the Brillouin diagram the wavefronts are as far apart as possible consistent with making the construction of the figure possible. This amounts to being a plane wave traveling at right angles to the waveguide and bouncing back and forth between the side walls. Such a wave has infinite phase velocity and zero group velocity. Since the fields at opposite sides of the waveguide are exactly out of phase, in the limiting case we have

$$\frac{\lambda}{2} = a, \quad \text{or} \quad \omega = c\frac{\pi}{a}; \tag{6.50}$$

the geometric construction therefore agrees with the analytic value (6.47) for the cut-off wavelength of the TM_{10} mode;

$$\omega_c(TM_{10}) = c\frac{\pi}{a}. \tag{6.51}$$

Clearly β is the wave number corresponding to the previously introduced guide wavelength λ_g which is related to the wave angle θ as shown in Figure 6.5;

$$\lambda_g = \frac{2\pi}{\beta} = \frac{\lambda_0}{\cos\theta}. \tag{6.52}$$

(To avoid clash of symbols with the conventional relativistic factor β, it would be clearer to use a symbol such as β_g or k_z, but the present notation is fairly standard.)

Since accelerator guides are generally circular, perhaps more time than is justified has already been devoted to rectangular guides and parallel planes in particular. However, there are a few more points which can illustrate the main ideas in this simple case, free from the intrusion of complicated geometry and mathematics. These have to do with power flow and power dissipation in the walls.

The non-vanishing transverse field components between infinite parallel planes, corresponding to E_z given by Eq. (6.46) are

$$E_x = -\frac{\gamma a^2}{\pi}\frac{\partial E_z}{\partial x} = -\frac{\gamma a}{\pi}C\cos\left(\frac{\pi x}{a}\right)e^{-\gamma z},$$

$$H_y = -j\frac{\omega\epsilon_0 a^2}{\pi^2}\frac{\partial E_z}{\partial x} = -j\frac{\omega\epsilon_0 a}{\pi}C\cos\left(\frac{\pi x}{a}\right)e^{-\gamma z}. \tag{6.53}$$

The (averaged) power flow P_{tr} is obtained by integrating the average Poynting vector over a transverse plane

$$P_{tr} = \frac{1}{2}\Re\int E_x H_y^* \, dx \, dy. \tag{6.54}$$

Since we are assuming there is no y-dependence it is not necessary to satisfy boundary conditions on bottom and top walls. Taking an artificial value b as the waveguide height, with γ taken as pure imaginary, we get

$$P_{tr} = C^2 \frac{ab}{4\pi^2} \epsilon_0 \omega a^2 \beta. \tag{6.55}$$

The standard method for calculating power loss in the walls (an issue of dominant importance in linear accelerators) is to regard the fields obtained so far (which have assumed infinite conductivity) as fields "incident" on the medium having the actual conductivity. In our case the loss occurs in the side walls where, by Ampére's law, the longitudinal surface current is

$$I_z = H_y b. \tag{6.56}$$

Although the model is only heuristic, one can assume this current flows in a resistive layer of "skin depth" δ given by

$$\delta = \left(\frac{2}{\mu_0 \sigma \omega}\right)^{1/2}, \tag{6.57}$$

where $\mu_0 = 4\pi \times 10^{-7}\,\mathrm{H\,m^{-1}}$ and σ is the conductivity in $\Omega^{-1}\mathrm{m}^{-1}$. For example, for copper, $1/\sigma = 1.68 \times 10^{-8}\,\Omega\,\mathrm{m}$. The precise definition of δ has been arranged so that using formula (6.57) gives the exact (or at least very accurate) power loss when the current flow is assumed uniform throughout the layer. Treating the layer as infinitely thin, its skin resistance (per square) is $R_s = (\sigma \delta)^{-1}$. Using Eq. (6.56), the average power lost per unit length along the guide is therefore given by

$$\frac{dP_{lost}}{dz} = \frac{|H_y b|^2}{\sigma \delta b} = \frac{\omega^2 \epsilon_0^2 a^2 C^2 b}{\sigma \delta \pi^2}. \tag{6.58}$$

To account for this loss, the propagation constant γ must be assigned a small real part given by

$$\alpha = \frac{dP_{lost}/dz}{2 P_{tr}} = \frac{1}{\beta a} \frac{4\pi}{\lambda_0} \frac{R_s}{\sqrt{\mu_0/\epsilon_0}}. \tag{6.59}$$

Essentially the same information can be quantified by introducing W_{st} for energy stored per unit length and defining a Q-factor for the guide by

$$Q = \frac{\omega W_{st}}{dP_{lost}/dz}. \tag{6.60}$$

The stored energy per unit length at $z = 0, t = 0$, can be calculated from the maximum electric field by averaging transversely and longitudinally (with α

neglected) and doubling to account for magnetic energy;

$$W_{st} = \frac{\epsilon_0}{4}\left(|E_x|^2 + |E_z|^2\right)ab$$

$$= \frac{\epsilon_0}{4}\left(\left(\frac{\beta a}{\pi}\right)^2 + 1\right)C^2 ab,$$

$$= \frac{\epsilon_0}{4}C^2\frac{\omega^2}{c^2}\frac{a^2}{\pi^2} ab. \tag{6.51}$$

which yields

$$Q = \frac{\pi\left((\beta a/\pi)^2 + 1\right)}{8\,R_s/\sqrt{\mu_0/\epsilon_0}}\frac{\lambda_0}{a}. \tag{6.62}$$

Finally there are other important relations describing propagation along the guide. Relation (6.48) can also be recognized to be the "dispersion relation" for wave propagation along the guide. Differentiating this relation yields for the group velocity

$$v_g = \frac{d\omega}{d\beta} = c^2\frac{\beta}{\omega} = \frac{c^2}{v_p}. \tag{6.63}$$

The facts that the phase velocity is greater than and the group velocity less than the speed of light can be inferred graphically from Figure 6.7 since the former is the slope of the line from the origin (times c) and the latter is the slope of the local tangent to the hyperbolic curve (times c). The tangent slope vanishes at the cutoff frequency and approaches 1 at high frequency. The velocity of propagation of energy along the guide is

$$v_E = \frac{P_{tr}}{W_s} = v_g, \tag{6.64}$$

where the last equality comes from Eqs. (6.55) and (6.61). This shows that, as expected, energy flows with the group velocity.

All of the quantities introduced in this section have to be calculated for any practical accelerator structure. This is usually quite complicated and will not be attempted here. Conceptually, however, these quantities are no different from what has been calculated so far. The exact formulas look the same as far as dimensional factors are concerned but, of course, the numerical values depend on the exact geometry.

6.3.4
Circular Waveguide

Most linear accelerator structures have circular cross sections. To simplify matters let us initially assume there is a single boundary at $r = b$. Limiting ourselves to TM modes having no θ dependence, the Helmholtz equation corresponding to Eq. (6.44) is

$$\left(\frac{\partial^2}{\partial r^2} + \frac{1}{r}\frac{\partial}{\partial r} + \gamma^2\right) E_z = -\frac{\omega^2}{c^2} E_z. \tag{6.65}$$

The other fields can be obtained from E_z using

$$H_\theta = -j\frac{\omega\epsilon_0}{k_c^2}\frac{\partial E_z}{\partial r}, \quad E_r = -\frac{\gamma}{k_c^2}\frac{\partial E_z}{\partial r},$$

$$\text{where} \quad k_c^2 = \gamma^2 + \frac{\omega^2}{c^2}. \tag{6.66}$$

The simplest solution of this type is the so-called TM$_{01}$ mode, where the first subscript indicates zero θ-variation and the second indicates one radial variation. For this mode the fields are

$$E_z = E_0 J_0(k_c r)\, e^{-\gamma z},$$
$$H_\theta = \frac{j\omega\epsilon_0}{k_c} E_0 J_1(k_c r)\, e^{-\gamma z},$$
$$E_r = \frac{\gamma}{k_c} E_0 J_1(k_c r)\, e^{-\gamma z}. \tag{6.67}$$

Comparing these expressions with Eqs. (6.46) and (6.53), one sees that the main changes amount to having replaced sine and cosine functions by Bessel functions. The waveguide radius enters through to the requirement $E_r(b) = 0$. This means that $k_c b$ has to be equal to the first zero of $J_0(k_c r)$;

$$k_c b = 2.405. \tag{6.68}$$

The guide dispersion relation takes the form of Eq. (6.48) if the cut-off frequency is defined by

$$\omega_c = k_c c = c\,\frac{2.405}{b}. \tag{6.69}$$

Note that, as expected, this expression differs from the corresponding relation (6.49) only by a numerical factor, in this case 2.405 instead of π. All the other waveguide properties such as P_{tr}, P_{lost}, α, Q, can be calculated as before. This will not be done since the phase velocity will again be greater than c, making the wave incapable of providing indefinitely cumulative acceleration. However the circular geometry discussed in this section is appropriate for analysing the cylindrical resonator shown in Figure 6.1. This geometry is very close to the geometry used in room temperature electron linacs. The RF cavities of superconducting linacs are more nearly spherical, but their field configurations are very much like the field in a cylindrical resonator.

6.3.5
Cylindrical "Pill-box" Resonator

In a circular waveguide, dispersion relation (6.48) relates propagation constant β and frequency ω;

$$\omega^2 = c^2\beta^2 + \omega_c^2. \tag{6.70}$$

This is a continuum relation in the sense that, at least within a band of frequencies, the guide wavelength ω can be adjusted smoothly by varying ω. Suppose that the waveguide is terminated at some point by a conducting plane. Such a termination reflects the wave. Suppose the waveguide has length d by virtue of being capped at both ends by conducting planes. One can envisage a wave bouncing back and forth along the guide. There will be a discrete set of frequencies for which the guide length d is an odd multiple of $\lambda_g/2$. Such a configuration is known as a resonator. (Neglecting the small holes in both ends) each individual cavity in Figure 6.1, with g now replaced by d, forms such a resonator. For the TM_{01} cylindrical waveguide mode analysed in the previous section the cut-off frequency ω_c was given by Eq. (6.69). Using this result in Eq. (6.70), the lowest resonance occurs for

$$\omega_r(TM_{011}) = c\sqrt{\frac{\pi^2}{d^2} + \frac{2.405^2}{b^2}}. \tag{6.71}$$

In the expression for E_r in Eq. (6.67), the electric boundary condition on the end surfaces is satisfied by using the relation $(\exp(i\beta z) - \exp(-i\beta z))/(2i) = \sin \beta z$ in forming a superposition of traveling waves. This is the so-called TM_{011} resonant mode because there is one longitudinal variation.

It turns out though that there is an even lower frequency mode—it is the mode used in the Alvarez drift tube linac. For this mode E_r vanishes not just at the ends but everywhere. In this mode the guide wavelength λ_g is infinite; in Eq. (6.70) this yields resonant frequency

$$\omega_r(TM_{010}) = \frac{2.405}{b} c. \tag{6.72}$$

The non-vanishing field components in this mode are

$$E_z = E_0 J_0(k_c r), \quad H_\theta = -j\frac{E_0}{\eta} J_1(k_c r), \tag{6.73}$$

where $\eta = \sqrt{\mu_0/\epsilon_0} = 377\,\Omega$. This TM_{010} is "degenerate" in the sense that it is not constructed from repeated reflections of propagating waveguide modes. Or rather, if one prefers, TM_{010} corresponds to the TM_{01} at cut-off, making λ_g infinite. Alternatively the mode can be viewed as a cylindrical wave emerging from the axis and reflecting back from the cylindrical boundary.

The total stored energy for this mode is

$$W_{st} = \pi \epsilon_0 d\, E_0^2 \frac{b^2}{2} J_1^2(k_c b), \tag{6.74}$$

and the wall power is

$$P_{lost} = \frac{\pi R_s b E_0^2}{\eta^2} J_1^2(k_c b)(d+b). \tag{6.75}$$

Using Eq. (6.73) to obtain E_z and the maximum voltage drop V_0, the cavity shunt impedance R_{lin} can be calculated using Eq. (6.7). The result is

$$R_{lin} = 2R_{cir} = \frac{\eta^2 d}{\pi R_s b(1+b/d) J_1(k_c b)}. \tag{6.76}$$

(As explained above there are two definitions of "shunt resistance.) The Q-value of the cavity is obtained using Eq. (6.60);

$$Q_0 = \frac{\eta}{R_s} \frac{2.045/2}{1+b/d}. \tag{6.77}$$

as the so-called "unloaded Q" of the cavity with no external inputs or outputs. The R_{lin}/Q value is approximately $230\, d/b\, \Omega$.

Problem 6.3.1 *The formulas for R_{lin} and Q_0 just derived are approximately applicable both to proton accelerators, for which, say, $\omega = 200$ MHz, $d = 8$ m and electron accelerators, for which the frequency is, say, $\omega = 2856$ MHz, $d = 3.5$ cm. For these choices evaluate R_{lin}, Q_0, and R_{lin}/Q, to gain a quantitative comparison between these two extremes of linac operation. Also work out the transit time factor T_{tr} for both cases.*

6.3.6
Lumped Constant Model for One Cavity Resonance

RF power is coupled into waveguides by input loops or probes. There is radiation into all propagating modes of the guide, and this leads to continuous extraction of power from the input. If the waveguide is effectively endless in both directions this power flows away and does not "react back" on the source. There is also local excitation of non-propagating modes, as needed to match boundary conditions, but this is negligible as far as power flow is concerned. Because of the absence of reaction, calculating radiation into a waveguide is straightforward, though complicated if there are many propagating modes at the drive frequency.

The situation with a resonant cavity driven from a single input loop or probe is very different. For a high-Q cavity the modes are well separated in frequency so a mono-frequency power source produces significant excitation in

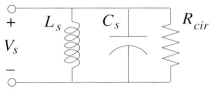

Fig. 6.8 Lumped circuit model for a single cavity resonant mode. The subscript "s" standing for "secondary" is intended to suggest the cavity forms the secondary circuit of a transformer.

only the mode whose frequency is closest. As the drive frequency approaches the resonant frequency, excitation builds up strongly and reacts back on the input source. For analysing this situation it is convenient to represent the cavity by a lumped constant model such as exhibited in Figure 6.8.

The parameters in the lumped model are not directly measurable. Rather, they have to be assigned consistent with the known cavity quantities ω_r, P_{lost}, W_{st}, and Q_0. When worked out in terms of the lumped parameters these quantities are given by

$$\omega_r = \frac{1}{\sqrt{L_s C_s}},$$
$$P_{lost} = \frac{|V_s|^2}{2R_{cir}},$$
$$W_{st} = \frac{1}{2} C_s |V_s|^2 = C_s P_{lost} R_{cir} = \frac{Q_0 P_{lost}}{\omega_r},$$
$$Q_0 = \frac{R_{cir}}{\sqrt{L_s/C_s}}. \tag{6.78}$$

Curiously enough, there is still freedom to vary $|V_s|$ without invalidating these equations. One can see this, for example, by doubling $|V_s|$ and altering the unmeasurable quantities C_s, L_s, and R_{cir} while leaving the measurable quantities fixed. A natural definition would be to set $|V_s|$ equal to the maximum voltage drop along the axis; $|V_s| = \int_0^d E_z dz$ ($= E_0 d$ for the TM$_{010}$ mode of the pill-box cavity.) Alternatively, one might choose to include the transit time correction so that $q|V_s|$ would be the actual energy gain of a particle traveling along the axis at the optimal phase. Any definition flagrantly different from either of these would likely be very misleading. Perhaps the lack of unanimity in the definition of shunt impedance derives from this ambiguity?

In any case, accepting R_{cir} as having been defined, we obtain

$$L_s = \frac{R_{cir}/Q_0}{\omega_r}, \quad C_s = \frac{Q_0/R_{cir}}{\omega_r}. \tag{6.79}$$

Then the impedance Z_s of the resonant circuit is given by

$$\frac{1}{Z_s} = \frac{1}{j\omega L_s} + j\omega C + \frac{1}{R_{cir}} = -j\frac{Q_0}{R_{cir}}\left(\frac{\omega_r}{\omega} - \frac{\omega}{\omega_r}\right) + \frac{1}{R_{cir}}, \quad (6.80)$$

which, except for the above-mentioned ambiguity, depends on nothing but measurable quantities.

6.3.7
Cavity Excitation

Accurate calculation of the coupling of energy into a resonant cavity requires numerical field calculation. But a semi-quantitative calculation will give the general idea.

The diagram in Figure 6.9(a) is a lumped model for the left pill-box cavity in Figure 6.1. The cavity is excited by current flowing in the semicircular loop of area S. This loop is placed in a location where the resonant magnetic field is maximum and its plane is normal to the magnetic field direction in order to maximize the coupling. In the circuit diagram this loop is represented by "primary" inductance L_p. Current i_p is driven through this loop by the RF power source.

The magnetic field of the half-ring protruding through the effectively infinite conducting plane that is the cavity wall is very much like the magnetic field of a full ring in free space. The loop self-inductance is therefore approximately half that of a ring of radius a_p made with wire of diameter $2b_p$; this gives

$$L_p \approx (1/2)\mu_0 a_p \left(\ln(8a_p/b_p) - 7/4\right). \quad (6.81)$$

For purposes of estimation $L_p \approx \mu_0 a_p$.

There is mutual inductance M between primary and secondary. The ratio M/L_s is approximately equal to the area fraction $S/(bd)$, but weighted by the relative value of H_ϕ at the loop; i.e. at $r \approx b$;

$$\frac{M}{L_s} = \frac{S}{d}\frac{J_1(2.405)}{\int_0^b J_1\left(\frac{2.405 r}{b}\right) dr} = \frac{S}{bd}\frac{2.405 J_1(2.405)}{\int_0^{2.405} J_1(x)\, dx} \approx 1.25\frac{S}{bd}. \quad (6.82)$$

We define two subsidiary dimensionless quantities,

$$K = \frac{M}{\sqrt{L_p L_s}}, \quad \alpha = \frac{M}{L_s}. \quad (6.83)$$

The quantities introduced so far satisfy inequalities:

$$L_p \ll L_s, \quad M \ll L_s, \quad \alpha \ll 1, \quad K \ll 1. \quad (6.84)$$

Together the primary and secondary inductances form a transformer. It is customary to define a "turns ratio" or "step-up factor" n for a transformer. In this case since both primary and secondary are best visualized as having a single turn it seems to me that naming a factor that brings to mind multiple turns is not heuristically helpful. Instead we will use output/input voltage ratio $\alpha(=1/n)$ to characterize the transformer. Of course the whole point of using a high Q resonant cavity is to take advantage of a gigantic voltage step up factor to get strong acceleration from a power supply of modest output voltage. In the circuit diagram the rest of the cavity impedance is represented by "load" Z_L. The circuit equations for circuit Figure 6.9(a) are

$$v_p = j\omega L_p i_p - j\omega M i_s,$$
$$0 = -j\omega M i_p + j\omega L_s i_s + Z_L i_s. \tag{6.85}$$

The corresponding equations for Figure 6.9(b) are

$$v_p = j\omega L_p i_p - j\omega L_p K^2 \frac{i_s}{\alpha},$$
$$0 = -j\omega K^2 L_p i_p + j\omega K^2 L_p \frac{i_s}{\alpha} + \alpha^2 Z_L \frac{i_s}{\alpha}. \tag{6.86}$$

With K and α given by Eq. (6.83) these equations reduce to Eqs. (6.85). The upper two circuits of Figure 6.9(a) and (b) are therefore equivalent.

We know that the impedance so far called Z_L is primarily $1/(j\omega C_s)$ where C_s is an equivalent lumped capacity, but Z_L also includes a resistive element R_{cir} to model wall loss. The output circuit is primarily a weakly damped LC circuit. In the equivalent circuit model the resonant frequency is

$$\omega_r = \frac{1}{\sqrt{(K^2 L_p)(C_s/\alpha^2)}} = \frac{1}{\sqrt{L_s C_s}}. \tag{6.87}$$

(This result has neglected both wall losses and the loading of the primary circuit. In any case, being resistive, these extra elements perturb the natural frequency very little.) Of course this result is consistent with expectation and only corroborates the equivalence of the circuit models. On the other hand it is clear that the (essentially resistive) input circuit will lower the Q of the configuration, producing the so-called "loaded Q". Neglecting the inductance in series with Z_0 (valid for sufficiently small loop size) Q_L is given by

$$\frac{1}{Q_L} = \frac{1}{Q_0} + \frac{1}{Q_{ext}}, \quad \text{where} \quad Q_{ext} = \frac{Z_0}{\alpha^2 \sqrt{L_s/C_s}}. \tag{6.88}$$

(Including finite input inductance would give a larger value for Q_L).

If the cavity is driven exactly on resonance the reactive impedance is infinite, the input voltage divides between Z_0 and $\alpha^2 R_{cir}$ (in the equivalent circuit)

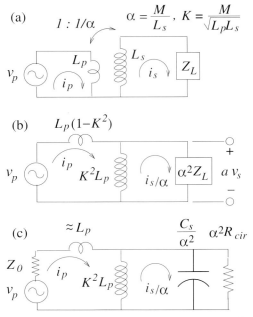

Fig. 6.9 (a) Representation of cavity excitation as a transformer coupled circuit. (b) An equivalent, transformer-free, circuit. (c) The load impedance is represented by capacitor and resistor and the source has been assigned output impedance which is probably the characteristic impedance Z_0 of the transmission line connecting power source to cavity.

and the (actual circuit) output excitation is

$$\left.\frac{v_s}{v_p}\right|_{\omega_r} = \frac{\alpha R_{cir}}{Z_0 + \alpha^2 R_{cir}} \quad \left(\approx \frac{1}{\alpha}, \quad \text{for small } Z_0.\right) \tag{6.89}$$

The input loop inductance has again been neglected.

For these calculations we have neglected the central, on-axis, beam holes, and this is a good approximation for individual cells. But a linear accelerator consists of stringing many such cavities together. In such a chain the main electrical effect of the holes is to couple fields from cell to cell. Even at high frequencies where the full electromagnetic field theory is required a good semi-quantitative model of the effects of such coupling can be obtained from lumped-constant models. That is the next topic.

6.3.8
Wave Propagation in Coupled Resonator Chain

A good lumped representation of the pill-box TM_{010} resonator is a simple LC-circuit, with capacity C_s ("s" for series) representing the capacity between the

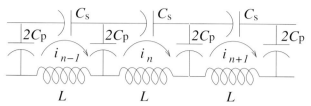

Fig. 6.10 Lumped constant model of successive linac cells as lightly coupled oscillators.

end surfaces, especially close to the axis where the electric field is dominant, and L_s representing the inductance of the toroidal outer volume where the magnetic field predominates. A lumped parameter model for a few linac cells is shown in Figure 6.10. Coupling between cells is represented by shunt or "parallel" capacitance $2C_p$.

In the limit of vanishing hole radius the value $2C_p$ would be infinite, and there would be no cell-to-cell coupling. To ensure we are close to that situation we require

$$C_p \gg C_s. \tag{6.90}$$

One qualitative feature of the true pill-box geometry that is masked by the lumped-constant model is the rotational symmetry around the beam axis. As described already, both C_s and L_s represent similarly symmetric field patterns. Thinking of the axis as a "virtual ground" helps in visualizing the representation of cell-to-cell coupling as shunt capacity C_p.

It was L. Brillouin who first systematized the analysis of such periodic chains. All the essential results we require can be derived in a few lines. Applying Kirchoff's laws ot the central loop in Figure 6.10 yields

$$\left(\frac{2}{j\omega 2C_p} + \frac{1}{j\omega C_s} + j\omega L\right) i_n - \frac{1}{j\omega 2C_p}(i_{n+1} + i_{n-1}) = 0,$$

$$\text{or} \quad i_{n+1} - 2\cos(\beta d) i_n + i_{n-1} = 0, \tag{6.91}$$

$$\text{where} \quad \cos(\beta d) = 1 + \frac{C_p}{C_s} - \omega^2 C_p L. \tag{6.92}$$

From the point of view of the present discussion we have simply introduced a new quantity βd which is a function of ω and circuit parameters. But this choice of symbol anticipates identifying this function with the previously defined quantities β and d.

Two solutions of difference equation (6.92) are

$$i_n = \genfrac{}{}{0pt}{}{\cos}{\sin}(n\beta d + \phi_0), \quad n = 0, \pm 1, \pm 2, \ldots \tag{6.93}$$

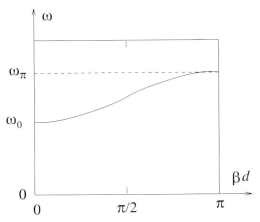

Fig. 6.11 Brillouin diagram for the chain of coupled LC circuits of Figure 6.10.

as can be checked with standard trigonometric identities. If βd, as defined by Eq. (6.92), is a real angle, then both sequences i_n represent oscillatory standing wave solutions of the chain of resonators. Remember that it is implicit in an impedance equation like Eq. (6.92) that the time dependence $e^{j\omega t}$ is assumed and the real part to be taken later. By superimposing the solutions (6.93) one can produce unattenuated "traveling waves" moving in either the direction of increasing or decreasing n-value.

Depending on the values of ω and the circuit parameters, βd can be real or complex. The condition for β to be real is

$$-1 \leq \cos(\beta d) \leq 1. \tag{6.94}$$

This can be re-expressed as a band of frequencies,

$$\omega_0 \leq \omega \leq \omega_\pi. \tag{6.95}$$

The frequencies ω_0 and ω_π defining the ends of this "propagating band" are given by

$$\omega_0 = \frac{1}{\sqrt{LC_s}}, \quad \omega_\pi \approx \frac{1}{\sqrt{LC_s}}\left(1 + \frac{C_s}{C_p}\right); \tag{6.96}$$

Condition (6.90) was used in obtaining ω_π. Note that the subscripts on ω_0 and ω_π are the corresponding values of the angle βd at the band edges.

The Brillouin diagram for this chain of resonators is plotted in Figure 6.11; this is a graph of Eq. (6.92). The band of propagating frequencies includes the resonant frequency $1/\sqrt{LC_s}$ of the uncoupled oscillators. For weak coupling (where condition (6.90) is especially true) the band (of frequencies) is very

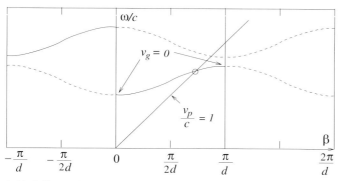

Fig. 6.12 Brillouin diagram for a chain of cells, each having two modes.

narrow, and its width increases as C_p decreases. The "band width" $\Delta \omega$ is given by

$$\frac{\Delta \omega}{\omega_0} = \frac{\omega_0 - \omega_\pi}{\omega_0} = \frac{C_s}{C_p}. \tag{6.97}$$

Certain features of this Brillouin diagram are important properties of periodic structures in general. The plot is an even function of βd as well as being periodic in βd with period 2π, since these are both properties of the function $\cos \beta d$. The broken line extensions convey these properties. As exhibited in Figure 6.12, these properties imply that the dispersion function $\omega(\beta)$ is horizontal at all multiples of $\beta = \pi/d$, and this implies the group velocity v_g vanishes at $\beta = 0$ and $\beta = \pi/d$. The frequencies at these end points are ω_0 and ω_π.

For our one mode model, these extensions of the range of βd are superfluous, since the replacement $\beta d \to \beta d + 2\pi$ in Eq. (6.93) reproduces the same solutions. This is related to the fact that our model has a discrete, independent index n rather than a continuous coordinate z. An actual cavity has higher modes, one of which is indicated as the upper curve in Figure 6.12. On general grounds this curve cannot intersect the lower curve and it has the same periodicity. One consequence is the "frequency gap" between modes at $\beta = \pi/d$. Also, comparing with Figure 6.7, one sees that there is necessarily a point, indicated by an open circle in Figure 6.12, at which there is an intersection with the curve $v_p/c = 1$ on which the phase velocity is equal to the speed of light. As explained earlier, this is the condition that the electromagnetic wave must satisfy to give cumulative particle acceleration.

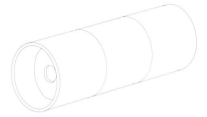

Fig. 6.13 The iris-loaded slow-wave structure used for most room temperature electron linacs.

6.3.9
Periodically Loaded Structures

It necessary to face up to the fact that a linear accelerator is more complicated than a chain of weakly coupled LC circuits. It is more nearly a periodically-loaded waveguide and the task is to design and analyse the loading in order to tailor the Brillouin diagram for effective acceleration. For a chain of pill-box resonators it is easy to meet the phase requirement. If the resonators are individually powered it is straightforward to adjust the successive cell phases appropriately, especially in the TM_{010} case in which, according to Eq. (6.72), the resonant frequency is independent of d.

But it is economically not feasible to power individually the large number of cells needed for electron acceleration. Rather, the natural evolution away from pill-box geometry is to the iris-loaded (or disk-loaded) slow-wave structure illustrated in Figure 6.13. This can be viewed as a chain of pill-box cavities butted together, but the field pattern for realistic excitation frequency is rather different than was true for the TM_{010} pill-box mode, as can be seen from Figure 6.14. Instead of treating the structure as a sequence of pill-boxes one can treat the structure as a waveguide with periodic irises. The former approach is appropriate for small hole radius r, the latter for r more nearly comparable with outer radius b.

Treating an iris as a perturbations, one expects its leading effect on a propagating mode to be generation of a reflected wave. Then, in a periodic structure, the important question is the degree to which reflected waves interfere constructively. For some frequencies the reflections from successive reflections interfere destructively and propagation is somewhat insensitive to the presence of the irises. But there is a kind of Bragg scattering condition, the condition for which is

$$\beta = \frac{\pi}{d} n, \quad n = 0, 1, 2, \ldots, \tag{6.98}$$

at which successive reflections are in phase and propagation is very sensitive to the presence of the irises. The frequencies at which this condition is satisfied are labeled $\omega_{n\pi}$. At these frequencies the wave number is very

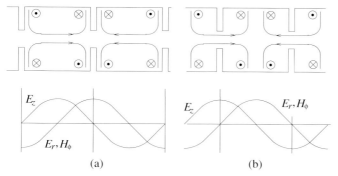

Fig. 6.14 Heuristic standing wave configurations of opposite symmetry relative to the irises of an iris-loaded linac. With the frequency chosen near the ω_π stop band, the guide wavelength matches the iris biperiodicity; $\beta = \pi/d$. In the upper figure the solid lines with arrows are electric field lines and the magnetic fields are represented by arrow heads or tails.

sensitive to the frequency or, conversely, the group velocity approaches zero; $v_g = d\beta/d\omega \approx 0$. In the mode illustrated in Figure 6.14 there is a complete phase reversal from iris to iris; this corresponds to the so-called "π-mode" at frequency ω_π.

To investigate behavior near ω_π consider the two modes illustrated in Figure 6.14. With the irises absent these two modes would be degenerate, differing only in phase. But the irises have greater perturbative effect in case (a) because the (unperturbed) radial electric field is nonvanishing tangential to the iris surface, maximally violating the boundary condition on that surface. Currents flowing in the irises cause the standing wave of wavelength $2d$ to have a different (actually lower because of increased C_s) frequency in case (a), while we expect that in case (b) the relation between ω and β will more nearly resemble the iris-free dispersion relation.

These comments are illustrated in Figure 6.15. As β increases from zero, the frequency initially follows a hyperbolic curve like that for the unloaded wave guide of Figure 6.7. But the curve flattens out and becomes horizontal at $\beta = \pi/d$, where the group velocity vanishes. We have just seen that there is another, slightly higher, frequency, that falls almost on the hyperbola for the unloaded guide. This accounts for the circled point just above the ω_π stop-band. Increasing the frequency from that point moves the wave number along that branch from $\beta = \pi/d$ to $\beta = 2\pi/d$. Similar considerations lead to successively higher branches. Only points lying on the 45° diagonal that remain synchronous with relativistic particles are relevant for an electron acceleration. In practice the intersection with the lowest branch, labeled β_{op}, ω_{op} in the figure, is chosen for operation.

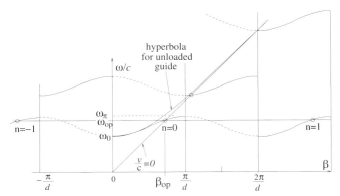

Fig. 6.15 Extension of the Brillouin diagram for an iris-loaded waveguide.

6.3.10
Space Harmonics

A more quantitative understanding of, and completion of, Figure 6.15 entails finding a method to calculate the exact dispersion relation and then, for a given operating frequency ω_{op}, to calculate the corresponding electric and magnetic fields for the slow-wave structure. Concentrating on the configuration of Figure 6.16, one requirement to be met is that of periodicity. This does not require the fields to be periodic but it does constrain monochromatic fields, according to Floquet's theorem, to be altered only by the same multiplicative factor, complex in general, when z is increased by d; call the factor $e^{-j\beta_0 d}$. This applies to any two planes separated by d, such as those indicated in the figure. This theorem is discussed, for example, in Landau and Lifshitz, *Mechanics*. Without loss of generality β_0 can be chosen in the range $0 \leq \beta_0 \leq \pi/d$. An electric field propagating in the positive z direction can tentatively be expressed by

$$E(r,z,t) = F(r,z)\, e^{-j\beta_0 z}\, e^{j\omega t}. \tag{6.99}$$

Through any single starting structure period this is a completely general expression. To extend it outside this period one requires $F(r,z)$ to be periodic in z with period d. For the adjacent period

$$E(r, z_1 + d, t) = e^{-j\beta_0 d}\, E(r, z_1, t), \tag{6.100}$$

for any z_1. Assumption (6.99) therefore meets the Floquet requirement. Field patterns for various values of $\beta_0 d$ are illustrated in Figure 6.17.

Since $F(r,z)$ is periodic it can be expanded in a Fourier series

$$F(r,z) = \sum_{n=-\infty}^{\infty} C a_n(r)\, e^{-j\frac{2\pi n}{d} z}, \tag{6.101}$$

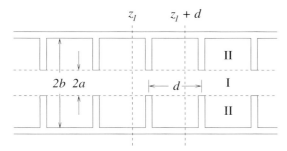

Fig. 6.16 Example of Floquet, single-period separated planes for the iris-loaded linac. Also shown is a possible subdivision of the cavity volume into regions I and II. As well as matching the boundary conditions, separate series expansions in these regions have to be matched along the broken lines.

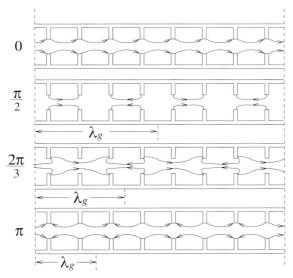

Fig. 6.17 Fixed time snapshots of electric field configurations for the iris-loaded linac. Phase advances per period, $\beta_0 d$ are indicated on the left.

where C is a normalization constant, and the $a_n(r)$ are radial functions to be determined. After substituting series (6.101) into Eq. (6.99), each term in the result has to satisfy the wave equation. This constrains the r-dependence of each coefficient

$$\frac{d^2 a_n(r)}{dr^2} + \frac{1}{r}\frac{d a_n(r)}{dr} - \left(\frac{\omega^2}{c^2} - \left(\beta_0 + \frac{2\pi n}{d}\right)^2\right) a_n(r) = 0. \qquad (6.102)$$

With $k = \omega/c$, introducing wave number k_{rn} by

$$k_{rn}^2 = k^2 - \beta_n^2, \quad \text{where} \quad \beta_n = \beta_0 + \frac{2\pi n}{d}, \tag{6.103}$$

the r-dependent factors are proportional to $J_0(k_{rn}r)$ and Eq. (6.99) becomes

$$E(r,z,t) = \sum_{n=-\infty}^{\infty} C\, a_n\, J_0(k_{rn}r)\, e^{j(\omega t - (\beta_0 + \frac{2\pi n}{d})z)}. \tag{6.104}$$

The constant a_n is called the amplitude of the nth "space-harmonic".

Individual terms in expansion (6.104) are much like the propagating modes in a uniform cylindrical waveguide; the dispersion relation for each of the terms is given by Eq. (6.103). The wavenumbers β_n can be recognized as the intersections of the horizontal line in Figure 6.15 with the lower dispersion curve (extended). The phase velocities of these intersections fall at

$$v_{pn} = \frac{\omega_{op}}{\beta_0 + 2\pi n/d}. \tag{6.105}$$

Several observations can be made concerning the space harmonics:

- The operating frequency will normally be chosen so that the $n = 0$ term provides the dominant acceleration, and has the required phase velocity, $v_{p0} = c$.

- According to Eq. (6.105) and Figure 6.15 it is then only the $n = 0$ term that meets the requirement for cumulative acceleration.

- The intersections discussed so far are with portions of the dispersion curve shown as solid in Figure 6.15. By the periodicity requirement the slopes at these points are all identical, and positive, meaning their group velocities are positive. Intersections with portions of the curve drawn with broken lines represent points with negative group velocities. One expects power to enter at the same end as the electrons so the linac wave energy travels in the same direction as the electrons. This corresponds to the solid curve intersections. But, in principle, the linac can be driven from the downstream end, in which case the broken sections of curve would be appropriate. Or there can be energy flowing in both directions, due, for example, to reflection at the downstream end, then all intersections would be relevant.

- To apply the definition of shunt impedance it is appropriate to adjust the constant a_0 in Eq. (6.104) to match V_0/d in Eq. (6.7).

- In general the longitudinal acceleration is a function of the radial position r of the beam being accelerated. For real k_{rn}, $|E_z|$ falls with increasing r. But the factor k_{rn}^2 in Eq. (6.103) will be negative for certain values

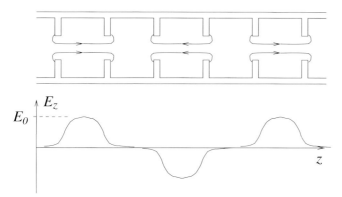

Fig. 6.18 On-axis, fixed-time snapshot of longitudinal dependence of electric field pattern for $\pi/2$ mode.

of n other than $n = 0$. For these solutions the J_0 Bessel function is replaced by a modified Bessel function I_0; this causes the field amplitude to increase with, rather than decrease with increasing r.

- But, for the most important case, namely $n = 0$, with v_g adjusted to be equal to c, there is a very fortuitous result; namely, $k_{r0} = k^2 - \beta_0^2 = 0$. Then the radial function is proportional to $J_0(0) = 1$, and the accelerating field is independent of r.

Nowadays all these calculations can be performed numerically. In the early days of linacs it was necessary to perform the calculations using analytical, series expansion approximations. A procedure developed by Walkingshaw in 1948 is suggested by Figure 6.16. In the region labeled I, the general electric field can be represented by a standard uniform circular wave guide expansion based on J_n Bessel functions. In regions II a similar expansion (but different because other Bessel functions are allowed) expansion holds. As well as matching the boundary conditions at the conducting boundaries, these series have to be matched along the boundary between I and II. An example field pattern for the $\pi/2$ mode is exhibited in Figure 6.18.

A complication that has not even been mentioned is the possible existence of fields in the cavities that are oscillating at frequencies other than the drive frequency. Such oscillations can occur only because they are driven by the beam itself. They are known as "higher modes" and are represented in the Brillouin diagram by higher pass band curves. Though unimportant for small beam currents, because their drive is proportional to the beam current, they become increasingly important with increasing beam current. They may set the limit of maximum possible beam current, for example by causing transverse emittance growth.

7
Undulator Radiation

7.1
Preview

An undulator (or wiggler) is a magnet with periodic field reversals along its length. Being a storage ring "insertion device", its purpose is to produce large on-axis radiation intensity while causing negligible net deflection of the electron beam. An undulator/wiggler can be characterized by three parameters: λ_w, the longitudinal wavelength of the magnetic field pattern; N_w, the number of periods; and K, the "undulator parameter", where $\Theta = K/\gamma$ is the maximum angle a passing electron makes with the undulator axis. The radiation pattern depends strongly on K and is considerably simpler to analyze for $K \lesssim 1$; magnets operating in this region have come to be known as "undulators" while magnets for which $K \gg 1$ are known as "wigglers". To me this distinction seems silly since, just by cranking up its magnetic field, an undulator becomes a wiggler. (So the magnets being described will usually be referred to as wiggler magnets, which is why the subscripts used are "w".) Furthermore the equations governing wigglers and undulators are the same and the essential physical phenomenon (coherence between radiation from different poles) is common to both devices.

What can be true, however, is that the mathematical approximations used to make the formulas manageable may be different for wigglers and undulators. The early sections of this chapter assume arbitrary K values, but this material is intended more to be instructive than to provide definitive formulas on which designs can be based. Problems based on this material are intended to provide insight into the essence of undulator radiation without being encumbered by intractable mathematical complication. Following this are analytic results which are more soundly based and accurate, but valid only in the $K \lesssim 1$ undulator regime. Finally will come analytical formulas that are valid for arbitrary values of K. They are simple in the sense of being analytic expressions in closed form, but complicated in the sense that the expressions are too complicated to be evaluated "by hand".

An important ingredient of the distinction between undulators and wigglers is the distinction (made in Eq. (3.26)) between "short" and "long" mag-

Accelerator X-Ray Sources. Richard Talman
Copyright © 2006 WILEY-VCH Verlag GmbH & Co. KGaA, Weinheim
ISBN: 3-527-40590-9

nets. For a "long" magnet the full radiation "headlight" pattern sweeps past an observation point while the electron remains in the same magnet. For a "short" magnet this is not the case. So an undulator consists of "short" magnets while a wiggler consists of "long" magnets. As a result the wiggler radiation more nearly resembles the radiation from arc magnets analysed in Chapters 3 and 5. But interference between the field components emitted from different poles, which is fundamental to undulator operation, also influences wiggler radiation. This makes it sensible to analyse wigglers and undulators together. As time goes on, because of their cleaner beams, more and more x-ray beamlines are being based on undulators. This will be especially true of fourth generation light sources in which only undulator radiation provides the desired high degree of coherence.

As in other parts of the text, for pedagogical purposes, some of the assumptions are extreme enough to approach oversimplification. However, starting with Section 7.10, the formulas are, as far as I know, as accurate and as general as are known, short of detailed numerical integration over measured field profiles.

One surprising feature of undulator radiation is that it is "the same as" Compton scattered radiation. This connection is spelled out in Chapter 12. Apart from developing the equivalence of Compton scattering and undulator radiation, emphasis in that chapter is on the use of "laser wire" apparatus that detects scattered visible laser light to diagnose ultra-small emittance electron beams. In laser wire setups, even with high power lasers, the wavelengths are so short that the undulator parameter K is much less than 1. This is the limit in which accurate analytical undulator radiation formulas are fairly easy to obtain. Many formulas in the present chapter therefore apply to Compton scattering (which is usually referred to as Thomson scattering in this long wavelength limit.)

7.2
Introduction

Some useful undulator references are Attwood, Halbach, and Kim [1], and Jackson [2] (who, though he treats ideal undulator radiation thoroughly, restricts treatment to the $K \ll 1$ case), and Kim [4].

The word "frequency" (or its corresponding wavelength) seems most apt when discussing interference of x-rays, while the word "energy" seems most apt when discussing the detection of individual x-ray photons. Of course the Planck formula guarantees the essential equivalence (except for units) of frequency and energy, and the two terms will be used interchangeably.

The instantaneously radiated power P_γ from a charged particle in a magnetic field has been given earlier in Eq. (5.2). This same formula is valid instantaneously within every single millimeter of an undulator or wiggler. Furthermore, in gross terms, the typical angles and typical energies of the emitted radiation are much the same as for the radiation from the arc magnets of the storage ring. But, instead of the "white" spectrum from a long magnet, interference effects cause the radiated power from an undulator to be concentrated into a "line spectrum" consisting of a fundamental frequency and its harmonics. Since the fractional bandwidths of these bands are of order $1/N_w$, the pattern evolves from a continuum spectrum to a spectrum of narrow lines as N_w increases from $1/2$ to large N_w. Most x-ray experiments require the incident radiation to be nearly monochromatic. The line spectrum produced by an undulator is one thing that makes the undulator an attractive insertion device for these experiments.

The range of wavelengths usefully produced at a light source is vast and the choice of topics likely to be of interest depends on what portion of this range is of special interest. For now I assume that x-ray diffraction is the subject of greatest interest. There is a fairly narrow band of energies that is ideal for x-ray diffraction. The band is limited on the high energy side by difficulty in making optical elements in that range, by excessive heating, by long term radiation damage, and by unwelcome backgrounds. The low energy limit is due to excessive attenuation in vacuum windows, protective covers and thick samples—the attenuation length of a 1 keV photon is so short as to cause unacceptable attenuation in anything other than vacuum. But, because of the extremely rapid energy dependence of attenuation length, a modest energy increase largely overcomes this problem. One therefore seeks a photon beam centered on (to pick an arbitrary but convenient nominal value) $\mathcal{E}_\gamma = 12.4\,\text{keV}$,[1] as brilliant as possible, consistent with being as monochromatic as possible.

With electron energy $\mathcal{E}_e$ limited to some maximum value by cost considerations, making the undulator period short enough can concentrate the beam power near this useful spectral range. But operational storage ring practicalities set a lower limit on λ_w—a magnetic undulator with ideal radiation properties commonly has a gap height too small for satisfactory operation at existing storage rings. This, in turn, sets an upper limit to the energy of the fundamental undulator harmonic. By increasing K, the intensity going into high orders can be increased but, as K increases into the wiggler regime, although the radiation pattern continues to consist only of lines, hundreds of significant lines emerge, so closely spaced that, for most purposes, the spectrum has again become effectively a continuum. Furthermore the resulting beam has undesirably high power relative to the flux of x-rays actually being used.

1) The choice of $\mathcal{E}_\gamma = 12.4\,\text{keV}$ as nominal energy corresponds to a wavelength $\lambda_{1,\text{edge}} = 1\,\text{Å}$, and to the (mnemonic) approximation
1Å → 12345 eV.

7 Undulator Radiation

The qualitative idea behind undulator radiation is familiar from the pattern produced by an optical diffraction grating having a large number $2N_w$ of identical slits. Let us recall that subject. The individual slits of a diffraction grating are so narrow that their diffraction patterns are very broad and the slits are so closely spaced that the diffraction patterns are very nearly superimposed. For most emergent angles the large spread of phases of the individual components causes almost complete destructive interference. But there are particular angles for which all components are in phase, causing narrow, intense, constructive maxima. The result is a spectrum of a few peaks having widths much less than their separations. The angular widths of these peaks are proportional to $1/N_w$ and the total energy coming through the slits is proportional to N_w. All this energy has to go somewhere, and somewhere, in this case, is the few lines. For fixed slit spacing the spacing of these lines is independent of the number of slits. The flux into each line is therefore proportional to N_w. This is the factor appropriate for estimating the total rate of photons in one grating order. Since the line width is proportional to $1/N_w$, the intensity at the line center is proportional to N_w^2. This factor can be understood as the resultunt N_w electric field vectors, all in phase, and squaring the result to obtain an intensity.

Undulator radiation exhibits analogous behavior. The primary element of a conventional undulator is a magnet having a large number $2N_w$ of magnetic poles, alternately north and south, with period λ_w. The trajectory of an electron through this magnet oscillates transversely about a straight central line, and this transverse acceleration of the electron results in synchrotron radiation. Though the radiation from any individual electron is incoherent with the radiation from any other, the waves emitted from any one electron during deflections at different locations interfere coherently. The fundamental interference maximum occurs when (because of the electron's speed deficit relative to c) the electron lags the radiated field by exactly one wavelength in passing through one period of the undulator. (Neglecting angular dependence and path length excess) this yields a condition (to be derived and refined below)

$$\lambda_{1,\text{edge}} = \frac{\lambda_w}{2\gamma^2}, \tag{7.1}$$

giving $\lambda_{1,\text{edge}}$, the short wavelength edge of the first order diffraction maximum, in the $K \ll 1$ limit of ideal undulator operation. For numerical estimates in this chapter the value $\gamma = 10^4$, corresponding to 5.1 GeV operation, will be assumed. Then the choice $\lambda_w = 2\,\text{cm}$ yields $\lambda_{1,\text{edge}} = 1.0\,\text{Å}$, or about 12.4 keV.

Because of the inevitable fringing between the poles of a magnet and a correspondingly too-small gap height requirement, it is difficult for λ_w to be as small as required by Eq. (7.1). One can contemplate using higher order interference maxima but, since the electron's trajectory through a standard undulator is essentially sinusoidal, the higher orders are extremely weak unless $K > 1$.

Formula (7.1) can also be obtained using an elementary relativistic argument. In the rest frame of the electron the undulator period is foreshortened by a factor γ, thereby increasing the electron's transverse oscillation frequency, and hence the energy of the photons it radiates by the same factor. When these photons are observed in the laboratory, their wavelengths are further foreshortened by a factor having maximum value 2γ in the forward direction. The result is Eq. (7.1). In the laboratory the radiation pattern extends from almost zero frequency up to the frequency corresponding to wavelength $\lambda_{1,\text{edge}}$.

This "scattering" approach is taken in Chapter 12. Working in the rest frame of the electron, where the undulator provides a sinusoidally varying transverse electric field, the electron's response is simple harmonic. In the electron rest frame the resulting (electric dipole radiation) is exactly monochromatic and, more or less isotropic. When viewed in the lab this radiation is no longer monochromatic. In fact it is distributed more or less uniformly in energy, from the peak value given above in the forward direction, down to essentially zero in the backward direction. But, because of the one-to-one relationship between energy and angle, the radiation viewed at fixed angle is monochromatic.

There is a widely shared impression that the radiation from $2N_w$ poles of an undulator is more forward-peaked (angular width proportional, say, to $1/N_w$) than would be the radiation that would come from any one of the poles in isolation. If the discussion of the previous paragraph is correct (and it is) then this impression is incorrect. The number of poles did not even enter that discussion.

Nevertheless, increasing N_w has a major beneficial effect for x-ray experiments. Dependence on N_w enters because, as viewed in the electron's frame, the undulator field is switched on for only a finite time—precisely N_w cycles in fact. Expressing the field as a wave packet, its fractional frequency width is approximately $1/N_w$. This fractional spread is inherited by the energy spectrum of the photons radiated in the electron's rest frame and passed on to the laboratory x-ray spectrum. The fractional energy spread at fixed angle in the lab inherits this same value. Though this might seem to be quite narrow, especially for large N_w, it is still far larger than the energy bandwidth required for typical x-ray experiments. To obtain such narrow bandwidth, the beam is passed through a "monochromator" (see Chapter 9). When the pass band of the monochromator is adjusted to coincide with the energy at the undulator resonance maximum, only an "undulator spike" is transmitted. This peak can be very narrow (with cone angle proportional to $1/N_w$. This is the source of the above-mentioned impression that the angular distribution of synchrotron radiation is made narrow by an undulator.

Another (equivalent) approach to the study of undulator radiation is to treat the undulator as a beam of "virtual photons" which Thomson scatter from the electrons in the electron beam. This approach is also fully discussed in Chapter 12.

The approach taken in this chapter, like that in Chapter 3 and Chapter 5, is to apply the retarded time formalism, working entirely in the laboratory frame of reference.

7.3
Electron Orbit in a Wiggler or Undulator

The magnetic field of an undulator has the form $B(z)\hat{\mathbf{y}}$, where $B(z)$ is sinusoidal with period λ_w.[2] An electron trajectory in this field is illustrated in Figure 7.1. The equation of motion in this field, of a particle whose orbit lies in the x, z plane is

$$m\gamma \frac{\mathrm{d}}{\mathrm{d}t}(v_x \hat{\mathbf{x}} + v_z \hat{\mathbf{z}}) = qB(z)(v_x \hat{\mathbf{x}} + v_z \hat{\mathbf{z}}) \times \hat{\mathbf{y}}, \qquad (7.2)$$

which yields

$$\frac{\mathrm{d}v_x}{\mathrm{d}t} = \frac{-qB(z)}{m\gamma} v_z, \quad \frac{\mathrm{d}v_z}{\mathrm{d}t} = \frac{qB(z)}{m\gamma} v_x. \qquad (7.3)$$

The constancy of γ can be expressed as $v_x dv_x + v_z dv_z = 0$, which is consistent with these equations. By using z instead of t as independent variable, Eqs. (7.3) become

$$\frac{\mathrm{d}v_x}{\mathrm{d}z} = \frac{-qB(z)}{m\gamma}, \quad \frac{\mathrm{d}v_z^2}{\mathrm{d}z} = \frac{2qB(z)}{m\gamma} v_x. \qquad (7.4)$$

Integrating the first of these, with the origin placed at a position z_0 where $v_x(z_0) = v_{x0}$, yields

$$v_x = v_{x0} - \frac{q}{m\gamma} \int_{z_0}^{z} B(z')\,\mathrm{d}z', \quad v_z^2 = v^2 - v_x^2. \qquad (7.5)$$

In practical wigglers the magnetic profile varies from "square saw-tooth" for large λ_w, to essentially sinusoidal for small λ_w. The undulator parameter K is traditionally defined assuming sinusoidal magnetic field, and related to the

---

2) It is more valid to say that the undulator magnetic field is periodic rather than sinusoidal, but a customary approximation is to treat the field as sinusoidal. The discussion of wiggler-dominated rings in Section 5.7 should provide warning that this approximation can introduce serious error.

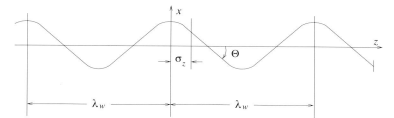

Fig. 7.1 The trajectory of an electron passing through an undulator magnet. For approximate purposes the deflecting force can be treated as a sequence of alternating sign impulses, each characterized by approximate half-width $\sigma_z = \lambda_w/(2\pi)$ and each bending through angle 2Θ. In the "ideal undulator limit", Θ is small compared to the half-angle of a cone containing most of the synchrotron radiation; i. e. $K \ll 1$.

peak magnetic field B_0. Then, performing the integral in Eq. (7.5) over a half-period in which the field is positive, yields

$$2\Theta \equiv \frac{2K}{\gamma} = \frac{\Delta v_x}{c} = \frac{e}{mc\gamma} B_0 \frac{\lambda_w}{2} \frac{2}{\pi}, \tag{7.6}$$

which reduces to

$$K = \frac{eB_0}{mc} \frac{\lambda_w}{2\pi} = 93.4 \, B_0[\text{T}] \lambda_w[\text{m}]. \tag{7.7}$$

One reason K is such a useful parameter is that it is independent of electron energy. The K value of a given undulator remains constant as the machine energy changes, or even if the undulator is moved from one accelerator to another. For electromagnets the K value is altered by controlling B_0 via the supplied current. For permanent magnets the K value is altered by controlling B_0 via the gap height.

It will simplify some calculations (especially in cases where the total undulator length is comparable with the distance to the observation point) if we can suppose that the deflecting element is "very short", in the sense defined earlier. (See Eq. (3.26).) In fact, as well as being short, (to avoid the need for step functions which would in any case be unachievable physically) the deflecting field of a single pole can be approximated as having a Gaussian longitudinal profile such that the curvature (inverse bending radius) is given by[3]

$$\frac{1}{R(z)} = \frac{1}{R_w} \exp\left(-\frac{z^2}{2\sigma_z^2}\right). \tag{7.8}$$

[3] What with fringe fields being inevitable, treating the field shape of a short magnet as Gaussian could provide an accurate approximation, but we are intending to apply these formulas eventually to a sinusoidal undulator orbit.

In passing one quarter of a wiggler period, an electron's angular deflection is

$$\Theta = \frac{1}{R_w} \int_0^\infty \exp\left(-\frac{z^2}{2\sigma_z^2}\right) dz = \sqrt{\frac{\pi}{2}} \frac{\sigma_z}{R_w} \equiv \frac{K_G}{\gamma}, \tag{7.9}$$

where the "effective wiggler strength parameter" K_G, with "G" for Gaussian, has been introduced to distinguish it from the sinusoid-based wiggler parameter K introduced above. Though K_G is approximately equal to K for the same maximum angle Θ, K and K_G are numerically interchangeable only for semi-quantitative purposes.

A sinusoidal orbit having maximum slope K/γ is

$$x = \frac{\lambda_w}{2\pi} \frac{K}{\gamma} \cos \frac{2\pi z}{\lambda_w}, \tag{7.10}$$

This curve can be matched approximately by choosing R_w and σ_z appropriately, using the approximation

$$\sin x \approx -\sum_{i=-\infty}^{\infty} (-1)^i \exp\left(-\frac{(x+(2i+1)\pi/2)^2}{2}\right). \tag{7.11}$$

This is illustrated in Figure 7.2, and the parameters are related by

$$\sigma_z = \frac{\lambda_w}{2\pi}. \tag{7.12}$$

For our nominal $\lambda_w = 2$ cm wavelength, $\sigma_z = 3.2$ mm.

With this (quite unconventional) approximation, the ends of the undulator can be represented by simple truncation of the sum in Eq. (7.11). Also, after having sliced the undulator longitudinally in this way, coherent superposition can be handled by the vector addition of phasors, one per deflection arc—$2N_w$ in all. By using a Gaussian shape, the artificial high frequency components that would accompany using truncated half-sinusoids are largely suppressed. For long undulators it may be appropriate to incorporate longitudinal dependence by making the phasor magnitude depend on longitudinal position.[4]

Problem 7.3.1 *With radius R_w corresponding to peak field B_0, if the Gaussian width parameter in Eq. (7.8) is taken to be $\sigma_z = \lambda_w/(2\pi)$, K is defined as in Eq. (7.6), and K_G by Eq. (7.9), show that $K_G \approx \sqrt{\pi/2}\, K$. So $K_G \approx K$ for semi-quantitative purposes.*

[4] The magnet representation by a phasor sum over single-pole contributions will prove inappropriate for analyzing wiggler radiation. For $K > 1$ it will prove to be more appropriate, and mathematically simpler, to first work out the amplitude from a full wiggler period. Then the radiation from N_w periods can be worked out by an appropriately different phasor summation.

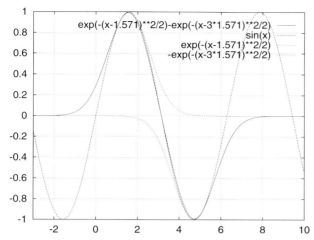

Fig. 7.2 Plot illustrating a sinusoid matched by a series of alternate sign Gaussians, of which only two are shown. See key in upper right for analytic forms. The standard deviations are related to undulator wavelength by $2\pi\sigma_z = \lambda_w$ in order to match curvatures at the peaks when $K \ll 1$. Though the true trajectory is sinusoidal, radiation integrals are dominated by high curvature regions, so the radiation deficiency or excess from the tail regions will be relatively insignificant.

Problem 7.3.2 *Show that trajectory (7.10) leads to a more-realistic-than-Eq. (7.8) version of the curvature of a sinusoidal orbit*

$$\frac{1}{R(z)} = \frac{2\pi}{\lambda_w} \frac{K}{\gamma} \frac{\cos(2\pi z/\lambda_w)}{(1 + (K/\gamma)^2 \sin^2(2\pi z/\lambda_w))^{3/2}}. \tag{7.13}$$

Also show that the maximum curvature matches the maximum curvature of Eq. (7.8) if Eq. (7.12) holds and the K^2 term in the denominator can be neglected. That is $K \ll 1$.

7.4
Energy Radiated From one Wiggler Pole

According to Sands [6] (and our Eq. (5.2) is equivalent) the energy dissipated per unit length in a region with bending radius R_0 is given by

$$\frac{dU}{dz} = \frac{q^2 \gamma^4}{6\pi\epsilon_0 R_0^2}. \tag{7.14}$$

This is sometimes known as "Schott's formula", though it is due to Liénard. If the electron trajectory is modeled as a sequence of circular arcs this formula can be used directly to give the energy radiated per pole to be $U_1 = \frac{dU}{dz} \frac{\lambda_w}{2}$.

But, not accounting for the reduced field between poles, this could give a substantial over-estimate. An accurate value of U_1 is easily obtained by numerical integration of the actual electron orbit. The energy U_{G1} radiated by an electron in traversing our thin "Gaussian single-pole" is given by

$$U_{G1} = \frac{q^2 \gamma^4}{6\pi \epsilon_0 R_w^2} \int_{-\infty}^{\infty} \exp\left(-\frac{z^2}{\sigma_z^2}\right) dz = \frac{1}{6\sqrt{\pi}} \frac{q^2 \gamma^4}{\epsilon_0} \frac{\sigma_z}{R_w^2}. \tag{7.15}$$

Problem 7.4.1 *Assuming $K \ll 1$, show that the energy U_1' radiated from a single undulator pole for which the curvature is given by Eq. (7.13), is*

$$U_1' = \frac{\pi}{24} \frac{q^2 \gamma^4}{\epsilon_0} \frac{\sigma_z}{R_w^2} + O(K^4), \tag{7.16}$$

where λ_w has been replaced using Eq. (7.12), and the last of Eqs. (7.9) has been used. Comparing with U_{G1} as given by Eq. (7.15), the ratio of the results is $(1/6\sqrt{\pi})/(\pi/24)$, which works out to about 0.72.. This discrepancy (which is at least partially due to ambiguity in the definition of K) means the Gaussian pulse value underestimates total energy radiated by some 28%.

To relate the energy emitted from a single pole to familiar quantities, U_{G1} can be expressed either as a fraction of U_0 (the energy radiated as an electron travels in a complete circle of radius R_0)[10] or as a function of undulator parameter K;

$$U_{G1} = \frac{1}{2\sqrt{\pi}} \frac{R_0 \sigma_z}{R_w^2} U_0 = \frac{1}{\pi^{3/2}} \frac{U_0 R_0}{\sigma_z} \Theta^2 = \frac{1}{\pi^{3/2}} \frac{C_\gamma \mathcal{E}_e^4}{\lambda_w/(2\pi)} \left(\frac{K}{\gamma}\right)^2. \tag{7.17}$$

The essential qualitative feature of this formula is that the radiation comes in $2N_w$ pulses of energy, each with energy given by Eq. (7.17). All that remains is to determine how this energy is distributed in direction and wavelength (or, preferably, into photon energies.) The total energy radiated from the wiggler/undulator is approximately $2N_w U_{G1}$, irrespective of interference effects.

7.5
Spectral Analysis for Arbitrary Longitudinal Field Profile

The orbit and detection point P geometry is illustrated in Figure 7.3. At $t_r = 0$ the electron passes O and its velocity vector points toward the origin O' in the

---

10) Application of Eq. (7.14), for example to obtain the energy radiated as a 5.1 GeV electron completes a full circle of radius $R_0 = 89$ m, yields $U_0 = 0.67$ MeV. Other numerical estimates can be scaled to U_0. Though this is artificial, it has mnemonic value, since it relates quantities to that feature of synchrotron radiation which imposes itself most emphatically upon the operation of storage rings—the average energy loss. For accurate calculation one should use $U_0 R_0 = C_\gamma \mathcal{E}_e^4$, where $C_\gamma = 0.885 \times 10^{-4}$ m GeV^{-3}; this formula gives $U_0 R_0$ in units of m GeV.

7.5 Spectral Analysis for Arbitrary Longitudinal Field Profile

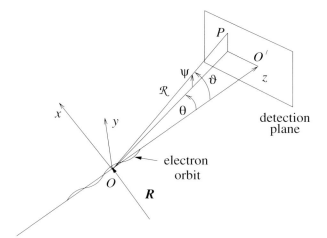

Fig. 7.3 Orbit geometry and definition of horizontal angle θ, vertical angle ψ, and polar angle ϑ, locating the detector position P relative to the $t_r = 0$ point on the central axis.

plane of detection. As in Chapter 4, the zero of time t at observation point P is adjusted to correspond to arrival time at P of a photon emitted at $t_r = 0$ at emission point O.

The detection point P is displaced from O' by horizontal angle θ and vertical angle ψ. The instantaneous radius of curvature is $R(z)$. The most important difference of the geometry here, from the bending magnet configuration of Chapter 4, concerns the placement of the origin O' in the detector plane. Now it lies on the undulator centerline extended. In Chapter 4 the origin O' lay on a tangent to the circular orbit, at arbitrarily chosen angle θ_h, and the observation point P lay directly above O', subtending angle ψ at O. Now the detector point P subtends vertical angle ψ and horizontal angle θ at O. So the undulator intensity depends on both ψ and θ. This differs from the case of pure circular motion in which the intensity depends on ψ but is independent of θ_h.

The formalism to be employed has been described in Chapter 3. The fundamental relationship governing synchrotron radiation is Eq. (3.3), which relates observation time t to "retarded time" t_r;

$$t = t_r + \frac{\mathcal{R}}{c}, \tag{7.18}$$

where $\mathcal{R}$ is the distance from source point to observation point P. Other formulas needed from Chapter 3 are

$$t \approx t_r \left(\frac{1}{2\gamma^2} + \frac{\vartheta^2}{2} + \frac{c^2 t_r^2}{6R^2} \right), \tag{7.19}$$

where $\vartheta = \sqrt{\theta^2 + \psi^2}$ is the angle between $\mathbf{v}$ and $\widehat{\mathcal{R}}$, and

$$\frac{dt}{dt_r} = 1 - \widehat{\mathcal{R}} \cdot \frac{\mathbf{v}(t_r)}{c}. \tag{7.20}$$

In terms of the instantaneous radius of curvature R, the electric field is given by Eq. (3.14);

$$\mathbf{E}(\mathbf{r},t) = \frac{q}{4\pi\epsilon_0 c} \frac{1}{\mathcal{R}} \left(\frac{\widehat{\mathcal{R}} \times ((\widehat{\mathcal{R}} - \mathbf{v}/c) \times \dot{\mathbf{v}}/c)}{(1 - \widehat{\mathcal{R}} \cdot \mathbf{v}/c)^3} \right)_{\text{ret.}} \tag{7.21}$$

Using Eq. (3.35), its transform is given by

$$\widetilde{\mathbf{E}}(\omega) = \frac{q}{4\pi\epsilon_0 c \mathcal{R}} \frac{1}{\sqrt{2\pi}} \int_{-\infty}^{\infty} e^{i\omega t} \left(\frac{\widehat{\mathcal{R}} \times ((\widehat{\mathcal{R}} - \mathbf{v}/c) \times \dot{\mathbf{v}}/c)}{(1 - \widehat{\mathcal{R}} \cdot \mathbf{v}/c)^3} \right)_{\text{ret.}} dt. \tag{7.22}$$

Changing the integration variable from t to t_r and replacing the dt/dt_r factor using Eq. (7.20) yields

$$\widetilde{\mathbf{E}}(\omega) = \frac{q}{4\pi\epsilon_0 c \mathcal{R}} \frac{1}{\sqrt{2\pi}} \int_{-\infty}^{\infty} e^{i\omega t(t_r)} \frac{\widehat{\mathcal{R}} \times ((\widehat{\mathcal{R}} - \mathbf{v}/c) \times \frac{\dot{\mathbf{v}}}{c})}{(1 - \widehat{\mathcal{R}} \cdot \frac{\mathbf{v}}{c})^2} dt_r. \tag{7.23}$$

We will use the result

$$\frac{\widehat{\mathcal{R}} \times ((\widehat{\mathcal{R}} - \mathbf{v}/c) \times \dot{\mathbf{v}}/c)}{(1 - \widehat{\mathcal{R}} \cdot \mathbf{v}/c)^2} = \frac{d}{dt_r} \frac{\widehat{\mathcal{R}} \times (\widehat{\mathcal{R}} \times \mathbf{v}/c)}{1 - \widehat{\mathcal{R}} \cdot \mathbf{v}/c} \tag{7.24}$$

Since this formula is critical, I exhibit the main step in its derivation:

$$\frac{d}{dt_r} \frac{\widehat{\mathcal{R}} \times (\widehat{\mathcal{R}} \times \mathbf{v})}{1 - \widehat{\mathcal{R}} \cdot \mathbf{v}} = \frac{\widehat{\mathcal{R}} \times (\widehat{\mathcal{R}} \times \dot{\mathbf{v}})}{1 - \widehat{\mathcal{R}} \cdot \mathbf{v}} + \frac{\widehat{\mathcal{R}} \times (\widehat{\mathcal{R}} \times \mathbf{v})}{(1 - \widehat{\mathcal{R}} \cdot \mathbf{v})^2} \widehat{\mathcal{R}} \cdot \dot{\mathbf{v}}$$

$$= \frac{\widehat{\mathcal{R}} \times (\widehat{\mathcal{R}} \times \dot{\mathbf{v}}) + (\mathbf{v} \cdot \widehat{\mathcal{R}}) \dot{\mathbf{v}} - (\dot{\mathbf{v}} \cdot \widehat{\mathcal{R}}) \mathbf{v}}{(1 - \widehat{\mathcal{R}} \cdot \mathbf{v})^2}, \tag{7.25}$$

where two terms have cancelled after expanding the triple cross product. Combined with the relation $\widehat{\mathcal{R}} \times (\mathbf{v} \times \dot{\mathbf{v}}) = (\widehat{\mathcal{R}} \cdot \dot{\mathbf{v}}) \mathbf{v} - (\widehat{\mathcal{R}} \cdot \mathbf{v}) \dot{\mathbf{v}}$, this yields Eq. (7.24). Since $\widehat{\mathcal{R}}$ is being held fixed in this calculation, we are neglecting the displacement of the electron off the undulator axis in the determination of $\widehat{\mathcal{R}}$. This assumes that $K\lambda_w \ll \mathcal{R}$, which will always be valid in practice. Substituting into Eq. (7.23), we obtain

$$\widetilde{\mathbf{E}}(\omega) = \frac{q}{4\pi\epsilon_0 c \mathcal{R}} \frac{1}{\sqrt{2\pi}} \int_{-\infty}^{\infty} e^{i\omega t(t_r)} \frac{d}{dt_r} \left(\frac{-\mathbf{v}_\perp/c}{1 - \widehat{\mathcal{R}} \cdot \mathbf{v}/c} \right) dt_r, \tag{7.26}$$

with $t(t_r)$ given by Eq. (7.19). This integral can be further simplified using integration by parts to yield

$$\frac{\widetilde{\mathbf{E}}(\omega, \theta, \psi)}{\frac{q}{4\pi\epsilon_0 c \mathcal{R}}} = \frac{i\omega}{\sqrt{2\pi}} \int_{-\infty}^{\infty} \exp\left(i\omega t_r \left(\frac{1}{2\gamma^2} + \frac{\vartheta^2}{2} + \frac{c^2 t_r^2}{6R^2} \right) \right) \frac{\mathbf{v}_\perp}{c} dt_r. \tag{7.27}$$

In this step the factor dt/dt_r has been replaced using Eq. (7.20). In spite of its relatively simple appearance this formula is (a) valid in considerable generality and (b) quite difficult to evaluate. Being valid for the radiation from a magnetic field of arbitrary longitudinal profile, it applies to undulators and wigglers having an arbitrary number of poles and arbitrary K-values. It will be the basis for calculating the spectrum from a single pole, in the next section, and from a complete undulator/wiggler in a later section.

It has been mentioned earlier that it is the frequency spectrum that is the main feature distinguishing undulator radiation from single magnet radiation. Nevertheless we will continue to pretend that a single pole can be analysed in isolation, and start by calculating the frequency spectrum of its radiation.

7.6
Spectrum of the Radiation from a Single Pole

7.6.1
Orbit Treated as Arc of Circle

Since radiation from an electron following a circular orbit has already been analysed in Chapters 3 and 5, this section is somewhat redundant. Proceeding as in Eq. (3.17), some of the quantities entering Eq. (7.27) are

$$\widehat{\mathcal{R}} \approx \theta \hat{\mathbf{x}} + \psi \hat{\mathbf{y}} + \left(1 - \frac{\theta^2}{2}\right) \hat{\mathbf{z}},$$

$$\frac{\mathbf{v}}{c} \approx -\sin \frac{vt_r}{R} \hat{\mathbf{x}} + \cos \frac{vt_r}{R} \hat{\mathbf{z}},$$

$$\widehat{\mathcal{R}} \cdot \frac{\mathbf{v}}{c} \approx -\theta \sin \frac{vt_r}{R} + \cos \frac{vt_r}{R},$$

$$\frac{\mathbf{v}_\perp}{c} = \frac{\mathbf{v}}{c} - \left(\widehat{\mathcal{R}} \cdot \frac{\mathbf{v}}{c}\right) \widehat{\mathcal{R}} \approx -\left(\sin \frac{vt_r}{R} + \theta \cos \frac{vt_r}{R}\right) \hat{\mathbf{x}} - \psi \cos \frac{vt_r}{R} \hat{\mathbf{y}}. \quad (7.28)$$

The only important difference from Eq. (3.17) concerns the introduction of θ as the non-centered position of observation point P.

It seems to be conventional to call the x-component the "σ-mode" and the y-component the "π-mode", but I will continue to use x and y. From Eq. (7.27), making small angle approximations, one obtains

$$\frac{\widetilde{E}_x(\omega,\theta,\psi)}{\frac{-i\omega q}{\sqrt{2\pi}4\pi\epsilon_0 c \mathcal{R}}} = \int_{t_{r,\text{in}}}^{t_{r,\text{out}}} dt_r \exp\left(i\omega t_r \left(\frac{1}{2\gamma^2} + \frac{\vartheta^2}{2} + \frac{c^2 t_r^2}{6R^2}\right)\right) \left(\sin\frac{vt_r}{R} + \theta\right)$$

$$\approx \int_{t_{r,\text{in}}}^{t_{r,\text{out}}} dt_r \left(\frac{ct_r}{R(t_r)} + \theta\right) \exp\left(i\omega t_r \left(\frac{1}{2\gamma^2} + \frac{\vartheta^2}{2}\right)\right)$$

$$\frac{\widetilde{E}_y(\omega,\theta,\psi)}{\frac{-i\omega q}{\sqrt{2\pi}4\pi\epsilon_0 c \mathcal{R}}} = \int_{t_{r,\text{in}}}^{t_{r,\text{out}}} dt_r \exp\left(i\omega t_r \left(\frac{1}{2\gamma^2} + \frac{\vartheta^2}{2} + \frac{c^2 t_r^2}{6R^2}\right)\right) \psi \cos\frac{vt_r}{R}.$$

$$\approx \int_{t_{r,\text{in}}}^{t_{r,\text{out}}} dt_r\, \psi \exp\left(i\omega t_r \left(\frac{1}{2\gamma^2} + \frac{\vartheta^2}{2}\right)\right), \qquad (7.29)$$

where $t_{r,\text{in}}$, $t_{r,\text{out}}$ are input and output retarded times.

Because the integrals are thoroughly dominated by regions in which $vt_r \ll R$ the trigonometric factors have been approximated by their leading terms. As written, for use in an approximate evaluation in the next section, the formulas admit the posibility of variable bending radius $R(t_r)$. This constitutes an approximation to be discussed below. It is consistent with the assumed circular orbit only if $K < 1$. (This is the only significant difference from Jackson's formulas Eqs. (14.79).) Otherwise the formulas are valid in general. For a short pole, with $K \ll 1$, the cubic retarded time correction terms can be dropped as shown in Eqs. (7.29). In the integrals of Eq. (7.29) the factors θ and ψ are constant, and could be moved outside the integral signs, but they have been left inside for ease of comparison.

A curious feature of the formulas (which is shared by Jackson's Eq. (14.79), is that the integrals proportional to θ and to ψ do not vanish in the limit $R \to \infty$. Though for practical parameter values these contributions are small compared to the ct_r/R contribution, this appears to predict radiation even with the magnet turned off. The explanation for this is that, if the magnet *were* turned off, the orbit would pass very close to the observation point which invalidates the radiation zone approximation which went into the original derivation.

7.6.2
Radiation from a Single, Short, Isolated, Magnet, $K \ll 1$

For our diffraction grating analogy we wish to treat an undulator as a sequence of alternating-sign single poles—one undulator pole will be the analog of one grating slit. If the magnetic field of an undulator is treated as a true square wave, with field reversing instantaneously at the interfaces, the formula derived in the previous section could be used to calculate the radiation field from a single pole. There are two things actually wrong with this, and one complication. Such a magnetic field violates Maxwell's equations and there would be unphysical high frequency components in the radiation corresponding to the discontinuities at the pole ends. (These unphysical contributions would, in principle, cancel when the fields of adjacent poles were added, but that would require mathematical care we are unwilling to commit to.) The complication is that the formulas are very complicated.

We therefore proceed to model the orbit through the undulator as a sequence of alternating-sign, bell-shaped curves, as is suggested by Figure 7.2. This overcomes both of the problems and the complication mentioned in the previous paragraph, though at the cost of being applicable only in the $K \ll 1$ ideal undulator limit. (See Problem 7.3.2.)

Substituting into Eq. (7.29), retaining only the dominant mode E_x and, accepting the short magnet ($K \ll 1$) limitation, yields

$$\frac{\tilde{E}_x(\omega, \vartheta)}{\frac{q}{4\pi\epsilon_0 R}} \approx -\frac{i\omega}{R_w}\sqrt{\frac{2}{\pi}} \int_0^\infty t_r \exp\left(-\frac{t_r^2}{2\sigma_z^2/c^2}\right) \sin\left(\omega t_r \left(\frac{1}{2\gamma^2} + \frac{\vartheta^2}{2}\right)\right) dt_r$$

$$= -\frac{i\omega}{R_w}\sqrt{\frac{2}{\pi}} \left(\frac{1}{2\gamma^2} + \frac{\vartheta^2}{2}\right)^{-2} \int_0^\infty t \exp\left(\frac{-t^2}{2\sigma_z^2/c^2 \left(\frac{1}{2\gamma^2} + \frac{\vartheta^2}{2}\right)^2}\right) \sin \omega t\, dt$$

$$= -\frac{i\omega}{R_w}\sqrt{\frac{2}{\pi}} \frac{\sigma_z^2}{c^2} a_\vartheta^2 \int_0^\infty t\, e^{-\frac{a_\vartheta^2 t^2}{2}} \sin \omega t\, dt$$

$$= -\frac{i}{R_w} \frac{\sigma_z^2}{c^2} a_\vartheta \frac{\omega^2}{a_\vartheta^2} \exp\left(\frac{-\omega^2}{2 a_\vartheta^2}\right) \quad (7.30)$$

where a function a_ϑ, with dimensions of frequency, has been defined by

$$a_\vartheta = \frac{a_0}{1 + \gamma^2 \vartheta^2} \quad \text{where} \quad a_0 = \frac{2\gamma^2 c}{\sigma_z} = \frac{4\pi \gamma^2 c}{\lambda_w}. \quad (7.31)$$

The spectrum given by Eq. (7.30), with $\vartheta = 0$, is plotted (modulo a multiplicative factor) in Figure 7.4. The radiated power is obtained from the Poynting vector which is proportional to the square of the function plotted. This gives the radiated power one has "to work with". As in an optical diffraction grating, though interference effects can "concentrate" the power into the form of photons centered on one or more diffraction peaks, such effects cannot alter the total power.

The parameter a_0 is a frequency typical of the radiation from a single deflection of r.m.s. length σ_z. Superficially a_0 might seem to play a role much like the "critical frequency", $\omega_c = (3/2) c \gamma^3 / R_0$, traditionally defined for synchrotron radiation from bend radius R_0, but this is misleading. Because of its cubic dependence on γ, one might be misled into believing that ω_c corresponds to a more "Lorentz contracted" and hence shorter wavelength than the wavelength $\lambda_0 = 2\pi c / a_0$, but this is wrong. In fact, the "short magnet

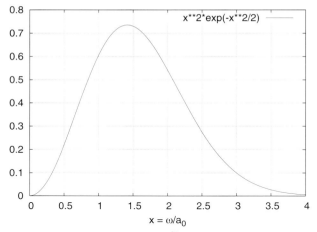

Fig. 7.4 (Shape of) Fourier transform $\tilde{E}_x(\omega, \vartheta = 0)$. As given by Eq. (7.30), the shape is independent of ϑ provided the horizontal axis is interpreted as ω/a_ϑ. There is appreciable response only up to about $\omega = 3a_\vartheta$. In the analogy between undulator and diffraction grating this frequency spectrum is the analog of the angular pattern from a single slit.

effect" makes the opposite true in a true undulator. Note also that, unlike ω_c, for which half the energy is above and half below, far more than half of the undulator energy comes in the form of photons with frequency higher than a_0. For a magnetic wiggler, though ω_c is indicative of the maximum magnetic field, and hence the total power radiated, the bandwidth of the single pole spectrum is independent of ω_c and depends only on a_0, which depends only on the pole width. The maximum of the single deflector Fourier transform can be seen to be located at $\omega = 1.45\, a_0$. This defines a_ϑ to be a kind of "typical" x-ray frequency at angle ϑ. It will prove to be significant that the spectral shape is a universal function of the ratio ω/a_ϑ. The energy radiated from a single undulator pole is distributed according to

$$\begin{aligned}\frac{d^2 U_{G1}}{d\omega d\Omega} &= \frac{2\mathcal{R}^2}{\mu_0 c} \tilde{E}_x(\omega)\tilde{E}_x(-\omega) \\ &= \frac{2\mathcal{R}^2}{\mu_0 c}\left(\frac{q}{4\pi\epsilon_0 \mathcal{R}}\right)^2 \left(\frac{\sigma_z^2}{R_w c^2}\right)^2 a_\vartheta^2 \frac{\omega^4}{a_\vartheta^4} \exp\left(-\frac{\omega^2}{a_\vartheta^2}\right) \\ &= \frac{1}{\pi^2 c^2}\frac{q^2}{4\pi\epsilon_0 c}\left(\frac{\lambda_w}{2\pi}\right)^2 \left(\frac{K}{\gamma}\right)^2 a_\vartheta^2 \frac{\omega^4}{a_\vartheta^4}\exp\left(-\frac{\omega^2}{a_\vartheta^2}\right) \end{aligned} \quad (7.32)$$

where the initial factor 2 accounts for the restriction of ω to positive values.

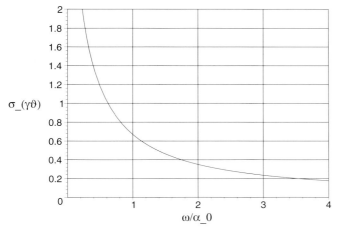

Fig. 7.5 The r.m.s. production angle $\sigma_{\gamma\vartheta}$ plotted versus normalized frequency ω/a_0 for the radiation from a single pole of an undulator, as given by Eq. (7.33). This graph is to be checked by Problem 7.6.1.

Problem 7.6.1 *The second moment of ϑ for fixed ω is defined by*

$$\sigma_\vartheta^2(\omega) = \frac{\int d\Omega\, \vartheta^2 \frac{d^2 U_{G1}}{d\omega d\Omega}}{\int d\Omega\, \frac{d^2 U_{G1}}{d\omega d\Omega}}. \tag{7.33}$$

Starting from Eq. (7.32), evaluate this ratio numerically and compare your result with Figure 7.5. To exploit inequality $\gamma \gg 1$, it is appropriate to extend the ϑ integration range from π to infinity.

Problem 7.6.2 *Check the consistency of the formulas by integrating $d^2 U_{G1}/d\omega d\Omega$, as given by Eq. (7.32), over ω and Ω to calculate the total energy radiated per deflection to be*

$$U_{G2} = \frac{3}{16\sqrt{\pi}} \frac{q^2 \gamma^4}{\epsilon_0} \frac{\sigma_z}{R_w^2}. \tag{7.34}$$

(Using ω/a_ϑ as the first integration variable seems to be the best approach. Also the relation $\gamma \gg 1$ should be exploited to extend the integration range of ϑ from π to ∞. Numerically $3/(16\sqrt{\pi}) = 0.1058$, not quite the same as $1/(6\sqrt{\pi}) = 0.0940$, which is the corresponding coefficient of U_{G1}. This disagreement is comparable with the inaccuracy of our other results. Note also that Eq. (7.34) exhibits the correct proportionality to K^2 for large K, in spite of its derivation's having assumed $K < 1$.

With the intended purpose of the undulator being to produce x-rays, it is appropriate to estimate typical wavelengths. The wavelength corresponding to a_0 is

$$\lambda_0 = 2\pi \frac{c}{a_0} = \frac{\lambda_w}{2\gamma^2}, \tag{7.35}$$

and comparison with Eq. (7.1) shows that $\lambda_0 = \lambda_{1,\mathrm{edge}}$. Order of magnitude equality of these quantities was to be expected, but *exact* equality is a coincidence. $\lambda_{1,\mathrm{edge}}$ is a characteristic of the radiation from a full periodic structure, while λ_0 is characteristic of the radiation from one half-wiggle.[6] With the wiggler field shape being treated as a sinusoid with wavelength equal to the wiggler period the approximate equality of λ_0 and $\lambda_{1,\mathrm{edge}}$ is assured. One consequence of their having comparable values is that there is a substantial flux of photons having wavelengths capable of constructive interference with the radiation from all the other "poles" of the wiggler.

Because positive and negative wiggles are being treated independently, the diffraction maxima occur at $\omega = n a_0$, where $n = 1, 3, 5, \ldots$. (This will be confirmed below.) From Figure 7.4 one sees that there is appreciable amplitude only up to two or three times a_0. One concludes that the amplitudes of diffraction maxima of order higher than, say, $n = 3$, will be negligible (for the case $K < 1$ being analysed.)

Wavelength λ_0 (corresponding to frequency a_0) has the remarkable feature of being independent of R_w, the central deflecting radius of curvature. The first person to emphasize the experimental significance of this feature was apparently Coisson [7]. Unlike regular arc radiation, the short magnet spectrum extends to high photon energies even if the deflection angle is arbitrarily small. (This refers to the spectral *shape*; of course the total intensity is miniscule for small deflections.) The theory has been amply corroborated at CERN, as part of diagnostics of the SPS, a 400 GeV proton accelerator, by Bossart et al. [8].

7.7
Coherence from Multiple Deflections

The coherence of amplitudes from different poles of an undulator depends critically on the average velocity of the electron. The electron's orbit will otherwise, for the time being, be treated as a straight line, with longitudinal velocity altered to account for the increased arc length of the actual (approximately sinusoidal) orbit;

$$\frac{\overline{v_z}}{c} \approx \sqrt{\frac{v^2}{c^2} - \overline{(\Theta \cos k_w z)^2}} \approx 1 - \frac{1}{2\gamma^2} - \frac{\Theta^2}{4}. \tag{7.36}$$

---
6) In principle the length of half-wiggle sections of a wiggler could be very short compared to the wiggler period. Then we would have $\lambda_0 \ll \lambda_{1,\mathrm{edge}}$, and high order diffraction maxima would become significant. The same would be true if the wiggler field shape were more nearly an ideal square wave.

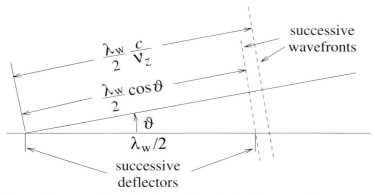

Fig. 7.6 Geometry illustrating the condition for interference maxima observed at angle ϑ.

A single electron is subject to $2N_w$ undulator pulses, of alternating polarity, with each pulse having r.m.s. (retarded time) duration given roughly by $2\sigma_z/c$. All radiation sources are centered on the same straight line. Consider a component of the radiation having wavelength λ and direction ϑ. A reference wavefront is defined to be the plane passing through the emission point, perpendicular the photon's direction. As the electron advances the distance $\lambda_w/2$ from one deflection to the next, its travel time is $(\lambda_w/2)/\bar{v}_z$. Meanwhile the reference wavefront (traveling at the speed of light in the photon's direction) has traveled a distance $(\lambda_w/2)(c/\bar{v}_z)$. Referring to Figure 7.6, consider another wavefront which is parallel to the original wavefront, but emerges from the new emission point. The distance of this wavefront from the first emitter is $(\lambda_w/2)\cos\vartheta \approx (\lambda_w/2)(1-\vartheta^2/2)$. The phase difference between these two wavefronts is

$$\Delta\phi(\vartheta) = 2\pi\frac{\lambda_w}{2}\frac{c/\bar{v}_z - 1 + \vartheta^2/2}{\lambda} = \pi\frac{\lambda_w/(2\gamma^2)}{\lambda}(1+K^2/2+\gamma^2\vartheta^2). \quad (7.37)$$

In a "Fraunhofer approximation", in which all emission at angle ϑ is "focused at infinity" the condition for the two waves to interfere constructively is that this phase shift be πn where n is any *odd* positive integer. That is, for $n = 1, 3, 5, \ldots$,

$$\begin{aligned}\lambda_n(\vartheta) &= \frac{\lambda_w/(2\gamma^2)}{n}(1+K^2/2+\gamma^2\vartheta^2), \quad \text{or} \\ \omega_n(\vartheta) &= \frac{n a_0}{1+K^2/2+\gamma^2\vartheta^2}.\end{aligned} \quad (7.38)$$

(The value of $\lambda_n(0)$ with $K=0$, $n=1$ has previously been denoted $\lambda_{1,\text{edge}}$.)

We see now that there is a trade-off between intensity and maximum energy edge. According to Eq. (7.17), the total energy radiated is proportional to K^2 while, according to Eq. (7.38), the resonant frequency tends toward proportionality to $1/K^2$ for large K. To get both high energy photons and high intensity requires going to values of n all the higher because of the $K^2/2$ term in the denominator of Eq. (7.38). This road leads to "wiggler" operation.

Due to betatron oscillation, the electron's angle θ_e, relative to the central axis, though small, is not zero, and there is a longer effective wiggler period, given by $\lambda_w \to \lambda_w/\cos\theta_e$. But, because λ_w appears only as a multiplicative factor in Eq. (7.38), this is a relatively insignificant effect.[7] For the same reason, though the term $K^2/2$ gives an (undesirable) shift to reduced energy, it does not cause the (typically more significant) energy broadening[8] and resultant brilliance reduction caused by the $\vartheta^2/2$ term.

It is not common, in practice, to demand a strictly ideal undulator $K \ll 1$ spectrum. To do so, and to limit the shift of interference maxima, it might be appropriate to require

$$\Theta \lesssim \frac{1}{2\gamma}, \quad \text{or} \quad K \lesssim \frac{1}{2}, \tag{7.39}$$

which limits the K-dependent energy shift to the 10% level. For a conventional magnetic undulator this condition would typically correspond to a low, and hence easily achievable, magnetic field. For radio frequency or laser "undulators" practical power considerations surely cause condition (7.39) to be easily satisfied. When undulator theory is applied to Compton scattering the ideal undulator approximation is even more valid.

In the limit $K \ll 1$, from Eq. (7.38), the relation between production angle and frequency of the fundamental $n = 1$ line is

$$\omega_1(\vartheta)\big|_{K \ll 1} = \frac{a_0}{1 + \gamma^2 \vartheta^2}, \tag{7.40}$$

This function is plotted in Figure 7.7. Note, from this equation and Eq. (7.31), that the angular patterns of $\omega_1(\vartheta)$ (maximum of the multisource pattern) and

---

7) Of course, the angular divergence of the radiated photon beam cannot be less than the angular divergence of the electron beam. This would only be possible if the radiation from different electrons were coherent. This can be true only at very long wavelengths, as in a free electron laser.

8) Because there is a functional relation between production angle and wavelength, the beam brilliance could, in principle, be infinite, in spite of the "Doppler" spread, since the detection apparatus could be designed to exploit this correlation. For example, if the beam is shone directly on a crystal, without having passed through a monochromator or other filter, the program analysing the diffraction pattern could exploit its full knowledge of the correlation. In practice the detection apparatus will usually sum incoherently over a finite range of ϑ, which will smear the energies and reduce the brilliance.

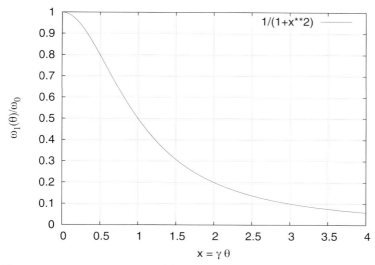

Fig. 7.7 Angular dependence $\omega_1(\theta)/\omega_0$, the frequency of the fundamental undulator line with $K \ll 1$. This angular dependence is sometimes described as "Doppler shifted", based on the frequency shift from its monochromatic electron frame value.

of a_ϑ (characteristic frequency of the single source pattern) are the same. This causes the diffraction maximum to retain its same position relative to the single source spectrum, independent of ϑ.

7.8 Phasor Summation for $K \ll 1$

We continue with the diffraction grating analogy for undulator radiation. To calculate the multiple source interference pattern we sum the amplitudes from the $2N_w$ undulator poles, using the phasor construction of Figure 7.8. (The construction works even if $2N_w$ is an odd integer, as in the figure, but $2N_w$ is normally even.) The phase slip per deflection can be expressed in terms of ω by substituting the second of Eqs. (7.38) back into Eq. (7.37);

$$\Delta\phi(\vartheta) = \pi n \frac{\omega}{\omega_n(\vartheta)} = \pi n \nu, \tag{7.41}$$

where relative energy variable

$$\nu = \frac{\omega}{\omega_n(\vartheta)} \tag{7.42}$$

has been introduced. For example, with $n = 1$, at $\omega = \omega_1(\vartheta)$, the phase slip per pole is π. It would have been appropriate to designate ν as $\nu_n(\vartheta)$ since ν

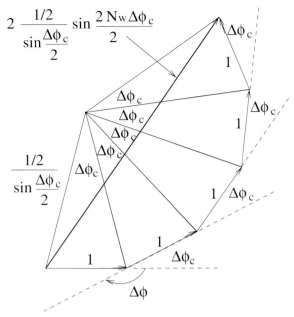

Fig. 7.8 Phasor diagram with $2N_w$ arrows to calculate the coherent sum of waves from $2N_w$ sources, or N_w undulator periods. The directions of alternate phasors are reversed to account for the half period phase shift; the complementary angle, $\Delta\phi_c = \pi - \Delta\phi$, where $\Delta\phi$ is the phase advance per half period of the undulator, is labeled in the figure. Normally there would be far more and, more likely, an even number, of phasors. This diagram is valid only for $K < 1$.

is the fractional energy *displacement* relative to the energy of the *n*th harmonic peak, as viewed at angle ϑ. But these dependences have been suppressed for brevity. A variable that is equivalent to ν, but is more useful for expressing deviation from a *nearby* resonance peak, is

$$\Delta\nu = \nu - 1 = \frac{\omega - \omega_n(\vartheta)}{\omega_n(\vartheta)} \tag{7.43}$$

The result of the phasor summation exhibited pictorially in Figure 7.8, is a "grating amplitude function"

$$G(2N_w, \Delta\phi) = \frac{\sin(2N_w(\pi/2 - \Delta\phi/2))}{\sin(\pi/2 - \Delta\phi/2)}. \tag{7.44}$$

The intensity is proportional to the square of this function. Though functions resembling G arise in a variety of multisource situations, its detailed interpretation requires a certain amount of care. Here we substitute for $\Delta\phi$ from

Eq. (7.41), and then for ν from Eq. (7.42);

$$G(2N_w, \Delta\phi) = \frac{\sin(N_w\pi(1-n\nu))}{\sin(\pi/2 - \pi n\nu/2)} = \pm\frac{\sin(N_w\pi n\nu)}{\cos(\pi n\nu/2)} \qquad (7.45)$$

$$= \pm\frac{\sin(N_w\pi n(1+\Delta\nu))}{\cos(\pi n(1+\Delta\nu)/2)}$$

$$= \begin{cases} \pm\frac{\sin(N_w\pi n\Delta\nu)}{\cos(\pi n\Delta\nu/2)} & n = 0, 2, 4, \ldots \\ \pm\frac{\sin(N_w\pi n\Delta\nu)}{\sin(\pi n\Delta\nu/2)} & n = 1, 3, 5, \ldots \end{cases}$$

$$\xrightarrow{\Delta\nu\to 0} \begin{cases} \pm\sin(N_w\pi n\Delta\nu) & n = 0, 2, 4, \ldots \\ \pm 2N_w \frac{\sin(N_w\pi n\Delta\nu)}{N_w\pi n\Delta\nu} & n = 1, 3, 5, \ldots \end{cases} \qquad (7.46)$$

The simplification in the last step is applicable only for $\pi\Delta\nu \ll 1/n$, which is to say, for energies very close to an interference maximum. (It is not necessary to keep track of the signs more carefully since the result will be squared. Furthermore a more numerically accurate formula will be derived later.) From this equation one sees that the only constructive interference maxima occur for $n = 1, 3, 5, \ldots$. To x-ray scientists this dominance of odd-integer undulator lines has become second nature.

For $2N_w = 20$, the function $G^2(20, \Delta\phi)\tilde{\mathbf{E}}_x^2(\omega, 0)$ is plotted in Figure 7.9 with, as yet, arbitrary units for the vertical scale. A logarithmic scale is used, to make the third harmonic peak more visible, and to show that its amplitude is small compared to the fundamental peak, but the absolute widths of first and third harmonic are the same. This makes the third three times narrower in relative terms. Like a grating spectrometer, the resolution is better in higher order. The replacement of the true transverse deflection profile by a Gaussian profile (Eq. (7.8)) has caused the higher order peak to be underestimated, but not by a large factor. No line spreading due to finite vertical acceptance is included in this spectrum.

The (squared) grating function, as approximated by the last expression in Eq. (7.46), near the $n = 1, 3, 5, \ldots$ resonance peaks, is a narrow positive definite function, having area $4N_w/n$. It is conveniently further approximated by a Dirac delta function or by a Gaussian;

$$G^2(2N_w, \Delta\phi) \approx \frac{4N_w}{n}\delta(\Delta\nu) \approx 4N_w^2 \exp\left(-\frac{\Delta\nu^2}{2(\sqrt{2\pi}N_w n)^{-2}}\right). \qquad (7.47)$$

For the Gaussian approximation, the central value has been matched to the central value and the r.m.s. width, $\sigma_{\delta\nu} = 1/(\sqrt{2\pi}N_w n)$, has been adjusted to give the correct area. An alternative definition of "width" is the value of $\Delta\nu$ at the first diffraction minimum—this would yield $\Delta\nu_{\text{width}} = 1/(N_w n)$. Whatever way it is defined, the relative width of the energy peak observed at fixed angle is inversely proportional to n and to N_w.

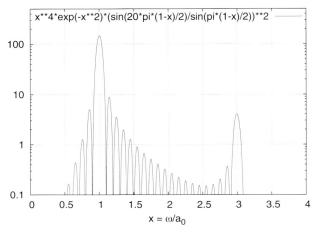

Fig. 7.9 Energy spectrum for undulator with $N_w = 10$ undulator periods, with $\vartheta = \Theta = 0$. Vertical axis units are arbitrary.

We are now in a position to complete the analogy between undulator radiation and diffraction gratings. The single source spectrum is given by Eq. (7.30), and plotted in Fig 7.4, and its corresponding intensity distribution is given by Eq. (7.32). The grating function is given by Eq. (7.46). From elementary physical optics, for example in the form of the Huyghens construction, one obtains the overall intensity pattern as the product of the squared amplitudes. In mathematical terms the result being used is that the Fourier transform of a convolution is the product of the individual Fourier transforms. The distribution of radiated energy from a single electron passing through the undulator into the n'th harmonic, for $n = 1, 3, 5, \ldots$, is given, therefore, by

$$\frac{d^2 U_n}{d\omega d\Omega} = \frac{4K^2\gamma^2}{\pi^2} \frac{q^2}{4\pi\epsilon_0 c} \frac{\omega_n^4(\vartheta)}{a_0^2 a_\vartheta^2} \exp\left(-\frac{\omega_n^2(\vartheta)}{a_\vartheta^2}\right) G^2(2N_w, \Delta\phi). \tag{7.48}$$

Equation (7.35) was used to eliminate λ_w, and $\Delta\phi$ is to be obtained from Eq. (7.41). This has assumed $K \ll 1$. This restriction came about not from the convolution procedure, but during the calculation of the single source spectrum. To remove the restriction it will be necessary to calculate the exact spectrum from a full undulator period before performing the phasor construction. Then there will be only half as many phasors.

7.9
Photon Energy Distributions

7.9.1
Energy Distribution from the $n = 1$ Undulator Fundamental

For ideal $K \ll 1$ undulator operation essentially all the power goes into the $n = 1$ resonance. For $K > 1$ the total radiated power can be calculated from a formula already derived, namely Eq. (7.14), but we have not yet learned how this energy is divided up into the various n-values. We are, however, in a position to calculate the distribution in energy of the photons associated with any particular n-value, especially $n = 1$.

The number spectrum of emitted photons for $K \ll 1$ undulator radiation is studied in Chapter 12. Photons are shown there to be "more-or-less" uniformly distributed from zero laboratory energy up to the maximum energy $\mathcal{E}_{1,\mathrm{edge}}$ of the undulator harmonic. This estimate is based on a picture in which, in the electron's rest frame, the undulator field resembles an incident plane wave to which the electron responds and hence radiates. Treating the radiation pattern as isotropic in the electron's frame, the uniform laboratory energy distribution results. The actual (normalized to 1) distribution is given by Eq. (12.110)

$$\frac{dN}{d\nu_J} = \frac{3}{2}(1 - 2\nu_J + 2\nu_J^2), \quad \text{where} \quad \nu_J = \frac{\mathcal{E}}{\mathcal{E}_{\max}}, \tag{7.49}$$

and plotted in Figure 12.8. The variable ν_J (with "J" for Jackson), differs from our variable ν. ν_J is normalized to the maximum energy at $\vartheta = 0$; ν is normalized to the maximum energy at the actual value of ϑ. The two definitions are the same, therefore, only for the $\vartheta = 0$ direction. The distribution of radiated power corresponding to Eq. (7.49) is obtained by multiplying by ν_J to account for the energy per photon, and normalizing to total power P;

$$\frac{dP}{d\nu_J} = 3P(\nu_J - 2\nu_J^2 + 2\nu_J^3). \tag{7.50}$$

This is plotted in Figure 12.8.

Implications of this spectrum for beam line design are pursued in Chapter 9. Operation near the "edge" of an undulator harmonic is usually favored because of the maximum in the power spectrum at that point. For a beam line accepting only photons in the narrow range $1 - |\Delta\nu_J| < \nu_J < 1$ the fraction of the power from the harmonic peak is given by

$$\frac{\Delta P}{P} = 3|\Delta\nu_J|. \tag{7.51}$$

Since the factor $|\Delta\nu_J|$ can be as small as 10^{-4}, or even smaller, the fraction of the total radiation that is present at the end of the external beamline is miniscule even under ideal conditions.

According to Eq. (7.38), for the undulator fundamental, the photon energy and photon laboratory angle are one-to-one related according to

$$\nu_J(\vartheta) = \frac{\omega_1(\vartheta)}{\omega_1(0)} = \frac{1 + K^2/2}{1 + K^2/2 + \gamma^2\vartheta^2}. \tag{7.52}$$

7.10
Undulator Radiation for Arbitrary K Value

7.10.1
Analytic Formulation

Since the formulas in this section are to be evaluated numerically, approximations will be avoided as far as possible. Also, at the expense of repetition, to emphasize the surprisingly small number of steps needed to derive the main result, some intermediate formulas already given in earlier sections will be repeated. In writing this section I have profited from conversations with Lewis Kotredes [3], and from his numerical calculations using Maple.

The approach so far has been to represent a single pole by one phasor and to evaluate the total amplitude as the vector sum of $2N_w$ such phasors. We should now note that what was called the "long magnet condition" in Eq. (3.26), namely $L > 2R/\gamma$, in undulator terminology corresponds to $K > 1$. As K increases from zero it becomes increasingly less valid to represent the radiation from a single pole by a single phasor amplitude. Rather, one should segment the entire undulator into sufficiently short intervals, then form their phasor sum. For large N_w this would be a formidable task, requiring the summation over large numbers of nearly-canceling quantities. Fortunately, as mentioned earlier, it will be sufficient to limit this segmentation to one period of the wiggler, since the superposition of all periods can be handled by phasor summation.

Because the magnetic field pattern is most naturally specified in terms of z we will proceed to change the integration variable from t_r to z. The extreme dependence of t_r upon t has been noted previously, and now the dependence of t_r upon z becomes progressively more influential as K increases.

For a purely sinusoidal orbit, using Eq. (7.5), the velocity components are given by

$$\frac{v_x}{c} = \frac{v}{c}\Theta\cos k_w z, \quad \frac{v_z}{c} = \frac{1}{c}\frac{dz}{dt_r} \approx 1 - \frac{1}{2\gamma^2} - \frac{\Theta^2}{4}(1 + \cos 2k_w z). \tag{7.53}$$

This formula gives v_z the correct maximum value, $v/c \approx 1 - 1/(2\gamma^2)$, and the correct average value as given by Eq. (7.36), and corresponds to having chosen the z-origin at a point where v_z is minimum. Inverting the second equation to give dt_r/dz, then integrating and setting $t_r = 0$ at $z = 0$, yields

$$t_r \approx \left(1 + \frac{1 + K^2/2}{2\gamma^2}\right)\frac{z}{c} + \frac{(K/\gamma)^2}{8k_w c}\sin(2k_w z). \tag{7.54}$$

This gives the laboratory frame time of arrival of a reference electron at z. The main inference to be drawn from this equation is that

$$t_r = \frac{z}{c} + O\left(\frac{1}{\gamma^2}\right). \tag{7.55}$$

Referring again to Figure 7.3, dropping factors that are cubic or higher in small quantities, expressed in terms of z, the factors governing radiation are

$$\widehat{\mathcal{R}} \approx \theta\hat{\mathbf{x}} + \psi\hat{\mathbf{y}} + \left(1 - \frac{\vartheta^2}{2}\right)\hat{\mathbf{z}}, \tag{7.56}$$

$$\frac{\mathbf{v}}{c} = \frac{v}{c}\Theta\cos k_w z\,\hat{\mathbf{x}} + \left(1 - \frac{1}{2\gamma^2} - \frac{\Theta^2}{2}\cos^2 k_w z\right)\hat{\mathbf{z}},$$

$$1 - \widehat{\mathcal{R}}\cdot\frac{\mathbf{v}}{c} \approx -\theta\Theta\cos k_w z + \frac{1}{2\gamma^2} + \frac{\vartheta^2}{2} + \frac{\Theta^2}{4}(1 + \cos 2k_w z),$$

$$\frac{\mathbf{v}_\perp}{c} = \frac{\mathbf{v}}{c} - \left(\widehat{\mathcal{R}}\cdot\frac{\mathbf{v}}{c}\right)\widehat{\mathcal{R}} \approx (\Theta\cos k_w z - \theta)\hat{\mathbf{x}} - \psi\hat{\mathbf{y}} + (-\theta\Theta\cos k_w z + \vartheta^2)\hat{\mathbf{z}}.$$

where $\vartheta^2 = \theta^2 + \psi^2$. The transverse displacement of the electron in the undulator has been neglected in these relations; this is a Fraunhofer approximation that becomes progressively more valid as the distance to the observation point is increased. From Eq. (7.20) and the third of Eqs. (7.56),

$$\frac{d(ct)}{dz} = \frac{dt}{dt_r} = -\theta\Theta\cos k_w z + \frac{1 + K^2/2}{2\gamma^2} + \frac{\vartheta^2}{2} + \frac{\Theta^2}{4}\cos 2k_w z. \tag{7.57}$$

Only the first term in Eq. (7.55) has been retained for t_r. Where z appears in the argument of $\cos k_w z$ or $\cos 2k_w z$, the dropped terms are smaller by another factor of $1/\gamma^2$. This amounts to noting that it is unnecessary to distinguish between t_r and z/c. But this *does not* mean that the phase shift dependence on the longitudinal position of the electron can be neglected. Integrating Eq. (7.57) and requiring $t = 0$ at $z = 0$ yields

$$ct = \frac{1}{2\gamma^2}\left(1 + \frac{K^2}{2} + \gamma^2\vartheta^2\right)z - \frac{\theta\Theta}{k_w}\sin k_w z + \frac{\Theta^2}{8k_w}\sin 2k_w z. \tag{7.58}$$

Position, velocity and acceleration components, as well as the trigonometric

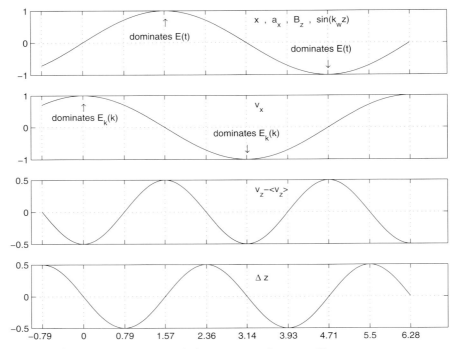

Fig. 7.10 Graphs showing the correlations among various quantities as functions of longitudinal position in the undulator. Vertical scales can be reconstructed using formulas in the text.

terms appearing in Eq. (7.58) are plotted as functions of z in Figure 7.10. Regions of the trajectory that dominate $E(t)$ (because B_y and hence transverse acceleration a_x is large) are indicated in the top figure. These regions are centered alternately on the north and south magnet poles.

The task is to evaluate the integral appearing in Eq. (7.27). The procedure will be to express the full undulator field as a sequence of paired north-south poles. After generating the amplitude from one pair, the full undulator amplitude will be obtained by phasor summation.

Following Kim [4] (though not in detail) as well as (and more closely) Als-Nielsen and McMorrow [9], to interpret Eq. (7.58) it is useful to re-arrange it so that the leading term is a uniformly advancing phase angle that matches the arguments of the perturbing trigonometric factors. Toward that end, copying from Eqs. (7.31) and (7.38), we introduce

$$\omega_n(\vartheta) = \frac{2n\gamma^2}{1 + K^2/2 + \gamma^2\vartheta^2} ck_w, \quad \phi_t = \omega_1(\vartheta)\, t, \quad \text{and} \quad \phi_z = k_w z \approx k_w c t_r. \tag{7.59}$$

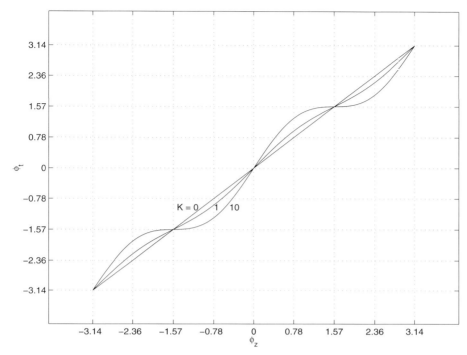

Fig. 7.11 Observation phase ϕ_t versus emission phase ϕ_z, as given by Eq. (7.60), for $K = 0, 1, 10$; $\vartheta = 0$. Where the curves are almost horizontal, a large range of ϕ_z corresponds to a small range of ϕ_t.

The newly introduced quantity ϕ_t is the observation time expressed as a phase angle, where the phase is referred to the $n = 1$ undulator resonance frequency at the particular angle ϑ. Then we obtain[9]

$$\phi_t = \phi_z - \frac{2\gamma\theta K}{1 + K^2/2 + \gamma^2\vartheta^2} \sin\phi_z + \frac{K^2/4}{1 + K^2/2 + \gamma^2\vartheta^2} \sin 2\phi_z$$
$$\equiv \phi_z + p \sin\phi_z + q \sin 2\phi_z. \qquad (7.60)$$

Examples of this dependence are plotted in Figure 7.11. The right-hand side of Eq. (7.60) is a periodic function of ϕ_z with period 2π. As ϕ_z advances by 2π, the first term on the right-hand side of Eq. (7.60) advances by 2π while the other terms return to their original values. Therefore ϕ_t also advances

---

9) Apart from different symbols, my p and q need to be multiplied by $\mp\omega/\omega_1(\vartheta)$ respectively to be the same as Kim's p and q. It appears that Kim's Eq. (4.23) has a typographical error—in his factor $\phi/K - \cos\xi$ the first term is of order $1/\gamma$ relative to the second, and hence would be neglible according to the usual approximation. To be consistent with my formulas his K should be replaced by Θ. Kim corrects this error in the equation below his Eq. (4.51).

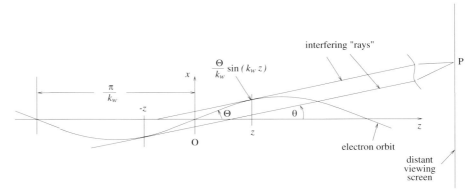

Fig. 7.12 The contributions to the radiated amplitude at any angle are greatest from regions near the two points of tangency. Because of their opposite curvatures these amplitudes interfere destructively at low frequencies. The $n = 1$ resonance condition is satisfied when the electron delay (relative to a photon) first converts this to constructive interference. As K increases from zero, the two contributions become increasingly out of phase due to the electron's speed deficit and non-straight path.

by 2π. Since $E_P(t)$, the electric field at observation point P at time t, must be a periodic function when expressed in terms of ϕ_t, the period must be 2π. But $E_P(t)$ is *not* a pure sinusoid. In fact, as K increases the electric field becomes progressively more peaked and therefore has progressively higher frequency components. Note also that, because $\omega_1(\vartheta)$ depends on ϑ, the frequencies are multiples of a fundamental frequency that depends on ϑ.

The leading cancellation within a single undulator period as K increases toward 1 and beyond is illustrated in Figure 7.12. At any point on the viewing screen[10] there are contributions to the amplitude for every value of z but, especially as K increases, the greatest contributions at any angle come from regions near the points of tangency with the electron's trajectory. One sees from the figure that there are two such points per undulator period. The procedure suggested by this figure is to calculate the amplitude from a full period (instead of from a half-period as previously) by summing the integrands from paired points before completing the integration. Then the complete undulator amplitude will be obtained as the sum of N_w (instead of the previous $2N_w$) phasors.

Another feature that can be inferred from Figure 7.12 is that for $\theta \neq 0$ (as in the figure) the spacing between the two major contributors deviates from $\lambda_w/2$. This defeats the usual cancellation on even radiation harmonics. As a result there are non-zero radiation peaks at even harmonics, in addition to the dominant, odd harmonic, peaks.

10) Recall that, in a Fraunhofer picture, parallel rays converge to a single point on the viewing screen.

7.10 Undulator Radiation for Arbitrary K Value

Applying Eq. (7.27), the frequency domain amplitude from the entire undulator is given by

$$\frac{\tilde{\mathbf{E}}(\omega, \theta, \psi)}{\frac{q}{4\pi\epsilon_0 c \mathcal{R}}} = \frac{i\omega}{\sqrt{2\pi}} \int_{-\frac{N_w \lambda_w}{2c}}^{\frac{N_w \lambda_w}{2c}} e^{i\omega t(t_r)} \frac{\mathbf{v}_\perp}{c} dt_r. \tag{7.61}$$

In this step we have restricted the range of integration to the actual length of the undulator which is $N_w \lambda_w$. Also, so that there will be a "central period" of the undulator, the inessential assumption has been made that N_w is an *odd* integer. In component form, from Eq. (7.56), the transverse velocity is given by[11]

$$\frac{\mathbf{v}_\perp}{c} = \begin{pmatrix} \Theta \cos k_w z - \theta \\ -\psi \end{pmatrix}. \tag{7.62}$$

Substituting this into Eq. (7.61), and changing integration variable from t_r to ϕ_z using Eq. (7.60), yields

$$\frac{\tilde{\mathbf{E}}(\omega, \theta, \psi)}{\frac{q}{4\pi\epsilon_0 c \mathcal{R}}} \tag{7.63}$$

$$= \frac{i\omega}{\sqrt{2\pi}} \int_{-N_w \pi}^{N_w \pi} \exp\left(i \frac{\omega}{\omega_1(\vartheta)} (\phi_z + p \sin \phi_z + q \sin 2\phi_z)\right) \begin{pmatrix} \Theta \cos \phi_z - \theta \\ -\psi \end{pmatrix} \frac{d\phi_z}{k_w c},$$

where N_w is assumed to be odd. We can now exploit the periodic nature of the exponent to represent the integral as a sum;

$$\frac{\tilde{\mathbf{E}}(\omega, \theta, \psi)}{\frac{q}{4\pi\epsilon_0 c \mathcal{R}}} = \frac{i}{\sqrt{2\pi}} \frac{\omega}{k_w c} \left(\sum_{j=-(N_w-1)/2}^{(N_w-1)/2} \exp\left(i \frac{\omega}{\omega_1(\vartheta)} 2\pi j\right) \right) \tag{7.64}$$

$$\times \int_{-\pi}^{\pi} \exp\left(i \frac{\omega}{\omega_1(\vartheta)} (\phi_z + p \sin \phi_z + q \sin 2\phi_z)\right) \begin{pmatrix} \Theta \cos \phi_z - \theta \\ -\psi \end{pmatrix} d\phi_z.$$

This is the main formula of the theory. Since the integral in the second line is independent of N_w, the result can be seen to be the product of a "phasor sum"

---

11) Because both components of $\mathbf{v}_\perp$ are even functions of z, their Fourier transforms are relatively real, which means they are in phase in all directions. This causes the radiation to be linearly polarized. This contrasts with the radiation from arc magnets, which has earlier been seen to be elliptically polarized.

part and a "single period amplitude". The phasor sum is

$$\sum_{j=-(N_w-1)/2}^{(N_w-1)/2} \exp\left(2\pi i \frac{\omega}{\omega_1(\vartheta)} j\right)$$

$$= \frac{\exp\left(-\pi i \frac{\omega}{\omega_1(\vartheta)}(N_w-1)\right) - \exp\left(\pi i \frac{\omega}{\omega_1(\vartheta)}(N_w+1)\right)}{1 - \exp\left(2\pi i \frac{\omega}{\omega_1(\vartheta)}\right)}$$

$$= \frac{\sin(N_w \pi \omega / \omega_1(\vartheta))}{\sin(\pi \omega / \omega_1(\vartheta))}. \tag{7.65}$$

The single period amplitude (which will, in general, have to be evaluated numerically) is

$$\frac{i}{\sqrt{2\pi}} \frac{\omega}{k_w c} \int_{-\pi}^{\pi} \exp\left(i \frac{\omega}{\omega_1(\vartheta)}(\phi_z + p \sin \phi_z + q \sin 2\phi_z)\right) \begin{pmatrix} \Theta \cos \phi_z - \theta \\ -\psi \end{pmatrix} d\phi_z. \tag{7.66}$$

For frequencies close to resonance, defining

$$\nu = \frac{1}{n} \frac{\omega}{\omega_1(\vartheta)} = 1 + \Delta\nu, \tag{7.67}$$

as in Eqs. (7.42) and (7.43), the phasor sum can be approximated, as in Eq. (7.46). But the details are quite different now, because the integral now covers a full period of the undulator. One result of this is that both numerator and denominator vanish at $\Delta\nu = 0$ for values of n that are either even or odd. Exactly on resonance the value of the sum is

$$\sum\bigg|_{\Delta\nu=0} = \frac{d/d\Delta\nu \sin(N_w \pi n(1+\Delta\nu))}{d/d\Delta\nu \sin(\pi n(1+\Delta\nu))}\bigg|_{\Delta\nu=0} \tag{7.68}$$

$$= N_w (-1)^{(N_w-1)n} = N_w;$$

the last step follows because N_w has been required to be odd. Near all resonances, for n either even or odd, we have

$$\sum \approx N_w \frac{\sin(N_w \pi n \Delta\nu)}{N_w \pi n \Delta\nu} \equiv N_w \operatorname{sinc}(N_w \pi n \Delta\nu), \quad n = 0, 1, 2, 3, \ldots. \tag{7.69}$$

(To recover the exact result from this approximation one need only replace the N_w sinc factor by expression (7.65).)

It is valid to suppose, near any undulator resonance, that the dependence of $\widetilde{\mathbf{E}}(\omega, \theta, \psi)$ on ω (which is to say the dependence on $\Delta\nu$ for $\Delta\nu \ll 1$) is dominated by the phasor factor (7.68). (Since the approximate vanishing of the field for even n harmonics is only implicit in the single period integral, for the dominant contribution we can approximate only the odd n harmonics in this way.) Then, in integral (7.66), we make the replacement $\omega = n\omega_1(\vartheta)$ and get

$$\frac{i}{\sqrt{2\pi}} \frac{n\omega_1(\vartheta)}{k_w c} \int_{-\pi}^{\pi} \exp\left(i n \left(\phi_z + p \sin\phi_z + q \sin 2\phi_z\right)\right) \begin{pmatrix} \Theta \cos\phi_z - \theta \\ -\psi \end{pmatrix} d\phi_z$$

$$= i\sqrt{\frac{2}{\pi}} \frac{n\omega_1(\vartheta)}{k_w c} \int_0^{\pi} \cos\left(n \left(\phi_z + p \sin\phi_z + q \sin 2\phi_z\right)\right) \begin{pmatrix} \Theta \cos\phi_z - \theta \\ -\psi \end{pmatrix} d\phi_z. \quad (7.70)$$

For $\theta = 0$ the dominant (upper) integral can be evaluated analytically;

$$\int_0^{\pi} \Big(\cos\left((n+1)\phi_z + nq \sin 2\phi_z\right) + \cos\left((n-1)\phi_z + nq \sin 2\phi_z\right)\Big) d\phi_z$$

$$= \frac{1}{2} \int_0^{2\pi} \left(\cos\left(\frac{n+1}{2}\xi + nq \sin\xi\right) + \cos\left(\frac{n-1}{2}\xi + nq \sin\xi\right)\right) d\xi$$

$$= \pi \left(J_{\frac{n+1}{2}}(-nq) + J_{\frac{n-1}{2}}(-nq)\right). \quad (7.71)$$

Recombining factors we obtain, for $\theta = 0$,

$$\frac{\tilde{E}_{x,n}(\omega,0,\psi)}{\frac{q}{4\pi\epsilon_0 c R}} \approx i\sqrt{\frac{\pi}{2}} \frac{n\omega_1(\vartheta)}{k_w c} \frac{K}{\gamma} N_w \operatorname{sinc}(N_w \pi n \Delta \nu) \left(J_{\frac{n+1}{2}}(-nq) + J_{\frac{n-1}{2}}(-nq)\right), \quad (7.72)$$

which is valid only for odd n and for $\Delta\nu \ll 1$. The lower component of Eq. (7.70) can be evaluated for even n. It yields

$$\frac{\tilde{E}_{y,n}(\omega,0,\psi)}{\frac{q}{4\pi\epsilon_0 c R}} \approx i\sqrt{2\pi} \frac{n\omega_1(\vartheta)}{k_w c} (-\psi) N_w \operatorname{sinc}(N_w \pi n \Delta \nu) J_{\frac{n}{2}}(-nq). \quad (7.73)$$

These formulas can serve to check the numerical evaluation of integral (7.70) in the $\theta = 0$ limit.

Recapitulating, the unapproximated, general K, undulator radiation formula derived in this section is

$$\frac{\tilde{\mathbf{E}}(\omega,\theta,\psi)}{\frac{q}{4\pi\epsilon_0 c R}} = i\sqrt{\frac{2}{\pi}} \frac{\omega}{k_w c} \frac{\sin\left(N_w \pi \omega/\omega_1(\vartheta)\right)}{\sin\left(\pi\omega/\omega_1(\vartheta)\right)} \times \quad (7.74)$$

$$\int_0^{\pi} \cos\left(\frac{\omega}{\omega_1(\vartheta)}\phi_z - \frac{\theta K}{\gamma} \frac{\omega}{ck_w} \sin\phi_z + \frac{K^2}{8\gamma^2} \frac{\omega}{ck_w} \sin 2\phi_z\right) \begin{pmatrix} \Theta \cos\phi_z - \theta \\ -\psi \end{pmatrix} d\phi_z.$$

The factorization into phasor part and single period part is analogous to the similar factorization in Eq. (7.48). In particular the single period integral in the lower line is independent of N_w. The only dependence on N_w enters via the phasor factor. Note that the suppression of even harmonics, previously coming from the phasor diagram, now comes from the pole-pair formula.

Furthermore, from the $N_w \operatorname{sinc}(N_w \pi n \Delta \nu)$ approximation to the phasor part, one sees that the pattern is independent of angles once the frequency is expressed via the offset variable $\Delta \nu$. For practical wigglers, having say $N_w > 10$, this *sinc* approximation is valid in regions where the radiation is significantly large. In these regions the replacement $\omega/\omega_1(\vartheta) \to n$ in the integrand (as in Eq. (7.70)) is also legitimate, yielding

$$\tilde{\mathbf{E}}(\omega,\theta,\psi) \approx \sum_{n=1,3,5,\ldots} \tilde{\mathbf{E}}_n(\theta,\psi)\, n N_w \operatorname{sinc}\left(N_w \pi \left(\frac{\omega}{\omega_1(\vartheta)} - n\right)\right), \tag{7.75}$$

where

$$\tilde{\mathbf{E}}_n(\theta,\psi) = i\frac{q}{4\pi\epsilon_0 c \mathcal{R}} \sqrt{\frac{2}{\pi}} \frac{2\gamma^2}{1+K^2/2+\gamma^2\vartheta^2} \times \tag{7.76}$$

$$\int_0^\pi \cos\left(n\left(\phi_z - \frac{2\gamma\theta K \sin\phi_z}{1+K^2/2+\gamma^2\vartheta^2} + \frac{(K^2/4)\sin 2\phi_z}{1+K^2/2+\gamma^2\vartheta^2}\right)\right) \begin{pmatrix} \Theta\cos\phi_z - \theta \\ -\psi \end{pmatrix} d\phi_z.$$

Referring back to Eq. (3.43), the radiation distribution in this approximation is given by

$$\frac{dU_0}{d\Omega d\omega} = \frac{2}{\mu_0 c}\mathcal{R}^2 |\mathbf{E}|^2 \tag{7.77}$$

$$= \frac{2}{\mu_0 c}\mathcal{R}^2 \left| \sum_{n=1,3,5,\ldots} \tilde{\mathbf{E}}_n(\theta,\psi)\, n N_w \operatorname{sinc}\left(N_w \pi \left(\frac{\omega}{\omega_1(\vartheta)} - n\right)\right) \right|^2.$$

7.10.2
Diffraction Grating Analogy

The analogy between undulator and diffraction grating has to be revised now that the pattern is based on N_w-fold repetition of a pole-pair instead of the $2N_w$-fold alternating repetition of individual poles. For $K \ll 1$ the pictures are equivalent, but for $K > 1$ treating the pole-pair as a unit is obligatory. With visible light, consider a double slit in which a half-wave plate has been placed in front of one of the two slits. The two-slit diffraction pattern would be unchanged except for the half-fringe translation corresponding to the vanishing of the forward amplitude. A full grating made from many of these pairs would be much the same as an ordinary grating except the modulating pattern would be a two-slit rather than a one-slit pattern. The most striking feature would be the absence of the $n = 0$ (forward) line. The character of undulator radiation is much like this, especially in the respect that the $n = 0$ line is missing in the forward direction.

In the following sections the pole-pair pattern will be investigated numerically. Unlike a single pole pattern, this pattern exhibits minima and maxima. These features are rather broad, though narrower than the $1/\gamma$ angular spread characteristic of bending magnet radiation—the angular spacings of the pole-pair pattern are of order $1/(n\gamma)$.

The grating-type, multiple-pole resonance lines are very narrow in energy, when viewed at fixed direction, with fractional energy widths of order $\Delta E/E \approx 1/N_w$. These lines are modulated by the pole-pair pattern. Because of the functional dependence of energy on angle (for a particular undulator resonance order) after passage through an appropriately-tuned monochromator, an undulator line observed in the forward direction can be very narrow compared to both $1/\gamma$ and to the pole-pair angular separation—with appropriately tuned monochromator the angular width can be of order $1/(nN_w\gamma)$.

7.10.3
Numerical/Graphical Representation of Undulator Radiation

This section starts by showing coarse undulator distributions obtained by numerical integration of Eq. (7.66) for the (moderately low) values of $n = 3$ and $n = 5$. Since the *sinc* factor is *not* included in these plots, they exhibit an N_w-independent, "modulating pole-pair profile" that is to be multiplied later by an N_w-dependent *sinc* factor. This multiplication will have the effect, for given direction (θ, ψ) (and hence given ϑ) of defining a narrow band of frequencies close to $\omega_n(\vartheta)$ as given by Eq. (7.59). The angular intensity pattern is proportional to $|\widetilde{E}_{n,x}|^2(\theta,\psi)$, which is given by Eq. (7.76). It is exhibited for $K = 1.5$, $n = 3, 5$, in Figure 7.13 and Figure 7.15, along with a grayscale plot that can be used to infer the photon frequency corresponding to the particular n value in the particular direction.

Each histogram tower represents both $|\widetilde{E}_{3,x}|^2_{K=1.5}(\theta,\psi)$ and the ratio $\omega(\theta,\psi)/\omega_3(0)$. The intensity is represented by the height of the tower and the photon energy by its grayscale value. The bin widths are $2\gamma\Delta\theta = \gamma\Delta\psi = 0.15$. A detector having these acceptances and centered, say, at the origin, would count only photons of frequency $\omega = \omega_3(0)$ (which makes the tower pure white.) Assuming the detector accepts all photons close to this energy (so the sinc-factor of Eq. (7.76) can be treated as a δ-function) the rate can be read off from the vertical axis, but still needs multiplication by the factor nN_w from Eq. (7.76) and by the factor $\Delta\theta\,\Delta\psi$. If the detector's fractional frequency acceptance is small compared to $1/N_w$ it is necessary to include the sinc-factor dependence of Eq. (7.76). The range of $\gamma\theta$ over which the intensity is appreciable is roughly equal to K, and n "two-source" interference fringes are visible within this range. The structure of Figure 7.15 can also be understood qualitatively. The angular separation $\Delta\theta$ of fringe maxima is approximately $0.375/\gamma$.

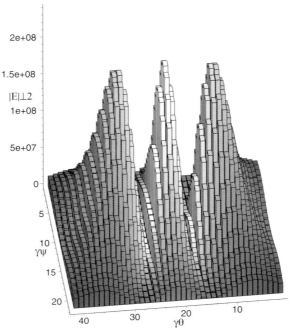

Fig. 7.13 Histogram representation of undulator radiation: the height of each bin gives the value of $|\tilde{E}_{3,x}|^2_{K=1.5}(\theta, \psi)$, as given by Eq. (7.76) (with factor $q/(4\pi\epsilon_0 c\mathcal{R})$. suppressed). The photon frequency ω is a function of the same independent-variable pair (θ, ψ) (Eq. (7.59)) and can be exhibited as a ratio $\omega/\omega_3(0)$ which is coded by the grayscale shading. The forward direction is marked by the highest tower which has $\omega/\omega_3(0) = 1$ and is therefore pure white. The bin widths are $2\gamma\Delta\theta = \gamma\Delta\psi = 0.15$. This "pole-pair modulating profile" is independent of N_w.

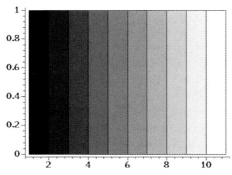

Fig. 7.14 Grayscale to be used for graphs in this section.

The factor $\sin k_w z$ advances from 0 to 1 between north and south magnet poles, so the second term on the right-hand side of Eq. (7.58) gives a phase advance (with $n = 5$, $K \equiv \Theta\gamma = 1.5$) equal to $5 \times 2\gamma \times 1.5 \times 0.375/\gamma \approx 2\pi$.

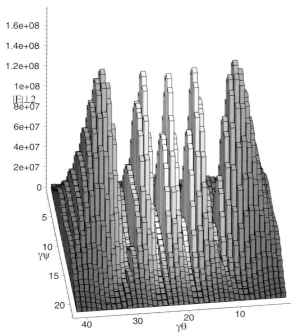

Fig. 7.15 Same as previous figure except now $n = 5$. The θ-angular separation of the fringe maxima is approximately $5 \times 0.075/\gamma = 0.375/\gamma$. The factor $\sin k_w z$ advances from 0 to 1 between north and south magnet poles, so the second term on the right-hand side of Eq. (7.58) gives a phase advance equal to $5 \times 2\gamma \times 1.5 \times 0.375/\gamma \approx 2\pi$. This is why angular fringes are visible even with $N_w = 1$. For $N_w > 1$ it is only the central maximum that necessarily coincides with a multiple-pole maximum. A diffraction grating constructed from repeated double slits having a quarter wave plate in front of one of the slits could give a similar pattern.

This is why angular fringes are visible even with $N_w = 1$. This θ-dependent pattern is analogous to the pattern observed with visible light passing through a double-slit. With $N_w \gg 1$ and the sinc-factor included, the peaks are much narrower and acquire ψ-dependent structure.

There appears to be no universally accepted distinction between "undulator operation" and "wiggler operation", though $K = 1$ is commonly quoted as the dividing line. Here is a suggested spectral signature. When the spectra corresponding to two (or more) n-values are superimposed (such as the $n = 3$ and $n = 5$ spectra illustrated in the figures) the $N_w = 1$ fringe structure is washed out. As K increases beyond 1, more and more n-values contribute significantly and the $N_w = 1$ fringe structure is more and more washed out. A figure below (Figure 7.18) will show, for given K, the relative contributions from different n-values. There are significant contributions from several n values for K-values greater than 1 or so. However, when the beam is passed through a monochromator, intense, large-N_w resonance lines will emerge from

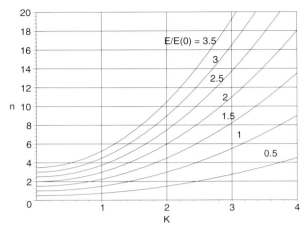

Fig. 7.16 In the (K, n) plane, the plot shows contours of constant photon energy, expressed as a ratio to the energy at the fundamental undulator edge, $\mathcal{E}_\gamma/\mathcal{E}_\gamma(0) = \omega_n(\vartheta = 0)|_K/\omega_1(0)|_0$ derived from Eq. (7.78).

the fog even in large-K wiggler operation. This will be illustrated in a later section.

There is an inevitable trade-off in which forward-going undulator frequency $\omega_n(0)$ and intensity $|\mathbf{E}|^2$ are made as high as possible consistent with the machine energy γ and undulator wave number k_w being as low as possible. For "cleaner" operation one prefers both the undulator parameter K and the harmonic number n to be as low as possible. (Until accelerator physics issues intrude, the bigger the better for N_w.)

The first formula governing this trade-off is Eq. (7.59);

$$\frac{\omega_n(0)}{ck_w 2\gamma^2} = \frac{n}{1 + K^2/2}. \tag{7.78}$$

This formula shows that the resonant frequency can be increased by increasing n but is unavoidably decreased by increasing K. Equal energy (that is, frequency) contours are plotted in Figure 7.16.

The second result governing the trade-off, from Eq. (7.72), with q (which is *not* the charge) defined by Eq. (7.60), still suppressing the sinc factor and the factor $q/(4\pi\epsilon_0 c\mathcal{R})$, is

$$\frac{|\mathbf{E}_n|^2(0)}{\gamma^2} = \frac{\pi}{2}\left(\frac{2Kn}{1+K^2/2}\right)^2 \left(J_{\frac{n+1}{2}}\left(\frac{-nK^2/4}{1+K^2/2}\right) + J_{\frac{n-1}{2}}\left(\frac{-nK^2/4}{1+K^2/2}\right)\right)^2. \tag{7.79}$$

These formulas are evaluated for ranges of n and K, and the results exhibited numerically in the following table and graphically in Figure 7.17.

7.10 Undulator Radiation for Arbitrary K Value

Tab. 7.1 Column headings are K-values. The upper entry is intensity as given by Eq. (7.79) and the lower is frequency as given by Eq. (7.78).

n	0.2	0.4	0.6	0.8	1.0	1.2	1.4	1.6	1.8	2.0	2.4	2.6	2.8	3.0
1	0.25	0.78	1.5	2.1	2.2	2.3	2.3	2.2	2.0	1.8	1.6	1.4	1.3	1.2
	1.0	0.91	0.83	0.77	0.67	0.59	0.50	0.43	0.38	0.33	0.29	0.26	0.23	0.20
3	0.49e-3	0.022	0.14	0.61	1.1	1.7	2.2	2.4	2.6	2.7	2.4	2.4	2.2	2.1
	3.0	2.7	2.5	2.3	2.0	1.8	1.5	1.3	1.2	1.0	0.88	0.79	0.68	0.60
5	0.60e-6	0.30e-3	0.012	0.11	0.31	0.60	1.3	1.9	2.6	2.6	2.8	2.7	2.7	2.6
	5.0	4.5	4.2	3.8	3.3	2.9	2.5	2.2	1.9	1.7	1.5	1.3	1.1	1.0
7	0.60e-9	0.39e-5	0.60e-3	0.015	0.11	0.25	0.86	1.4	1.8	1.9	2.7	3.1	2.8	2.6
	7.0	6.4	5.8	5.4	4.7	4.1	3.5	3.0	2.7	2.3	2.1	1.8	1.6	1.4
9	0.58e-12	0.46e-7	0.33e-4	0.0019	0.023	0.11	0.34	0.72	1.4	1.7	2.0	1.9	2.5	2.2
	9.0	8.2	7.5	6.9	6.0	5.3	4.5	3.9	3.5	3.0	2.6	2.4	2.0	1.8
11	0.54e-15	0.61e-9	0.15e-5	0.39e-3	0.0044	0.029	0.22	0.60	0.55	1.8	1.9	1.9	2.6	2.9
	11.	10.	9.2	8.5	7.3	6.5	5.5	4.8	4.2	3.7	3.2	2.9	2.5	2.2

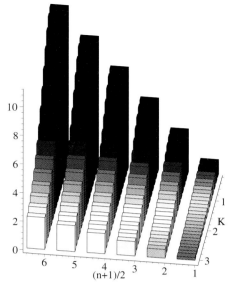

Fig. 7.17 Histograms illustrating the dependence of undulator frequency, indicated by tower height, with the unlabeled axis being the lower entry in Table 7.1. The other parameters are undulator parameter $0 < K < 3.0$ and resonance order $n = 1, 3, 5, 7, 9, 11$. The on-resonance intensity is indicated by tower grayscale, the same as in Figure 7.13, and is normalized to 1 (white) in the lower left-hand corner. Black regions have negligible flux.

An example may help to interpret this data. Consider the series of points $(n, K, \text{intensity}, \text{frequency}) = (1, 0.2, 0.25, 1.0), (3, 0.6, 0.14, 2.5), (5, 0.8, 0.11, 3.8), (7, 1.0, 0.11, 4.7), (9, 1.2, 0.11, 5.3)$. These points all have roughly the same intensities but, by increasing n and K together, it is possible to increase the beam frequency from 1.0 to 5.3 (in units of $ck_w 2\gamma^2$.) Being proportional to K^2, the total radiated beam power increases by a factor of 36 over this range. The beam is therefore much "cleaner" for low n-values than for high. Calculations like this are useful in fixing the major storage ring and undulator parameters to achieve high brilliance.

Equation (7.79) is also plotted in Figure 7.18 though this time the calculation was performed using the computer program described in the next section. For arbitrary angles the special functions entering the calculation are Weber and Anger functions. On axis, this analytic form reduces to Eq. (7.79). This can be used as a test of the validity of the numerical calculation, especially because Figure 7.18 can be compared with Figure 10 in a paper concerning undulator radiation by Walker [10]. The same data is plotted in Figure 7.19 in a way that makes clear, for fixed K-values, the breakdown of the forward intensity into the different harmonic numbers n.

7.10 Undulator Radiation for Arbitrary K Value

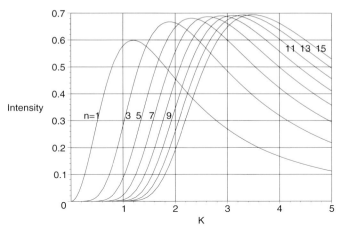

Fig. 7.18 On-axis intensity of undulator radiation as calculated using expansion (7.83), which reduces to Eq. (7.79) on axis. "Intensity" stands for the right-hand side of Eq. (7.79) which has a multiplicative factor $\pi/2$ compared to Walker's [10] function $F_n(K)$, plotted in his Figure 10. By Eq. (7.77), "Intensity" is proportional to $dU_0/(d\Omega d\omega)$.

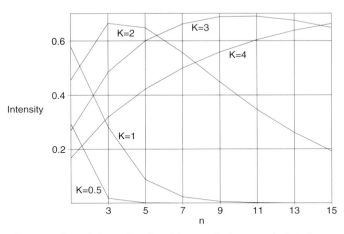

Fig. 7.19 On-axis intensity of undulator radiation as calculated using expansion (7.83), which reduces to Eq. (7.79) on axis. "Intensity" stands for the right-hand side of Eq. (7.79). This data is the same as Figure 7.18 except the forward contributions from various harmonics n are plotted for fixed values of K. By Eq. (7.77), "Intensity" is proportional to $dU_0/(d\Omega d\omega)$.

7.10.4
Approximation of the Integrals by Special Functions

The integrand in Eq. (7.74) can be Taylor expanded in terms of the (small) variable θ

$$\cos(n\phi + np\sin\phi + nq\sin 2\phi)$$
$$= \cos(n\phi + nq\sin 2\phi)\left(1 - \frac{n^2p^2}{4} + \frac{n^2p^2}{4}\cos 2\phi + \ldots\right)$$
$$- \sin(n\phi + nq\sin 2\phi)(np\sin\phi + \ldots), \qquad (7.80)$$

and further terms can be derived easily. Here, motivated by Eq. (7.76), we have generalized the meaning of variable n by making the substitution

$$\frac{\omega}{\omega_1(\vartheta)} = n, \qquad (7.81)$$

which means that n is now allowed to lie anywhere in the range $0 < n < \infty$ and, especially, to not necessarily be an integer. Nevertheless, especially for large N_w, the sinc factor suppresses the complete expression when n is not close to an integer, so n can be thought of as being close to an integer.

Using abbreviation $\mathcal{C} \equiv \cos(n\phi + nq\sin 2\phi)$ we define standard integrals

$$I_{C0} = \int_0^\pi \mathcal{C}\,d\phi, \quad I_{C1} = \int_0^\pi \mathcal{C}\cos\phi\,d\phi, \quad I_{C2} = \int_0^\pi \mathcal{C}\cos 2\phi\,d\phi, \qquad (7.82)$$

integrals $I_{S1}, I_{S2}, \ldots$, are defined similarly, but with $\mathcal{C}$ replaced by $\mathcal{S} \equiv \sin(n\phi + nq\sin 2\phi)$. The required integral is

$$\int_0^\pi d\phi \cos(n\phi + np\sin\phi + nq\sin 2\phi)\left(\begin{pmatrix}\Theta\\0\end{pmatrix}\cos\phi + \begin{pmatrix}-\theta\\-\psi\end{pmatrix}\right)$$
$$= \begin{pmatrix}\Theta(1-n^2p^2/8)\\0\end{pmatrix}I_{C1} + \begin{pmatrix}\Theta n^2p^2/8\\0\end{pmatrix}I_{C3} + \begin{pmatrix}-np\Theta/2\\0\end{pmatrix}I_{S2}$$
$$+ \begin{pmatrix}-\theta(1-n^2p^2/4)\\-\psi(1-n^2p^2/4)\end{pmatrix}I_{C0} + \begin{pmatrix}-\theta n^2p^2/4\\-\psi n^2p^2/4\end{pmatrix}I_{C2} + \begin{pmatrix}np\theta\\np\psi\end{pmatrix}I_{S1} + \ldots .$$
$$(7.83)$$

I_{C1} was evaluated for odd n in Eq. (7.71), but it, and the other integrals, can now be expressed for arbitrary n and for positive integers j as

$$I_{Cj} = \frac{1}{4}\int_0^{2\pi}\cos\left(\frac{n-j}{2}\xi + nq\sin\xi\right)d\xi + \frac{1}{4}\int_0^{2\pi}\cos\left(\frac{n+j}{2}\xi + nq\sin\xi\right)d\xi$$
$$I_{Sj} = \int_0^\pi \sin(n\phi + nq\sin 2\phi)\sin j\phi\,d\phi \qquad (7.84)$$
$$= \frac{1}{4}\int_0^{2\pi}\cos\left(\frac{n-j}{2}\xi + nq\sin\xi\right)d\xi - \frac{1}{4}\int_0^{2\pi}\cos\left(\frac{n+j}{2}\xi + nq\sin\xi\right)d\xi.$$

7.10 Undulator Radiation for Arbitrary K Value

All these integrals can be expressed in terms of the functions

$$\pi \mathbf{J}_\nu = \int_0^\pi \cos(\nu\theta - z\sin\theta)\, d\theta$$
$$\pi \mathbf{E}_\nu = \int_0^\pi \sin(\nu\theta - z\sin\theta)\, d\theta, \tag{7.85}$$

where $\mathbf{J}_i$ is known as an "Anger" function and $\mathbf{E}_i$ as a "Weber" function [11]. These definitions are valid for general values of ν, but we will mainly use only integers or half-integer indices. Both functions are known to Maple and are presumably rapidly calculable. Bisecting the range, and replacing θ by $2\pi - \theta$ in the second integral, Watson gives the formula

$$\int_0^{2\pi} \cos(\nu\theta - z\sin\theta)\, d\theta$$
$$= \int_0^\pi \left(\cos(\nu\theta - z\sin\theta) + \cos(2\nu\pi - \nu\theta + z\sin\theta) \right) d\theta$$
$$= 2\pi \cos^2 \nu\pi\, \mathbf{J}_\nu(z) + \pi \sin 2\nu\pi\, \mathbf{E}_\nu(z). \tag{7.86}$$

In terms of these functions the required integrals are

$$I_{Cj} = \frac{\pi}{2} \cos^2\left(\frac{n-j}{2}\pi\right) \mathbf{J}_{\frac{n-j}{2}}(-nq) + \frac{\pi}{4} \sin((n-j)\pi)\, \mathbf{E}_{\frac{n-j}{2}}(-nq),$$
$$+ \frac{\pi}{2} \cos^2\left(\frac{n+j}{2}\pi\right) \mathbf{J}_{\frac{n+j}{2}}(-nq) + \frac{\pi}{4} \sin((n+j)\pi)\, \mathbf{E}_{\frac{n+j}{2}}(-nq),$$
$$I_{Sj} = \frac{\pi}{2} \cos^2\left(\frac{n-j}{2}\pi\right) \mathbf{J}_{\frac{n-j}{2}}(-nq) + \frac{\pi}{4} \sin((n-j)\pi)\, \mathbf{E}_{\frac{n-j}{2}}(-nq),$$
$$- \frac{\pi}{2} \cos^2\left(\frac{n+j}{2}\pi\right) \mathbf{J}_{\frac{n+j}{2}}(-nq) - \frac{\pi}{4} \sin((n+j)\pi)\, \mathbf{E}_{\frac{n+j}{2}}(-nq). \tag{7.87}$$

7.10.5
Practical Evaluation of the Series

The Taylor expansion of Eq. (7.83) can be spelled out in general as follows:

$$f(m,j) = \frac{\text{binomial}(m,(m-j)/2)}{m!\,2^{(m-1)}}$$

$$a(\eta,j) = (-1)^{j/2} \sum_{0}^{i_{\max}} (-1)^i (\eta p)^{2i} f(2i,j), \quad j = 0,2,4,\ldots$$

$$a_0(\eta) = a(\eta,0)/2,$$

$$b(\eta,j) = (-1)^{(j-1)/2} \sum_{0}^{i_{\max}} (-1)^i (\eta p)^{(2i+1)} f(2i+1,j), \quad j = 1,3,5,\ldots$$

$$\gamma \tilde{E}_x(\eta) = \frac{K}{2}\left(2a_0(\eta) + a(\eta,2)\right) I_{C,1}$$

$$+ \frac{K}{2} \sum_{i'=1}^{i_{\max}-1} \left(a(\eta,2i') + a(\eta,2i'+2)\right) I_{C,2i'+1} + \frac{K}{2} a(\eta,2i_{\max}) I_{C,2i_{\max}+1}$$

$$- \frac{K}{2} \sum_{i'=0}^{i_{\max}-1} \left(b(\eta,2i'+1) + b(\eta,2i'+3)\right) I_{S,2i'+2} - \frac{K}{2} b(\eta,2i_{\max}+1) I_{S,2i_{\max}+2}$$

$$- \gamma\theta \left(a_0(\eta) I_{C,0} + \sum_{i'=1}^{i_{\max}} a(\eta,2i') I_{C,2i'} - \sum_{i'=0}^{i_{\max}} b(\eta,2i'+1) I_{S,2i'+1}\right)$$

$$\gamma \tilde{E}_y(\eta) = -\gamma\psi \left(a_0(\eta) I_{C,0} + \sum_{i'=1}^{i_{\max}} a(\eta,2i') I_{C,2i'} - \sum_{i'=0}^{i_{\max}} b(\eta,2i'+1) I_{S,2i'+1}\right). \tag{7.88}$$

These expressions still need to be multiplied by the factor

$$i\sqrt{\frac{2}{\pi}} \frac{\omega}{k_w c} \frac{\sin N_w \pi \eta}{\sin \pi \eta}. \tag{7.89}$$

This factor (except for the i) is included for the following graphs. The maximum power of p retained in the expansion is $2i_{\max} + 1$. For constant accuracy $i_{\max}$ has to increase with increasing $\gamma\vartheta$. At fixed K the ring structure depends only on $\gamma\vartheta$. Because an oscillatory function is being fit by a power series, the number of well fit cycles is probably proportional to the highest power retained. From the example we have studied most carefully ($K = 1.35$) a suggested rule of thumb is to choose $i_{\max}$ to be about four (or more) times the number of rings to be faithfully calculated. This should be investigated in each case. Figures 7.20(a) and 7.20(b) illustrate these comments. The former evaluates the integral in Eq. (7.66) numerically, the latter shows the deviation from this result when the same quantity is calculated using Eqs. (7.88). The fractional accuracy is excellent over the full range $-1/\gamma < \theta < 1/\gamma$ which

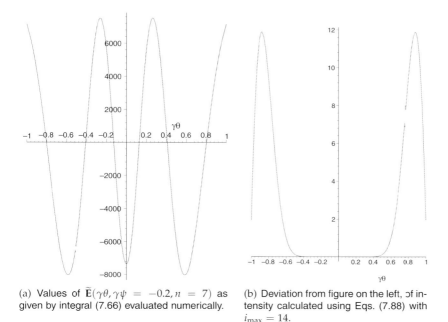

(a) Values of $\widetilde{\mathbf{E}}(\gamma\theta, \gamma\psi = -0.2, n = 7)$ as given by integral (7.66) evaluated numerically.

(b) Deviation from figure on the left, of intensity calculated using Eqs. (7.88) with $i_{max} = 14$.

Fig. 7.20 Comparison results obtained using Eq. (7.66) and using the method of Section 7.10.5.

includes essentially all the radiation. To obtain excellent accuracy in the experimentally relevant region of the central peak it is sufficient to use only the terms exhibited explicitly in Eq. (7.83).

Since θ, but not ψ, has been assumed small in deriving Eq. (7.83), its region of validity should be a narrow band centered on the ψ-axis. From Figure 7.13 and Figure 7.15 one knows that the radiation pattern is made up of parallel valleys separated by long mountains that are aligned with the ψ-axis. Since the variation with θ (at fixed ψ) is roughly sinusoidal (squared) one cannot expect a power series truncated to the terms shown explicitly in Eq. (7.83) to remain accurate outside the central three mountains. Nevertheless this truncated (and hence quite simple) form should be useful in practice because it it is the central mountain that is mainly used in most applications of undulator radiation.

7.11
Post-monochromator Profile

To calculate either "flux" or "brilliance" it is necessary to integrate $c^2 U / d\omega \, d\Omega$, as given, for example, by Eq. (7.48), over an actual detection apparatus. A δ-function replacement, as in Equation (7.47), can only be used for apertures that are large compared to the natural spreads, but not otherwise, since for example, an arbitrarily small aperture could lead to infinite brilliance. Similarly, since the Gaussian approximation does not convey the true diffraction pattern, it can only be used to *estimate* the brilliance if the aperture size is broad compared to this diffractive structure.

This cautionary sentence notwithstanding, there is a picture of the forward distribution of photons, due to Attwood et al. [1] based on a formula very similar to the Gaussian approximation of Eqs. (7.47). I find this picture confusing and misleading, but it seems to be in common use and mis-use by workers in the field, so I will try to reconstruct the conditions for which the calculation is applicable.

Imagine an apparatus whose central energy is tuneable and which accepts only an infinitesimal band of photon energies. Let us suppose that the apparatus accepts a broad range of production angles but measures the direction of each photon with perfect angular resolution. This apparatus can therefore be used to measure the dependence of radiated power on production angle ϑ; (and azimuthal dependence as well, should that be of interest.) A popular setting for such an apparatus has its energy acceptance tuned to be centered on $n a_0/(1 + K^2/2)$, which is the central frequency of the peak, in the forward direction, for the nth harmonic. Substituting the second of Eqs. (7.38) into Eq. (7.43) yields the fractional deviation from resonance of this detector, at angle ϑ, to be

$$\Delta \nu(\vartheta) = -\frac{\gamma^2 \vartheta^2}{1 + K^2/2}. \tag{7.90}$$

Using Eq. (7.47), the angular dependence of photons striking the detector is given by

$$\frac{d^2 U_n}{d\omega d\Omega} \sim \exp(-r^4 \vartheta^4) \quad \text{where} \quad r^2 = \frac{\gamma^2}{1 + K^2/2} \sqrt{\pi} N_w n. \tag{7.91}$$

Since this formula has a "hyper-Gaussian" dependence, with ϑ^4 in the exponent, it disagrees with Kim, whose exponent is proportional to ϑ^2. Nevertheless we can use it to "calculate" an angular variance;

$$\sigma_\vartheta^2 = \frac{\int_0^\infty \vartheta^3 \exp(-r^4 \vartheta^4) \, d\vartheta}{\int_0^\infty \vartheta \exp(-r^4 \vartheta^4) \, d\vartheta} = \frac{1}{\Gamma(1/2)} \frac{1}{r^2} = \frac{1}{\pi} \frac{1 + K^2/2}{\gamma^2} \frac{1}{N_w n}. \tag{7.92}$$

This agrees with Kim's formula Eq. (4.33) giving the variance (square of r.m.s. value) of ϑ, except the factor $1/\pi$ is replaced by $1/4$ in his formula. This discrepancy may be due to slightly different definition of the parameters and, in any case, constitutes good agreement within the spirit of the approximation. The same formula appears in *X-Ray Data Booklet* [12], and in the Wiedemann contribution, to the *Handbook of Accelerator Physics and Engineering* [13]. See, for example, page 189 of Wiedemann (or pages 4-7, Kim). Equivalent formulas also occur in Eq. (2.9) of Mills' book [14] and on p. 21 of Krinsky's article in the same Mills-edited book [14].

The reason I disapprove of this analysis is that it is so easily misinterpreted as showing that the average production angle of undulator-produced photons decreases with increasing N_w and with increasing n. One hears it said that the production angle "narrows" inversely with nN_w. In fact it is only the production angle of photons through a monochromator, carefully set up as described in the previous paragraphs, that "narrows". In fact, if the full energy spread is accepted at every angle, the angular pattern from the whole undulator is very nearly the same as would come from any one of its poles in isolation.

To understand this behavior it is necessary to understand the source of the curious $\exp(-r^4\vartheta^4)$ angular dependence in Eq. (7.91). The undulator energy width at fixed ϑ is quite narrow; according to the last of Eqs. (7.48) the fractional energy spread is of order $1/N_w$. This might be about 1%. This would be narrow enough to pass for an "experimentalist's delta function" in many contexts—but *not* for the present apparatus, whose fractional energy acceptance is 10^{-4}, a hundred times narrower. By using the Gaussian representation of the diffraction peak one improves on the simple delta function representation but, in truth, the discontinuity at the diffraction edge deserves a more careful mathematical treatment. This topic is pursued further in the next section (Figure 7.21) and in Chapter 9 where beamline design is discussed.

7.11.1
Monochromatic Annular Rings

Some of the points that have been raised are illustrated in Figure 7.21 which shows little parallellograms (at intersections of undulator curves and monochromator lines) within which undulator resonance energies are equal to the monochromator energy. In greater detail, the parallelograms are defined above and below by the monochromator energy limits and on the sloping sides by the edges of the undulator peaks. At fixed energy, lines from undulator harmonics appear at angles increasing monotonically with n. Grating structure $G(2N_w, \Delta\phi)$, as defined in Eq. (7.44)) is not shown. The effect of this dependence would be to produce fringes on the sloping sides of the little acceptance parallelograms. The complete pattern observed by a detector with large angular acceptance would consist of rings; Figure 7.21 just shows the intersections of these rings with one coordinate plane. The radii of the annular rings increase monotonically with n. If the energy window were reduced slightly from the value shown, even the $n=1$ resonance would give an annular ring, and if the window were raised the central spot would vanish.

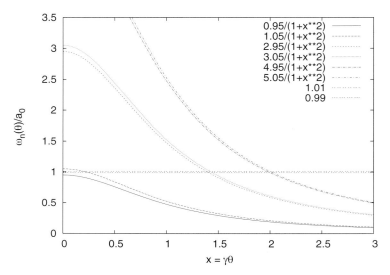

Fig. 7.21 The photon flux through a monochromator with narrow energy acceptance and full angular acceptance is proportional to the areas of the little "parallelograms" shown. Their sloping sides are contours of constant n resonance (Eq. (7.38)) and their bottoms and tops define the monochromator acceptance. There is no significance to the solid, broken, etc. curves other than to correlate with the mathematical form shown in the key. The strong "hyper-Gaussian", $\exp(-\vartheta^4)$ dependence in Eq. (7.91) comes about because the narrow $n = 1$ undulator band diverges so rapidly from the even narrower monochromator band.

The setting shown, with the energy window centered on the $n = 1$ resonance, is optimal for achieving maximum brilliance. It is also the configuration for which formula (7.92) is most nearly valid. But one sees that the vanishing of $d\omega_n/d\vartheta$ at this point complicates the mathematics. It would be unreasonable to suppose that, in such a situation, a formula for the variance (Eq. (7.92)) will yield anything better than a semi-quantitative indication of the nature of the distribution near this point. An equation providing the same qualitative content could just as well have been derived by finding the intersection of the constant energy line with the undulator band edge. This amounts to finding the angle $\vartheta_{\text{lim.}}$ at which the numerator factor in the last form of Eq. (7.48) has fallen by the same amount the numerator factor is reduced for a typical value $\Delta \nu = 1/(2N_w n)$;

$$\vartheta_{\text{lim.}}^2 = \frac{1}{4\gamma^2 N_w n}. \tag{7.93}$$

This is in semi-quantitative agreement with Eq. (7.92), and with Kim's Eq. (4.33).

7.11.2
Numerical Investigation of Undulator Rings

An ideal monochromator passes only frequencies in a narrow band centered at, say,

$$\omega_{\text{mono.}} = n_{\text{mono.}}\, \omega_1(0), \tag{7.94}$$

where $n_{\text{mono.}}$ is set to an integer or to slightly below an integer. For the example to be worked out shortly, $n_{\text{mono.}} = 7.0$. Substituting Eq. (7.94) into Eq. (7.81) yields

$$\eta(\vartheta) = n_{\text{mono.}} \frac{1 + K^2/2 + \gamma^2 \vartheta^2}{1 + K^2/2} \tag{7.95}$$

as the appropriate parameter at which integral (7.83) is to be evaluated. For large N_w we know that the phasor factor will suppress the field unless η is close to an integer; call the integer $n_{\text{harm.}}$ where

$$n_{\text{mono.}} \leq n_{\text{harm.}} \leq n_{\text{max.}}, \tag{7.96}$$

where $n_{\text{max.}}$ is the highest undulator harmonic that is kinematically possible or some arbitrarily chosen maximum value. The analysis would be simplest for $n_{\text{mono.}} = n_{\text{max.}}$ but, in practice, it may be desirable to center the monochromator on an undulator resonance lower than the maximum possible. (For the example to be worked out shortly $n_{\text{max.}} = 10$.) When this is done the harmonics for which $n_{\text{harm.}} > n_{\text{mono.}}$ yield circular ring profiles centered on the undulator axis, at angle $\vartheta_{\text{harm.}}$ given by solving Eq. (7.95) to obtain

$$\gamma \vartheta_{\text{harm.}} = \sqrt{\left(1 + \frac{K^2}{2}\right) \frac{n_{\text{harm.}} - n_{\text{mono.}}}{n_{\text{mono.}}}}. \tag{7.97}$$

The convergence of series (7.83) is worst for $n_{\text{harm.}} = n_{\text{max.}}$. To save computer time it is sensible to calculate only at points for which the phasor factor is not negligibly small. The larger N_w, the slimmer the rings in which there is any appreciable response. For example in generating Figure 7.22 we have taken $\Delta n_{\text{harm.}} = \pm 1/N_w$ as the range over which the phasor factor is not negligible. This suppresses secondary diffraction rings having intensities in the percent range. The parameters for Fig. (7.22) are identical to the parameters used for the figure on the cover of the text.

7.11.3
Is the Forward Undulator Peak Subject to Angular Narrowing?

We start with a fallacious analogy that suggests there will be strong forward peaking of undulator radiation.

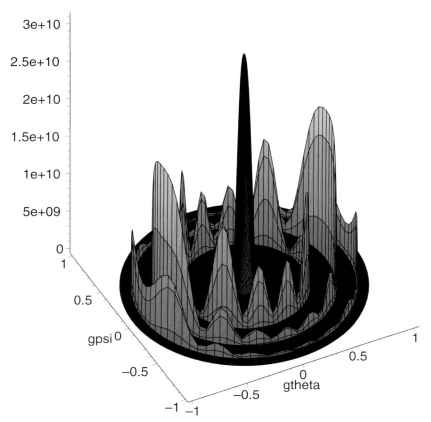

Fig. 7.22 Angular x-ray pattern for published ESRF configuration as recalculated using Eq. (7.83). Physical parameters were $E = 6\,\text{GeV}$, $N_w = 20$, $\lambda_w = 46\,\text{mm}$, $\mathcal{E}_\gamma = 27\,\text{keV}$. Calculational parameters are $N_w = 19$, $i_{\max} = 14$, $n_{\text{mono.}} = 7$. "gtheta" stands for $\gamma\theta$ and "gpsi" stands for $\gamma\psi$. A strong forward peak is visible because the monochromator energy has been matched to the $n = 7$ resonance energy in the forward direction. This figure can be directly compared with another calculation: www.esrf.fr/machine/support/ids/Public/CentralCone/CentralCone.html

The $2N_w$ undulator pulses might be thought to resemble the emission from a linear, phased array of transmitting antennae. Even though the individual elements in such an array radiate more or less isotropically, when they are phased correctly, a narrow beam parallel to the array can be produced. To produce such a beam with free-space wavelength λ, because the wavefronts propagate at the speed of light, successive radiators should be phase shifted by (an odd multiple of) $\pi\lambda_w/\lambda$ to give constructive interference in the direction parallel to the array. The angular width of the radiation pattern can be defined to be the angle of the first interference minimum. For emission at (small) angle ϑ there is a phase shift $N_w\lambda_w\vartheta^2\pi/\lambda$ between radiation from first and last radiators. The condition for the vanishing of the amplitude from all

$2N_w$ radiators is that this phase shift be 2π. For the smallest angle interference minimum this gives

$$\theta_{\min} = \sqrt{\frac{1}{N_w}}\sqrt{\frac{\lambda}{\lambda_w}} = \frac{1}{\gamma}\frac{1}{\sqrt{2N_w}}, \tag{7.98}$$

where, in the last step, to make the connection to undulator radiation, the relation $\lambda = \lambda_w/(2\gamma^2)$ has been used. $\theta_{\min}$ is indeed a small angle; for large N_w, it is much smaller than the cone angle $1/\gamma$ characterizing radiation from a single radiator. This narrowing has occurred through interference in the Fourier angular domain.

The argument in the previous paragraph is fallacious however, since it assumed the radiation to be monochromatic, with wavelength independent of angle. In fact, undulator radiation is spread more or less uniformly in energy all the way from zero to a maximum value. At any angle the interference of all contributions at angle ϑ has already been accounted for in Eqs. (7.37), and Eq. (7.47). Hence, though there are angular peaks of some sort in the undulator spectra, they are not analogous to those due to a linear array of phased, monochromatic transmitters.

This comment is certainly not intended to belittle the usefulness of undulators in general. The narrowing proportional to $1/N_w$ of the frequency spectrum at fixed angle, say in the forward direction, is both uncontroversial and invaluable. When combined with a narrow energy band monochromator centered just below an undulator harmonic edge, it can result in greatly increased brilliance.

Before leaving this topic one can contemplate it using the work-it-out-in-the-electron's-rest-system approach. Representing the undulator field as a monochromatic wave in the electron's frame is manifestly incorrect in the sense that, there, the wave has only N_w full wavelengths and should, therefore, be treated as a wave packet. The electrons subject to this wave execute simple harmonic motion in their own rest frame only during the "time window" during which the wiggler is flying by. The width in time of this window is $\Delta \bar{T}_w = N_w L_w/(\gamma c)$. (Even this is only approximate since the turn-on transient is ignored.) The dipole radiation due to this oscillation will therefore be gated on for a time interval of length $\Delta \bar{T}_w$. At any angle the fields will be the product of a pure sinusoid and a square pulse. The frequency domain spectrum will therefore be the convolution of a single line (from the sinusoid) and a $(\sin \omega)/\omega$ spectrum (from the square pulse). Qualitatively, the rest system frequency will be spread over a range $\Delta \bar{\omega}_w \approx 1/\Delta \bar{T}_w$.

In the absence of the spread just described, the electron rest system dipole radiation is monochromatic; it is only after transformation back into the laboratory system that energy variation results and, even then, there is a one-to-one relation between frequency and angle because the laboratory system angle increases monotonically with increasing rest system angle. At fixed laboratory angle the radiation is therefore not only monochromatic, but is a pure sine wave. The presence of frequency spread in the electron's rest system changes this. Fractionally, the shape of the laboratory frequency spectrum is the same as the shape of the rest system spectrum.

Yet one more point can be made. Analyses of undulator radiation usually employ the Fraunhofer picture, in which the detector is "at infinity". This assumption can be validated either by the image distance being large relative to all relevant source dimensions or by the presence of a parallel-to-point focusing lens. For undulator sources neither of these possibilities is fully available. X-ray lenses do not exist and focusing mirrors are problematic. And, especially with long undulators, the ratio of detector distance to undulator length may not be very large.

References

1 Attwood, D., Halbach, K., Kim, K. (1985) *Science*, **228**, 1265.

2 Jackson, J. (1999), *Classical Electrodynamics*, 3rd edn., John Wiley, New York.

3 Kotredes, L. (2001) *An Explicit Formula for Undulator Radiation,* Cornell Laboratory of Nuclear Studies Report, CBN01-14.

4 Kim, K. (1988), *Characteristics of Synchrotron Radiation*, in *Physics of Particle Accelerators*, M. Month, M. Dienes, (Eds.), American Institute of Physics, Vol. 184, New York.

5 Hoffmann, A. (1978) *Quasi-Monochromatic Synchrotron Radiation From Undulators*, Nucl. Instrum. Methods, **152**, 17.

6 Sands, M. (1971), *The Physics of Electron Storage Rings,* in *International School of Physics,* "Enrico Fermi", Academic Press, New York.

7 Coisson, R. (1977) *On Synchrotron Radiation in Nonuniform Magnetic Fields*, Opt. Commun., **22**, 135.

8 Bossart, (1979) *Observation of Visible Synchrotron Radiation Emitted by a High-Energy Proton Beam at the Edge of a Magnetic Field*, Nucl. Instrum. Methods, **164**, 375.

9 Als-Nielsen, J., McMorrow, D (2001) *Elements of Modern X-Ray Physics*, John Wiley, New York.

10 Walker, R. (1998) *Insertion Devices: Undulators and Wigglers,* in CERN 98-04, S. Turner, (Ed.).

11 Watson, G. (1995), *A Treatise on the Theory of Bessel Functions*, Cambridge University Press, Cambridge, p. 308.

12 Thompson, A. et al., (2001) *X-Ray Data Booklet,* 2nd edn., Berkeley.

13 Chao, A., Tigner, M., (Eds.), (2002) *Handbook of Accelerator Physics and Engineering*, World Scientific, Singapore.

14 Mills, D. (2002) *Third Generation Hard X-Ray Synchrotron Radiation Sources,* John Wiley, New York.

8
Undulator Magnets

8.1
Preview

The purpose of this chapter is to consider the design of undulator magnets and the issues that enter into the choice of their parameters.

Attractive properties of a light source are, low electron energy (for economy), high energy x-rays (to open up experimental possibilities), and high brightness (hence undulator) beams. Unfortunately these properties tend to be mutually exclusive. One thing that can facilitate the compromise among these competing properties is appropriate undulator design. This chapter starts with a (very much over-simplified) discussion of the way that the undulator gap height can be regarded as "determining" the energy of the storage ring in which it is used.

Storage ring practicalities require the chamber height $2g$ to exceed a value of order 1 cm. This sets a minimum value to the undulator wavelength λ_w, which, in turn, sets a practical upper limit to the energy $\mathcal{E}_\gamma$ of x-rays that can be produced with usefully large intensity from an undulator, especially if it is to be operated only on its fundamental resonance. This maximum energy is proportional to the square E_e^2 of the beam energy. For $\mathcal{E}_e = 2.9$ GeV a practical maximum is perhaps 10 keV. Undulator beams of higher energy than this require operation on higher undulator harmonics.

To motivate the discussion, the design of an undulator that maximizes the flux of high energy x-rays will be investigated. This route will lead to a hybrid, asymmetric, electro/permanent-magnet, undulator. As with some other topics in this text, this choice is made more for pedagogical reasons than for the purpose of developing a completely practical design. The single configuration to be investigated exhibits, all by itself, most of the problems to which the various, simpler, more conventional undulator designs are separately subject.

It should be clear from the previous chapters that getting an x-ray beam of maximally high energy requires the strongest possible magnetic field over the shortest possible field region. The idea, illustrated in Figure 8.3, is to produce very strong magnetic fields over very short intervals (spaced by somewhat larger intervals) in order to increase the "critical energy" without caus-

Accelerator X-Ray Sources. Richard Talman
Copyright © 2006 WILEY-VCH Verlag GmbH & Co. KGaA, Weinheim
ISBN: 3-527-40590-9

ing any net beam deflection. The inequality of these intervals corresponds to the "asymmetric" designation.

Though the asymmetric undulator can give a brilliant beam at high energy, this will have to correspond to operation on a quite high order undulator resonance since, because λ_w is relatively large, the fundamental is at a relatively low energy. From a symmetric undulator, because of the cancellation of adjacent pole amplitudes, only odd undulator harmonics contribute flux in the forward direction. For an asymmetric undulator this cancellation is largely suppressed, so the fluxes from even and odd harmonics will be comparable.

8.2
Considerations Governing Undulator Parameters

For centuries optical physicists had to be satisfied with the modest 2:1 dynamic wavelength range from red to violet, because light outside that range was invisible or their equipment was opaque. X-ray physicists have been reluctant to accept a similar restriction, say to the range from 5 keV to 20 keV, even though their equipment tends to be opaque at energies below 5 keV and energies above 20 keV are hard to produce, hard to focus, and hard to handle. For designing a storage ring to be used as a highly selective instrument it may therefore be sensible to adopt a nominal "useful" energy range such as this, and to optimize the accelerator design for that range.[1] Here it is assumed that the nominal range is centered on x-rays of wavelength equal to 1 Å.

Since the cleanest x-ray beams are produced by undulators, it is essential to fold undulator properties into the storage ring design. Apart from the number of poles N_w, the key undulator/wiggler parameter K can be crudely interpreted as

$$K = \frac{\text{maximum deflection angle in wiggler}}{\text{half-angle of radiation cone}}, \tag{8.1}$$

Because the denominator has only a semi-quantitative significance, the same can be said for this ratio, but it is the best we have. Because the deflection angle is proportional to the wiggler field B_0 and the radiated power is proportional to B^2, the radiated power is proportional to K^2. Until recent, third generation, light sources, this dependence has usually caused K to be run to high values (making the insertion device a "wiggler") in order to obtain high flux. As K increases the fundamental undulator resonance shifts to longer wavelengths,

---

[1] Since any electron storage ring produces x-rays at essentially all energies, even though the storage ring has been optimized for a particular energy, it can be used for x-ray experiments at any energy—though perhaps at unattractively low flux values.

so the radiation at a given energy has to come from a higher order harmonic, and the beam is less "clean".

For clean operation, one wants a major share of the power radiated from the undulator to be contained in a single, low order, resonance line. In terms of wiggler period λ_w, the lowest order is given by

$$\lambda_\gamma = \frac{\lambda_w}{2\gamma^2}\left(1 + \frac{K^2}{2} + \text{angular fall-off factor}\right). \tag{8.2}$$

The spectrum (integrated over "out-of-plane" angle) is shown in Figure 12.8. As a compromise between too-dirty operation at too-great wavelength shift, on the one hand, and too-low flux, on the other, the value $K = 1$ will be adopted. (This is typical.)

From the present perspective, the gap height $2g$ of the undulator turns out to be the most important parameter in the whole facility. One visualizes a magnetic field pattern that is independent of x and y and has a perfect square-sawtooth z-dependence, but the reality is far different. No matter what contortions one goes through in undulator design, because of the need to reduce the period λ_w, the on-axis magnetic field always ends up essentially sinusoidal; $B(x, y = 0, z) \sim \cos(2\pi z/\lambda_w)$. The assumed x-independence requires that the magnet poles have been made sufficiently wide. Using Maxwell's equations to match the off-axis behavior yields

$$B(x, y, z) = B_0 \cosh(2\pi y/\lambda_w) \cos(2\pi z/\lambda_w). \tag{8.3}$$

Though this formula is not strictly valid for $y \approx g$, it can be used to estimate the peak pole-tip field $B_{p,\max}$;

$$\frac{B_0}{B_{p,\max}} = \frac{1}{\cosh(\pi 2g/\lambda_w)}. \tag{8.4}$$

This function is plotted in Figure 8.1. Though one is probably willing to tolerate an appreciable reduction from the pole tip of the on-axis field, the presence of rapid dependence on y is undesirable, since it makes the radiated beam properties depend sensitively on steering through the undulator. For this reason, a requirement such as $\lambda_w \geq 8g$ is fairly conservative—the on-axis field is some 30% less than the pole-tip field. Combining the values $K = 1$ and $\lambda_w = 8g$ with Eq. (8.2) yields $\lambda_\gamma = 6g/\gamma^2$ or

$$\mathcal{E}_0 = \gamma mc^2 = mc^2 \sqrt{\frac{6g}{\lambda_\gamma}} = \sqrt{2g[\text{mm}]}\, 2.80\,\text{GeV}. \tag{8.5}$$

This function is plotted in Figure 8.2.

The model leading to Figure 8.2 has been based on crude scaling. The electron energies of actual existing light sources are perhaps a factor of two less than this scaling relation gives.

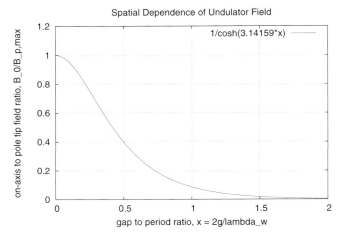

Fig. 8.1 On-axis magnetic field relative to pole tip field as a fraction of gap-height/wiggler-period, $2g/\lambda_w$.

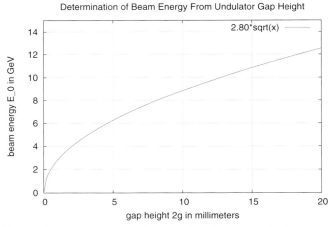

Fig. 8.2 Having chosen λ_γ, $K = 1$ and $\lambda_w/(2g) = 4$, the beam energy is "determined" as a function of $2g$, which is limited (from below) by beam steering and current dependent considerations. This is just a rough scaling illustration.

8.3
Simplified Radiation Formulas

For purposes of estimating the properties of the x-ray beam produced by an undulator it is convenient to specialize some previously derived formulas. The so-called "critical photon energy" u_c for electron energy $\mathcal{E}_e$ in magnetic field B is given by

$$u_c = \frac{3}{4\pi} hc\, \gamma^3 \frac{cB}{\gamma m_e c^2/e} = B[T]\left(\frac{\mathcal{E}_e[\text{GeV}]}{2.9[\text{GeV}]}\right)^2 5.59\,\text{keV}. \qquad (8.6)$$

Half of the radiated energy comes in the form of photons of energy exceeding u_c. For energies greater than u_c, the probability distribution of photon energies is given approximately by

$$\frac{dP}{d(u/u_c)} \approx 0.24\, \frac{\exp(-u/u_c)}{\sqrt{u/u_c}}. \qquad (8.7)$$

where the argument u/u_c is the photon energy in units of the critical energy. This factor has fallen by a factor of 35 for $u = 3u_c$, which makes $3u_c$ a kind of upper limit for what will be called "favorable" operation. Applied to an undulator, Eq. (8.7) provides the "envelope" of a "comb" of undulator resonances. In gross terms therefore (neglecting the comb structure), Eq. (8.7) describes the photon energy distribution, independent of whether the radiation comes from bending magnet, wiggler or undulator. The total energy radiated per electron per unit length, dU/dz, also depends only on the local magnetic field, not the entire insertion device;

$$\frac{dU}{dz} = \frac{e^2}{6\pi\epsilon_0} \gamma^4 \left(\frac{cB}{\gamma m_e c^2/e}\right)^2 = B^2[T^2]\left(\frac{\mathcal{E}_e[\text{GeV}]}{2.9[\text{GeV}]}\right)^2 10.6\,\text{keV}\,\text{m}^{-1}. \qquad (8.8)$$

From an undulator, some fraction of this energy is contained within a single resonance order and, potentially, most of that energy impinges on the scientific apparatus the beamline is intended to serve. From a bending magnet or wiggler the energy is spread in a broad fan, most of which must be removed by collimation.

For $B = 1\,\text{T}$, $\mathcal{E}_e = 2.9\,\text{GeV}$ the critical photon energy (5.59 keV) is, roughly speaking, at the dividing line above which x-rays are sufficiently energetic to pass through a vacuum window without unacceptable attenuation. For present purposes let us refer to photons with energies above 5 keV as "high energy x-rays". From Eqs. (8.6) and (8.7), if we require appreciable flux of high energy x-rays from a beam with energy $\mathcal{E}_e$ as low as 2.9 GeV it will be necessary to have magnetic fields of order 1 T or, preferably, greater. Such magnetic fields are at the high end, or beyond, of what is achievable with permanent magnet material. This requires the use of electromagnets, for which B can be as high as 2 T or superconducting magnets, which can have even higher fields.

Another important consideration is the extent to which "undulator" performance, in which the amplitudes from multiple poles sum coherently, can be achieved. In this regime essentially all the radiation is directed toward the physical apparatus using it (rather than in a broad spray) and, as well as the enhancement factor equal to the number of poles, there is strong forward peaking of the x-ray energies passed by a monochromator. These features make the undulator beam "clean". An important parameter characterizing this behavior is $E_{1,\text{edge}}^{(0)}$ the "edge energy" of the fundamental undulator resonance, for favorable, $K \ll 1$ undulator operation;

$$E_{1,\text{edge}}^{(0)} = hc\frac{\gamma^2}{\lambda_w/2} = \left(\frac{\mathcal{E}_e[\text{GeV}]}{2.9[\text{GeV}]}\right)^2 \left(\frac{1\,\text{cm}}{\lambda_w/2[\text{cm}]}\right) 3.99\,\text{keV}; \qquad (8.9)$$

(the factor 1/2 has been left explicitly in the factor $\lambda_w/2$ for later convenience in analysing an asymmetric undulator.) This edge energy is independent of B. So, surprisingly, the range of energies radiated coherently is independent of B (in the limit of "pure undulator operation" $K \ll 1$.) This is somewhat academic however, as the flux increases strongly with K and one essentially never runs in the limit of pure undulator operation. In fact, satisfactory undulator operation can be achieved up to $K = 2$ or somewhat higher, and say, $n = 10$ (actually 9 or 11) where n labels the undulator harmonic. There is appreciable intensity into the higher harmonics only for $K \approx 1$ or greater. The edge energies of general harmonics are given by

$$\mathcal{E}_{n,\text{edge}} = \frac{n}{1+K^2/2}\left(\frac{\mathcal{E}_e[\text{GeV}]}{2.9[\text{GeV}]}\right)^2 \left(\frac{1\,\text{cm}}{\lambda_w/2[\text{cm}]}\right) 3.99\,\text{keV}. \qquad (8.10)$$

Since the undulator parameter K is proportional to B, the denominator factor $1 + K^2/2$ causes "diminishing returns" to set in as one attempts to increase the x-ray energies by increasing B. This sets a practical upper limit on K of order 2 or slightly higher if "clean beam" undulator performance is to be achieved.

For a fixed value of $\mathcal{E}_e$, these consideration and the practicalities of magnet construction, set an upper x-ray energy limit for "clean" operation. It is necessary to reduce λ_w to the extent possible, but it is impractical for $\lambda_w/2$ to be less than the magnet gap height $2g$ which cannot be less than, say, 6 mm, without impairing storage ring operation. Also it is difficult to achieve small values of $\lambda_w/2$ using powered magnets requiring current-carrying coils around every pole. For the value $\mathcal{E}_e = 2.9\,\text{GeV}$ featured in the formulas given so far, one needs values $\lambda_w < 2\,\text{cm}$, $B > 1\,\text{T}$ to obtain undulator operation in the range from 5 keV (easy) to 20 keV (hard). Increasing $\mathcal{E}_e$ from 2.9 GeV would, of course, pay off handsomely in higher energy x-rays.

Formulas given in this section have assumed a "symmetric" undulator—equal pole lengths and equal but opposite field strengths. Relaxing this requirement makes the constraints more easily achievable.

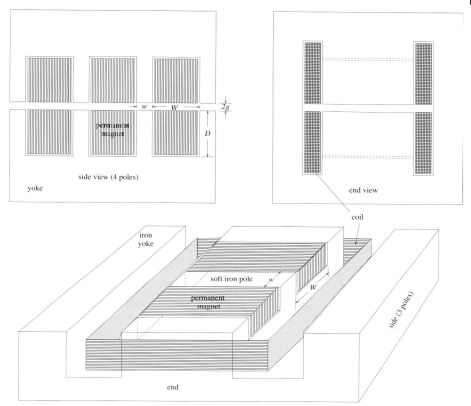

Fig. 8.3 Schematic illustration of coil, yoke, and permanent magnetic material in a hybrid, electro/permanent-magnet, asymmetric undulator. The mechanism to vary the gap height $2g$ is not shown. Only half of the magnet is shown in the lower figure, and only a few of the poles at that.

8.4
A Hybrid, Electo-permanent, Asymmetric Undulator

In this section analytic design formulas for both electromagnet and permanent magnet sections will be given and it is argued that the fields should approximately superimpose. But the interaction of electromagnet and permanent magnets is subtle so (obviously) a numerical study is needed when electromagnets and permanent magnets are present simultaneously. Certain undulator constraints can be relaxed by permitting the undulator to be "asymmetric", meaning that the magnetic field is strong and of one sign over a short section and weak with opposite sign over a long section; the length×strength products must be equal if the beams from all the poles are to superimpose, and the electron beam is to suffer no net deflection. It will be argued that u_c

(as given by Eq. (3.39) or by Eq. (8.6)) needs only to be large in the strong field regions. Then, for estimating the brilliance of an asymmetric undulator, $2w$, where w is the pole width, can be substituted for λ_w in formulas derived for symmetric undulators. The large values of B in high field regions can be provided by an electromagnet. The longer and weaker, opposite sign, fields can be provided by permanent magnets. Such a magnet design is described in this section.

The proposed magnet is illustrated in Figure 8.3. Its purpose is to produce a very strong vertical magnetic field in the short gap region between soft iron poles. The maximum x-ray energy that can be produced coherently from multiple magnet poles is dominated by the pole width w, which is therefore to be made as short as possible consistent with the product $B_0 w$ being sufficiently large. Since the maximum field B_0 is produced by an electromagnet, its value can approach 2 T. The purpose of the permanent magnet inserts is to produce magnetic field of opposite sign, needed because the undulator must produce no net bending. Most undulators are "symmetric" such that north and south pole widths are equal ($w = W$) and the field maxima are equal but opposite in sign. Here $w \ll W$ is being permitted. Since the maximum field achievable with permanent magnets is less than with electromagnets, and the permanent magnet has to overcome the (small) electromagnetic field between poles, the field in the permanent magnet sections will be much less than between poles, which is consistent with $w \ll W$. For assessing the x-ray energies emitted from a symmetric undulator, the undulator period λ_w is the most important parameter, but for our asymmetric undulator it will be w rather than period $w + W$ that governs the high energy cut-off of the radiation. The energy of the fundamental undulator resonance will, however, be governed by $w + W$. For improved field uniformity the gap height $2g$ should be as small as possible, but storage ring practicalities will make it impossible to reduce $2g$ below some minimum value such as 6 mm.

For purposes of first-cut design of such a magnet, a kind of superposition principle will be assumed. First the electromagnet field between soft iron poles will be calculated, ignoring the permanent magnet material. Then the field due to the permanent magnet material will be calculated ignoring the electromagnet (except for the iron return yoke). The two field components will then simply be summed. Since ferromagnetism is thoroughly nonlinear, such superposition can only be approximate. Two extreme simplifying assumptions will be that the iron permeablility is infinite and the permanent magnet permeability is equal to the free space value μ_0. Two features possibly invalidating this approach are demagnetization of the permanent magnets and saturation of the iron. Failure of superposition will be briefly considered below.

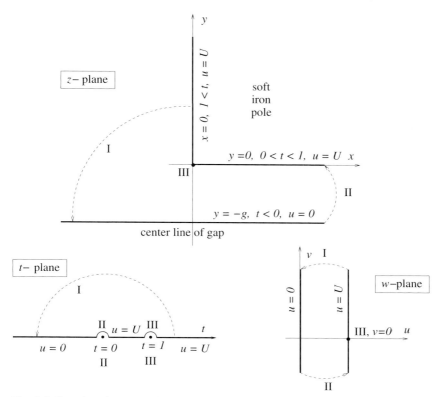

Fig. 8.4 Complex planes used to determine the magnetic field in the region between one half pole and the centerline of the gap. At point III, the corner of the pole, $t = 1$ and $x = y = v = 0$.

Though the magnet design to be studied is somewhat impractical, the analytic formulas describing its magnetic field profiles are intended to be very realistic. They should therefore be applicable to most undulator magnets.

8.4.1
Electromagnet Design

For the small values of gap height $2g$ that will undoubtedly be required, the magnetic field can be calculated by concentrating on the region between the iron pole and the magnet centerline. Half of this region is labeled "z-plane" in the upper part of Figure 8.4. The field will be calculated by conformal transformation, using the subsidiary complex w and t planes, also shown in Figure 8.4. These figures, and the subsequent analysis, are copied from Bewley [1]. In this approach any analytic function in any of the planes automatically satisfies the 2D Laplace equation in that plane, with the real and imaginary coordinates interpreted as Cartesian coordinates. $\Phi(w)$ is known as the "complex, magnetic scalar potential", and the magnetic field components are obtained from it by

$$-B_x + iB_y = \frac{d\Phi}{dz}. \tag{8.11}$$

This formula is valid because both divergence and curl of **B** vanish. In our case $\Phi \equiv w$.

The starting point for the calculation is the known (linearly increasing) potential between parallel plates in the $w = u + iv$-plane. In this plane $\Re(\Phi(w)) \equiv \Re(w)$ satisfies the boundary conditions $\Phi(0 + iv) = 0$, $\Phi(U + iv) = U$.

The concatenated analytic transformation $w \to t \to z$, produces the desired boundaries and the appropriate boundary conditions in the z-plane. The tranformation $w \to t$ is

$$t = -e^{-i\pi w/U} = e^{\pi v/U}\left(\cos\frac{U-u}{U}\pi + i\sin\frac{U-u}{U}\pi\right), \tag{8.12}$$

On the boundaries this simplifies to

$$t = \begin{cases} -e^{\pi v/U} & \text{for } u = 0, \\ e^{\pi v/U} & \text{for } u = U. \end{cases} \tag{8.13}$$

The inverse transformation is

$$\frac{w}{U} = 1 + \frac{i}{\pi}\ln t = 1 - \frac{\theta}{\pi} - \frac{i}{\pi}\ln\rho, \tag{8.14}$$

where $t = \rho e^{i\theta}$. One can check that the function $\Re(w(t))$ satisfies the boundary conditions shown in the figure (U on the positive real t-axis, 0 on the negative real t-axis.) The further transformation $t \to z$ is given by

$$\begin{aligned}x + iy &= \frac{-2g}{\pi}\left(\sqrt{1-t} - \frac{1}{2}\ln\frac{1+\sqrt{1-t}}{1-\sqrt{1-t}}\right) \\ &= i\frac{2g}{\pi}\left(\sqrt{t-1} - \tan^{-1}\sqrt{t-1}\right),\end{aligned} \tag{8.15}$$

where the formula option permits the argument of the square root always to be chosen positive. Eliminating t from these formulas using Eq. (8.12) gives transformation formulas $(u,v) \to (x,y)$. v varies on contours of constant "potential" u, while on (orthogonal) field lines u varies and v is constant. Eqs. (8.15) therefore give the equations of both sets of curves in parametric form. Explicitly, the upper formula becomes

$$\frac{x+iy}{g} = \frac{-2}{\pi}\sqrt{1+\exp(-i\pi w/U)} + \frac{1}{\pi}\ln\frac{1+\sqrt{1+\exp(-i\pi w/U)}}{1-\sqrt{1+\exp(-i\pi w/U)}}. \tag{8.16}$$

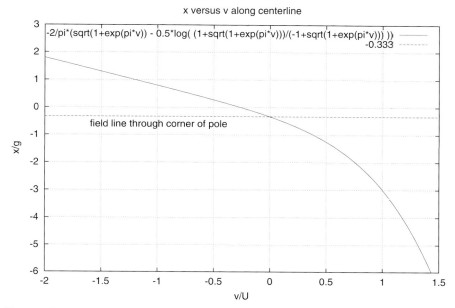

Fig. 8.5 Relation between x and v along the centerline. Note that the $v = 0$ contour (the one leaving the corner of the pole) intersects the centerline at a slightly negative value of $x = -0.333g$. As v becomes more positive, the constant v contour bulges out more and more (i. e. x becomes increasingly negative.)

On the boundaries, pole-side, pole-face, and centerline, the boundary conditions are, respectively,

$$\frac{y}{g} = \frac{2}{\pi}\sqrt{e^{\pi v/U} - 1} - \frac{2}{\pi}\tan^{-1}\sqrt{e^{\pi v/U} - 1} \quad 1 < t, 0 < v, w = U + iv,$$

$$\frac{x}{g} = \frac{-2}{\pi}\sqrt{1 - e^{\pi v/U}} + \frac{1}{\pi}\ln\frac{1 + \sqrt{1 - e^{\pi v/U}}}{1 - \sqrt{1 - e^{\pi v/U}}}, \quad 0 < t < 1, v < 0, w = U + iv,$$

$$= \frac{-2}{\pi}\sqrt{1 + e^{\pi v/U}} + \frac{1}{\pi}\ln\frac{1 + \sqrt{1 + e^{\pi v/U}}}{1 - \sqrt{1 + e^{\pi v/U}}}. \quad t < 0, w = iv. \tag{8.17}$$

The third of these relations is plotted in Figure 8.5 and Figure 8.6 shows a complete field plot. Using Eq. (8.11) the field is given by

$$B_x - iB_y = -\frac{dw}{dz}; \tag{8.18}$$

In the process of deriving these transformations these two derivatives appeared:

$$\frac{dz}{dt} = i\frac{g}{\pi}\frac{\sqrt{t-1}}{t}, \quad \frac{dw}{dt} = i\frac{U}{\pi}\frac{1}{t}. \tag{8.19}$$

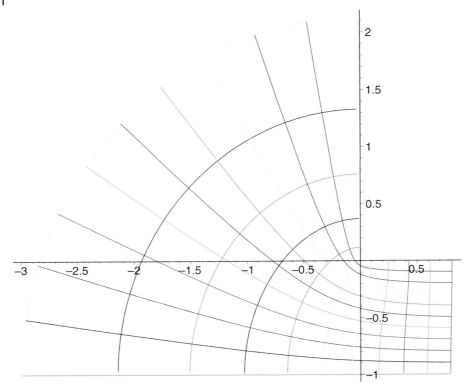

Fig. 8.6 Field plot for the ranges $0 < u < 1$, $-1 < v < 1$. Judging by the spacing between curves of constant v the field magnitude at $(x, y) = (0, 2g)$ is roughly one fifth of the field in the gap. This suggests that D should be greater than $2g$.

We therefore have

$$B_x + iB_y = -\left(\frac{dw}{dz}\right)^* = -\left(\frac{dw}{dt}\frac{dt}{dz}\right)^* = -\left(\frac{U}{g\sqrt{t-1}}\right)^*$$
$$= \frac{-U}{g}\left(e^{\pi v/U}\left(\cos\frac{U-u}{U}\pi - i\sin\frac{U-u}{U}\pi\right) - 1\right)^{-1/2}. \quad (8.20)$$

We are primarily interested in the value of B_y along the centerline where $u = 0$;

$$\frac{B_y g}{U} = \frac{1}{\sqrt{1 + e^{\pi v/U}}}, \quad (8.21)$$

and along the pole face where $u = U$;

$$\frac{B_y g}{U} = \frac{1}{\sqrt{1 - e^{\pi v/U}}}, \quad (8.22)$$

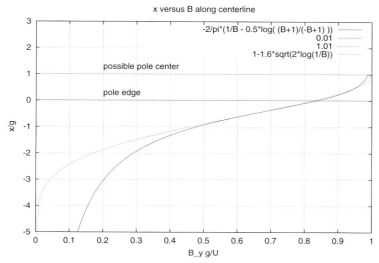

Fig. 8.7 Magnetic field profile along the centerline due to one half pole. The pole edge is at $x = 0$. The pole width can be as little as $2g$ (the case shown) with only 2% field reduction from the nominal (thick pole) value. Also shown is a Gaussian fit that matches fairly closely this particular choice of pole width.

The absolute value of the magnetic field is given by

$$\frac{|\mathbf{B}|g}{U} = \left(e^{2\pi v/U} - 2e^{\pi v/U}\cos\frac{U-u}{U}\pi + 1\right)^{-1/4}. \tag{8.23}$$

Substituting from Eq. (8.21) into the third of Eqs. (8.17) yields

$$\frac{x}{g} = \frac{-2}{\pi}\left(\frac{U}{B_y g} - \frac{1}{2}\ln\frac{B_y g/U + 1}{-B_y g/U + 1}\right), \quad t < 0, \tag{8.24}$$

along the centerline. The inverse of this formula gives the magnetic field profile along the centerline of the magnet; it is plotted in Figure 8.7. Because $x = 0$ corresponds to the pole edge, rather than the pole center, it is necessary to specify the position of the pole center. For a wide pole $B_y g/U$ approaches 1 in the interior of the pole. Since we want the pole to be as slender as possible a choice, such as shown in the figure, with full pole width equal to full gap $2g$ seems appropriate since the field is within 2% of its full value at $x = g$. In the spirit of making σ_z as small as possible this is an aggressively small choice. But an even somewhat smaller choice could be made without much loss of central field relative to nominal. Also plotted in Figure 8.7 is an empirical (Gaussian) fit of the form

$$B_y = \frac{(Bg)_{\text{nom.}}}{g} \exp\left(-\frac{(x-w/2)^2}{2(1.6)^2 g^2}\right); \tag{8.25}$$

This is a crude fit, that does not properly combine the pole width and field spreading in the gap except at the one point where the fit was made. So it can be expected to give the dependence on g over only a small range. More properly the parameters should be fit for various values of g holding w fixed. Of course saturation also invalidates the formula for small g. Nevertheless, for subsequent calculations, I will use the value

$$\sigma_z = 1.6\, g, \quad \left(\stackrel{\text{e.g.}}{=} 0.51\text{ cm for } 2g = 0.6\text{ cm.}\right) \tag{8.26}$$

8.4.2
Permanent Magnet Design—Small Gap Limit

The magnetic field in the permanent magnet sections can be estimated while referring to Figure 8.8. Assuming constant magnetization M (directed up in the figure) within the permanent magnet material, the permanent magnet block can be modelled by bound surface current densities M flowing in and out of the page. In this picture the material itself is replaced by free space, of permeability μ_0. The B-field will therefore resemble that of a rectangular solenoid, more or less uniform in the interior of these current sheets. Assuming $g \ll W$ the outward bowing of the fields in the gap region should be modest and will be neglected. As a result

$$B_g = B_m \tag{8.27}$$

where subscripts g and m stand for "gap" and "magnetic material". Treating the iron as ideal, there is no contribution within the iron region to the Ampèrian loop integral along the path shown so this loop links zero true current.[2]

---

2) In fact the Ampère loop also links the current in the main dipole coil, but I am assuming $D \gg g$, and hence that negligible flux from the main coil actually passes through the permanent magnet blocks. Of course this is wrong, especially adjacent to the corners of the iron poles where the "fringe fields" pass through the corners of the permanent magnet blocks. In the present approximation this is being neglected. It will be (crudely) accounted for later. The error becomes fractionally less significant for $W \gg w$. Since one will be striving to make B_w as large as possible it may be necessary, for obtainable magnet material, even to allow the magnet block corners to be demagnetized by the main dipole current. This may make it tricky to maintain zero field integral, because of hysteresis, for example, as the gap height is changed. At worst the beam steering caused by the device can be adjusted empirically to zero by varying the main dipole current while viewing the closed orbit in the storage ring.

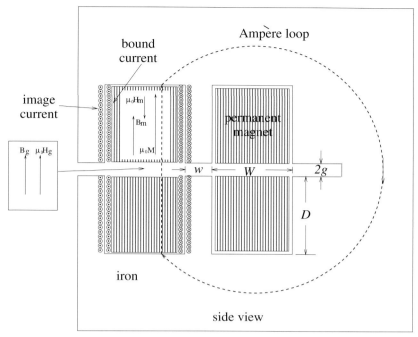

Fig. 8.8 Two adjacent permanent magnet sections, separated by an iron pole, are shown. The Ampère loop links the "bound" current of the permanent magnet and the "image" current in the iron, but no "free" current.

From this and Eq. (8.27) the equation of the "load line" is obtained;

$$2gH_g + 2DH_m = 0, \quad \text{so} \quad B_m = -\frac{D}{g}\mu_0 H_m. \tag{8.28}$$

Load lines for various values of D/g are plotted in Figure 8.9. Also shown are their intersections with the demagnetization curve for $SmCo_5$. This seems to be the material of choice because of its extremely large value of coercive force, which is the intersection of the dashed curve with the horizontal axis; in the units of the figure the intercept is at about 2 T. Maximizing this parameter will minimize the problem of irreversible demagnetization due to the electromagnet field.

Demagnetization curves for other materials are also shown. By using Nd-FeB, fields almost 50% higher could be achieved. But there seems to be little reason to push for the highest possible field, since this would only reduce the ratio W/w proportionally. This would permit more complete periods per unit length of undulator, and hence higher total flux, but it would also exaggerate the importance of the problematical regions close to the iron poles, both because of demagnification and their greater fractional importance. Figure 8.9(b)

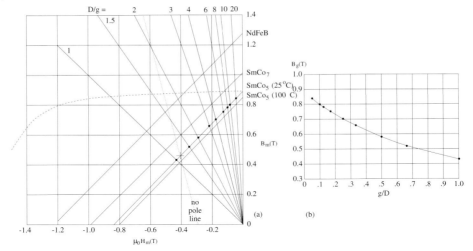

Fig. 8.9 (a) Demagnetization curves for various permanent magnet materials. From Campbell [2], Also shown are load lines given by Eq. (8.28) and their intersections with the SmCo$_5$ load line for various values of D/g. The dashed curve gives the magnetization $\mu_0 M$. (b) The corresponding dependence of B_g, the field in the gap, on g/D.

shows the dependence of B_g on g/D. A polynomial fit to the data is

$$B_g[T] = 0.8704 - 0.7359(g/D) + 0.2974(g/D)^2. \tag{8.29}$$

8.4.3
Combined Electro-/Permanent- magnet Design

Figure 8.9 suggests that the peak field at minimum g can be $B_g \approx 0.8\,\text{T}$, with $g/D = 0.1$. However the dependence of B_g on g is rather weak at this point; the dependence is slower than the $1/g$ dependence of the iron pole fields. This suggests that the operating point should be chosen closer to the center of the range shown in Figure 8.9(b). Tentatively I take $g/D = 0.25$ and therefore $B_g = 0.7$. Then, selecting $B_w = 1.6\,\text{T}$ would seem to require $W/w \approx 2.5$. But the "stray flux" through the permanent magnet material from the electromagnet has been ignored. A feeble attempt to estimate this effect is indicated by the dotted line labeled "no pole line" in Figure 8.9. If there were no iron poles then, effectively, the gap would be five times greater and the B-field five times less, or about 0.3 T. This reduces the permanent magnet field by roughly $1/3$. Furthermore the reduction in the corners is even greater. To compensate I select $W/w \approx 5$. The aspect ratios visible in Figure 8.10 correspond, more or less, to these tentative choices.

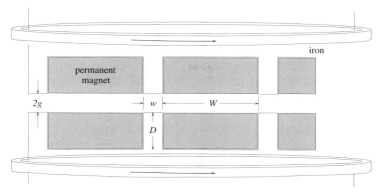

Fig. 8.10 Pole profile of the asymmetric undulator. The relative dimensions shown are a tentative starting point for a more careful analysis that accounts for the interaction of electro- and permanent-magnet effects. The electromagnet coils are shown schematically only as a reminder of their presence.

8.4.4
Estimated X-ray Flux

In the high energy region this undulator is intended to service, the flux will be dominated by the high field regions. According to Eq. (8.25) the magnetic field in these regions is given by

$$B(z) = B_0 \exp\left(-\frac{z^2}{2(1.6g)^2}\right). \tag{8.30}$$

I will neglect all fields other than this. From Eq. (8.6), assuming a field $B_0 = 1.6\,\text{T}$ can be achieved, the critical energy will be $u_c \approx 9\,\text{keV}$ at $\mathcal{E}_e = 2.9\,\text{GeV}$, and there will be "favorable" (defined above) flux up to about 27 keV.

One can introduce a kind of symmetric "comparison undulator" for which the fields within one half period more or less match Eq. (8.30). Its undulator wavelength would be

$$\lambda_{w,\text{comp.}} \approx 2\pi\sigma_z = 1.6\pi(2g), \quad (\stackrel{\text{e.g.}}{=} 3\,\text{cm for } 2g = 6\,\text{mm.}) \tag{8.31}$$

The comparison K value would then be

$$K_{\text{comp.}} = 0.934\, B_0[\text{T}]\, \lambda_{w,\text{comp.}}[\text{cm}] \,(\stackrel{\text{e.g.}}{=} 2.8\, B_0[\text{T}] \text{ for } 2g = 6\,\text{mm,})$$
$$(\stackrel{\text{e.g.}}{=} 4.6 \text{ for } B_0 = 1.6[\text{T}].) \tag{8.32}$$

This comfortably exceeds anything one would need in practice. Compared to the comparison undulator, the total power radiated from the asymmetric undulator of the same total length will be reduced by a factor equal to twice the ratio of periods,

$$\frac{N_{\text{asym.}}}{N_{\text{comp.}}} = \frac{1}{2}\frac{\lambda_{w,\text{comp.}}}{w+W} \approx \frac{1}{2}\frac{1.6\pi(2g)}{2g+10g} \approx 0.4. \tag{8.33}$$

As mentioned before, the fluxes into even and odd undulator harmonics will be comparable. This means the flux into any particular odd harmonic will be reduced by another factor of two over and above the reduction given by Eq. (8.33). In compensation for this, the presence of intense even harmonics will ameliorate the problem of "holes" in the spectrum—this is a nuisance associated with running in the low order harmonic region.

References

[1] Bewley, L. (1948), *Two-Dimensional Fields in Electrical Engineering*, Dover, New York.

[2] Campbell, P. (1994), *Permanent Magnet Materials and Their Applicatation*, Cambridge University Press, Cambridge.

9
X-Ray Beam Line Design

9.1
Preview

Most of this text is devoted to the accelerator end of x-ray beam lines, and very little is devoted to the detector end. One source discussing detection apparatus in detail is a book edited by Mills [1]. This chapter concentrates on the design of the beam lines used to connect the accelerator to the detection equipment. Hofmann's article [2] contains similar material.

Rapid advances are currently being made in x-ray optics—the rate rivals that of any field of modern science. This is at least partly due to recent rapid advances in *nanofabrication*. Probably the main impediment to x-ray beam line design has been, until recently, the absence of practical x-ray lenses. This is primarily because the ultraweak refraction of x-rays, coupled with strong absorption, makes the attenuation too great in an x-ray lens of useful strength. Fresnel-like modification is especially appropriate for x-rays because it can greatly reduce attenuation, by making the lens more or less uniformly thin. Evans-Lutterodt et al. [3] refer to this construction method as "removing 2π phase-shifting regions", but the extraordinarily small spacings needed for such lenses have made their fabrication very difficult. Modern nanofabrication techniques are enabling rapid advances in the area of zone plate growth and schemes for saw-tooth lens sculpturing that reduces absorption without seriously degrading focal properties.

As well as covering some tried-and-true techniques cursorily, a few more futuristic applications, that are likely to become useful only with fourth generation light sources, are covered in considerable detail in this chapter. X-ray lenses are inherently highly chromatic, but with the ultraslim electron beams anticipated from fourth generation electron beams it should become possible to compensate chromaticity using the well defined undulator radiation pattern. This applies especially to the use of x-ray lenses. The above-mentioned Fresnel-type lenses will not be emphasized however. Rather the analyses will assume simple refractive lenses, but the methods should apply satisfactorily to any sort of lens.

Accelerator X-Ray Sources. Richard Talman
Copyright © 2006 WILEY-VCH Verlag GmbH & Co. KGaA, Weinheim
ISBN: 3-527-40590-9

9.2
Beam Line Generalities

The typical length of an x-ray beam line is tens of meters, but the characteristic transverse dimension is in the micrometer to millimeter range. This calls for apparatus of very high precision. The angular spreads of x-ray beams are extraordinarily small. This, plus the fact that several physical phenomena independently establish small angle limits, complicates beam line design. The beam properties depend on the relative importance of the various small angle phenomena. The natural cone angle of synchrotron radiation is $1/\gamma$ which, though small itself, is typically large compared to the angular acceptances of collimators and monochromators, and to the horizontal and vertical angular spreads of the electron beam. Also the energy spread at fixed angle of undulator radiation is quite narrow. As one beam line parameter, such as photon energy, is varied continuously, it is not uncommon for another beam line property, such as flux, to change discontinuously when one constraint is replaced by another.

An electron beam passing through a magnet makes its presence known by the radiation it emits. By imaging this radiation it is possible to determine the transverse dimensions (and hence emittances) of the beam. A task of greater practical importance is to tailor the radiation optimally into a beam of photons to be used as the incident beam for an x-ray science experiment. These two tasks will be considered together, but concentrating on those features that will be of greatest significance in producing photon beams of high quality. Scrapers are to be avoided as far as possible. Certainly, for diagnosing the electron beam, scrapers must not be permitted to destroy the image.[1] For the design of actual beam lines the no-scraper limitation eventually has to be relaxed.

Most x-ray beam line design is predicated on providing a monochromatic beam that is as brilliant as possible. In first generation light sources the typical configuration resembled Figure 3.6; achievable fluxes and brilliance in that configuration were estimated in connection with that figure. By now, to obtain higher useful flux, the source of choice is an insertion device operated in the wiggler or, better yet, the undulator regime. Since a typical beam line is tens of meters in length, in spite of the small cone of synchrotron radiation, some means must usually be found to limit the spot size at the detector end of the beam line. To preserve flux this would preferably be done losslessly by means of lenses. For the visible part of the synchrotron spectrum this is practical, but x-ray lenses have, until recently, been somewhere between difficult and impossible to produce. A certain amount of focusing can also be achieved

---

[1] An example of beam collimation that *does not* destroy the beam image is the "pin hole camera", though its resolution leaves much to be desired. This, "lowest tech" diagnostic device there is, permits the beam to be imaged on a distant screen.

using curved mirrors. But, especially as the x-ray energy increases, only very glancing reflection is possible. This makes reflective optical design difficult and distortion-free optics impossible.

Commonly the first element of the beam line is an element that limits the beam transversely. This element may be a scraper/collimator that allows through only photons that are close to the axis. Another element that can limit the beam transversely is the monochromator, a device based on Bragg scattering from a crystal or crystals. Since achieving a monochromatic beam is usually a requirement, there will usually be a monochromator somewhere along the beam line. As well as limiting energy bandwidth this will limit angular spread in at least one of the two transverse planes. In some cases, though, it may be useful to suppose that the monochromator reflects in both transverse planes and therefore that it establishes angular apertures in both planes. Where appropriate one should also consider the possibility of restricting the beam transversely using scrapers. In fact, if a monochromator limits the beam transversely, a scraper placed nearby can establish a slightly larger aperture without intercepting any but stray photons.

Though considerable cleverness goes into the design of the line following an initial collimator, the properties of the beam line are largely already determined by the initial collimation. The term "aperture-defined" will be used to describe a beam with limited initial angular range and "aperture-free" for the alternative possibility. Incidentally, the terms "scraper", "collimator", "aperture", and "jaws" are used more or less interchangeably.

Viewed from the accelerator end, it is *brilliance*, because it is an invariant quantity, that is the most readily evaluated and the most useful measure of beam intensity. Crudely speaking, the challenge is to produce maximum brilliance at the upstream end of the beam line in order to make it possible to obtain maximum useful flux through the sample at the downstream end.

For designing a photon beam line it is necessary to have accurate descriptions of its elements. For visible light these elements include lenses, mirrors, prisms, gratings, filters etc., all of which are presumed to be well understood. The emphasis here will be on analogous elements for x-rays. The only "new" physics that this entails is Bragg scattering from crystals and the index of refraction (including absorption) of x-rays in homogeneous matter. The main beam line element using Bragg scattering is the monochromator. The beam line elements that depend on index of refraction are mirrors (for which only glancing reflections of x-rays are possible) and lenses (for which only very weak, very long focal length lenses are possible). Just enough of the physics of these devices will be given to support their phenomenological description.

After introducing those accelerator parameters that influence beam line properties, because monochromators are so important, this chapter continues with a brief discussion of Bragg scattering. A discussion of aperture-limited beams follows. After that some other possible beam elements are introduced, including mirrors and lenses. Following that is discussion of the possibilities opened up by lenses. A synchrotron light beam, either visible or x-ray, can also be used to "diagnose" the electron beam's properties. It is natural to consider such "beam cameras" along with scraper-free beam lines.

The design of an x-ray beam line depends critically on the sort of experiment for which the x-rays are intended. Some of the important parameters, such as flux, brightness, and brilliance, have been defined in Chapter 3. All assume a narrow energy bandwidth, 0.1%BW. Ultimately it is *flux through sample* that is most important, since that is what determines the data rate of the experiment. However, *flux through sample* becomes definite only along with a designation of the sample size. So, when only small samples are available, it becomes *flux density* (per unit area) that determines the data collection time. For some experimental set-ups, such as crystal diffraction, near parallelism is required, making *brilliance* the critical parameter. For other experiments, such as x-ray absorption or atomic excitation, the angular spread is relatively unimportant, so *flux* is the important rate-determining measure. There is a strong tendency for flux density to decrease with distance along the beam line. This tendency can only be reversed with some sort of focusing device.

9.3
Accelerator Parameters

To determine intensity, brilliance, distribution functions and other properties of the produced beam, it is necessary to address accelerator physics practicalities. For a start we assume the radiated power is small enough to make it legitimate to neglect degradation of the electron beam caused by the undulator deflections. This topic is discussed further in Section 5.7.

It is implicitly assumed in most discussions of synchrotron radiation (including this one) that the bend plane is horizontal and is designated as the x plane; the dominant field component is then E_x. (For the same reason) practical electron beams are usually ribbon-shaped, with transverse sigmas related by $\sigma_y \ll \sigma_x$. Because of this, it can turn out that vertical deflections, for example in monochromators, give superior performance for some purposes. I leave this as an open question, but continue to assume implicitly that bend planes are horizontal.

Some typical parameter values for the beta functions and emittances of an electron beam in a present day, second or third generation light source, at the center of the undulator are:[2]

$$\beta_x = 8\,\text{m}, \quad \beta_y = 4\,\text{m}, \quad \epsilon_x = 2 \times 10^{-8}\,\text{m r}, \quad \epsilon_y = 2 \times 10^{-9}\,\text{m r}. \tag{9.1}$$

The beam sizes are therefore

$$\sigma_x = \sqrt{\beta_x \epsilon_x} = 400\,\mu\text{m}, \quad \sigma_y = \sqrt{\beta_y \epsilon_y} = 90\,\mu\text{m},$$

$$\sigma'_x = \sqrt{\epsilon_x/\beta_x} = 50\,\mu\text{r}, \quad \sigma'_y = \sqrt{\epsilon_y/\beta_y} = 22\,\mu\text{r}. \tag{9.2}$$

In some cases the beta function can be varied. One can reduce either the transverse beam size or the angular spread, but only at the cost of increasing the other. The optimal choice depends on the intended application.

For a "fourth generation" source, such as the energy recovery linacs (ERL) (discussed in Chapter 10) or rapid cycling storage ring (discussed in Chapter 11) currently under development, the emittances will be much smaller, and probably equal in both planes; say

$$\beta = 8\,\text{m}, \quad \epsilon = 10^{-10}\,\text{m r}. \tag{9.3}$$

The spatial and angular spreads corresponding to these values are

$$\sigma = 28\,\mu\text{m}, \quad \sigma' = 3.5\,\mu\text{r}. \tag{9.4}$$

The angular spreads just calculated can be compared with the synchrotron radiation cone angle, which is roughly $1/\gamma$. Using the same value $\gamma = 5675$ used in previous calculations, this angular spread is 176 μr. If the x-ray beam is produced by an undulator, and has been passed through a monochromator, the beam cone is far narrower. The cone angle for a commonly used configuration has been determined in Eq. (7.92), to be

$$\sigma_\theta = \sqrt{\frac{1}{\pi} \frac{1 + K^2/2}{\gamma^2} \frac{1}{N_w n}}. \tag{9.5}$$

Variants of this formula (possibly due to Kim in the first place) are in common use. See, e.g. Mills Eq. (2.9) on p. 44, or Krinsky Eq. (1.58) on p. 21 of Mills' book [1]. This cone angle typically works out to be far less than $1/\gamma$. It is specific to the highly specialized situation (very relevant in the context of beam line design) of radiation passing through a monochromator which has been adjusted to "sit" precisely on the energy of the forward peak of a particular undulator harmonic. Should the monochromator be adjusted to lower energy the radiation would appear (with only somewhat reduced, but not easily usable, brilliance) in a hollow cone at a far larger angle of order $1/\gamma$. For third

---

[2] The vertical emittance is extremely conservative. The NSLS X-Ray Ring at Brookhaven reports $\epsilon_y = 0.7 \times 10^{-10}$ m at 2.6 Gev in a ring of circumference 170 m.

generation light sources this radiation cone is much greater than the spread of vertical electron angles. For fourth generation sources the same will be even more true in both planes.

9.4 Bragg Scattering and Darwin Width

There is more or less total scattering of x-rays incident on a crystal at the Bragg angle θ_B (unlike Snell's law, this angle is measured from the plane, not from the normal) which is given by

$$\sin \theta_B = \frac{\lambda}{2d}, \qquad (9.6)$$

where λ ($\stackrel{\text{e.g.}}{=} 1.0$ Å) and d is the spacing between the crystal planes involved in the scattering. Some Bragg scattering parameters are shown in Table 9.1 and Table 9.2 for some crystals commonly used for monochromators.

Figure 9.1 exhibits the theoretical dependence of reflectivity and phase of scattered beam (relative to incident beam) on "rocking angle" $\Delta\omega$, which is the angular deviation away from the exact Bragg orientation. The Darwin theory is explained by Als-Nielsen and McMorrow [4]. Figure 9.1 is adapted from their Figure 5.8. As the monochromator is rocked by tens of microradians the phase of the reflected beam shifts by 2π. The reflection is complete over a range of angles $\theta \pm \omega_D/2$ centered on the Bragg angle θ_B. The angular width ω_D is known as the "Darwin width". The reflectivity is essentially perfect for photons parallel to the plane orthogonal to the plane of incidence. The Bragg planes need not be parallel to the surface of the crystal but, if they are, the crystal behaves just like an ordinary plane mirror for appropriately directed photons. A "white beam" beam of parallel-traveling photons can be made monochromatic by reflecting it from the crystal. Alternatively, a diffuse beam of monochromatic photons, after Bragg reflection, consists entirely of parallel photons. A monochromator uses two or more crystal reflections to produce a beam that, as well as being monochromatic, consists of photons that are parallel in at least one transverse projection. By differentiating Eq. (9.6) one finds the fractional energy width corresponding to the Darwin width (assuming a perfectly collimated incident beam),

$$\frac{\Delta \mathcal{E}_\gamma}{\mathcal{E}_\gamma} = \frac{w_D}{\tan \theta_B}. \qquad (9.7)$$

This quantity is also shown in Table 9.1 and Table 9.2. One notes that ω_D is more or less proportional to λ while the fractional energy spread is independent of λ, which makes ΔE vary as $1/\lambda$. Large reflection angles are favorable

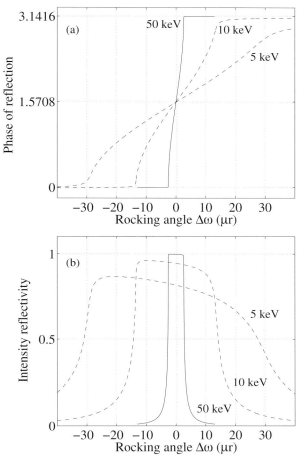

Fig. 9.1 Predictions of Darwin theory for (a) reflection phase and (b) intensity reflectivity, for Si(111) reflection. $\Delta\omega$ is the angular deviation away from the exact Bragg angle. These figures are adapted from Als-Nielsen and McMorrow.

for obtaining small energy spreads. For example, with $\mathcal{E}_\gamma = 12\,\text{keV}$, from C(400) an energy spread of roughly 0.1 eV is achieved. For reflection angles close to $\pi/2$ (i.e. "back scattering") the width can be far less than this. Of course, if the incident energy spread is large compared to $\Delta\mathcal{E}_\gamma$, as is usually the case, the reflected beam flux is inversely proportional to this width.

It was mentioned in the introduction that beam line properties depend critically on the relative size of small parameters. Two such small parameters are ω_D and $\sigma_{y'}$, the r.m.s. vertical angle of electrons in the beam. For the somewhat arbitrary choice $\sigma_{y'} = 22\,\mu\text{r}$ (from Eq. (9.2)) values of $2\sigma_{y'}/\omega_D$ are listed in one of the columns of the tables. One sees that the range of beam angles

Tab. 9.1 Darwin width ω_D for $\lambda = 1.0$ Å. Adapted from Mills [14]. Note: 1 arc-sec= 4.85×10^{-6} r. The choice $\sigma_{y'} = 22$ μr in the second last column is an arbitrarily chosen example value of the r.m.s. vertical electron beam angle for comparison with ω_D. N_D is defined in Eq. (9.58) and $\hat{\vartheta}$ is defined in Eq. (9.61). The entries in the last four columns are just specialized example ratios.

Crystal planes	d-spacing Å	tan θ_B	ω_D μ-rad.	$\frac{\Delta\mathcal{E}_\gamma}{\mathcal{E}_\gamma} = \frac{\omega_D}{\tan\theta_B}$ parts/10^6	$\frac{2\sigma_{y'}}{\omega_D}$	N_D	$\frac{2\hat{\vartheta}}{\omega_D}$	$N_D \frac{\Delta\mathcal{E}_\gamma}{\mathcal{E}_\gamma}$ parts/10^3
Si(111)	3.1355	0.1615	21.6	134	2.0	3.7	7.2	0.50
Ge(111)	3.2664	0.1549	49.0	316	0.9	1.6	3.2	0.51
C(111)	2.0593	0.2503	14.9	59.5	2.9	5.4	10.5	0.32
Ge(400)	1.4143	0.3779	22.0	58.2	2.0	3.6	7.1	0.21
Si(400)	1.3577	0.3961	9.2	23.2	4.8	8.7	17.0	0.20
C(400)	0.8917	0.6772	5.0	7.38	8.8	16	31.2	0.12

Tab. 9.2 Same as Table 9.1 except $\lambda = 0.2$ Å.

Crystal planes	d-spacing Å	tan θ_B	ω_D μ-rad.	$\frac{\Delta\mathcal{E}_\gamma}{\mathcal{E}_\gamma} = \frac{\omega_D}{\tan\theta_B}$ parts/10^6	$\frac{2\sigma_{y'}}{\omega_D}$	N_D	$\frac{2\hat{\vartheta}}{\omega_D}$	$N_D \frac{\Delta\mathcal{E}_\gamma}{\mathcal{E}_\gamma}$ parts/10^3
Ge(111)	3.2664	0.0306	10.0	326	4.4	8.0	3.5	2.6
Si(111)	3.1355	0.0317	4.2	133	10	19	8.3	2.5
C(111)	2.0593	0.0486	2.9	58.9	15	27	12.0	1.6
Ge(400)	1.4143	0.0709	4.27	60.2	10	19	8.2	1.1
Si(400)	1.3577	0.0739	1.70	23.0	26	47	20.5	1.1
C(400)	0.8917	0.1129	0.82	7.30	54	98	42.5	0.71

typically exceeds the Darwin width, especially for sub-Å wavelengths. However a fairly large value has been taken for $\sigma_{y'}$ compared to what can actually be achieved. On the other hand, for a third generation light source $\sigma_{x'} = 22$ μr would be a *small* horizontal angular spread.

9.5
Aperture-defined Beam Line Design

9.5.1
Undulator Radiation, $n = 1$, Negligible Electron Divergence

To concentrate on the effect of the angular radiation pattern let us first assume the angular divergence of the electron beam is negligibly small. A primary collimation, by monochromator or scrapers, takes place at a distance L_1, (perhaps 10 m) from the radiation source. Temporarily neglecting the electron spot size,[3] this sets horizontal and vertical angular limits $\pm \Delta\theta/2$ and $\pm \Delta\psi/2$. In practice the horizontal and vertical acceptances need not be equal but, to simplify the present discussion, let us assume the collimator passes only photons in a cone of half-angle $\vartheta_{\text{collim.}}$, thereby defining solid angle $\Delta\Omega_{\text{collim.}} = \pi \vartheta_{\text{collim.}}^2$.

The angular and energy distributions produced by undulators and wigglers are thoroughly analysed in Chapter 7 and Chapter 12. For any single undulator harmonic there is a maximum photon energy, and photons of that energy travel in the forward direction, presumably right through the center of the collimating device.

For operation on the $n = 1$ undulator fundamental the normalized-to-one photon number density as a function of fractional energy v_J, integrated over azimuthal production angle φ, was given in Eq. (7.49);

$$\frac{dN}{dv_J} = \frac{3}{2}(1 - 2v_J + 2v_J^2), \quad \text{for} \quad 0 < v_J < 1, \tag{9.8}$$

where $v_J = \mathcal{E}/\mathcal{E}_{\max}$. This spectrum is plotted in Figure 12.8. Recall that this number distribution is almost uniform because the angular distribution is almost isotropic in the electron rest system. Of course the discontinuous drop to zero at $v_J = 1$ is valid only as $N_w \to \infty$; for finite N_w the spectrum at fixed angle falls continuously over a range $\approx 1/N_w$.

Corresponding to Eq. (9.8), it was established in Eq. (12.112), that the fractional power ΔP at the undulator edge is related to fractional energy range

---

3) Since we are neglecting the angular divergence of the electron beam it would be physically inconsistent to neglect also the spot size. In spite of this, with collimator area visualized as large compared to spot size, all rays from source to collimator will be treated as if they have come from the center of the source. For the numerical example discussed shortly, there will be a (substantial) correction to compensate for this temporary over-simplification.

$\Delta\nu_J$ by[4]

$$\frac{\Delta P}{P} = 3\Delta\nu_J. \tag{9.9}$$

Recall, from Eq. (7.90), the one-to-one relation between ν_J and production angle ϑ;

$$\Delta\nu_J(\vartheta) = 1 - \nu_J(\vartheta) \approx \frac{\gamma^2 \vartheta^2}{1 + K^2/2}. \tag{9.10}$$

For the limits defined so far to be "matched", this requires collimator solid angle, fractional photon energy, and fractional beam power through the collimator to be related by

$$\frac{\gamma^2}{\pi} \frac{1}{1 + K^2/2} \Delta\Omega_{\text{collim.}} = \Delta\nu_J(\vartheta_{\text{collim.}}) = \frac{\Delta P}{3P}. \tag{9.11}$$

But the angular and energy acceptances of a monochromator are fixed by the design of the monochromator and will not, in general, provide energy spread consistent with the first of these two conditions. Too large monochromator solid angle acceptance would not be a problem because the collimation could be matched using scrapers. Too small monochromator solid angle acceptance could, perhaps, be fixed by monochromator re-design—from the discussion in Section 9.4, there is a fairly broad range of choice of crystals. In any case, to simplify the formulas, let us accept the collimation as being "matched" according to Eq. (9.11). Pictorially this requires threefold intersections of the curve, the (heavy) horizontal line, and the vertical lines in Figure 9.2.

By convention the "useful for x-ray physics" fractional energy range is taken to be $\Delta\nu_{\text{nom.}} = 0.001$. Equation (9.9) shows that only three parts in one thousand of the beam power in a pure ($n = 1$) undulator beam lie within the nominal energy bandwidth. The potentially "useful" power is therefore less than the total beam power by a factor of at least 300. Furthermore, for any particular experiment, the energy acceptance is typically controlled by a monochromator to a value $\Delta\nu_{\text{monochrom.}}$ which is typically much less even than $\Delta\nu_{\text{nom.}}$.

Because it is an undulator beam there is another energy-band-defining feature. By Eq. (7.47) the fractional energy spread at fixed angle is related to the number of undulator periods N_w by $\Delta\nu_{J,\text{und.}} \approx 1/N_w$. Since a typical value for N_w is not greater than 100, we expect

$$\Delta\nu_{\text{monochrom.}} < \Delta\nu_{\text{nom.}} < \Delta\nu_{\text{und.}}. \tag{9.12}$$

---

4) Since it is inconvenient for $\Delta\nu_J(\vartheta)$ to be negative, its sign has been reversed in this section compared to its earlier definition.

The quantities $\Delta\nu$, with various subscripts, will all be positive.

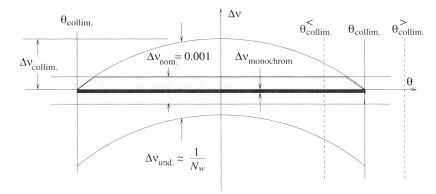

Fig. 9.2 In the ν_1, ϑ plane, apertures are defined by monochromatization, by collimation, and by the undulator characteristics (i.e. number of poles N_w.) For the relative sizes of these apertures suggested in the text, "flux" is conventionally defined by the lightly shaded region even though the monochromator-limited photons populate the darker region. The collimation angle can be reduced to the value $\vartheta_{\text{collim.}}$ without limiting the flux appreciably.

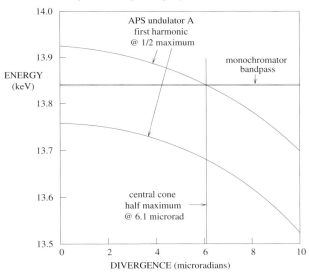

Fig. 9.3 Correlation between energy and angle for an undulator beam at APS. This figure is copied from Macrander [5].

The apertures that have been defined are illustrated schematically in Figure 9.2 and, with realistic numerical values for an APS undulator beam, in Figure 9.3, which is copied from Macrander [5]. Let us approximate the grating function to be uniform within the curved band shown in these figures. Although it is the photons inside the monochromator band that impinge on the experimental apparatus, the conventionally quoted "flux" for the beam line is the flux within the nominal, 0.1%BW fractional energy bandwidth; in Figure 9.2 this flux consists of the photons in the lightly shaded region. The logic behind this convention is probably that the experimenter is the one in the best position to determine $\Delta \nu_{\text{monochrom.}}$, and to normalize the measured data accordingly. When the term "flux" appears in the following discussion it will be the flux as defined in this conventional way.

We assume the monochromator energy is set exactly on the diffraction maximum. From Figure 9.3, the limiting collimator angle is determined by the equation

$$\Delta \nu_{\text{collim.}} \approx \frac{1}{2N_w}. \tag{9.13}$$

Collimation very much tighter than this would reduce the flux. One sees from Figure 9.2 that conventionally useful photons make up a fraction

$$\frac{\Delta \nu_{\text{nom.}}}{\Delta \nu_{\text{und.}}} \approx 10^{-3} N_w \tag{9.14}$$

of the power through the collimator. Combining formulas, the flux $\widetilde{\mathcal{F}}_1$ corresponding to total beam power P is

$$\widetilde{\mathcal{F}}_1 \approx \frac{3}{2} \frac{P}{\mathcal{E}_\gamma} 10^{-3}. \tag{9.15}$$

The tilde is a reminder that this is an "idealized flux" in which the angular cone angle of the electron beam has been set to zero and the subscript 1 indicates operation on the $n = 1$ peak. The relevant beam is spread over solid angle $\Delta \Omega_{\text{collim.}}$ which, from Eq. (9.11), is

$$\Delta \Omega_{\text{collim.}} \approx \frac{\pi}{2N_w} \left(1 + \frac{K^2}{2}\right) \frac{1}{\gamma^2} \tag{9.16}$$

Even though we have been treating ν_J and ϑ as perfectly correlated, for computing the beam brilliance let us start treating them as uncorrelated. (This will be valid if there is nothing present in the subsequent apparatus that is capable of exploiting correlation internal to such a slender beam.) The area of the photon beam at its source is equal to the area $\pi \sigma_x \sigma_y$ of the electron beam. The brilliance of the beam is therefore

$$\widetilde{\mathcal{B}}_1 = \frac{\widetilde{\mathcal{F}}_1}{\Delta \Omega_{\text{collim.}} \pi \sigma_x \sigma_y} \approx \frac{3 \times 10^{-3}}{\pi} \frac{N_w \gamma^2}{1 + K^2/2} \frac{P/\mathcal{E}_\gamma}{\pi \sigma_x \sigma_y}, \tag{9.17}$$

Reducing the product $\sigma_x \sigma_y$, for example by reducing beta-functions at the undulator, would increase $\widetilde{\mathcal{B}}_1$ proportionally. Of course, this dependence will be tempered by the effect of the angular divergence of the electron beam that has so far been neglected.

For $K \ll 1$, P is proportional to $N_w K^2$, so $\tilde{\mathcal{B}}_1$ is proportional to $N_w^2 K^2$. For $K > 1$, the denominator factor $1 + K^2/2$ tends to cancel the K^2 dependence. Furthermore, an effect not yet included in the formula is that the fraction of energy going into the fundamental decreases with increasing K. As a result the brilliance due to the $n = 1$ resonance falls rapidly as K increases past 1.

9.5.2
Effect of Electron Beam Emittances on Flux and Brilliance

To estimate the importance of electron beam spreads on the brilliance of x-ray beams the key comparison is between the angular spread of the monochromator-passing radiation pattern (Eq. (7.93)).

$$\vartheta_{\text{lim.}} = \frac{1}{2\gamma\sqrt{N_w n}} \quad \left(\stackrel{\text{e.g.}}{=} \frac{8.8\,\mu\text{r}}{\sqrt{n}} \quad \text{at} \quad \mathcal{E}_e = 2.9\,\text{GeV.}\right) \tag{9.18}$$

and the angular spreads of the electron beam, which, for the "third generation light source" example given in Eq. (9.2) are some tens of microradians. If the electron cone angle is large compared to the radiation cone angle the flux through the primary collimator will be multiplied by a factor roughly equal to the ratios of the cone solid angles, which is[5]

$$\frac{1/(4\gamma^2 N_w n)}{(\sigma_x/\beta_x)(\sigma_y/\beta_y)} \equiv \frac{n_{\text{norm.}}}{n}. \tag{9.19}$$

For want of a better symbol, the constant of proportionality in this relation has been called $n_{\text{norm.}}$, as a reminder that it is to be used only in the dimensionless ratio $n_{\text{norm.}}/n$. For numerical values given so far, $n_{\text{norm.}}$ works out to

$$n_{\text{norm.}} = \frac{(4 \times 5675^2 \times 100)^{-1}}{(400 \times 10^{-6}/8)(90 \times 10^{-6}/4)} = 0.069 \tag{9.20}$$

Setting $n = 1$ and multiplying the "idealized" brilliance given in Eq. (9.17) by factor (9.19) yields

$$\mathcal{B}_1 = \frac{3 \times 10^{-3}}{4\pi^2} \frac{1}{1 + K^2/2} \frac{P/\mathcal{E}_\gamma}{\epsilon_x \epsilon_y} \tag{9.21}$$

---

5) The angle at the collimator due to transverse displacement at the source has been neglected. The effect of this neglect could be compensated by increasing the effective electron angular divergence, $(\sigma_x/\beta_x) \to (\sigma_x/\beta_x)\sqrt{1 + (\beta_x/L_{\text{collim.}})^2}$ and $(\sigma_y/\beta_y) \to (\sigma_y/\beta_y)\sqrt{1 + (\beta_y/L_{\text{collim.}})^2}$. For the numerical example given in the text this correction reduces $n_{\text{norm.}}$ by a factor of perhaps two, depending on $L_{\text{collim.}}$.

An accurate determination of power P would require numerical integration along the true electron orbit. We can estimate P using Eq. (7.17),

$$U_{G1} = \frac{1}{\pi^{3/2}} \frac{C_\gamma \mathcal{E}_e^4}{\lambda_w/(2\pi)} \left(\frac{K}{\gamma}\right)^2 = \frac{1}{\pi^{1/2}} \frac{C_\gamma (m_e c)^4}{\lambda_w/(2\gamma^2)} K^2. \quad (9.22)$$

which gives the energy radiated by one electron passing one pole of the undulator. C_γ was defined in Eq. (3.32);

$$C_\gamma = \frac{e^2}{3\epsilon_0 (m_e c^2)^4}. \quad (9.23)$$

At average beam current I, the number of electrons passing the undulator per second is I/e. Since "brilliance" is defined with "per unit current" dimensionality, we should set to $I = 1$ which leads to

$$P = \frac{1}{e} \frac{2}{\pi^{1/2}} \frac{C_\gamma (m_e c^2)^4}{\lambda_w/(2\gamma^2)} N_w K^2, \quad (9.24)$$

The monochromator has been tuned to

$$\mathcal{E}_\gamma = hc \frac{2\gamma^2}{\lambda_w}. \quad (9.25)$$

Substituting these two expressions into Eq. (9.21) and using $e^2/(4\pi\epsilon_0 \hbar c) = 1/137.03$,

$$B_1 = \frac{1}{e} \frac{3 \times 10^{-3}}{2\pi^{5/2}} \frac{e^2}{3\epsilon_0 hc} \frac{1}{\epsilon_x \epsilon_y} \frac{N_w K^2}{1 + K^2/2} \approx 2.61 \times 10^{12} \frac{1}{\epsilon_x \epsilon_y} \frac{N_w K^2}{1 + K^2/2}. \quad (9.26)$$

Since the units used here are meters and radians, while brilliance traditionally uses millimeters and milliradians, the nominal-expressed brilliance is less by a factor 10^{12}. Then, if ϵ_x and ϵ_y are measured in units of nm, the brilliance is

$$B_1 = \frac{2.6 \times 10^{18}}{\epsilon_x(\text{nm})\epsilon_y(\text{nm})} \frac{N_w K^2}{1 + K^2/2} \left(\frac{\text{photons}}{\text{s mm}^2 \text{ mr}^2 \text{ 0.1\%BW A}}\right). \quad (9.27)$$

The coefficient is quoted to higher than justified precision only to permit numerical checking. Such crude approximations have gone into the derivation that the result could scarcely be accurate to better than a factor of 2. Krinsky [6] gives a formula identical to Eq. (9.27) but with 2.6 replaced by 2. Taking the grating function to be uniform (just below Eq. (9.12)) was probably an underestimate, since much of the range of integration (in this still-idealized model) passes exactly along the peak of the grating function. But there are other neglected effects, such as the effective increase in electron cone angle described in a recent footnote, which reduces the brilliance.

According to Eq. (9.20) the electron cone solid angle is some fifteen times greater than the cone of useful photons. If the aperture has been set by scraping (rather than by the monochromator), by increasing the collimator aperture, the flux through the collimator would increase almost proportionally until the collimation solid angle becomes comparable with the electron cone (or the monochromator angular acceptance is exceeded). But, because of the denominator factor $\Delta\Omega_{\text{collim.}}$ in Eq. (9.27), the resulting brilliance will be somewhat less than that given by Eq. (9.27). Furthermore, subsequent elements in the beam line, such as mirror focusing devices, may have transverse acceptance limits making them unable to tolerate a wider beam.

9.5.3
Brilliance with $K > 1$ and $n > 1$

With electron energy and wiggler fixed, striving for higher resonant photon energy, one can increase n. Striving for more brilliance, one is naturally tempted to increase the wiggler magnetic field, thereby increasing K. But, not only does the $n = 1$ energy drop as a result of increasing K, soon the $n = 1$ brilliance falls also. To hold $\mathcal{E}_\gamma$ constant while increasing K it is necessary to use $n > 1$. So far, calculation of the brilliance has been based entirely on theoretical Eqs. (9.8), (9.9), and (9.10). For $K > 1$ or $n > 1$ it is necessary to use calculations from Section 7.10.3. The dependence of forward intensity on n and K is given in Eq. (7.79). Results from this formula are tabulated in Table 7.1, exhibited as histograms in Figure 7.17, and plotted in Figure 7.18 and Figure 7.19.

For convenience in the present context the same data, except with forward intensity divided by n, is plotted in Figure 9.4 and, with logarithmic scale, Figure 9.5. This allows for the n-dependence of $\vartheta_{\text{lim.}}$ as given in Eq. (9.18). This factor causes the solid angle passed by the monochromator to be proportional to $1/n$. It is shown in Eq. (7.93) that the angular range of photons passed by a narrow band monochromator centered on a harmonic is proportional to $1/n$. For given K, these figures can therefore be used to estimate how the forward intensity is "effectively" distributed over n. That is, with K fixed, and monochromator tuned to n, to answer the question "what is the flux through the monochromator?"

To obtain high energy x-rays from an undulator it is appropriate to use a high n resonance. The same formulas can be used to obtain the flux. The ratio of beam energy to fundamental resonance energy is $n/(1 + K^2/2)$. Curves giving forward-intensity/n as a function of n and K are plotted in Figure 9.6, but with $n/(1 + K^2/2)$ as the horizontal axis.

Suppose we need x-rays having energy some three times higher than the $n = 1$, $K \ll 1$, peak energy of the undulator fundamental. For exam-

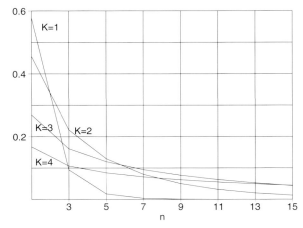

Fig. 9.4 The same data as in Figure 7.19 except the vertical axis is "forward-intensity/n". which is the forward intensity divided by harmonic number n. This is proportional to the flux through a narrow band monochromator centered on harmonic n, assuming the angular acceptance is proportional to $1/\sqrt{n}$, as given by Eq. (7.93).

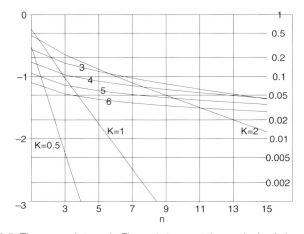

Fig. 9.5 The same data as in Figure 9.4 except the vertical axis is $\log 10$(forward-intensity/n). With the absolute $n = 1$, $K \ll 1$ brilliance having been established, these curves can be used to extrapolate to $n > 1$ and $K > 1$.

ple, with $K = 2$, $n = 9$, the appropriate point on the horizontal axis is at $9/(1 + 2^2/2) = 3$. From the graph the (relative) forward-intensity/n is 0.05. Comparing with the point $n = 1$, $K = 0.5$ in Figure 9.5, the energy ratio is $3/0.89 \approx 3.4$ and the ratio of intensity/n values is $0.05/0.25 \approx 0.2$. To convert an intensity ratio into a flux ratio it is necessary to include one power of photon energy. The ratio of $K = 2$, $n = 9$ flux to $K = 0.5$, $n = 1$ flux is therefore

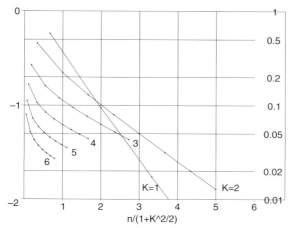

Fig. 9.6 The same data as in Figure 9.5 except the horizontal scale is $\omega_{n,\text{edge}}/\omega_{1,\text{edge}} = n/(1 + K^2/2)$, which is the ratio of the high energy edge of the undulator harmonic n at K to the $n = 1$ edge at $K = 0$. The vertical axis is $\log 10(\text{intensity}/n)$. Counting backwards the tiny black dots on the curves correspond to $n = 15, 13, 11, \ldots$.

$0.2/3.4 \approx 0.06$. This comparison has assumed optimally-matched collimation angles for both configurations. Also the effect of electron beam divergence has not yet been included.

As n increases it becomes progressively more difficult to experimentally isolate a given n value, though values at least as high as 9 have been used effectively. To estimate the brilliance it is necessary to specify the collimation explicitly. Because of the $1/\sqrt{n}$ variation of $\vartheta_{\text{lim.}}$, and continuing to assume zero electron beam divergence, the collimation solid angle is implicitly being reduced to the value given by Eq. (9.18). For $n = 9$, barring other complications, the collimation angle has therefore been reduced from 8.8 µr to 3.0 µr with no loss of flux. As a result the brilliance has increased by a factor of 9. Continuing the numerical comparison of the previous paragraph, the $n = 3$, $K = 2$ brilliance would be roughly half as great as the $n = 1$, $K = 0.5$ brilliance, in spite of the reduced flux and 3.4-times greater photon energy.

Unfortunately the assumptions of the previous paragraph may be unrealistic. A collimation angle of 3.0 µr at a distance of, say, 10 m corresponds to a collimation radius of 30 µm, which may be too small. Furthermore, the effect of electron beam divergence has not yet been included.

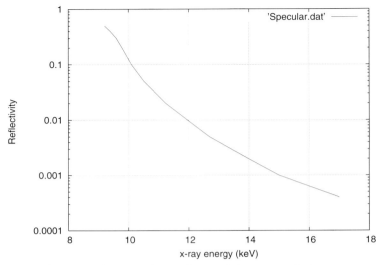

Fig. 9.7 Specular reflectivity of x-rays from a platinum-coated mirror at grazing incidence angle of 8 mr [7]. For angles below about 5 mr the reflectivity of longer than Å wavelength photons is close to 100%.

9.6
X-ray Mirrors

9.6.1
Specular Reflection of X-rays

X-rays can reflect specularly from a metallic surface, but only at extremely glancing angles. One is familiar with total *internal* reflection at an interface from, say, glass to a medium with lower index of refraction such as air. Because the index of refraction for x-rays is slightly lower than one, the same phenomenon becomes one of total *external* reflection. Figure 9.7 shows the reflectivity from a platinum-coated mirror.

9.6.2
Elliptical Mirrors

A common mirror system for focusing the x-ray beam to a small spot is the so-called "Kirkpatrick–Baez mirror pair" shown in Figure 9.8. Because of the glancing incidence such mirrors tend to be extremely long, and correspondingly difficult to produce with optical grade surface contours. It is apparently easier to make two cylindrical lenses than one ellipsoidal lens. Being (highly eccentric) elliptical in shape, these lenses focus a source at the distant focal

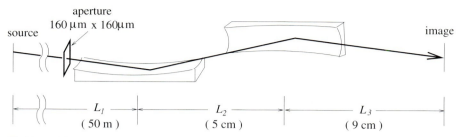

Fig. 9.8 A Kirkpatrick–Baez mirror pair for focusing an x-ray beam to a small spot. The dimensions shown are for Spring-8 apparatus of Hayakawa et al. [8].

point of the ellipse onto an image at the nearby focal point. For the example shown in Figure 9.8 the lens to focus distance L_3 was made very short in order to de-emphasize magnet surface slope errors. To limit geometric aberation, a 160 μm-square collimator was placed at the entrance. Being 50 m from the source, this represents an angular aperture of ± 1.6 μr. Since the electron beam angular size is ± 56 μr, the angle ratio is $1.6/56 = 0.029$. This represents a substantial loss of flux (especially if the same factor applies to both transverse planes.) This was a price the designers were prepared to pay to obtain good quality optics (i. e. a small spot size.) This beam line achieved a 10 keV photon flux of 10^{10} photons per second into a 2 μm $\times$ 4 μm spot.

9.6.3
Hyperbolic Mirrors

To form a real image of a real object using a mirror requires the mirror shape to be elliptical. Since x-ray beams are rarely converging toward a focus it is only elliptical mirrors that have customarily been used for x-ray lines. But an already convergent beam can be focused to a real image using a hyperbolic mirror.

One device that relies on reflections to concentrate an x-ray beam is the so-called "capillary tube" of gradually diminishing radius aiming toward a tiny sample area. X-rays can be visualized to be bouncing repeatedly along the reflective inner surface of the "light pipe" toward the sample. But Liouville's theorem demands conservation of phase space density and Snell's reflection law confirms this. So the angular divergence increases with each reflection and the incidence angle soon exceeds the critical angle of the metallic coating. Also, to limit beam spreading, the sample plane must be kept inconveniently close to the end of the capillary tube.

A more disciplined variant of the capillary tube is to be analysed next—it is a tube of (exactly) hyperboloidal shape in which x-rays bounce precisely once. For a beam that is already converging toward a point focus a hyperbolic

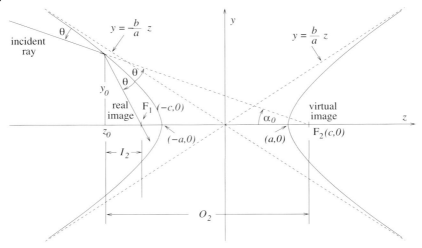

Fig. 9.9 An incident ray aimed toward a virtual image point at the far focal point F_2 of a hyperbola pair is reflected through the near focal point F_1. For this to be thought of as a "one-bounce capillary tube" the hyperbolas are so squashed that $c = \sqrt{a^2 + b^2} \approx a$. To reflect high energy x-rays the angle θ cannot exceed several milliradians.

mirror with one focal point at that (virtual) focal point can focus the beam to the local focal point. The geometry is shown in Figure 9.9. Figure 9.10 shows a blown up region near the focus with rays impinging on such a mirror and being brought to a common focus. The mirror equation is

$$\frac{y^2}{a^2} - \frac{z^2}{b^2} = 1, \tag{9.28}$$

and other equations relating object and image geometry to the mirror geometry are

$$c^2 = a^2 + b^2 = \left(\frac{O_2 - I_2}{2}\right)^2, \quad y_0^2 = O_2^2 \tan^2 \alpha_0, \quad z_0^2 = \left(\frac{O_2 + I_2}{2}\right)^2. \tag{9.29}$$

Eliminating b^2 and solving for a^2 yields the equation

$$a^4 + Ba^2 + C = 0, \quad \text{where} \quad B = -\left(\frac{O_2^2}{2} + \frac{I_2^2}{2} + y_0^2\right), \quad C = \left(\frac{O_2^2}{4} - \frac{I_2^2}{4}\right), \tag{9.30}$$

which, choosing the root that makes b^2 positive, yields

$$a^2 = -\frac{B}{2} - \sqrt{\left(\frac{B}{2}\right)^2 - C}. \tag{9.31}$$

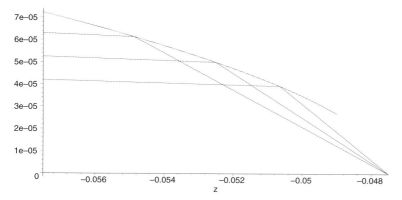

Fig. 9.10 Realistic hyperboloid focusing geometry. Note though that the vertical scale is still expanded by two orders of magnitude compared to the horizontal scale. All coordinates shown are measured in meters. The hyperbola parameters are $a = 0.0474999$ m, $b = 0.00010622$ m, $c = \sqrt{a^2 + b^2} = 0.047500$. The origin is off the page to the right and the focal points are at $z = \pm c$. The mirror terminates about 2 mm short of the focal plane to leave room for secondary detection apparatus.

Figure 9.10 shows more nearly realistic geometry, though the transverse scale is still much expanded compared to the longitudinal scale. The hyperbola parameters are given in the caption. Taking as the mirror position the intersection with the mirror of the central ray in Figure 9.10 (for which $c_0 = 0.0005$), the object and image distances, demagnification factor, and the coordinates of the reflection point are

$$O_2 = 0.1, \quad I_2 = 0.005, \quad \frac{I_2}{O_2} = \frac{1}{20}, \quad (9.32)$$
$$(y_0, z_0) = (49.999 \times 10^{-6}, -0.05249950).$$

For such an extreme surface the magnification factor is approximately

$$M_m \approx \frac{z_0 + c}{z_0 - c}. \quad (9.33)$$

The angles of incidence of the three rays shown in Figure 9.10 are 3.9 mr, 4.7 mr, and 6.0 mr which are small enough for good reflectivity of medium energy x-rays. The mirror is shown to be terminated short of the focal point to make "stay clear" room in front of the sample for slits or a window or other purposes. In any case the angle of incidence increases rapidly there, so the reflectivity becomes poor as the focal point is approached

9.7
X-ray Lenses

9.7.1
Monochromatic X-ray Lens

The only fundamental difference between lens design for monochromatic x-rays and for visible light is that the absorption of x-rays in the lens medium cannot be neglected. An ordinary converging glass lens could serve as an x-ray lens with the differences that the focal length would be extraordinarily long because the index of refraction is so close to 1 for x-rays, and the lens would be divergent because the index is slightly *less than* 1 for x-rays. For the glass lens to be usefully convergent its thickness would have to be so great that the x-ray absorption would be intolerable. As it happens, even though it is opaque for visible light, aluminum could be used instead of glass—the first demonstration of x-ray lensing employed aluminum—but the absorption is still unacceptably large. So neither glass nor aluminum is an appropriate medium to be used for x-ray lenses. To improve the ratio of index of refraction (or rather its deviation from 1) to absorption, low Z is required. If it were easier to work with, lithium would be ideal. For practical reasons beryllium seems like a good compromise.

As with visible light, the issue of chromatic aberation in lenses is important, that is, the focal length depends on the wavelength of the radiation. But, for the x-rays that have passed through a monochromator, chromatic aberation is less of a problem because the wavelength bandwidth is so narrow.

9.7.2
Focusing a Monochromatic Undulator Radiation Ring

With a high brilliance x-ray source the electron beam can be treated, to a good approximation, as having vanishing emittances. Even so, the radiated x-rays have substantial angular spreads. Even from a single, harmonic order n, resonance, the angular spread is of order $1/\gamma$. But, because of the one-to-one relation between angle and energy, it is possible, in principle, for a single lens to focus all radiation for that n value to a single point.

Reviewing results obtained previously, the radiation comes in bands for which the dependence of energy on production angle ϑ is given by

$$\mathcal{E}_{\gamma,n}(\vartheta) = n \frac{2\gamma^2 hck_w}{1 + K^2/2 + \gamma^2\vartheta^2} = \frac{n\mathcal{E}_{\gamma,1}(0)}{1 + \frac{\gamma^2\vartheta^2}{1+K^2/2}} = \frac{n\mathcal{E}_{\gamma,1}(0)}{1 + x^2}, \quad (9.34)$$

where $k_w = 2\pi/\lambda_w$ and dimensionless radial coordinate x has been defined

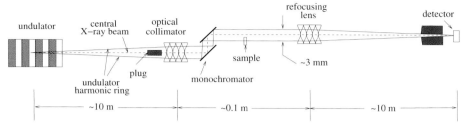

Fig. 9.11 Top view of beam line configuration for focusing the radiation from a single undulator resonance to a point. This configuration can also serve as an interferometer. A phase-contrast image of a sample placed in one of the side lobes is formed at the detector.

by

$$x = \sqrt{\frac{\gamma^2 \vartheta^2}{1 + K^2/2}} \equiv \frac{\gamma r}{\ell} \frac{1}{\sqrt{1 + K^2/2}}. \tag{9.35}$$

x is, on the one hand, proportional to the transverse ray displacement r at longitudinal distance ℓ and, on the other hand, scaled so that the half-energy point occurs at $x = 1$. The variable x is proportional to r and differs from $\gamma \vartheta$ only by a factor depending on K that is not very different from 1 for small K.

The pattern of angular rings from the undulator has been plotted in Figure 7.21. An apparatus intended to exploit this pattern is shown in Figure 9.11. Some particular value of n, say $n = 5$, can be isolated by adjusting the monochromator energy acceptance. To eliminate flux from other lines the plug shown can be adjusted to intercept photons from lower values of n. Only odd harmonics are significant as the even harmonics are weak in the forward direction. For reference, in Figure 7.21, the monochromator is set exactly on the $n = 1$ resonance and the ring to be focused could be $n_{\text{harm.}} = 3$ or $n_{\text{harm.}} = 5$.

Possible parameter choices for this configuration are tabulated in Table 9.3. This table gives the radii of monochromatized undulator rings for the following parameters: $\lambda_w = 20$ mm, $\mathcal{E}_e = 2.9$ GeV, $\gamma = 5675$. $\mathcal{E}_\gamma$ is the monochromator energy. For this table the monochromator is tuned to match the forward peak of the undulator line labeled $n_{\text{mono.}}$ (a number close to an integer) and Δn is the harmonic number relative to this of the harmonic to be viewed. Underlined values in the first row indicate a configuration appropriate for producing a ring of 10.65 keV x-rays from the $n_{\text{harm.}} = 5$ undulator harmonic, with the monochromator energy set to just exclude the forward, $n = 3$ peak. The peak energy of this undulator line is 17.7 keV.

Tab. 9.3 Various parameter choices for post-monochromator undulator harmonic rings with monochomator set near energy of the forward peak of the $\eta_{mono.}$ harmonic.

K	$\eta_{mono.}$	Δn	$\gamma\vartheta_{harm.}$	$\vartheta_{harm.}$ µr	$\mathcal{E}_\gamma$ keV	$\mathcal{E}_{\gamma,harm.}(0)$ keV
0.5	3	0,1,2	0, 0.612, 0.866	0, 108, 153	10.65	10.7, 14.2, 17.7
	5	0,1,2	0, 0.474, 0.671	0, 84, 118	17.75	17.8, 21.3, 24.8
	7	0,1,2	0, 0.401, 0.567	0, 71, 100	24.85	24.9, 28.4, 31.9
1.0	3	0,1,2	0, 0.707, 1.000	0, 125, 176	7.99	8.0, 10.6, 13.3
	5	0,1,2	0, 0.548, 0.775	0, 97, 137	13.31	13.3, 16.0, 18.6
	7	0,1,2	0, 0.463, 0.655	0, 82, 176	18.63	18.6, 21.3, 24.0
1.5	3	0,1,2	0, 0.842, 1.190	0, 148, 210	5.64	5.6, 7.5, 9.4
	5	0,1,2	0, 0.652, 0.922	0, 115, 162	9.40	9.4, 11.3, 13.2
	7	0,1,2	0, 0.551, 0.779	0, 97, 137	13.15	13.1, 15.0, 16.9
1.0	2	0,1	0, 0.866	0, 153	5.32	5.32, 7.99
	4	0,1	0, 0.612	0, 108	10.65	10.7, 13.3
	6	0,1	0, 0.50	0, 88	15.97	16.0, 18.6
	8	0,1	0, 0.433	0, 76	21.30	21.3, 24.0

9.7.3
Undulator-specific X-ray Lens

In this section a so-called "undulator-specific" lens is described. An achromatic lens could focus a parallel beam with large momentum spread to a single point. The undulator-specific lens is *not* achromatic in this sense. Rather it compensates for the energy dependence on the angle of undulator radiation to, for example, convert into a parallel beam, all x-rays from the nth order undulator resonance, or to focus a broad bandwidth to a single point. The undulator-specific lens would only be appropriate for angle-insensitive experiments requiring very intense broad band radiation. Also, being so reliant on the undulator radiation pattern, the lens would only be applicable with an extremely low emittance beam for which the spread of electron angles is much less than the undulator angular pattern.

For the configuration of Figure 9.11, but with the monochromator removed, the lens being discussed concentrates the undulator radiation for a given n value onto a point. For concreteness let us concentrate on focusing x-rays which have energy equal to 10.65 keV produced at angle 153 µr. These photons belong to the $n = 5$ line, for which the forward energy is 17.7 keV; that is $\lambda_5(0) = 0.70$ Å. (See Table 9.3).

The index of refraction for x-rays in matter is given by

$$n(\lambda) \equiv 1 - \delta + i\beta = 1 - \frac{r_0 Z}{2\pi}\lambda^2 + i\frac{\mu\lambda}{4\pi}, \qquad (9.36)$$

where $r_0 = 2.82 \times 10^{-15}$ m is the classical electron radius, Z is the electron density, λ is the photon wavelength, and μ is the total attenuation coefficient. (All quantities are in SI units.) For rough calculations, taking $Z/A = 1/2$, in terms of mass density ρ, one obtains

$$\delta(\rho, \lambda) \approx 1.3 \times 10^{11} \rho \lambda^2 \qquad (9.37)$$

This value of δ is extremely small; for example, beryllium[6], with $\rho = 1.846 \times 10^3$ kg/m^3 has $\delta \approx 2.22 \times 10^{-6}$ at $\lambda = 1.0$ Å. The photoelectric and scattering (coherent plus incoherent) absorption lengths are equal at 12.2 keV [9]. Expressed in terms of photon energy[7]

$$\delta_{\text{Be}} \approx \frac{3.41 \times 10^{-4}}{\mathcal{E}_\gamma^2[\text{keV}^2]}. \qquad (9.38)$$

This is so small as to make x-ray lenses impractical for most purposes. But, by forming a compound lens consisting of many lenses in series, Snigerev et al. [9,10] have shown that it is possible to achieve substantial focusing, though only at the cost of substantial attenuation caused by the nonvanishing value of β in Eq. (9.36).

A compound undulator-specific lens is illustrated in Figure 9.12. Except for the detailed cavity contours (or possible sawtooth Fresnel-like contouring) any x-ray lens will be similar. For the "undulator-specific lens" application, there are two factors that make the achievable focusing adequate. One is that the required focusing is extremely weak; a focal length $f \gtrsim 10$ m is all that is required. Also important is the fact that, because of angle-energy relation (9.34), the bend angle required to "straighten out" a large angle photon tends to be compensated by its reduced energy and hence greater refraction. In this sense the x-ray lens is "matched" to undulator radiation.

The defining purpose of an undulator-specific lens is to steer all photons belonging to a single undulator harmonic through a single point. This is only possible because of the one-to-one relation between energy and angle for undulator radiation on a single harmonic. The undulator-specific lens could

---

6) Beryllium is a promising candidate for x-ray lenses. It has the (small) values $Z = 4$, $A = 9$, is robust, has excellent thermal conductivity (almost as good as aluminum) and is capable of being accurately formed and polished, though granularity poses a problem. Lithium may be superior, especially at lower energy, but, for concreteness, all calculations in this proposal are based on beryllium.

7) Equation (9.38) uses the correct density rather than the $Z/A = 1/2$ rule of thumb value mentioned earlier.

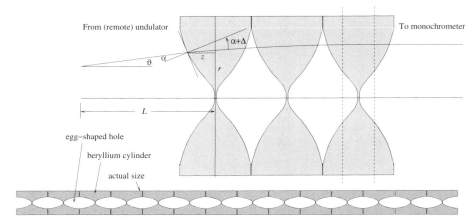

Fig. 9.12 An undulator-specific lens consists of one or more solids of revolution, shaped like miniature soup cans with pushed in ends, manufactured from low-Z material, and lined up along the undulator beam line. The exact shape has to be matched to the particular undulator configuration. The lens is "multichromatic" because the dependence of focal length on position matches the dependence of wavelength on position of the incident radiation. If rays close to the axis are unwanted a uniform slab (indicated by broken lines in the third lens) could be removed in order to reduce attenuation without affecting focusing. More complicated cutting away of material can also be performed to reduce absorption with little degradation of the focusing. A more or less actual size, full multi-element, "compound" lens is also shown.

therefore, if one wished, be used as a "harmonic filter" that passes only the photons coming from that particular harmonic. This could be implemented by placing a collimator at the focal point. Photons issuing from the collimator, by virtue of the one-to-one relation mentioned above would (ideally) occupy zero phase space volume, and would therefore, in spite of broad energy bandwidth, be extremely brilliant. This could be of interest for some purposes, such as maximizing local power density or for Laue diffraction with broadband illumination and subsequent large solid angle detection.

The undulator-specific lens can be said to be "multichromatic" in that photons of arbitrary wavelength are steered through the same point. This relies on the organization of the incident photons and is, of course, not the same as a lens being achromatic, which (conventionally defined) would mean that the lens focal length is independent of wavelength. Since the x-ray lens is made from many separate partial lenses, the optics can be kept in focus by extracting or inserting lenses as small changes are made in the photon energy.

For a conventional lens a simple spherical surface is adequate but the undulator-specific lens requires a more complicated surface. To obtain the lens profile we start by calculating the focusing due to the front edge of the first lens. Because the focusing is so weak, the focusing strength is proportional to the number of edges n_E (which could also be said to be the number of half lenses) and can be obtained as a simple multiple of the value now to be calculated. The angle $\vartheta = r/L$ is vastly exaggerated in Figure 9.12, and will, in fact, be neglected. Then, applying Snell's law, $\sin\alpha = (1-\delta)\sin(\alpha + \Delta)$, yields

$$\Delta = \delta\tan\alpha = \delta\frac{dz}{dr}, \tag{9.39}$$

where Δ is the angle of bending of the ray due to refraction, and $\tan(\alpha(r))$ is the slope of the lens at radius r. The requirement that the "undulator radiation" focal length of the lens be equal to f_{ur} for all values or r yields[8]

$$\frac{r}{f_{ur}} = n_E\,\delta(r)\,\frac{dz}{dr}. \tag{9.40}$$

where $\delta(r)$ is the index of refraction deviation appropriate for photons at radius r. Combining Eqs. (9.34) and Eq. (9.38), the index deviation from 1 is

$$\delta(r) = \frac{3.41\times 10^{-4}}{n^2\mathcal{E}_{\gamma,1}^2(0)}(1+x^2)^2. \tag{9.41}$$

Combining Eqs. (9.40), (9.41), and (9.35) The lens shape is governed by the differential equation

$$\frac{d(n_E z)}{dx} = \frac{n^2\mathcal{E}_{\gamma,1}^2(0)}{3.41\times 10^{-4}}\left(1+\frac{K^2}{2}\right)\frac{\ell^2}{f_{ur}\gamma^2}\frac{x}{(1+x^2)^2}. \tag{9.42}$$

Integrating this equation and setting $z=0$ at $x=0$ yields

$$n_E z = \frac{n^2\mathcal{E}_{\gamma,1}^2(0)}{3.41\times 10^{-4}}\left(1+\frac{K^2}{2}\right)\frac{\ell}{2f_{ur}}\frac{\ell}{\gamma^2}\frac{x^2}{1+x^2}. \tag{9.43}$$

as the shape of the lens surface. This shape is plotted in Figure 9.13 and can be seen to resemble the shape appearing in earlier figures.

Consider a numerical example: $n\mathcal{E}_{\gamma,1}(0) = 17.7\,\text{keV}$, $n\mathcal{E}_{\gamma,1}(\vartheta) = 10.7\,\text{keV}$, which corresponds to $x^2 = 0.654$. Taking $K = 0.5$ yields $\gamma\vartheta = 0.817$ which for $\gamma = 5670$ yields $\vartheta = 144\,\mu\text{r}$. For symmetric, unit magnification optics $\ell = 2f_{ur}$. Let us take $\ell = 10\,\text{m}$ and $f_{ur} = 5\,\text{m}$. Substituting into Eq. (9.43) yields $n_E z \approx 14\,\text{cm}$. At this thickness most of the photons will be absorbed so the effective lens acceptance is considerably less than the $\vartheta = 144\,\mu\text{r}$ value. But the lens thickness falls rapidly with decreasing x so there will be an appreciably large aperture. (Note also, as shown in Figure 9.12, that lens material can be cut away without affecting the focal properties outside the cut away region.) Since there is no particular advantage to minimizing the lens radius, the numbers

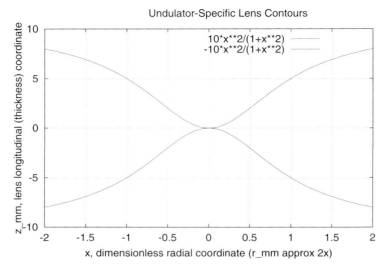

Fig. 9.13 Front and back contours of undulator-specific lens as given by Eq. (9.43) assuming $n_E = 28$ (i. e. 14 simple, concave-concave lenses stacked end to end, as shown at the bottom of Figure 9.12. Note though, the 90° relative rotation, and reversed aspect ratio, of this and that figure.) By Eq. (9.35), $r \approx 2.0x$, for the parameters given in the text. The lenses can be flattened at $z = \pm 5$ mm since the transmission at larger radius is negligible.

given in this paragraph have been used (approximately) to establish the scales of Figure 9.13.

Attenuation sets a fundamental limit on any x-ray lens. For beryllium the energy dependence of the total absorption coefficient is shown in Figure 9.14 [11]. This issue has been carefully analysed by Yang [12]. Though Yang analyses only monochromatic lenses, his analysis can be directly adapted to the omnichromatic lens under discussion. Since δ and β are roughly proportional to ρZ and ρZ^4 respectively, low-Z materials are strongly favored. Yang introduces a "critical Fresnel number"[9]

$$N_0 = \frac{\delta}{2\pi\beta} = \frac{2\delta}{\mu\lambda}, \qquad (9.44)$$

which is a ratio of the real to imaginary increments of the index of refraction defined in Eq. (9.36). N_0 depends only on the medium and the photon energy.

8) The term "focal length" (f_{ur}) is being used in an unconventional sense in that the angle energy relation specific to undulator radiation is assumed.

9) The Fresnel number indexes Fresnel zones starting at the origin. The *critical* Fresnel number is the value of the Fresnel number for which the lens thickness is equal to the attenuation length.

9.7 X-ray Lenses

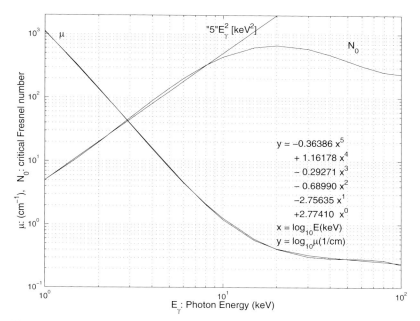

Fig. 9.14 Energy dependence of total absorption coefficient μ for beryllium. An empirical fit, accurate to about 5%, is indicated. Also shown (same scale) is the critical Fresnel number N_0 and the (valid over limited range) approximation "5"E_γ^2 [keV^2].

The energy dependence of N_0 for beryllium is shown in Figure 9.14. In terms of N_0, the absorption length is given by

$$t_{\text{abs.,Be}} = \frac{N_0 \lambda_\gamma}{2\delta_{\text{Be}}} = \frac{N_0 \lambda_\gamma}{2} \frac{\mathcal{E}_\gamma^2 [\text{keV}^2]}{3.41 \times 10^{-4}} \quad (\approx 1.03\,\text{cm}@10.65\,\text{keV}). \tag{9.45}$$

For a lens to be practical, its thickness must not be too large compared to $t_{\text{abs.}}$. Using Eqs. (9.41), (9.43), and the first of Eqs. (9.45), our lens thickness, measured in absorption lengths at the operative radius, is

$$\frac{n_E z}{t_{\text{abs.}}} = \left(1 + \frac{K^2}{2}\right) \frac{\ell}{2 f_{\text{ur}}} \frac{2\ell}{N_0 \lambda_\gamma} \frac{1}{\gamma^2} \left(\frac{\mathcal{E}_\gamma(0)}{\mathcal{E}_\gamma}\right)^2 \frac{x^2}{1+x^2}. \tag{9.46}$$

It is the cancellation of δ from this equation that justifies having introduced N_0. The equation makes it clear why choosing large-N_0 lens material is appropriate. Note, from Figure 9.14, that the product $N_0 \lambda$ in the denominator is more or less constant over the range from 5 keV to 20 keV.

9.7.4
Lens Quality

The Rayleigh [14] criterion (originally intended for glass lenses and with visible light) states that *optical path length* errors due to surface imperfection and incorrect shape should not exceed $\lambda_\gamma/4$. Errors less than this are dwarfed by diffractive effects. If its thickness had to be accurate to less than the $1/4\,\text{Å}$ it would be impossible to construct any x-ray lens; but this is far from necessary. For glass the index of refraction deviation $n_{\text{glass}} - 1$ is about 0.5 while for x-rays this deviation δ is some five orders of magnitude smaller. The *optical path length* deviation caused by a given surface imperfection is reduced by this factor of order 10^5 for x-rays as compared to visible light. Since this factor is comparable with the visible/x-ray wavelength ratio, the lens shape surface uniformity tolerances of x-ray and visible lenses are similar. The fact that an x-ray lens has more surfaces than a simple lens (factor $n_E/2$) leads to only a factor of $\sqrt{2N_E}$ more precise surface requirement (because of statistical averaging).

To make these considerations quantitative, using Eq. (9.38), we introduce an "effective wavelength" [13]

$$\lambda_{\text{eff.}} = \frac{\lambda_\gamma}{1-n} = \frac{\mathcal{E}_\gamma^2[\text{keV}^2]}{3.4 \times 10^{-4}} \lambda_\gamma \quad (= 0.19\,\mu\text{m at } \lambda_\gamma = 1\,\text{Å.}) \tag{9.47}$$

Let σ_z be the r.m.s. surface height variation over some appropriately chosen transverse distance (such as $100\,\mu\text{m}$) so the r.m.s. variation of a compound lens is $\sqrt{n_E}\sigma_z$ where n_E is the number of surfaces. Combining this with the Rayleigh formula we obtain a condition for surface quality

$$\sigma_z < \frac{1}{4\sqrt{n_E}} \frac{\mathcal{E}_\gamma^2[\text{keV}^2]}{3.4 \times 10^{-4}} \lambda_\gamma \quad (\stackrel{\text{e.g.}}{=} 0.02\,\mu\text{m at } \lambda_\gamma = 1\,\text{Å.}) \tag{9.48}$$

With proper methods (of course including proper safety precautions) it appears that beryllium can be machined and ground to accuracy comparable with other high-performance lens materials. Also, even though beryllium is opaque, aspheric surface dimensions can be measured by reflection to $\pm 0.01\,\mu\text{m}$ r.m.s. [15] This level of accuracy seems consistent with satisfying condition (9.48).

One could contemplate achieving the proposed focusing using mirrors rather than refractive lenses. A surface error of $0.02\,\mu\text{m}$ over a transverse distance of $100\,\mu\text{m}$ would cause a ray deflection of $4\,\text{mm}$ at $10\,\text{m}$. This is some twenty times greater than the diffraction limited size we have been assuming. It appears therefore that the construction tolerances are much easier to meet using refractive lenses than using curved mirrors. In spite of this difficulty, remarkably small spot sizes have, however, been obtained using KB mirrors.

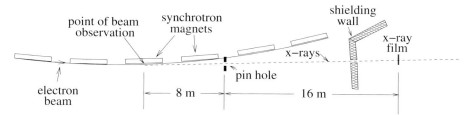

Fig. 9.15 Pin hole camera experiment performed at CEA to "photograph" the electron beam using x-rays.

9.8
Beam cameras

9.8.1
The Pin-hole Camera

From the preceeding discussion of the diffraction limit for beam cameras it is clear that x-rays, if they could be focused, would be appropriate for "photographing" electron beams of tiny transverse dimension. In the absence of focusing, if there is an abundance of flux, as there *is*, a pin-hole camera can be used.

Hofmann describes an experiment at the Cambridge Electron Accelerator [17] using apparatus illustrated in Figure 9.15. The beam image was captured by the remote x-ray film. (A modern implementation of the device would use an electronic detector such as a multi-pixel device or an x-ray-to-visible conversion medium followed by microscope and TV camera.) As well as dimensions shown in the figure, some of the parameters of this set-up were: magnet bending radius, $R = 26.2$ m, pin-hole diameter=70 μm, magnification=$-16/8 = -2$. The radiation in this set-up was "white", though wavelengths longer than 1 Å were absorbed in the exit window from the acccelerator vacuum chamber or in the air along the line. Also the sensitivity of x-ray film depends on wavelength. The experimenters estimated the average wavelength contributing to the photograph was about 0.5 Å. As a result the diffraction-limited spot size was about 10 μm which was negligible compared to the pin hole "shadow diameter" of 140 μm. Even so, the resolution was superior to what could have been obtained using visible light optics.

This set-up measures the transverse electron beam sizes σ_x and σ_y. The image is unaffected (except for flux, and hence exposure time) by the vertical angular spread σ'_y. The object depth is affected by σ'_x which influences the tangential electron path length over which radiation gets to the pin-hole. This effect is relatively unimportant in the pin-hole configuration just described but it can be quite important in more sophisticated modern devices.

The main operational use of a device like this is for monitoring the "decoupling" of the accelerator. As has been mentioned repeatedly, the vertical beam size (i.e. height) of the beam in an ideal electron storage ring is very nearly zero. The leading cause of beam height is the inadvertent presence of cross plane coupling elements such as "rolled" bending magnets or displaced quadrupoles.[10] Such coupling increases the vertical beam spreads both because coupled motion anywhere in the ring leads to increased vertical emittance and, if the local motion is coupled, the vertical component of horizontal eigenmotion contributes to the spreads.[11]

Skew quad elements are included in the ring to compensate for such errors. Since the location of the errors are unknown, the skew quad element strengths have to be set empirically. This operation depends on the ability to measure the beam height. That is the role of the pin-hole camera.

9.8.2
Imaging the Beam with Visible Light

Because of the ready availability of optical elements for visible light, most synchrotron radiation beam viewers work in the visible. They are primarily useful for measuring σ_x because diffraction prevents the accurate measurement of σ_y (except when the beam is "blown-up", either intentionally or unintentionally, by coupling elements in the ring.) For the casual visitor to an accelerator control room, and even for some of the operators, the most persuasive evidence for the presence of beam in the storage ring is the TV image of the beam image formed by the synchrotron radiation. In the days when intensities were much lower and safety standards more lax, even more persuasive evidence came from looking directly into a viewing port. (Actual blindness was prevented by the presence of mirrors which reflected mainly visible light.)

A feature of the synchrotron radiation pattern that influences the beam camera is σ_ψ which is the r.m.s. angle (from the beam plane) of the radiation. For the dominant (horizontally polarized) component of the radiation, this quantity was calculated approximately in Problem 3.7.2. Hofmann [16] gives formulas for the individual polarizations and for the total:

$$\sigma_{\psi,x} = 0.4097 \left(\frac{\lambda}{R}\right)^{1/3}, \quad \sigma_{\psi,y} = 0.5497 \left(\frac{\lambda}{R}\right)^{1/3}, \quad \sigma_\psi = 0.4488 \left(\frac{\lambda}{R}\right)^{1/3}, \quad (9.49)$$

10) Vertical dispersion due to vertically mis-steering of the orbit can also contribute appreciably to the vertical emittance.
11) A certain amount of vertical coupling is sometimes used to increase the Touschek, intrabeam scattering lifetime. An ideal design would arrange the coupling so that the nominally horizontal eigenmode is exactly horizontal at insertion devices for which mimimal vertical electron beam spreads are required.

which are valid at the low visible frequencies. It has been stated repeatedly that typical emission angles are of order $1/\gamma$. Let us reconcile this with Eqs. (9.49). Using Eq. (3.38) for critical frequency ω_c, and converting it to corresponding wavelength λ_c one obtains

$$\left(\frac{\lambda_c}{R}\right)^{1/3} = \left(\frac{4\pi}{3}\right)^{1/3} \frac{1}{\gamma} \approx \frac{1.6}{\gamma}. \tag{9.50}$$

Substituting this into Eq. (9.49) yields

$$\sigma_\psi \approx \frac{0.72}{\gamma} \left(\frac{\lambda}{\lambda_c}\right)^{1/3}. \tag{9.51}$$

This is a relatively weak dependence on λ, but the wavelength of visible light is some three orders of magnitude greater than λ_c, so the typical angle for visible light is roughly $7/\gamma$. Compared to x-rays, this ten-fold angular increase permits a ten-fold increase in lens diameter which gives a ten-fold improvement in the diffraction-limited resolution. But this only partially makes up for the thousand-fold loss of resolution because of the longer wavelength.

Figure 9.16 shows a visible light beam camera configuration. The portion $D_{\text{eff.}}$ of the lens that is illuminated is related to σ_ψ by $D_{\text{eff.}} \approx 2\sigma_\psi L$. With unit magnification, when in focus, the spot height $2d$ visible on the viewing screen is interpreted as the height of the electron beam. To estimate the smallest possible beam height that can be measured we could refer back to Section 2.4.2, for example Eq. (2.37), where behavior of an electromagnetic wave near a focus was discussed. Instead I will use a more qualitative (but presumably more or less equivalent) argument based on the uncertainty principle, which is a (classical) relation between the spatial width and wave number (inverse wave length) width. The relation is[12]

$$\Delta y \, \Delta k_y \gtrsim \pi, \tag{9.52}$$

where Δy is the height of the wave field, $k_y = 2\pi/\lambda$ is the vertical wave number, and Δk_y is the range of vertical wave numbers making up the photon beam. Knowing that k is proportional to photon momentum and k_y is proportional (with the same factor) to transverse momentum, the range Δk_y can be replaced by $(2\pi/\lambda)\Delta\theta$, where $\Delta\theta = D_{\text{eff.}}/L$ is the range of angles converging from the lens toward the focus. From Eq. (9.52) the half-height d satisfies

$$d \gtrsim \frac{\pi}{2k\Delta\theta} = \frac{\lambda}{4} \frac{L}{D_{\text{eff.}}} = \frac{\lambda}{8\sigma_\psi} \approx 0.3 \, (\lambda^2 R)^{1/3}. \tag{9.53}$$

Numerical examples for a few storage rings are shown in Table 9.4

[12] The numerical factor on the right-hand side of Eq. (9.52) depends on the detailed definition of the factors on the left. Taking the factor to be π yields conventional optical diffraction limits.

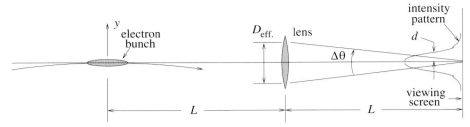

Fig. 9.16 Imaging the electron beam using its visible synchrotron radiation and an optical lens.

Tab. 9.4 Diffraction-limited spot diameter d for various electron rings. Missing entries are to be filled in as a problem.

Ring	R	λ	σ_ψ	d
	m	nm	mr	mm
EPA(CERN)	1.4	400	2.9	0.017
CLS	10	400		
CESR	90	400	0.74	0.068
LEP	3100	400	0.23	0.22

As Hofmann points out, there is a depth of field effect that artificially increases the image height by a comparable amount. As shown in Figure 9.17, the light getting to the screen is emitted over an appreciable tangential interval of length $\Delta z \approx 4\sigma_\psi R$. This discussion is continued as Problem 9.8.2.

Problem 9.8.1 *Fill in the CLS blanks in Table 9.4.*

Problem 9.8.2 *For visible light the longitudinal range over which light gets to the observation point P is far greater than the $2R/\gamma$ length we have been assuming. Rather we should use the appropriate horizontal spread of the "headlight" sweeping past the*

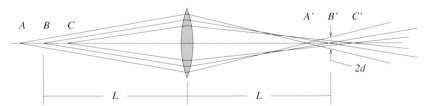

Fig. 9.17 Geometric construction illustrating the artificial electron beam height $2d$ that results because radiation emitted along an appreciable longitudinal source region contributes to the image at B'. Point B, at the tangent point of the orbit, is "in focus" at point B', but the points A and C focus to A' and C' as shown. An electron beam with vanishing actual height would appear to have height $2d$.

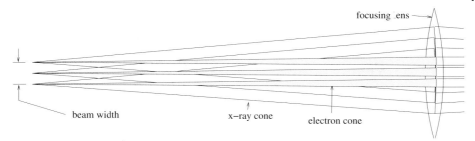

Fig. 9.18 A rudimentary beam camera. Emitted from a transversely spread source region, the cones of photon angles are typically much greater than the cones of electron angles. The lens produces an image of the beam on a screen off the page to the right.

observation point. For the horizontally polarized component this can be estimated to be the same as the vertical spread $2\sigma_{\psi,x} = 0.82(\lambda/R)^{1/3}$ given by Eq. (9.49). Estimate the distance AC in Figure 9.17 and complete the geometry to show that "spurious" half-height d is given by

$$d \approx \sigma_{\psi,x}^2 R. \tag{9.54}$$

(Hofmann gives this formula on p. 318, but I question the arithmetic in which it evaluates to $0.34(\lambda^2 R)^{1/3}$. In any case the functional dependence of the depth of field limit and the diffraction limit are the same, and the magnitudes are roughly the same.)

9.8.3
Practicality of Lens-based, X-ray Beam Camera?

The relationship between the cone of electron angles and the cone of photon angles is shown qualitatively in Figure 9.18. The lens brings the rays to a distant focus (not shown).[13] Especially for the small electron beam sizes anticipated in the near future, a beam camera resolution as small as 1 micrometer would be desirable. The diffraction-limited image spot size radius of such a camera, if Å x-rays could be used, would be given roughly by

$$r_{\text{diff.}} \approx \frac{\lambda_\gamma \ell}{2D} \approx \frac{10^{-10} \times 20}{2 \times 10^{-3}} \approx 1\,\mu\text{m}, \tag{9.55}$$

where D is the lens diameter and $\ell = 20\,\text{m}$ is the image distance from the lens. The numerical estimate given in Eq. (9.55) assumes that it is possible to focus x-rays having wavelength 1 Å, and the lens diameter has been taken to be 1 mm. For visible light, since the wavelength is greater by four orders of

---

[13] A lens such as shown in Figure 9.18 is clearly practical for visible light, but it remains to be seen whether such a lens is practical for x-rays.

magnitude, to obtain the same resolution at the same distance would require D to be increased by the same four orders of magnitude. Of course this is impossible and, in fact, even by increasing D and decreasing ℓ, as discussed in an earlier section, it is impossible to obtain resolutions better than several tens of µm using visible light. This makes it attractive to investigate the feasibility of x-ray cameras.

Spot radius (9.55) can be compared with the following "no-focusing" quantities (drawn from Section 9.3) that would be observed at a screen 20 m from the source *with no lens interposed*:

$$\text{spot-radius of projected ERL electrons} \approx 3.5 \times 10^{-6} \times 20 = 70\,\mu\text{m},$$
$$\text{half-height of projected existing-ring electrons} \approx 22 \times 10^{-6} \times 20 = 440\,\mu\text{m},$$
$$\text{half-width of projected existing-ring electrons} \approx 50 \times 10^{-6} \times 20 = 1000\,\mu\text{m},$$
$$\text{spot-radius of radiated photons} \approx 176 \times 10^{-6} \times 20 = 3500\,\mu\text{m}, \quad (9.56)$$

These estimates are illustrated qualitatively in Figure 9.18, though the beam width is much exaggerated and the lens distance much foreshortened. The final entry (3500 µm) shows that the image produced by a photon beam on a screen at 20 m distance, with no lens in the path is much larger than the (e. g. 70 µm) spot that would be produced if the electrons could simply be allowed to continue in straight lines to the screen. So detection of photons in this set-up would not serve as a useful emittance determining diagnostic, even for present day electron beams.

On the other hand, with perfect, unit-magnification optics, the image of an ideal electron beam (no transverse extent) could be as small as the diffraction limit of Eq. (9.55). But practical optics are not perfect. The emitted radiation (a) is not monochromatic and (b) is radiated over a finite longitudinal range. Also the lens (c) is not achromatic (d) is not astigmatic (focal length depends on where the ray passes through lens) and (e) is very absorptive in all wavelength ranges except the visible.

Of these effects, (e) has usually been used to rule out lensing at most wavelengths shorter than visible, but operation in the hard x-ray region is possible, though problematical. Rays very close to the axis are passed with little attenuation but other rays will be strongly absorbed in any conventional lens. Actually this effect was already factored into estimate Eq. (9.55) in that the lens diameter was estimated to be only 1 mm. The effect of absorption is to "stop down" the lens to a very small aperture, and the diffraction-limited spot size already accounted for this. A further consequence is that the limitation (a) is rather unimportant since only a very narrow band of x-ray energies will be used. For the same reasons effects (c) and (d) are fairly unimportant for an x-ray lens being used in a beam line for which the energy spread has been narrowed by monochromatization. But changing the photon energy appreciably

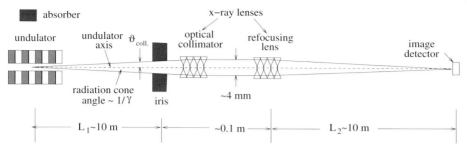

Fig. 9.19 Side view of beam camera set-up using x-ray lenses to focus most of the radiation from electrons passing through the same point in the undulator onto a single point.

would require the lens to be changed, for example by inserting or extracting partial lenses.

Effect (b) cannot be dispensed with so quickly. There is a "depth of focus" effect that causes a spot radius proportional to the length interval over which emitted photons are detected. For example

$$r_{\text{d.of.f.}} \approx \frac{\ell_u}{4} \frac{D/2}{\ell} \approx \frac{1}{4} \frac{0.25 \times 10^{-3}}{10} \approx 6\,\mu\text{m}, \qquad (9.57)$$

where the source length ℓ_u has been taken to be 1 m which is a possible undulator length, but which could be made shorter to improve the resolution. This calculated resolution is some 5 or 10 times smaller than the previously-mentioned beam size measurements using visible light optics. The object-depth-caused spot size could be reduced by reducing the source length, but that seems to be unnecessary as long as the expected value is substantially less than the anticipated spot size to be measured.

From the discussion so far we conclude that an x-ray lens shows promise for the purpose of "photographing" present-day and next-generation electron beams. By optimizing the configuration, for example by optimal choice of lattice beta functions and optimal choice of photon wavelengths, it should be possible to obtain a minimally broadened image of the beam on a remote screen.

The essential requirement for an x-ray camera is to focus a substantial fraction of the flux from each point in the undulator (or bending magnet) onto a small spot in the image plane. A rudimentary configuration is shown in Figure 9.19. Because there will be ample flux the lens can be "stopped down" for improved resolution using an iris close to the lens. Realistically, absorption in the lens provides an effective aperture stop.

9.9
Aperture-free X-ray Beam Line Design

9.9.1
Aperture-free Rationale

Most experiments require nearly monochromatic, nearly parallel, beams of very small transverse dimensions. Such beams can be produced by scraping, but there is always a trade-off between beam flux and the spread of relevant beam parameters. This trade-off is usually best analysed using the brilliance measure of beam intensity introduced in Chapter 3. Recall that the brilliance parameter is preserved along paraxial, lossless beam lines. In an electron beam, to avoid density dilution by filamentation, transfer lines have to be matched to the source characteristics. It has been much emphasized, especially in Chapters 1 and 2, that the optics of photon lines and electron lines are closely analogous. So the task to be addressed now is to "match" the optics of the photon line to the characteristics of the source, which is an electron beam in an undulator.

To simplify the discussion the entire beam will be treated initially as emerging from a point source, only later considering the issue of combining the contributions from an extended source.

In practice all x-ray beams are "extravagant" in that they convey only a tiny fraction of the radiation being produced onto the experimental apparatus. Synchrotron radiation is so intense that, for some applications, it is possible to be wrecklessly extravagant in beam scraping, while still retaining enough flux for the intended task; in any case, a given exposure can be accomplished, even with small flux, by exposing for a sufficiently long time. But most modern experiments cannot afford such extravagance. One must therefore maximize the flux of photons satisfying the beam quality limits specified for each experiment. Certainly a scraper can only reduce photon flux. But an ideally placed scraper has the property that cutting the flux in half also cuts in half the width of the particular beam parameter the scraper is intended to refine. By first analysing the beam properties with no beam scraping it should be possible to determine favorable locations for later beam tailoring using scrapers.

This section has a pedagogical purpose as well as a fundamental design purpose. The pedagogical purpose is to illustrate how the tools that have been described can be used in an x-ray beam line. The fundamental physics ingredient concerns the design of a beam line using the physics of synchrotron radiation from an undulator discussed in earlier chapters. This is intended to re-confirm the importance of "optical matching", in this case matching photon line properties to the electron beam and synchrotron radiation patterns. Scrapers are not to be used. Examples of *Scraper-defined* lines have already been analysed in Section 3.11 and Section 9.5.

Building a beam line without scrapers is about as practical as building a human eye without an iris. In any optical system the ability to absorb aberrant rays can greatly improve the optical performance. But, much as one should be reluctant to improve a class average by expelling all the students with low grades, one should be reluctant to improve beam quality by eliminating rays that are less than ideal. In the end one has to do it, but it is best deferred as long as possible.

A microbeam is an x-ray beam contained within a restricted region of six-dimensional phase space. For time-resolved experiments the bunch length (which is just inherited from the electron bunch length in the accelerator) is important. These experiments are fairly rare, however, and they are very much in the domain of FELs. So our discussion will ignore the bunch length, though certainly not the energy spread, of the x-ray beam. Many experiments require high photon number density in the remaining five-dimensional phase space. That is, transverse photon "emittances" in both planes and the energy spread should all be small. For these experiments, assuming the beam widths can be more or less matched to the sample size, it is the brilliance parameter that gives the most useful measure. Some experiments are fairly insensitive to the incident photon direction. For them it is the flux density (a density in transverse area) that is important. The beam line design to be described next emphasizes this application by attempting to maximize flux density over a small sample, but conclusions about the brilliance will also be drawn. For the sake of definiteness, let us shoot for one micrometer square as a size and shape of the radiation spot.

The beam line to be described next is too futuristic to be just like any existing line except in the elements it uses, which are standard components in the hundreds of x-ray beam lines currently in existence at electron storage rings. In designing the line certain idealizations will be made. Whether the idealization can be realized in practice and whether or not the beam line is practical will remain to be seen. The requirements placed on the individual elements in the beam line are very much like the requirements placed on the elements in any beam line. So the following material is intended to be relevant to beam line design, whether or not the actual beam line being described is practical for any particular purpose.

9.9.2
Aperture-free Microbeam Line Based on Lenses

The microbeam design being proposed here was motivated by a conversation with Ronald Cavell concerning microbeam requirements and by discussions concerning x-ray lenses with Alex Deyhim, Basil Blank, and Sarvjit Shastri. I also profited from conversations with Alexander Moewes, even though x-ray lenses are really not applicable to the sub-keV energy range appropriate for his fluorescence experiments.

This section contemplates the use of lenses for designing x-ray lines. If x-ray lenses were anywhere near as ideal as lenses for visible light, it is obvious that their use would greatly improve the performance of x-ray beam lines. But x-ray lenses are highly chromatic (not serious at each energy setting because of narrow bandwidth, but requiring addition or subtraction of focal power when changing x-ray energy) and highly absorptive (which is the fundamental problem). After designing a line using ideal lenses, the task will be to find whether absorption overwhelms whatever benefit came from the lens action. There are only three grounds for optimism; the nominal (10 keV) x-ray energy to be emphasized is optimal for beryllium lenses, and only very weak lenses are needed to give large flux increase. Thirdly, Fresnel lenses could be used.

In concept the beam line to be described is a "camera" that forms an image of a source (electron beam in insertion device) on a user apparatus in the image plane. In an ordinary camera, improved resolution can be achieved (at the cost of longer exposure time) by "stopping down" the lens. In the x-ray lens, the "lens stop" is controlled not by the photographer but by physics. Furthermore the "f-number" is strongly dependent on the x-ray wavelength. So, after designing the camera assuming ideally transparent lenses, the task will be to determine the loss of flux accompanying the physics-imposed stopping down of the lens.

The present design can be contrasted with the lens-free design of Section 9.5. Because of point-to-point focusing of the lens-defined line, all flux from a given point in the source contributes to the flux at the image, independent of the angular spreads of the electron beam and the radiation pattern. This is opposite to the collimator-defined line in which the flux through the collimator depends inversely on the angular spreads at the source. This will seem less paradoxical later when the above-mentioned "stopping down" of the lens is taken into account.

The guiding principle for the design will be "look after the flux and the brilliance will look after itself." Using idealized components, if we refuse to introduce anything that stops photons and manage to convey all the physically useful photons onto a small spot at the experimental apparatus, we will clearly have established an upper limit to the useful beam flux. As well as being maximum, if no flux has been lost, the flux will also be unambiguously calculable. Unfortunately, the elements required to achieve this goal are not physically realizable. Still, to the extent that realism leads to a calculable loss of photons, even though the flux has been reduced, it will remain calculable.

The proposed beam line is shown schematically in Figure 9.20. In simplest terms the optics forms an image of the source onto a distant image plane, where the x-rays impinge on some tiny target for some scientific purpose. As

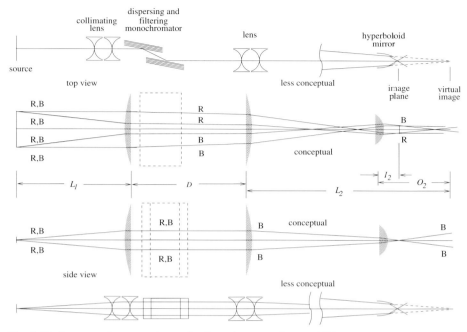

Fig. 9.20 A scraper-free x-ray beam line based on x-ray lenses. With ideal (but not necessarily achievable) optics, the unattenuated, "white" photon beam is, except for source position dependence, parallel at the monochromator location. Demagnification to a spot having the same aspect ratio as the electron beam occurs at the hyperboloidal mirror. The illumination is dispersed horizontally by "color".

shown, the line contains lenses, monochromators, and a hyperboloidal mirror. As already stated, the question of whether such elements are, in fact, physically realizable, will be deferred, but it is mainly the lenses that are problematical.[14] For visible light these elements are routine, but for x-rays they are certainly not. In any case the beam line will be described initially under the hypothesis that all elements are realistic. Since monochromatization occurs in a region where the photon beam is parallel, the monochromator itself does not limit the transverse aperture. With the lens magnification being approximately -1, the beam size at the detector is controlled by the magnification $0 < M_h \ll 1$ produced by the hyperboloidal mirror. Apart from increased flux, a sometimes-stated purpose for undulators is to produce x-rays that are more nearly monochromatic than the x-rays produced in a regular bending magnet. Nevertheless, for the time being, we will treat the radiation

14) It has to be admitted that I am told, by someone experienced in their use (AM), that the miniature hyperboloidal mirrors are unconscionably expensive and all but impossible to align.

as "white". In the figure, rays are labeled "R" for red, and "B" for blue. Of course, since it is x-rays under discussion, neither of these names is apt, and, furthermore, the labels are not supposed to indicate that the spread of wavelengths is anything like the two to one spread from blue to red. Rather "R" is supposed to mean "shifted toward red" and "B" means "shifted toward blue". Even more specifically "R"/"B" will be used to stand for the longest/shortest wavelength making it to the image plane.

Because the electron beam transverse dimensions are at least tens of microns and the desired image dimensions are as little as one micron, the image needs to be demagnified by a factor of at least 20. The aspect ratio of the source cannot be ignored though. The physics of storage rings permits the beam height to be almost arbitrarily small. I will assume that the r.m.s. beam height given previously is way too conservative and can be assumed to be less by a factor of at least three compared to the value given in Section 9.3, say to "object height" $\sigma_{y,o} = 20\,\mu\text{m}$. With forty-fold demagnification the vertical "image height" will then be $2\sigma_{y,1} \approx 1\,\mu\text{m}$, as desired. From here on I will therefore neglect the source height, and will defer discussion of the r.m.s. angular spread $\sigma_{y',o}$.

Unfortunately the electron beam width cannot be reduced below some undesirably large value—for the sake of definiteness let us take the r.m.s. source width to be $\sigma_{x,o} = 400\,\mu\text{m}$, as given in Section 9.3. To image this width to an acceptably small spot size would require unrealistically great demagnification. This consideration dominates the design of the beam line under discussion.

As illustrated in Figure 9.20, the crystal reflections provide a monochromatization in which the light emitted from a given point in the source contributes to the image of that point only if its color lies in a narrow band. The same comment applies to every point in the source, but the transmitted color depends on the horizontal source position. So the imaging is geometrically accurate but the line image is rainbow colored. (As previously stressed the terminology being used here here is entirely metaphorical; the fractional spread of wavelengths from one end to the other of the image will typically be less than the nominal, one part in a thousand, energy bandwidth that enters the conventional definitions of flux and brilliance. The full energy bandwidth is obtained by multiplying the entry in the ω_D column in Table 9.1 or Table 9.2 by the entry in the N_D column. The products are shown in the final column.) For experiments requiring narrower bandwidth further tailoring will be required.

Though all rays from a single source point have been made parallel by the optical collimator, there is an appreciable spread, $\pm\sigma_{x,o}/L_1$, of angles incident on the monochromator. The Bragg condition will be satisfied uniformly for all rays from the same source point, but the central Bragg energy $\mathcal{E}_\gamma$ will be different for different source points, yielding a one-to-one correlation between $\mathcal{E}_\gamma$ and x_o. From the white incident illumination, the monochromator will therefore generate parallel beams from each source point, with color correlated to x_o. We can express the electron beam width (taking it to be $2\sigma_x (\stackrel{e.g.}{=} 800\,\mu m)$) as a "number of Darwin widths" N_D given by

$$N_D = \frac{2\sigma_{x,o}/L_1}{w_D}, \quad \left(\stackrel{e.g.}{=} \frac{80}{w_D[\mu r]} \quad @L_1 = 10\,m. \right) \tag{9.58}$$

The N_D columns of Table 9.1 and Table 9.2 make the arbitrary choice $L_1 = 10\,m$ and $\sigma_{x,o} = 400\,\mu m$. For purposes of qualitative discussion it can be noted that numerical values of N_D are equal to the photon energy in keV multiplied by a number of order 0.5.

Taking the magnification to be $M \stackrel{e.g.}{=} -1/40$ in both planes, what has been achieved is illumination along a horizontal line of length about $2M\sigma_x \stackrel{e.g.}{=} 20\,\mu m$ and height about $2M\sigma_x \stackrel{e.g.}{=} 1\,\mu m$. But the illumination is "dispersed", with the wavelength $\lambda_\gamma(x_i)$ being a function of the horizontal position x_i. For some experiments, both the ribbon-shaped beam and the full wavelength band may be usable as-is. In principle, some experiments could be designed to exploit the position-energy correlation to effectively narrow the bandwidth without loss of flux. But experiments that need a narrower bandwidth would have to scrape the beam sideways with perhaps a tenfold narrowing of bandwidth and corresponding loss of flux. Yet another possibility, for experiments demanding the micrometer-squared spot size that was our initial goal, the final hyperboloidal lens could, at least in principle, be made highly asymmetric, to provide further demagnification in the horizontal plane. Of course the angular divergence would increase correspondingly, so the brilliance would be unchanged.

9.9.3
Effective Lens Stop Caused by Absorption

Absorption in the lenses is a fundamental limitation of the proposed beam line. This issue has been discussed in Section 9.7.3. The important parameter introduced there was the critical Fresnel number N_0, the Fresnel number of the zone at which the attenuation is $1/e$. From Figure 9.14, an empirical formula for N_0 for beryllium, valid roughly over the energy range from 1 to 100 keV, is

$$N_0 = \begin{cases} 5.0\,\mathcal{E}_\gamma^2\,[keV^2] & \text{if } \mathcal{E}_\gamma < 10\,keV, \\ 500 & \text{if } 10\,keV < \mathcal{E}_\gamma. \end{cases} \tag{9.59}$$

For a lens with focal length f, the Fresnel number at radius r is $r^2/(f\lambda_\gamma)$. The radius $\hat{r}$ (in meters) at which attenuation to e^{-1} occurs is given by

$$\hat{r} = \sqrt{N_0 f \lambda_\gamma} = 10^{-5}\sqrt{N_0 f \frac{12.4}{\mathcal{E}_\gamma[\text{keV}]}}$$

$$= \begin{cases} 0.79 \times 10^{-4} \sqrt{f\mathcal{E}_\gamma[\text{keV}]} \text{ m} & \text{if } \mathcal{E}_\gamma < 10\,\text{keV}, \\ 0.79 \times 10^{-3} \sqrt{\frac{f}{\mathcal{E}_\gamma[\text{keV}]}} \text{ m} & \text{if } 10\,\text{keV} < \mathcal{E}_\gamma. \end{cases} \quad (9.60)$$

The tentative choice used so far for the distance from source to collimating lens has been[15] $L_1 = f$, so the effective collimation cone half-angle is

$$\hat{\vartheta} = \begin{cases} 79\,\mu\text{r}\sqrt{\frac{\mathcal{E}_\gamma[\text{keV}]}{L_1}} & \text{if } \mathcal{E}_\gamma < 10\,\text{keV}, \\ 790\,\mu\text{r}\sqrt{\frac{1}{L_1 \mathcal{E}_\gamma[\text{keV}]}} & \text{if } 10\,\text{keV} < \mathcal{E}_\gamma. \end{cases} \quad (9.61)$$

This is a fairly small effective collimation angle. But note the final columns of Table 9.1 and Table 9.2 which, for $L_1 = 10$ m, give ratios of cone full-angle to Darwin width. The effect of the lensing has been to increase the angular acceptance of the monochromator by this ratio. For 1 Å photons the ratio is of order 10.

For any photon making it through this collimation, the probability of making it to the final image is proportional to the monochromator energy band given by Eq. (9.7). This is the same probability as would apply to any photon incident on the same monochromator. This observation allows an easy comparison of the flux in this beam line with the flux in a line with monochromator-governed angular acceptance. Allowing a loss factor of two for absorption even within the lens stop, if this same enhancement factor applies to both planes, (as it usually will) the lens will have given a flux enhancement of order 50.

9.9.4
Choice of Undulator Parameter K

For cleanest operation one would like to run on the $n = 1$ undulator harmonic. But, with electron energy of 2.9 GeV, one sees that for the wavelenth range, say, (1/3) Å to 3 Å and realistic undulator wavelength, this is impossible; with an already so-small-as-to-be-difficult-to-achieve value $\lambda_w = 2.0$ cm, $\mathcal{E}_{1,\text{edge}} = 4.0$ keV. From Figure 9.6, with $n = 5$, $K = 2$, there is appreciable flux up $n/(1 + K^2/2) = 5$, which is to say, up to 20 keV.

This is sufficiently far into the wiggler regime that, for estimating rates, one may as well revert to the use of thick element radiation formulas, perhaps

---

15) As mentioned before, though the bandwidth at each energy setting is narrow, because the x-ray lens is so chromatic, a special lens is required for each central setting of $\mathcal{E}_\gamma$.

with K-value even larger than 2. The optimal choice of K will depend on the photon energy range required, and the required beam "cleanliness". The wiggler magnetic field required can be estimated by comparing to the critical energy $\mathcal{E}_\gamma$ of magnets in the arcs of the storage ring. (For CLS, $\mathcal{E}_\gamma = 7.4\,\text{keV}$ at $B = 1.33\,\text{T}$ so a magnetic field of, say, $B = 2\,\text{T}$, will be required to get large flux at 20 keV.

9.9.5
Estimated Flux

Let us treat the radiation using the bending magnet formulas of Chapter 3. According to Eq. (3.62) the total number of photons emitted from every electron, every turn, from the $2N_w$ poles of the wiggler is

$$\mathcal{N}_{2N_w} = 0.0662\gamma \frac{2K/\gamma}{2\pi} 2N_w = 0.042\,KN_w; \tag{9.62}$$

(in words "a few".) To obtain the rate making it through the lens stop this rate will be multiplied by a horizontal probability and a vertical probability. Then the rate through the monochromator will be obtained by multiplying by a factor obtained from the normalized distribution function of photon energies multiplied by the monochromator energy pass bandwidth. These factors, symbolized by P_h, P_v, and P_E are probabilities, *not* powers.

Assuming $K > 1$, the $\mathcal{N}_{2N_w}$ photons are spread more or less uniformly over the horizontal angular range $-K/\gamma < \theta < K/\gamma$. This is obvious from the "locomotive headlight" picture. For confirmation or a more quantitative determination one can look at Figure 2-7 of the X-Ray Data Booklet [18] or graph 3.16 of Elleaume [19] Expressed in γ-scaled angles the effective collimation angle of Eq. (9.61) is

$$\gamma\hat{\vartheta} = \begin{cases} 0.45\sqrt{\frac{\mathcal{E}_\gamma[\text{keV}]}{L_1}} & \text{if } \mathcal{E}_\gamma < 10\,\text{keV}, \\ 4.5\sqrt{\frac{1}{L_1\mathcal{E}_\gamma[\text{keV}]}} & \text{if } 10\,\text{keV} < \mathcal{E}_\gamma. \end{cases} \tag{9.63}$$

The horizontal fraction of the emitted photons within this range is therefore

$$P_h = \begin{cases} \frac{0.45}{K}\sqrt{\frac{\mathcal{E}_\gamma[\text{keV}]}{L_1}} & \text{if } \mathcal{E}_\gamma < 10\,\text{keV}, \\ \frac{4.5}{K}\sqrt{\frac{1}{L_1\mathcal{E}_\gamma[\text{keV}]}} & \text{if } 10\,\text{keV} < \mathcal{E}_\gamma; \end{cases} \tag{9.64}$$

(which is of order 0.1.) The vertical photon angular distribution depends on the photon energy, but only mildly. For accurate work one should account for this dependence but, for purposes of estimation, let us ignore it. (This will overestimate low energies and underestimate high energies, since the high energies are more forward-peaked.) Then, according to Eq. (3.33), if the dependence on γ-scaled vertical angle $\gamma\psi$ is expressed by normalized differential

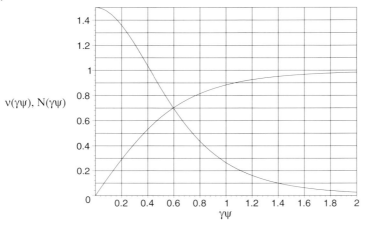

Fig. 9.21 Differential number distribution $\nu_{\gamma\psi}(\gamma\psi)$ and cumulative number distribution $N_{\gamma\psi}(\gamma\psi)$, for ($\gamma$-scaled) vertical angle $\gamma\psi$ of bending magnet radiation. Dependence on $\mathcal{E}_\gamma$ has been neglected. For confirmation or a more quantitative determination one can look at Figure 2-7 of the X-Ray Data Booklet [18] or graph 3.17 of Elleaume [19].

probability distribution $\nu_{\gamma\psi}(\gamma\psi)$ and corresponding cumulative distribution $N_{\gamma\psi}(\gamma\psi)$,

$$\nu_{\gamma\psi}(\gamma\psi) = \frac{3/2}{(1+\gamma^2\psi^2)^{5/2}}, \quad N_{\gamma\psi}(\gamma\psi) = \frac{1.5\gamma\psi + \gamma^3\psi^3}{(1+\gamma^2\psi^2)^{3/2}}. \tag{9.65}$$

(The shape expressed by this formula agrees with Hofmann's equation on page 28. Presumably θ in this formula is our ψ.) These distributions are plotted in Figure 9.21.

From Figure 9.21 one sees that the vertical angular distribution is reasonably constant out to $\hat{\vartheta}$, even at its maximum at $\mathcal{E}_\gamma = 10\,\text{keV}$; there the cumulative probability corresponding to the range $0 < \psi < \hat{\vartheta}$ is 0.56. So, as a function of $\mathcal{E}_\gamma$, the cumulative probability defined vertically by the lens stop is approximately

$$P_v(\mathcal{E}_\gamma) = \begin{cases} 0.56\sqrt{\frac{\mathcal{E}_\gamma[\text{keV}]}{L_1}} & \text{if } \mathcal{E}_\gamma < 10\,\text{keV}, \\ 5.6\sqrt{\frac{1}{L_1\mathcal{E}_\gamma[\text{keV}]}} & \text{if } 10\,\text{keV} < \mathcal{E}_\gamma; \end{cases} \tag{9.66}$$

(which is of order 0.3.) The total flux in the beam line is proportional to this probability. The flux is also proportional to the bending magnet number density given in terms of critical energy $\mathcal{E}_c$ by Eq. (5.14), multiplied by the monochromator energy acceptance $\Delta\mathcal{E}_\gamma = \mathcal{E}_\gamma\omega_D/\tan\theta_B$, as given by Eq. (9.7);

$$P_E = n_{\mathcal{E}_\gamma}(\mathcal{E}_\gamma)\frac{\mathcal{E}_\gamma\omega_D}{\tan\theta_B} = 0.3264\left(\frac{\mathcal{E}_\gamma}{\mathcal{E}_c}\right)^{0.275}\frac{\omega_D}{\tan\theta_B}\exp\left(-0.965\frac{\mathcal{E}_\gamma}{\mathcal{E}_c}\right); \tag{9.67}$$

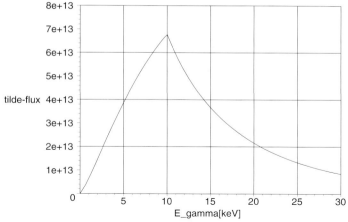

Fig. 9.22 Dependence on $\mathcal{E}_\gamma$ of "flux" $\widetilde{\mathcal{F}}(\mathcal{E}_\gamma)$ for $L_1 = 10$ m, $N_w = 50$, $\mathcal{E}_c = 15$ keV, and monochromator values for Si(1,1,1) from Table 9.1.

(in the range 10^{-5} to 10^{-4}.) In this expression, though ω_D and $\tan\theta_B$ are individually strongly dependent on $\mathcal{E}_\gamma$, their ratio is not. Combining all these factors, and including a factor I/e for the number of electron passages per second, we get the flux $\widetilde{\mathcal{F}}$ at the detector, per ampere of beam, to be

$$\widetilde{\mathcal{F}} = \frac{1}{e} \mathcal{N}_{2N_w} P_h P_v P_E$$

$$= \frac{0.0137}{1.6\mathrm{e}{-19}} \left(\frac{\mathcal{E}_\gamma}{\mathcal{E}_c}\right)^{0.275} \frac{N_w \omega_D}{\tan\theta_B}$$

$$\times \exp\left(-0.965 \frac{\mathcal{E}_\gamma}{\mathcal{E}_c}\right) \begin{cases} 0.252 \frac{\mathcal{E}_\gamma[\mathrm{keV}]}{L_1} & \text{if } \mathcal{E}_\gamma < 10\,\mathrm{keV}, \\ 25.2 \frac{1}{L_1 \mathcal{E}_\gamma[\mathrm{keV}]} & \text{if } 10\,\mathrm{keV} < \mathcal{E}_\gamma \end{cases} \quad (9.68)$$

This is a rate of perhaps $10^{-5} \times 10^{19}$ photons per ampere. The tilde on $\widetilde{\mathcal{F}}$ is a reminder that it is not yet quite the same as "flux" conventionally defined. The flux is plotted as a function of $\mathcal{E}_\gamma$ in Figure 9.22 for parameter values given in the caption.

9.9.6
Estimated Brilliance and Qualifying Comments

Various considerations will alter the flux shown in Figure 9.22. No allowance has been made for absorption of on-axis x-rays due to the impossibility of zero lens thickness. This could lower the flux by a factor of two or so. A factor going the other way is that no allowance has been made for the energy peaking due to the discrete, line-like nature of the spectrum. What has been calculated is an average flux. By tuning onto a local maximum the flux could perhaps be increased appreciably. Flux $\widetilde{\mathcal{F}}$ given by Eq. (9.68) seems (to me) like the quantity of greatest interest to experimenters but, for comparison with other beam line designs, the conventional "flux" $\mathcal{F}$ (flux per 0.001 BW) should perhaps be quoted, perhaps using $\mathcal{F} = \widetilde{\mathcal{F}}/(N_D \Delta \mathcal{E}_\gamma / \mathcal{E}_\gamma)$, where the denominator factor can be read from Table 9.1 and Table 9.2. This factor is not very different from 1.

It is convenient to calculate the brilliance directly at the source. The source area is

$$\pi \sigma_x \sigma_y = \pi \times 10^{-12} \times 400 \times 20 = 2.51 \times 10^{-8}\,\mathrm{m}^2, \tag{9.69}$$

and, at $\mathcal{E}_\gamma = 10\,\mathrm{keV}$ the effective solid angle is

$$\frac{\pi}{L_1^2} \sqrt{\hat{r}^2 + \sigma_{\hat{x}}^2} \sqrt{\hat{r}^2 + \sigma_{\hat{x}}^2} = \frac{10^{-12}\pi}{100} \sqrt{790^2 + 400^2} \sqrt{790^2 + 20^2} = 2.20 \times 10^{-8}. \tag{9.70}$$

Including a factor 10^{-12} for conversions from m to mm and from r to mr units, the brilliance (in conventional units) is therefore

$$\widetilde{\mathcal{B}} = \frac{10^{-12} L_1^2 \widetilde{\mathcal{F}}}{\pi^2 \sqrt{\hat{r}^2 + \sigma_{\hat{x}}^2} \sqrt{\hat{r}^2 + \sigma_{\hat{x}}^2}\, \sigma_x \sigma_y} \quad (\approx 1.2 \times 10^{17} \text{ at } \mathcal{E}_\gamma = 10\,\mathrm{keV}.) \tag{9.71}$$

This is some two orders of magnitude less brilliance at the same x-ray energy as can be achieved from the same undulator, using an electron beam energy of, say, 6 GeV. The most important factor entering this comparison is the critical energy $\mathcal{E}_c$ which, at fixed magnetic field, is proportional to γ^2. A factor of two in electron energy gives a factor of four in $\mathcal{E}_c$ which permits hard x-ray emission even on the peaks of a low order undulator resonances at the higher value of γ.

Most of the reduction of the flux from its maximum at 10 keV was due to reduction of angular aperture. Since this effectively reduces the angular beam spread, the brilliance is more nearly independent of energy than is the flux. Comments made a few paragraphs back qualifying the meaning of the quoted flux apply also to the brilliance.

The beam line that has been described is really quite simple and therefore, presumably cheap and reliable. Like all x-ray lines it will be hypersensitive to transverse alignment. As the energy is changed, it will be necessary to insert and extract lens elements, which will undoubtedly introduce some missteering that will have to be corrected. Adjusting the monochromator will also cause beam translation which will force the second lens position to be compensated accordingly, but monochromator sensitivities like this are common to most lines. Also, though almost trivial in principle, the hyperboloidal mirror can be expected to be difficult to control. Of course other focusing systems, like KB mirrors, are also difficult to control.

References

1. D. Mills (2002), *Third Generation Hard X-Ray Synchrotron Radiation Sources*, John Wiley, New York.

2. A. Hofmann (1998), *Diagnostics with synchrotron radiation*, in CERN Acclelerator School on Synchrotron Radiation and Free Electron Lasers, S. Turner (Ed.), CERN 98-04.

3. K. Evans-Lutterodt et al. (2003), *Single-element elliptical hard x-ray micro-optics*, Opt. Express, **11**, p. 919.

4. J. Als-Nielsen, D. McMorrow (2000), *Elements of Modern X-Ray Physics*, John Wiley, New York.

5. A. Macrander et al. (1992), *Emittance, Brilliance, and Bandpass Issues Related to an Inclined Crystal Monochromator*, in *Optics for High-Brightness Synchrotron Radiation Beam Lines*, J. Arthur, (Ed.), SPIE Proc., **1740**, 2.

6. S. Krinsky (2002), *Fundamentals of Hard X-Ray Synchrotron Radiation Sources*, in *Third Generation Hard X-Ray Synchrotron Radiation Sources*, D. Mills (Ed.), John Wiley, New York.

7. D. Bilderback(1981), SPIE, **315**, 90

8. S. Hayakawa et al. (2001), *Generation of an X-ray microbeam for spectromicroscopy at Spring-8 BL39XU*, J. Synchrotron Rad., **8**, 328.

9. B. Lengeler et al. (1998), Journ. Appl. Phys. **84**, 5855.

10. A. Snigerev et al. (1996), *A compound refractive lens for focusing high-energy x-rays*, Nature, **384**, 49. A. Snigerev et al. (1998), Appl. Opt., **37**, 653.

11. J. Hubbell and S. Seltzer (1996), *X-ray Attenuation Coefficients*, Nat. Inst. of Stand. and Tech., Gaithersburg, MD, USA.

12. B. X. Yang (1993), *Fresnel and refractive lenses for x-rays*, Nucl. Instrum. Methods Phys. Res., **A,328**, 578.

13. E. Church, P. Takacs (1993), Appl. Opt., **32**, 3344, 3346.

14. Lord Rayleigh, *Scientific Papers*, (1964), Vol. I, 436, Dover, New York.

15. E. Durand, J.-M. Bacchus(1992), *Phase Detection Deflectometry*, in *Specification and Measurement of Optical Systems*, L. Baker, Ed., SPIE **1781**, 249.

16. A. Hofmann (1998), *Characteristics of Synchrotron Radiation*, in CERN Acclelerator School on Synchrotron Radiation and Free Electron Lasers, S. Turner (Ed.), CERN 98-04.

17. A. Hofmann, K. Robinson (1971), *Measurement of a high-energy electron beam by means of the x-ray portion of the synchrotron radiation*, Proc. 1971 Particle Accel. Conference, IEEE Trans. Nucl. Sci. NS 18-3, 973.

18. A. Thompson, D. Vaughan (2001), *X-Ray Data Booklet*, Lawrence Berkeley National laboratory, LBNL/PUB-490, Rev. 2.

19. P. Elleaume (2003), *Undulator Radiation*, in *Undulators, Wigglers and Their Applications*, H. Onuki, P. Elleaume, (Eds.), Taylor and Francis, London.

10
The Energy Recovery Linac X-Ray Source

10.1
Preview

To obtain x-ray beams having high brilliance it is necessary to have an electron beam with high current and small emittance. There are two potential ways of achieving low emittance: one can start with arbitrary emittance and "cool" or one can start with, and then preserve, low emittance. As described in earlier chapters, the so-called "conventional" light sources exploit the cooling made possible by synchrotron radiation to obtain low emittance electron beams. Energy recovery linacs (ERL) take the other approach, starting with, and preserving, low emittance electron beams. Such a bright beam is made possible by the extremely small emittance initially available from a linear accelerator.

Ideally one would simply send the high quality linac beam through one or more undulators to produce the desired x-ray beams and, after that, dump the surviving beam. Unfortunately, for the high beam current (such as 0.1 A) and high energy (such as 5 GeV) needed for high brightness, the resultant power (500 MW) is some two orders of magnitude greater than can be afforded. In the energy recovery scheme the beam, after having been used to produce x-rays, is sent on a second pass through the linac(s). With this pass being at opposite RF phase, the energy in the beam is put back into the linac structure. Linacs having been analysed previously, the ERL approach is the subject of this chapter.

Historically the preservation of emittance has been crucial for hadron accelerators, because of the near absence of damping, but unimportant for electron accelerators, because of strong synchrotron radiation induced damping. In this respect an ERL more nearly resembles a hadron accelerator. Therefore, to begin, emittances, geometric, invariant, and normalized (the latter two of which are usually regarded as equivalent) are here defined and discussed. Then a simplified x-ray source based on energy recovery linacs is described. For compactness the design energy of 5 GeV is achieved with two 2.5 GeV linacs, one going, one returning, joined by circular arcs at each end to form a ring shaped like a bicycle chain.

Accelerator X-Ray Sources. Richard Talman
Copyright © 2006 WILEY-VCH Verlag GmbH & Co. KGaA, Weinheim
ISBN: 3-527-40590-9

One of the novel potentialities of an ERL x-ray source is the production of ultrashort (femtosecond scale) x-ray pulses. To support this feature, especially at multiple locations, the circular arcs of the ERL need to be made up of isochronous, achromatic, sections. These have to be matched to each other and to the straight sections. The major new complication to be faced in this, or any other, energy recovering design is that electrons have to traverse the linac straight sections twice; once during acceleration (where betatron oscillations are adiabatically damped) and once during deceleration (where they are anti-damped). First cut designs are given for these various components.

Because the success of the ERL is so dependent on the preservation of emittance, that is a parallel theme of the chapter.

10.2
Introduction

The beam emittance at the output of an electron linac is, to a large extent, inherited from the distribution of momenta of the electrons at the cathode of the gun, where they are born with velocities that are only thermal. Emittance preservation is especially easy to analyse for DC guns operating at charge per bunch small enough for space charge forces to be negligible. An electric field, constant in both space and time, is applied in the region between cathode and anode—this field is maximized in order to minimize emittance growth due to space charge forces. One purpose of this chapter is to define "emittance" in a way that is valid throughout the full range of velocities, from thermal to relativistic.

The "Courant–Snyder invariant" ϵ_{CS} of a particle in a storage ring is a positive-definite quadratic function of its position x and momentum p. The very essence of betatron dynamics is the conservation of ϵ_{CS}. A related quantity, but one which is a bunch property rather than a particle property, is the "emittance" ϵ_B, which is (roughly speaking) the average value of ϵ_{CS} for all the particles in the bunch. As such (at least after equilibration) ϵ_B is also conserved as a bunch circulates in a linear lattice. In its purest form this constancy depends on the constancy of the energy of every particle, which is certainly valid in a ring made up only of magnets. The *exact* constancy of ϵ_{CS} is a Hamiltonian property of linearized, or Gaussian, beam optics. The *derivation* of this result is invalidated by the presence of momentum offset which, in the presence of RF cavities, causes the momentum to vary. In a conventional ring, since the fractional momentum deviations are *tiny*, the fractional deviation of ϵ_B would be expected to be similarly small. In fact, however, for reasons to be discussed, the constancy of ϵ_B is assured to even much higher accuracy. The emittance defined so far is sometimes referred to as the "geometric emittance", to distinguish it from the quantity to be introduced next.

During acceleration to full energy the momentum of an electron varies by a large factor, perhaps a few orders of magnitude. The "normalized emittance" of a bunch of relativistic particles is defined by $\epsilon^{(N)} \approx \gamma \epsilon_B$, where γ is the usual relativistic factor. (Note that, for reasons to be discussed in this chapter, this definition is not yet expressed as an exact equality.) In the next section the careful definition of $\epsilon^{(N)}$ is based on its evolution in the simplest possible accelerator, namely a constant electric field.

For high energy accelerators, because $\beta \approx 1$, it is customary to neglect the distinction between γ and $\beta\gamma$ and to define the normalized emittance as above. A further velocity-dependent factor $\beta = v/c$ is sometimes included when less than fully relativistic particles are being discussed. In an energy recovery linac the electron kinetic energy varies from "thermal", a fraction of one eV, where the velocity ≈ 0, up through multi-GeV, and back down again to several MeV. This makes it essential to introduce an emittance definition that remains valid through the entire nonrelativistic/relativistic range.

10.3
Emittance Evolution in a DC Electron Gun

To study emittance evolution, consider a particle having charge e and mass m released with positive transverse momentum P_{y0} perpendicular to an electric field E that is parallel to the z-axis [1]. (Uppercase letter P is being used for momentum so that, later, lower case p can be used for fractional momentum.) The particle travels with uniformly increasing energy along the z-axis. Emittance conservation in this special case will later generalize to more realistic configurations having spatially and temporally varying longitudinal accelerating fields. Evolution only in the transverse y plane will be described here, but the x-motion is equivalent.

The initial longitudinal momentum P_{z0} can, to adequate accuracy, be assumed to vanish, and the particle energy is $\mathcal{E} = \gamma mc^2$. The initial particle energy therefore depends only on the initial transverse momentum P_{y0} and is

$$\mathcal{E}_0 = c\sqrt{m^2c^2 + P_{y0}^2}. \tag{10.1}$$

The momenta evolve with time t according to

$$P_y = P_{y0}, \quad P_z = eEt, \tag{10.2}$$

so the energy at time t is given by

$$\mathcal{E} = \sqrt{\mathcal{E}_0^2 + (ceEt)^2}. \tag{10.3}$$

Combining these results the velocity components evolve as

$$\frac{dy}{dt} = c\frac{cP_y}{\mathcal{E}} = \frac{P_{y0}c^2}{\sqrt{\mathcal{E}_0^2 + (ceEt)^2}},$$

$$\frac{dz}{dt} = c\frac{ceEt}{\mathcal{E}} = \frac{c^2 eEt}{\sqrt{\mathcal{E}_0^2 + (ceEt)^2}}. \quad (10.4)$$

Integrating these equations, arbitrarily setting the starting longitudinal position to be $z_0 = \mathcal{E}_0/(eE)$ at $t = 0$, and allowing for initial transverse offset y_0, the particle position coordinates are given by

$$y = y_0 + \frac{P_{y0}c}{eE}\sinh^{-1}\left(\frac{ceEt}{\mathcal{E}_0}\right),$$

$$z = \frac{1}{eE}\sqrt{\mathcal{E}_0^2 + (ceEt)^2}. \quad (10.5)$$

Dividing the first of Eqs. (10.4) by the second, and expressing t in terms of z using the first of Eqs. (10.5), produces the differential equation satisfied by the spatial curve $y = y(z)$. This equation can be integrated immediately to yield

$$\tilde{z} = \cosh(\tilde{y} - \tilde{y}_0), \quad \text{or} \quad \tilde{y} = \tilde{y}_0 + \cosh^{-1}\tilde{z}. \quad (10.6)$$

where dimensionless coordinates have been defined by

$$\tilde{z} = \frac{eEz}{\mathcal{E}_0}, \quad \tilde{y} = \frac{eEy}{P_{y0}c}, \quad (10.7)$$

In these coordinates the origin is at $\tilde{z}_0 = 1$. At thermal electron velocities it is an excellent approximation to assume $\mathcal{E}_0 = mc^2$. Using this and

$$(\gamma - 1)mc^2 = eE(z - z_0) = (\tilde{z} - 1)mc^2, \quad (10.8)$$

one obtains

$$\gamma = \tilde{z}, \quad (10.9)$$

which describes the energy evolution accurately all the way from thermal to fully relativistic energies. Dimensionless "phase space coordinates" can be expressed as

$$\tilde{y} = \tilde{y}_0 + \cosh^{-1}\gamma, \quad \frac{d\tilde{y}}{d\tilde{z}} = \frac{1}{\sqrt{\gamma^2 - 1}}. \quad (10.10)$$

The singular behavior of the slope at $\gamma = 1$ reflects the purely transverse motion that has been assumed at the starting position. In the paraxial limit

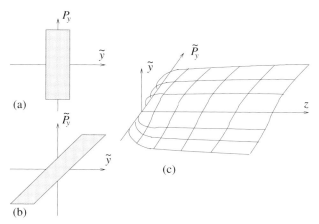

Fig. 10.1 Evolution of transverse phase space from initial condition (a) to final condition (b). Since both $\tilde{P}_y$ and the vertical position spread at fixed $\tilde{P}_y$ are conserved, the enclosed area is conserved (even if slanted edges are not as straight as they are shown). (c) illustrates transverse dependence on z for $y_0 = 0$ for a spread of values of P_{y0}. The beam height increases only logarithmically with increasing z.

the fractional particle momentum is equal to this slope. Clearly the paraxial approximation is invalid near $\gamma = 1$.

Suppose the particle just analysed is one of many particles in a bunch of particles. Figure 10.1 exhibits the evolution of an initially upright rectangle of particles in phase space. Since momentum P_y is conserved and an initial displacement along the y-axis is also preserved downstream, the are of the region containing the original particles is conserved. Here intrabeam interactions are being neglected and, since the motion is Hamiltonian, this area conservation is a manifestation of Liouville's theorem—the density of particles in (y, P_y) "true phase space" is conserved. This is another expression of emittance conservation.

In a photoelectric gun, electrons are ejected from the cathode by an incident laser beam. At the gun cathode, the spot size is governed by the area of this laser illumination. The transverse momentum distribution is governed (one assumes) by statistical physics at temperature T. Without being fussy about the actual physics we can speculate that the average transverse energy is kT, one half of which corresponds to the y-motion under study. This would correspond to an (assumed Gaussian) transverse momentum distribution

$$N_P(P_{yG}) = \frac{1}{\sqrt{2\pi}\,\sigma_{P_{yG}}} \exp\left(-\frac{P_{yG}^2}{2\sigma_{P_{yG}}^2}\right), \tag{10.11}$$

where

$$\sigma_{P_{yG}} \approx \sqrt{mkT}. \tag{10.12}$$

Here the approximation sign signifies uncertainty about the thermal distribution of momenta, and does not refer to uncertainty concerning phase space evolution during acceleration, which is the topic under study, and can be treated exactly. From here on we will use this equation without acknowledging its only approximate initial status.

With spot size at the gun characterized by r.m.s. value σ_{yG}, the "geometric emittance" is defined, to within a temporarily arbitrary constant factor, by

$$\epsilon_y^{(G)} = \sigma_{yG} \frac{\sigma_{P_{yG}}}{\mathcal{E}_0/c} = \sigma_{yG} \frac{\sigma_{P_{yG}}}{mc} = \sigma_{yG} \sqrt{\frac{kT}{mc^2}}, \tag{10.13}$$

where superscript G stands for "gun". ϵ_y is a *true phase space area* (meaning an area in a graph with canonically conjugate axes). Note though that the momentum coordinate has been scaled to units of mc so that this "emittance" has dimensions of length. Being evaluated at the origin, this quantity is, by definition, an *invariant* property of the beam. In defining emittance as the beam energy evolves toward a general, typically relativistic, energy $\mathcal{E}$, it is conventional to generalize Eq. (10.13) by defining "invariant emittance"

$$\epsilon_y^{(I)} = \sigma_y \frac{\sigma_{P_y}}{\mathcal{E}/c} \tag{10.14}$$

which reduces to the geometric emittance at the source as $\mathcal{E} \to \mathcal{E}_0$. Since the area factor $\sigma_y \sigma_{P_y}$ has previously been seen to be conserved, the numerical value of $\epsilon_y^{(I)}$ defined by this equation is the same as the numerical value of $\epsilon^{(G)}$ defined in Eq. (10.13) except for a factor $\mathcal{E}/\mathcal{E}_0 = \gamma$. (Strictly speaking this neglects the initial transverse as well as the initial longitudinal momentum—this is amply accurate in practice.)

Once the beam has become nearly relativistic, the responsibility for describing the propagation of individual particles has to be handed over to an accelerator physics code such as MAD or UAL. Like most such codes these codes "geometricize" (i.e. make independent of reference momentum) the lattice description, by expressing the transverse momentum components as fractions of a longitudinal momentum. For example the vertical momentum component p_y is defined by the equation

$$p_y = \frac{P_y}{P_0} = \frac{P_y}{\beta \gamma mc}. \tag{10.15}$$

where P_0 is the momentum of a central "reference" particle. (It would simplify the present discussion if this normalization were with respect to energy instead of momentum, but there is a good reason (symplicity) why the momentum normalization is chosen.) In the relativistic regime, in these codes, the *invariant emittance* $\epsilon_y^{(I)}$ can be obtained from the *geometric emittance* $\epsilon = \sigma_y \sigma_{p_y}$ by

$$\epsilon_y^{(I)} = \epsilon \frac{\beta \mathcal{E}}{mc^2} = \sigma_y \frac{\sigma_{p_y}}{\gamma mc} \frac{\mathcal{E}}{mc^2} = \sigma_y \frac{\sigma_{p_y}}{mc}. \qquad (10.16)$$

Since the final factor can again be recognized to be the conserved, *true phase space* area, this definition agrees with earlier definitions of $\epsilon_y^{(I)}$.

This definition does, however, have the curious property that it does not evolve smoothly back to the nonrelativistic regime. In that limit, as $\beta \to 0$, it is necessary that $\epsilon \to \infty$ for the result to remain finite. This behavior is related to the previously noted fact that the paraxial approximation cannot be applied to beam evolution in the nonrelativistic gun region. This inelegance is somewhat academic however as one rarely applies the accelerator lattice formalism for $\beta \ll 1$.

There is an even less seemly practice, but which is also of little more than academic importance. For high energy accelerators one also defines a so-called "normalized emittance"

$$\epsilon_y^{(N)} = \epsilon \frac{\mathcal{E}}{mc^2}, \qquad (10.17)$$

which differs from definition (10.16) only by the missing β factor. Since, typically, $\beta \approx 1$, the presence or absence of this factor makes little numerical difference. What makes this "unseemly" is that one rarely distinguishes between $\epsilon_y^{(N)}$ and $\epsilon_y^{(I)}$. In fact it is common practice for Eq. (10.17) to be used as the defining relation for "invariant emittance". With this terminology the missing factor of β makes it impossible to join the relativistic and nonrelativistic description. In spite of this it is common, regrettably even in this text, to blur the distinction between these two definitions. The fractional uncertainty induced by this is typically in the several percent range which is usually small compared to other uncertainties and ambiguities. This is especially true when the emittance is first reliably measured after the beam has become relativistic.

We now return to the task of obtaining the beam emittance at high energy from its value at the source. Combining Eqs. (10.13) and (10.16)

$$\epsilon_y = \sigma_y \sigma_{p_y} = \sigma_{yG} \sqrt{\frac{kT}{mc^2} \frac{1}{\beta \gamma}}. \qquad (10.18)$$

A curious feature of this equation is that it contains the source temperature T, whose meaning is only unambiguous at the source location, in an equation that makes no sense in the nonrelativistic regime because $\beta \approx 0$.

This seeming paradox has been adequately resolved by the discussion just completed.

The main purpose of this discussion has been to justify Eq. (10.18), and in particular, the absolute emittance it predicts. For room temperature, slipping back to the notation earlier referred to as "unseemly", the normalized emittance is given approximately by

$$\epsilon^{(N)} = \sigma_{yG} \sqrt{\frac{kT}{mc^2}} \quad \left(\stackrel{\text{e.g.}}{=} 2 \times 10^{-3} \sqrt{\frac{1/40}{5 \times 10^5}} \approx 0.5\,\mu\text{m.} \right) \quad (10.19)$$

This agrees with an approximate formula given by Groening [7]. Taken at face value, much smaller emittance could be obtained by cooling the cathode. Conversely, if the cathode gets hot, as seems likely to happen at high beam current, the emittance will be greater. In practice, because of various uncertainties, the invariant emittance calculated from first principles as in Eq. (10.19) serves only as a goal; a more reliable value is obtained by measurement.

10.4
Qualitative Description of Lattice Design Issues

Possible ERL designs that take advantage of a large (750 m) circumference existing tunnel are shown schematically in Figure 10.2 and Figure 10.3. The so-called "sprocket" design strives to be a minimal evolutionary upgrade of the existing Cornell University CESR storage ring facility into a "fourth generation" x-ray source. The rest of this chapter is devoted to describing the design and the expected properties of this configuration. The "dog-bone" design, shown in Figure 10.3, increases the overall circumference in order to support more flexible x-ray beamlines. Since lattice design issues are similar for the two designs, only the sprocket design is detailed here.

Because the beam emittance available from a linac is so small, it might be thought that the overall focusing optics is not very critical. What makes the design tricky, in spite of small emittances, is that the beam passes through both linacs twice, once during acceleration and once during deceleration. For the accelerating phase the far-end quadrupoles want to be strong, the near-end quads weak. For the decelerating phase the opposite is true.

The decelerating phases are at a disadvantage in this competition. The adiabatic damping accompanying *acceleration* may decrease the need for focusing appreciably. But the adiabatic anti-damping during *deceleration* makes focusing in the linac sections all the more important. That is the purpose of the quadrupoles (shown as triplets) labeled 2a, 2b, 5a, etc., only a few of which are shown. The labels 2a and 2b actually refer to the same triplet, but on the two passes at different energies. The same applies to 5a and 5b, and so on. The focusing is ascribed to triplets only to emphasize that there must be focusing

10.4 Qualitative Description of Lattice Design Issues

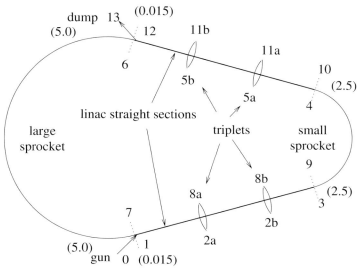

Fig. 10.2 The sprocket ERL design consists of a 5 GeV arc, somewhat greater than a semicircle (in a pre-existing tunnel), two long, linac-containing, straight sections in the same (new) tunnel, and a new, slightly less than semicircular (new) 2.5 GeV arc. Key points in the lattice are numbered sequentially, starting with 0 at the electron gun source and ending with the dump at location 13. Numbers in parentheses are energies in GeV [2].

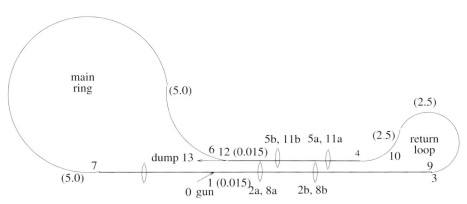

Fig. 10.3 The dog-bone ERL design consists of a major 5 GeV arc in the same pre-existing tunnel, two long linac-containing straight sections, and a 2.5 GeV "return loop. Key points in the lattice are numbered sequentially; the points they mark correspond to points with the same numbers in the sprocket figure, as do the energies, given in GeV.

in both planes. This is a natural property of quadrupole triplets. Later calculations will indicate that quadrupoles in FODO configuration are probably more appropriate than triplets.

If worse comes to worst, the increasing and decreasing energy beams could be separated at the quadrupole locations, so that ideal matching of all sections could be achieved, but, since the focusing elements already have to share space needed primarily for accelerating elements, one works to avoid this complication. The presence of such separation regions would increase the straight section lengths appreciably, especially considering the need to shield the superconducting linac sections from the synchrotron radiation emitted from the separation magnets. Furthermore, lengthening the straight sections further tends to enlarge the beam envelope, even with well matched optics.

Ideally the linac sections would be so short that no external focusing would be required. With no lenses, the too-strong during first passage/too-weak during second passage dilemma would not present itself. But the linac sections are too long to make this realistic. We assume that some focusing will be required, though the number and pattern of quadrupoles remains to be fixed. To idealize the theoretical analysis of the configuration we assume the existence of lenses that simultaneously focus both horizontally and vertically. Of course this is impossible with single quadrupoles, but something approximating it can be accomplished with quadrupole triplets. Even less uniform focusing in both planes is achievable with FODO optics. In the figure the focusing is symbolized by conventional, *thin* optical lenses, even though, in practice, quadrupole triplets are far from thin.

Solenoidal fields have the attraction of "focusing" in both transverse directions. Furthermore, since one has tended to visualize the ERL beam as being more or less round, the coupling produced by solenoids might be thought to be harmless. But this line of reasoning may be misguided. Eventually one will wish to send the beam through small gap height undulators. This already has implications for the shape of the laser beam illuminating the cathode of the electron gun. It may be appropriate for this illumination to be rectangular, much wider than high. Then, to preserve this aspect ratio, it will be appropriate to have decoupled optics through the whole system, just as in any conventional storage ring. In any case solenoidal focusing is probably too weak to be appropriate at the high electron energies of interest. (Incidentally, even if technically possible, imbedding the superconducting RF linac in a solenoidal field would spoil its superconductivity.)

Two principles are helpul in conceptualizing the "two passage" focusing problem. One is that, as mentioned above, the design should be predicated on the decelerating beam, because of its inherent emittance growth. This suggests that the optical design should concentrate on, and favor, the 7–9 and the 10–12 sections indicated in Figure 10.2. At first glance there is a kind of symmetry between these two linac sections. But a section decelerating from 2.5 GeV to 10 MeV (which is the nominal dump energy) has a vastly greater dynamic range than the section going from 5 GeV to 2.5 GeV and, loosely speaking, the difficulty of two-pass optics is proportional to the dynamic range. By this logic, it is the 1–3 and 10–12 sections that are critical.

One sees that the 10–12 section is doubly disadvantaged; in it the beam *decelerates* through a *large dynamic range*. This is the section that will limit the ERL operation. Rather than dumping the beam at, say, 10 MeV, one may entertain dumping it at, say, 20 MeV (greatly increasing the power bill by reducing the energy recoverable). Or, conversely, if dumping the beam at 10 MeV is feasible, then injecting the beam at, say, 5 MeV may be feasible. However there may be little benefit from this—the excess of dumped energy over injected energy has to have been provided by the main linac without being energy-recovered. To state the obvious: these considerations deserve careful compromise because they impact so directly on the economics of ERL operation. We will assume that setting injection and extraction energies equal is close to optimal.

To design the ring optics one can start with a well-defined beam condition at a single point in the lattice and, from there, match each successive lattice sector to the sector preceeding it. The complete two-sprocket, two-straight-section ring can be matched as a storage ring in this way, albeit with dynamic range of energy vastly greater than any conventional storage ring. Then, with the freedom available because it is external, the injection line can, at least in principle, be matched to this "circular" lattice. With this much design the beam can be counted on to advance with controlled, and acceptably small, transverse size from the source at location 0 to the start of the final linac passage at location 10.

Getting the beam from location 10 to the beam dump promises to be the real adventure. During acceleration to full energy there has been, theoretically, a 10/5000 reduction in geometric emittance. If the lattice design, because of mismatches or insertion devices or chicanes, or any other "beam heating" effects, fails to achieve this emittance reduction during acceleration, then the 5000/10 emittance growth certain to occur during deceleration may cause transverse beam dimensions large enough to give unacceptably large beam loss before the beam gets to the final dump.

10.5
Isochronous Arc Design

An important attraction of an ERL is its potential for producing beam bunches short enough to study dynamical evolution of molecular structures on ultrashort time scales. For bunches circulating in conventional light sources the r.m.s. bunch lengths are one, or a few, centimeters (perhaps 50 ps in time units). In so-called "pump–probe" experiments an ultrashort laser pulse excites a localized structure that is subsequently probed by an ultrashort x-ray pulse. With 50 ps x-ray pulses the dynamics time scale studied cannot be less than about 100 ps. One would like to reduce this time by two orders of magnitude or more, but producing and handling femtosecond-scale bunches is not easy.

As in other chapters, much of the emphasis will be on fitting lattice sections into a pre-existing more-or-less-circular tunnel. Let T (typical value 1 µs, which is about a half turn around the ring) be the time spent by electrons of momentum p_0 as they traverse an arbitrary arc of the ring. A momentum spread $\Delta p/p$ (typical value 10^{-3}) will result in a flight time spread ΔT given by

$$\frac{\Delta T}{T} = \left(\alpha - \frac{1}{\gamma^2}\right)\frac{\Delta p}{p_0} \approx \alpha \frac{\Delta p}{p_0}, \tag{10.20}$$

where $\gamma \approx 10^4$ and α is the so-called "momentum compaction factor". (Though α is conventionally regarded as a parameter of an entire ring we are generalizing its definition here to apply to a partial sector of a ring.) For present purposes the final, approximate version of Eq. (10.20) is adequate since, for a ring with tune Q dominated by FODO arcs, $\alpha_{\text{typ.}} = 1/Q^2 \approx 0.01 \gg 1/\gamma^2$. Inserting the typical numerical values given so far yields

$$\Delta T \sim 0.01 \times 10^{-3} \times 10^{-6} = 10\,\text{ps}. \tag{10.21}$$

This shows that the length of a zero length bunch will grow to 10 ps in a half turn around the ring. It follows that, even if the bunches are arbitrarily short as they leave the linac, they will become unacceptably long (for femtosecond experiments) after traversing any appreciable FODO arc.

At this time it is not clear how many beamlines needing ultrashort bunches are justified, or where in a ring they should be located. Once the difficulty of conveying short bunches is fully appreciated it seems likely that the ultrashort bunch region will be restricted to a relatively short sector. Over most of the ring the bunch length will be considerably longer. Even so, to facilitate bunch-shortening chicanery, it seems appropriate to design the arcs to be approximately isochronous ($\alpha = 0$), but with provision for adjusting the local values of α.

The equation for α applicable to an arc of length C is

$$\alpha = \frac{1}{C}\int_C \frac{D(s)}{\rho(s)}\,ds, \tag{10.22}$$

where the "dispersion" $D(s)$ satisfies the equation

$$\frac{d^2 D}{ds^2} + K_x(s) D(s) = \frac{1}{\rho}. \tag{10.23}$$

Here $1/\rho$, the curvature (inverse radius of curvature) acts as a "source" of dispersion, and K_x is the coefficient of the force that "focuses" $D(s)$. Propagation of D in bend-free regions is the same as a particle trajectory. Because ρ is normally positive the value of $D(s)$ tends to be positive, with the result, from Eq. (10.22), that α is positive. There are two ways to overcome this tendency. One is to use reverse bends so that $\rho < 0$. This is rather unattractive however, since the main function of the arcs is to turn the electrons through 360 degrees, which requires the bends to be predominantly positive. The more favorable possibility is to adjust the horizontal focusing in such a way as to make D_s negative in some sectors of the ring. Since this has the concomitant effect of increasing $D(s)$ in other sectors of the ring it is necessary to concentrate the bending in regions where $D(s)$ is negative. There have been several papers developing this approach [3–5].

A natural approach, possibly due to Guignard [3], is to design a triplet of FODO cells, symmetric about the center cell, with negative dispersion in the central cell and positive dispersion in the outer cells. Then one increases the steering in the negative dispersion region and decreases it in the positive dispersion region. This reduces the momentum compaction. Since the beta functions are reasonably insensitive to the distribution of steering, these steering manipulations have little effect on the lattice optics. In the designs shown next the three-cell isochronous modules are trimmed to also be achromatic using two weak negative bends per module.

Figure 10.4 shows the beta functions through a big sprocket triplet sector, and Figure 10.5 shows the dispersion through the same sector. Somewhat arbitrarily, the matching conditions at the ends of this sector (as well as small sprocket sectors) have been chosen to be, $\beta_x = 33.0$ m, $\alpha_x = 0$, $\beta_y = 10.0$ m, $\alpha_y = 0$. The optics in a small sprocket sector are essentially the same, with the same α and β values. The dispersion is matched to zero at all junctions and throughout both straight sections. Because the beam energy in the small sprocket has half the value in the large sprocket, the bend per dipole is roughly twice as great as in the small sprocket. Figure 10.6 shows a fully matched ring. Figure 10.7 shows the corresponding dispersion function.

312 | 10 The Energy Recovery Linac X-Ray Source

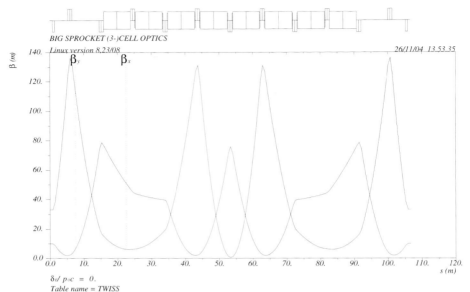

Fig. 10.4 Beta functions for one three-cell isochronous module of the big sprocket.

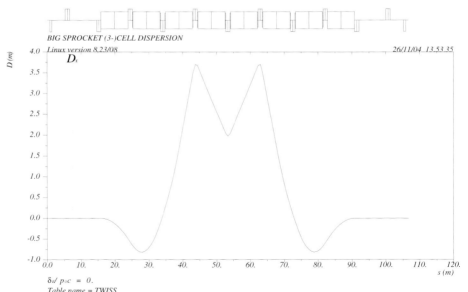

Fig. 10.5 Dispersion function for one three-cell isochronous module of the big sprocket.

10.5 Isochronous Arc Design

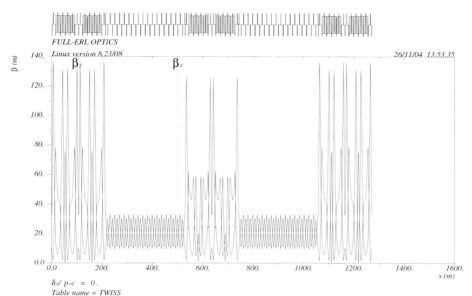

Fig. 10.6 Beta functions for sprocket ERL configuration matched as a constant energy storage ring.

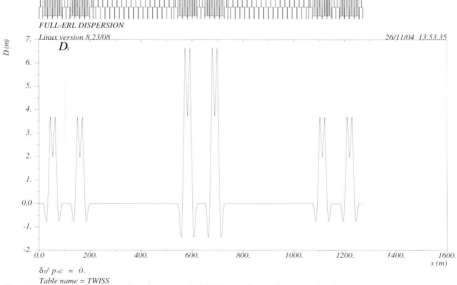

Fig. 10.7 Dispersion function for sprocket ERL configuration matched as a constant energy storage ring.

This is not quite practical as ERL optics since acceleration in the straight sections is absent. To enable full lattice matching, FODO cells have been artificially included in the straight sections as place holders. The resulting optics would not be practical for ERL operation because the beam energy is the same in both sprocket arcs. Such a configuration would presumably be tuned up, with linacs turned off, as a prelude to commissioning full ERL operation. The remaining design amounts to replacing these FODO sections by accelerating sections (which is, of course, made difficult by the two-pass problem discussed earlier.)

Much of the rest of the chapter is devoted to analysis of what has been referred to as the "two pass" problem. The problem will be idealized by assuming uniform (not lumpy) focusing through the linac sections. The only free parameter will therefore be the strength k_0 of this focusing.

10.6
Evolution of Betatron Amplitudes through the Linac Sections

10.6.1
Deceleration through Linac Section to Dump

Acceleration starting from thermal energies has been described in Section 10.3. That analysis of transverse, or betatron, motion is too specialized to be applied here because it assumed both constant accelerating field and zero initial longitudinal momentum. Here, because the initial longitudinal momentum is relativistic and the paraxial approximation is valid, it is not necessary to be as careful as in the nonrelativistic region. The orbit can be determined just using transverse momentum conservation. With $dz \approx \beta c\, dt$, and $p_y \approx \beta \gamma mc\, dy/dz$, and neglecting *variation* of β, though allowing modest deviation of β from 1, this reduces to

$$\frac{d}{dz}\left(\gamma(z)\frac{dy}{dz}\right) = 0. \tag{10.24}$$

With $\mathcal{E}$ being the uniformly varying total energy this becomes

$$\frac{d^2y}{dz^2} + \frac{\mathcal{E}'_D}{\mathcal{E}}\frac{dy}{dz} = 0, \quad \mathcal{E} = \mathcal{E}_0 + \mathcal{E}'_D z. \tag{10.25}$$

The "D" subscript signifies *deceleration*, which implies $\mathcal{E}'_D < 0$, causing the first derivative term to give *anti-damping*. For energy evolving uniformly from $\mathcal{E}_0$ to $\mathcal{E}(z)$, following Minty [6] by adding another term to this equation to allow for focusing, we have

$$\frac{d^2y}{dz^2} + \frac{1}{z + \mathcal{E}_0/\mathcal{E}_D'} \frac{dy}{dz} + k_0^2 y = 0. \tag{10.26}$$

where $k_0^2 y$ gives the focusing force provided by external focusing elements. Solutions of this equation can be expressed in terms of Bessel functions;[1]

$$y = A J_0(k_0(z + \mathcal{E}_0/\mathcal{E}_D')) + B Y_0(k_0(z + \mathcal{E}_0/\mathcal{E}_D')),$$
$$y' = -A k_0 J_1(k_0(z + \mathcal{E}_0/\mathcal{E}_D')) - B k_0 Y_1(k_0(z + \mathcal{E}_0/\mathcal{E}_D')). \tag{10.27}$$

The constants A and B are available to match initial conditions. The fact that the $Y_0(z)$ function is singular at $z = 0$ makes it inappropriate to identify the two terms in this solution as cosine-like and sine-like. Sine-like $S(z)$ and cosine-like $C(z)$ solutions can be obtained by solving the simple algebraic equations (for example using Maple);

$$y(A,B; z=0) = 1, y'(A,B; z=0) = 0, \longrightarrow C(z),$$
$$y(A,B; z=0) = 0, y'(A,B; z=0) = 1, \longrightarrow S(z). \tag{10.28}$$

Equation (10.26) looks simpler than it really is. The coefficient k_0 is a *geometric* focusing parameter (meaning the absolute momentum has been "factored out".) For k_0 to be constant the absolute restoring fields need to be proportional to beam energy, which depends on z. Especially for linacs with large energy increase, this requires strong increase of the absolute quadrupole strengths along the length of the linac. Since the final deceleration section is the critical one in the ERL we will assume that the focusing has been optimized for deceleration. Solutions (10.28) are also deceptively simple. Though A and B are individually complex, $C(z)$ and $S(z)$ are real.

Furthermore, the focusing needs to be lumped and cannot be distributed uniformly as is implied by Eq. (10.26). The plan is to break the 10-12 sector into multiple subsectors, as shown in Figure 10.2. In the absence of a more systematic approach it seems the optical choices of segment lengths, intermediate energies, and focal lengths have to be obtained iteratively, by trial and error. With the focusing strengths adjusted for the deceleration pass they are mismatched for the acceleration phase. Encouragement can be drawn from the fact that the focusing that predominantly restrains the beam size during deceleration (say from 2.5 GeV to 16 MeV) will be weak and not very influential during acceleration (say from 2.5 GeV to 5 GeV). Unfortunately, since the adiabatic focusing during the latter, fractionally small, energy increase is modest, there needs to be some focusing even on the accelerating pass. Practical focusing issues will be discussed later. For now we assume the focusing

---

1) The solution given in Eq. (10.27) differs from Minty's more approximate solution.

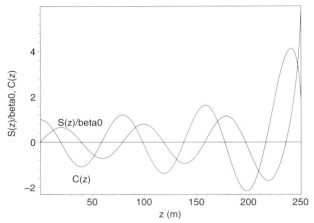

Fig. 10.8 Evolution of matrix elements $S(z)/\beta_0$ and $C(z)$ in the final deceleration section from lattice position 10 to 12, within which the beam energy $E(z)$ falls from 2.5 GeV to 15 MeV. A (uniform) focusing coefficient $k_0 = 6.34\pi/250\,\mathrm{m}^{-1}$ is assumed. The energy variation is also shown.

coefficient k_0 is to be optimized based on the final deceleration section with no concern whatsoever paid to the acceleration pass to high energy.

Similar considerations will eventually have to be applied to the design of the sector from lattice position 1 to lattice position 3. But, as discussed earlier, that sector is less demanding because the emittances have had less chance to blow up before the large dynamic range pass.

For now the value of k_0 is some kind of average over the eventual quadrupole focusing. The matrix elements $S(z)$ and $C(z)$ are plotted in Figure 10.8 for $k_0 = 6.34\pi/250\,\mathrm{m}^{-1}$. (Solving Eq. (10.26) numerically gives the same curves, which confirms Eqs. (10.27) and (10.28).) The main effect of deceleration is that $S(z)$ achieves a very large value (which varies inversely with k_0). Any starting slope increases rapidly with decreasing particle energy, but the growth is eventually restrained by the quadrupole focusing. The value of $C(z)$ remains of order 1 because, for the relatively weak quadrupole focusing, the transverse motion depends only weakly on the initial displacement.

It may be qualitatively more instructive to multiply the sine-like and cosine-like functions by typical values of $x'(0)$ and $x(0)$ which, being actual transverse displacements, are measured in meters. This is done in Figure 10.9.

The formula for lattice Twiss function evolution is

$$\begin{pmatrix} \beta \\ \alpha \\ \gamma \end{pmatrix} = \frac{E(z)}{E(z_0)} \begin{pmatrix} C^2 & -2CS & S^2 \\ -CC' & CS'+SC' & -SS' \\ C'^2 & -2C'S' & S'^2 \end{pmatrix} \begin{pmatrix} \beta_0 \\ \alpha_0 \\ \gamma_0 \end{pmatrix}, \qquad (10.29)$$

where C, S, C', S' are the elements of the 2×2 transfer matrix derived in

10.6 Evolution of Betatron Amplitudes through the Linac Sections

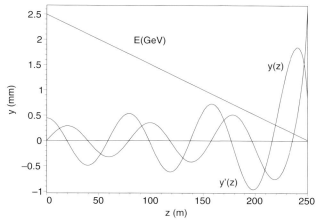

Fig. 10.9 Evolution of $y(z) = \sqrt{\epsilon \beta}\, C(z)$ and $y'(z) = \sqrt{\epsilon/\beta}\, S(z)$ trajectories in the final deceleration section from lattice position 10 to 12. A fairly large value, $\epsilon_x = 10^{-8}$ m, is assumed to allow for possible previous emittance growth.

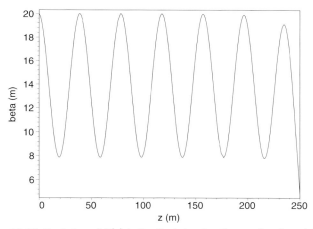

Fig. 10.10 Evolution of $\beta(z)$ in the final deceleration section from lattice position 10 to 12, assuming $\alpha = 0$, $\beta = 20$ m at position 10.

Eq. (10.28). This differs from the standard (constant energy) evolution formula only by the leading $E(z)/E(z_0)$ factor. The extreme reduction in Twiss function values caused by this factor is, to a large extent, compensated for by very large values of the matrix elements, especially $S(z)$ and $S'(z)$. Figure 10.10 shows the evolution of $\beta(z)$ calculated using the upper row of Eq. (10.29).

10.6.2
Triplet Design

Since quadrupole triplets focus simulataneously in both planes it is attractive to contemplate their application to the "two pass" problem. Other considerations suggesting this are:

- The focusing strengths of a triplet (unlike a doublet) are approximately the same for the x and y transverse coordinates. This is appropriate for beams that are more or less round.
- A triplet through which both a low and a high energy beam pass can be designed primarily with the low energy in mind. For the high energy beam this lens will be more or less inert; i. e. its focal length will be long.
- Because the individual lenses in a triplet "fight each other" their strengths have to be extravagantly large for many purposes. But this behavior is ameliorated in the present application by the relatively low energy for which the focal lengths have to be short.

We therefore digress into a brief discussion of triplet design. Wollnik [8] gives handy design formulas for quadrupole triplets. His formulas allow for the quadrupoles to have finite thickness but, for simplicity, we will treat them as thin, as shown in Figure 10.11. The triplet is designed to produce a parallel beam after the triplet from a point source at the origin the focal lengths. The focal lengths are therefore

$$f_x = f_y = l + d \approx l. \tag{10.30}$$

The focal lengths are given by

$$f_1 = -\sqrt{2ld}\sqrt{\frac{1}{1+d/l}}, \quad f_2 = \sqrt{\frac{ld}{2}}\sqrt{1+d/l}, \quad f_3 = -\sqrt{2ld}\sqrt{1+d/l}. \tag{10.31}$$

In the present application, since the triplet thickness is to be much less than the focal length, $d \ll l$, and the final factors will all be replaced by 1 for a first cut design, let $f_0 \approx l$ be the "nominal" focal length of the triplet at momentum p_0. The problem, as has been emphasized, is that the triplet will have a different focal length f at momentum p. By assuming $d \ll l$ and manipulating Eqs. (10.31), one finds

$$f \approx f_0 \left(\frac{p}{p_0}\right)^2. \tag{10.32}$$

This obviously gives a very strong chromatic focusing effect. It makes more quantitative our earlier statement that a lens designed for the low momentum pass is very weak on the high energy beam pass.

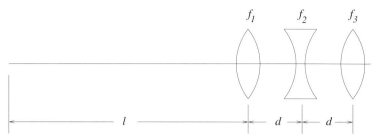

Fig. 10.11 Dimensions of thin element quadrupole triplet with point to parallel focusing and having focal lengths $f_x = f_y = l_1 + d$.

10.6.3
Acceleration through the High Energy Linac Section

Deceleration through the sector from lattice position 10 to 12 has been described earlier. This same section serves to accelerate the beam from 2.5 GeV to 5 GeV. Since this represents a rather small fractional increase in energy through a conventional linac its analysis should be straightforward. There is, however, the complication of passing through focusing elements designed for the deceleration pass. In the latter half of this section the focusing tends to be so weak as to be insignificant. This constitutes a problem however since, in a drift length of 100 m or more the β functions are sure to grow to a value of order 100 m or more.

Continuing to represent the focusing as being continuously distributed, on the accelerating pass differential equation (10.25) is replaced by

$$\frac{d^2y}{dz^2} - \frac{\mathcal{E}'_D}{\mathcal{E}} \frac{dy}{dz} + k_0^2 \left(\frac{\mathcal{E}_D(z)}{\mathcal{E}}\right)^n y = 0,$$
$$2.5 \,\text{GeV} < \mathcal{E} = 2.5 - \mathcal{E}'_D z < 5.0 \,\text{GeV}. \tag{10.33}$$

The coefficient of the second term has switched and is now numerically positive (which corresponds to damping) and $\mathcal{E}_D(z)$ is the energy of the beam at position z during the deceleration pass. The strength of the focusing term has been derated to account for the different energy on this pass compared to the deceleration pass and the same, but opposite sign, gradient $\mathcal{E}'_D$ is assumed. The exponent n has been left free to allow for different quadrupole configurations. For triplets n has the surprisingly high value of 4, which results from the coefficient being proportional to inverse focal length squared, and the effective strength of the triplet also being quadratic (as shown in Eq. (10.32).) Accounting for the momentum dependence of the focusing in this way is only a crude approximation, but it is most nearly valid near the beginning of the section where the focusing is strong. In this region the energies are not very different during acceleration and deceleration passes. Where the approxima-

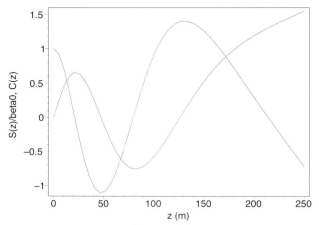

Fig. 10.12 Matrix elements $S(z)$ and $C(z)$ for the acceleration pass from lattice position 4 to 6 assuming chromatic index $n = 2$, which is appropriate for FODO (rather than triplet) optics. These functions have been determined by numerical solution of differential equation (10.33).

tion breaks down the focusing strengths have become weak. The value of n for FODO focusing can be closer to 2.

Curves for the accelerating pass are shown in Figure 10.12 and Figure 10.13. These are directly comparable with the curves shown earlier for the deceleration pass. For these curves the chromatic power index has been taken to be $n = 2$. Similar curves were obtained for the case $n = 4$ but, for brevity, they are not shown. In spite of arguments given earlier, recommending the use of triplet optics, the $n = 4$ performance seems to be far inferior to the $n = 2$ performance. The chief problem seems to be the β-blow up toward the end of the line that was mentioned earlier.

Recapitulating, beam evolution through the most difficult section of the ERL has been exhibited in the preceeding five graphs. The same focusing coefficient k_0 and the same accelerating gradient $\mathcal{E}'_D$ have been used in all of these graphs. The "fine" part of k_0 has been determined by the β-function match at the exit from the line. This selects one out of a sequence of "course" values of k_0 corresponding to different "integer tune advances" along the linac section. Other than choosing from this discrete set, all parameters have been determined.

Even for the idealized (unphysically simple) model that has been assumed here, the graphs describing betatron evolution in the ERL can only be described as *ugly*. This reflects the inelegance of sending beams of different energies through the same beam line. Replacing the uniform focusing by a realistic quadrupole distribution cannot turn this duckling into a swan. The degree of difficulty is a function of the dynamic ranges that have been assumed. Here

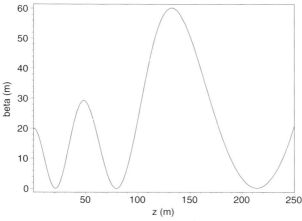

Fig. 10.13 Evolution of $\beta(z)$ in the acceleration section from lattice position 4 to 6, assuming $\alpha = 0$, $\beta = 20$ m at position 4. Quadrupole (rather than triplet) optics is assumed. The parameter k_0 has been chosen to match the beta functions at the output of the section to the value at the large sprocket input. (This match does not need to be perfect as a matching section could be inserted.)

the deceleration has been by a factor 2.5/0.015 and the acceleration factor has been 5/2.5. This is far larger than the 10 MeV to 90 Mev energy ratio that has been successfully achieved in 2005 at the Jefferson Laboratory ERL [9]. Exactly how much greater a ratio can be successfully handled remains to be established.

Though the graphs that have been exhibited meet the nominal requirements for ERL operation, it is my opinion that their behavior is too crude to be acceptable. The dynamic ranges assumed are simply too great. As mentioned earlier, however, the situation can be greatly improved (at modestly increased cost and complexity) by splitting the linac section just studied into a long followed by a short section, with the beams separated in the junction region on separate passes. For the sake of symmetry, the first half linac should probably be split similarly. By performing these splits all dynamic ranges could be kept closer to the value that has been shown to be achievable at Jefferson Lab.

10.7 Emittance Growth Due to CSR and Space Charge

For successful ERL operation it is essential that emittance growth be limited. Some sources of growth, such as steering jitter, can be ameliorated by precision regulation, or by feedback or, more likely, by feed-forward. These topics, though important, are too technical for discussion here.

Beam growth due to space charge effects is more complicated. All of Chapter 13 is devoted to this topic. The simulation described there subsumes *all* space charge effects, including Coulomb repulsion, Ampère attraction, the self-force associated with coherent synchrotron radiation (CSR) and the centrifugal space charge force (CSCF). The latter two effects are present only while the beam is in magnetic field regions.

Using the programs described in Chapter 13, some effects of CSR and CSRF on critical regions of an ERL are shown in the following figures: Figure 10.14, Figure 10.15, and Figure 10.16.

Figure 10.14 shows particle distributions in 6D phase space, before and after a single turn around the *return loop* of the *dog-bone* ERL. Some parameters for that series of graphs are: number of particles in bunch, qtot(C)=5e10, electron energy ee(GeV)=1.5, bunch half width xhW(micron)=50, bunch half height yhW(micron)=50, bunch half length cthW(mm)=0.5, etc. Other parameters can be read similarly from the data at the top of the figure.

Figure 10.15 displays emittance growth in the return loop of the dog-bone ERL. Different parameter sets are indicated in the key. Points plotted are *output* values of normalized emittance. For all points on any one curve the *input* value of normalized emittance was the same. It can therefore be inferred to be the *output* value at the left end of each curve, where the space charge forces are negligible. Because of the beam's relatively high energy in this ring, conventional Coulomb/Ampère forces are negligible. But, CSR and CSCF forces become strong as the charge per bunch Q increases toward 1 nC and beyond. Chapter 9 showed that the brilliance of produced x-ray beams varies proportionally to Q and inversely with $\epsilon^{(N)2}$. The data of this graph therefore shows there to be little benefit in increasing Q beyond, say, 1 nC. This relatively small value makes it important for the number of bunches circulating to be high. (The bunch repetition frequency is 1.3 GHz for the proposed Cornell ERL.)

Loss of brightness due to emittance growth is not the only thing limiting bunch charge Q. Growth of beam "tails" due to space charge, or anything else, can cause particle loss of large amplitude particles in intense bunches before the final beam dump. Note that, for long bunches, such as $\sigma_s = 1$ cm, that the emittance growth is negligible for the range of Q shown. This is consistent with the assumed dominance of CSR and CSCF, which weaken rapidly as the bunch length increases.

Figure 10.16 displays emittance growth in the injection merge region. (The large radius of curvature orbit of the high energy beam and the small radius of curvature injection orbit share the same tangent as they leave the last bending magnet of this merge.) Because of the low electron energy at this point, the space charge force is dominated by the Coulomb repulsion and it is strong enough to give emittance growth. This growth increases strongly with increasing charge per bunch. The "cross-over" near $Q = 20$ pC is probably a

10.7 Emittance Growth Due to CSR and Space Charge

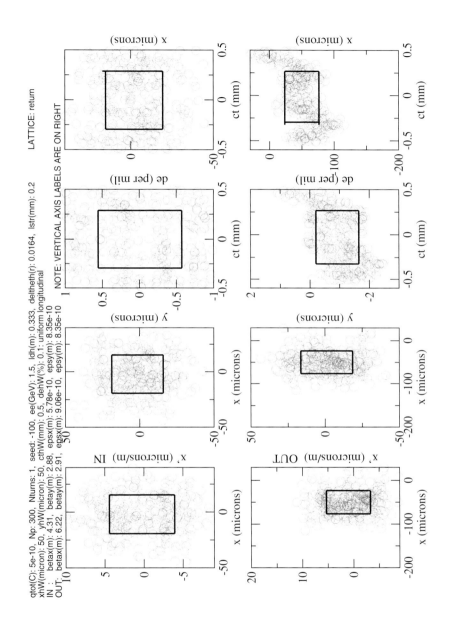

Fig. 10.14 Particle distributions in 6D phase space before (top row) and after (bottom row) a single turn around the *return loop* of the *dogbone* ERL.

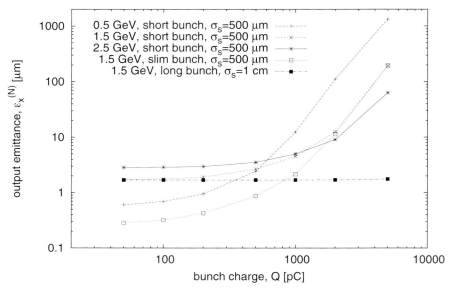

Fig. 10.15 Emittance growth caused by CSR and CSCF in the return loop (the region from lattice point 3 to 4 in Figure 10.3.) of the dog-bone ERL.

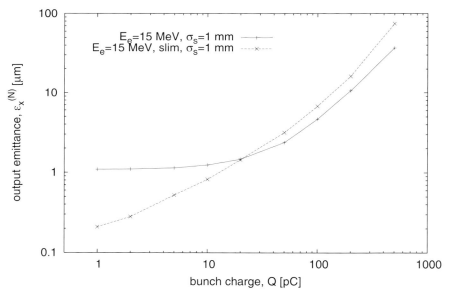

Fig. 10.16 Emittance growth caused by CSR and CSCF in the injection merge (near lattice point 1 in the figures) of ERL of either sprocket or dog-bone design.

statistical artifact of the limited number of macroparticles (namely 300) used in these simulations. It seems more appropriate to interpret the data as showing that there is little or no advantage—in reduced output emittance—from reducing the input invariant emittance below about 1 μm.

Parameters for the proposed Cornell ERL are $Q = 77\,\text{pC}$, $\epsilon^{(N)} = 1\,\mu\text{m}$. These values are just consistent with the emittance growth shown in the previous two graphs, but with little margin for error.

References

1 Landau, L., Lifshitz, E. (1989), *The Classical Theory of Fields*, Pergamon Press, New York.

2 Talman, R. (2002), *Energy Recovery Linac in the Wilson Tunnel*, Cornell Report, June, (2002).

3 Guiniard, G. (1989), *A Lattice With No Transition and Largest Aperture*, Proc. IEEE Part. Acc. Conf., **1989**, 915.

4 Trbojevic, D., Courant, E. (1994), *A Standard FODO Lattice With Adjustable Momentum Compaction*, Brookhaven report.

5 Lee, S., Ng, K., Trbojevic, D. (1993), *Minimizing Dispersion in Flexible Momentum Compaction Lattices*, Phys. Rev. E, **48**, page 3040

6 Minty, M. (2000), *Emittance Preservation in Linear Accelerators*, in *High Quality Beams*, Joint Accelerator School, S. Korokawa, et al. (Eds.), AIP Conf. Proc., **592**, 118.

7 Groening, L. (2000), *Dependence of Thermionic Gun Emittance on Beam Energy Calculated With EGUN*, CERN, February.

8 Wollnik, H. (1987), *Optics of Charged Particles*, Academic Press, New York, 76.

9 Jefferson Laboratory website, http://conferences.jlab.org/ERL.

11
A Fourth Generation, Fast Cycling, Conventional Light Source

11.1
Preview

"First generation" storage ring x-ray sources parasitically utilized the radiation produced by electron accelerators designed for, and in use by, elementary particle physics. "Second generation" x-ray sources, for the first time, derived beams from "insertion devices" introduced into lattices for the specific purpose of producing external x-ray beams. The designs of "third generation" storage rings have been optimized as x-ray sources and have been dedicated to that use. Mainly this means they have emittances far smaller than was useful for rings dedicated to elementary particle physics. Here we consider a "fourth generation" of sources made possible when a (large) tunnel and accelerator infrastructure is made available by the decommissioning of a colliding beam facility such as PEP, PETRA, CESR, and, eventually, even KEK.[1] The inherited assets yield very different optimization considerations than are applicable to the design of an x-ray facility built "from the ground up". Mainly this amounts to being able to take advantage of the "extavagantly" large circumferences of an inherited ring to produce major improvement of brilliance compared to existing third generation sources. In particular, this increases the optimal energy, which increases the achievable brilliance, especially for hard x-rays.

Ultrabright x-ray-producing rings have much in common with the damping rings needed for next-generation linear colliders, such as the International Linear Collider (ILC). Both produce intense, low emittance electron beams. These are essential in a damping ring (to produce low emittance) and in an x-ray source (to produce x-rays).

The main subject of this chapter then, is the detailed design, along conventional lines, of an ultra-brilliant x-ray source. This ring will be referred to as FCSR, for fast cycling storage ring. The main novel feature of the ring is its

---

1) At the time this is being written, projects converting colliding beam facilities into x-ray sources, similar in motivation to the design described in this chapter, have recently been completed (SPEAR) or begun (PETRA).

Accelerator X-Ray Sources. Richard Talman
Copyright © 2006 WILEY-VCH Verlag GmbH & Co. KGaA, Weinheim
ISBN: 3-527-40590-9

fast cycling. This feature allows the facility to tolerate the short beam lifetime caused by intrabeam scattering, which is known as the Touschek effect.

First, however, a general discussion of the design of low emittance light source lattices will be given.

11.2
Low Emittance Lattices

Many discussions of low emittance lattice design are available in the literature. Some useful articles are in the CERN Yellow Report 98-04 [1], and they include references to many earlier studies. The article most directly applicable to the present subject, *Lattices and Emittances*, is by Ropert [2]. She gives a thorough review of most of the light sources operating at the (1996) time of writing, with emphasis on designs intended to produce small emittance, and hence high brilliance.

The main lattice parameters governing the emittance of storage rings are β_x, the horizontal Twiss lattice function, and D, the horizontal dispersion. Ropert [1] gives a formula (our Eq. (5.76)) for the horizontal emittance,

$$\epsilon_x = \frac{C_q \gamma^2 <\mathcal{H}/\rho^3>}{J_x <1/\rho^2>}, \quad \text{where} \quad C_q = \frac{55}{32\sqrt{3}} \frac{h}{2\pi mc} = 3.84 \times 10^{-13} \text{ m}, \tag{11.1}$$

where ρ is the bending magnet radius and $\gamma = \mathcal{E}_e/mc^2$ is the usual relativistic factor and other factors will be defined in what follows. The factor $\mathcal{H}(s)$, depending on longitudinal position s, and known as Sands's "curly H", is given in terms of Twiss functions and dispersion function D by Eq. (5.65),

$$\mathcal{H}(s) = \gamma_x D^2 + 2\alpha_x DD' + \beta_x D'^2, \tag{11.2}$$

where $\gamma_x = (1 + \alpha_x^2)/\beta_x$. One noteworthy feature of these formulas is that bending magnets are the only contributors to beam emittance, since the inverse radius $1/\rho$ vanishes everywhere else. The main way, therefore, to minimize ϵ_x is to reduce $D(s)$ at the position of bending magnets. Furthermore, in spite of the γ_x factor in the first term of $\mathcal{H}$, it is advantageous to also minimize β_x at bending magnet locations—the D^2 numerator factor dominates the β_x factor in the denominator of the first term of $\mathcal{H}$.

In Figure 11.1 these lattice functions are plotted for the Canadian Light Source (CLS), an example of a double focusing achromat (DFA) cell. This is a variant of the "Chasman-Green" cell [3], first introduced for the Brookhaven NSLS.

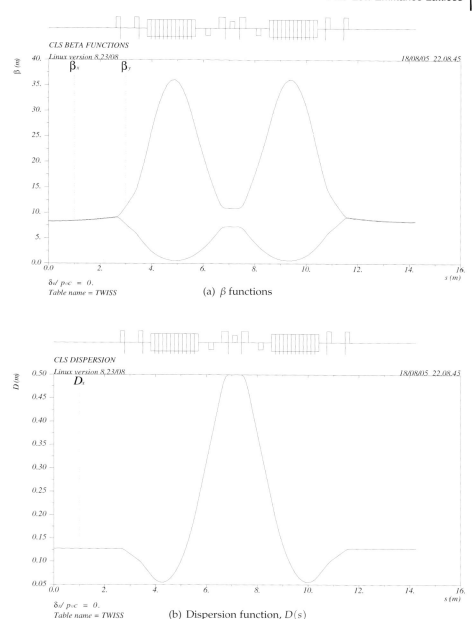

Fig. 11.1 Lattice functions for one cell of the Canadian Light Source (CLS), a double focusing achromat (DFA). The bending magnets are actually *combined function*, vertically focusing, field index $n = 20$, strength $\text{k1} \approx -0.4\,\text{m}^{-2}$. $\mathcal{E}_e = 2.9\,\text{GeV}$. The complete lattice consists of 12 such cells, which gives $\theta = 2\pi/24$.

In Figure 11.2 the same functions are shown for the Advanced Light Source (ALS), at Berkeley, which is an example of a triple bend achromat (TBA).

These plots have been produced using MAD8 [4]. The tops of these MAD plots identify the magnet types and locations in the cell. Bends are identified by up–down centered-boxes, horizontally (vertically) focusing quads by half-boxes above (below) the axis, and sextupoles are represented by less tall half-boxes. Features common to these (and all other low emittance lattices) are the presence of minimal values of both β_x and D at the positions of the bending magnets. The TBA design provides minima of these functions near the centers of the bending magnets.

An important issue that has to be faced is *chromaticity compensation*. To minimize the "tune footprint" of the beam, and to suppress the so-called "head–tail" effect, it is necessary for both chromaticities, Q'_x and Q'_y, to be approximately zero for the lattice as a whole. Especially for the strong focusing needed to minimize D and β_x, the "natural chromaticities" are large and negative. To compensate these chromaticities, sextupoles are inserted into the lattice. Since the impact on the chromaticity of a sextupole is proportional to D, the requirements for low emittance and zero chromaticity are in direct conflict. Strong focusing necessarily leads to strong sextupoles. Even apart from the fact that the required sextupoles may be expensive, or impossible, to build, the presence of strong sextupoles limits the so-called "dynamic aperture" of the lattice. Much of this chapter is devoted to this trade-off.

For the DFA lattice, Ropert quotes the equilibrium emittance as

$$\epsilon_x = \frac{C_q \gamma^2 \theta^3}{J_x} \frac{1}{4\sqrt{15}}. \tag{11.3}$$

One effect of the combined function, vertical focusing built into the CLS bending magnets is to increase J_x from 1.0 to 1.56, without invalidating Eq. (11.3). (Careful scrutiny of the magnet representations in Figure 11.1 shows how this focusing is modeled by thin interspersed, vertically focusing, quadrupole elements.) This yields a corresponding reduction in ϵ_x. This vertical focusing has the further beneficial effect of reducing the β_y maximum at the cell boundaries. Note that the CLS cell is *not*, in spite of its DFA heritage, achromatic—D does not quite vanish at the cell boundaries.

Applicable to the Advanced Light Source (ALS) lattice exhibited in Figure 11.2, Ropert gives the theoretical minimum emittance for the TBA lattice to be

$$\epsilon_x = \frac{C_q \gamma^2 \theta^3}{J_x} \frac{7}{36\sqrt{15}}. \tag{11.4}$$

Like CLS, there is combined-function vertical focusing in the ALS bending magnets, which increases J_x from 1. Also the cells are not quite achromatic.

11.2 Low Emittance Lattices

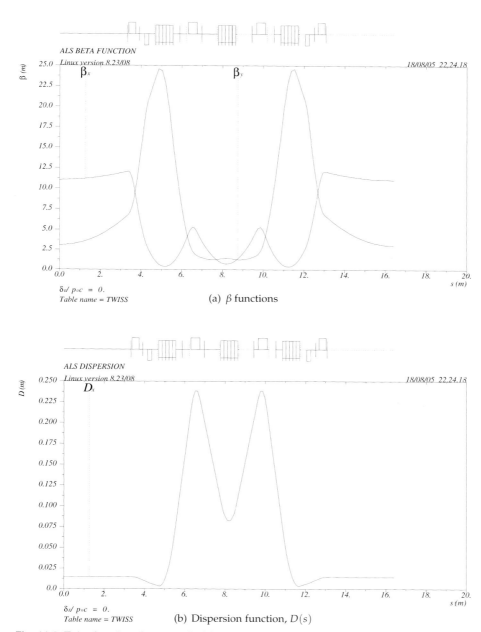

Fig. 11.2 Twiss functions for one cell of the Advanced Light Source (ALS), an example of a triple bend achromat (TBA). $\mathcal{E}_e = 1.5$ GeV. The superperiodicity is 12, giving 36 bending magnets, and $\theta = 2\pi/36$.

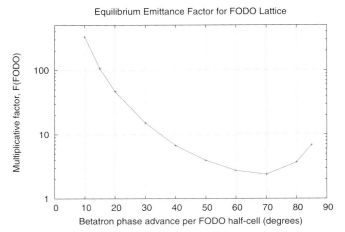

Fig. 11.3 The dimensionless factor $F_{\text{FODO}}(\phi)$ in Eq. (11.5) which gives the equilibrium emittance of FODO lattices.

The equilibrium emittances of these low emittance lattices can be compared with the emittance of a FODO lattice, which can be expressed as [5]

$$\epsilon_x = \frac{C_q \gamma^2 \theta^3}{J_x} F_{\text{FODO}}(\phi), \tag{11.5}$$

where ϕ is the phase advance per half-cell of the FODO lattice.

The function $F_{\text{FODO}}(\phi)$ is plotted in Figure 11.3. For a colliding beam lattice a typical value for the half-cell phase advance is 35° and the factor $4F(\text{FODO}) \approx 10$. One sees that lattice design optimized for x-ray production can potentially reduce ϵ_x by a substantial factor. Note though that, because the fraction of ring devoted to bending magnets is high in FODO designs, the bend per magnet θ can be relatively small. This causes the factor $F_{\text{FODO}}(\phi)$ to be unrealistically small when comparing FODO lattices with other designs, such as DFA or TBA. But reduction of ϵ_x by at least an order of magnitude is typical for the various high brilliance lattice designs. Furthermore, the brilliance will be seen later to be proportional to $1/\epsilon_x^2$.

11.3
Production of High Quality X-ray Beams

11.3.1
A Formula for Brilliance $\mathcal{B}$

The "brilliance" $\mathcal{B}$ of an x-ray source measures a (relativistically invariant) density of photons in six-dimensional phase space. As such $\mathcal{B}$ is the number of photons inside a six-dimensional volume $\mathcal{L}_3 \mathcal{P}_3$ (sufficiently small that all photon distribution functions are more or less constant throughout) divided by $\mathcal{L}_3 \mathcal{P}_3$. For x-rays produced in a storage ring, $\mathcal{B}$ is made large, at least partly by permitting $\mathcal{L}_3 \mathcal{P}_3$ to be chosen very small. Another figure of merit, the "flux" $\mathcal{F}$ is, crudely speaking, given by the product $\mathcal{F} \sim \mathcal{L}_3 \mathcal{P}_3 \mathcal{B}$.

When x-rays are used in a scientific experiment the volume of the sample being radiated can be compared to $\mathcal{L}_3 \mathcal{P}_3$. A storage ring is not well matched, for example, to making dental x-rays, because the sample size is very large and the large value of brightness over a small volume is inappropriate. In this case the flux $\mathcal{F}$ is a better measure than $\mathcal{B}$ of the usefulness of the source.

The real value of storage ring x-ray sources is in the opposite extreme, where the sample size is comparable to, or smaller than $\mathcal{L}_3 \mathcal{P}_3$. At the risk of annoying some fraction of the x-ray community, in this chapter I will refer to this as the "physically important" regime of storage ring x-ray operation. Whether or not this opinion is valid, everything said in this chapter applies only to this case. (In truth, important x-ray experiments span such a broad range that it is not really valid to say that $\mathcal{B}$ is always more important than $\mathcal{F}$. The extreme position taken here is mainly to avoid the constant repetition of qualifying statements in what follows.) As a result $\mathcal{B}$ will be taken as the figure of merit to be maximized and $\mathcal{F}$ will not be mentioned again.

A maxim heard in the x-ray field, "never trade flux for brilliance", may seem to contradict the previous paragraph, but it *does not*. When discussing an experiment in which x-rays are scattered from a *small* sample, the maxim could be re-expressed more carefully as "never trade more detected photons for fewer detected photons", which no one should disagree with. For experiments falling into this category the rate of detected photons is governed by $\mathcal{B}$ more than by $\mathcal{F}$. In this sense the use of the word "flux" in "never trade flux etc. etc. " is simply misleading.

The master formula governing the brilliance of the x-ray beam produced by electron beam current I from the fundamental $n = 1$ undulator resonance, in an undulator having N_w periods, with undulator parameter K (assumed less than 1), due to Krinsky [6] and possibly others, and our Eq. (9.27) is,[2]

$$\mathcal{B}_1 = \frac{2 \times 10^{18} \, I(\text{A})}{\epsilon_x(\text{nm})\epsilon_y(\text{nm})} \frac{N_w \, K^2}{1 + K^2/2} \frac{\text{photons}}{\text{s mm}^2 \, \text{mr}^2 \, 0.1\%\text{BW}}. \tag{11.6}$$

Here ϵ_x and ϵ_y are un-normalized emittances. This formula describes the $n = 1$, monochromatic x-ray beam whose wavelength depends, among other things, on the undulator period λ_w.[3]

In practice, especially for high energy x-rays, production from higher undulator resonances, $n = 3$, $n = 5$, or even higher, produces greater brilliance than does $n = 1$. To mask this dependence, from here on, the brilliance will be indicated by $\mathcal{B}$ rather than by $\mathcal{B}_1$. Achievable values of $\mathcal{B}_n$ are given later.

11.3.2
A Strategy to Maximize $\mathcal{B}$

With $\mathcal{B}$ declared to be of pre-eminent importance, it is clear from Eq. (11.6) that I should be maximized consistent with ϵ_x and ϵ_y being minimized. The brilliance is also influenced by *undulator* design and optimization, but those dependences are neglected here.

The theory of electron storage ring equilibration (due mainly to Sands) is well established, including formulas for ϵ_x and ϵ_y, along with lattice designs that minimize the emittances. One such design is spelled out in detail in a later section. Before going into analytic detail, only the following two effects are here promoted to *fundamental limitation* status:

- The basic approach to reducing ϵ_x and ϵ_y is to focus the beam more strongly. This stronger focusing inevitably leads to reduced dynamic acceptance. With care, the equilibrium beam size decreases more or less proportionally to the dynamic aperture, which seems to augur success in reducing emittances. The proportionality of beam size and dynamic aperture is demonstrated in the discussion accompanying Figure 11.10. Unfortunately the need to inject beam into the storage ring limits this approach. Either the beam is too large initially, or the betatron oscillations imparted by multiturn injection set a limit on maximum focusing strength.

---

2) The functional dependences in Eq. (11.6) are not controversial, at least for some conventional beamline designs, but the multiplying factor may vary by a factor of two or so, depending on assumptions.

3) Qualification: Eq. (11.6), especially its only-linear dependence on N_w, depends on relations between the angular spreads of the electron beam and the produced photon cone angle. These relations are likely to be different for horizontal and vertical emittances in the ranges of interest for this chapter.

- Another limit, almost as important, is set by the so-called Touschek effect, an intrabeam scattering process in which individual particles are scattered and lost. This limits the degree to which the emittances, especially ϵ_y, can be reduced. This consideration becomes relatively unimportant for energies above, say, 5 GeV.

With these two effects taken to be the only *fundamental* limitations, the approach taken here is to eliminate them both by taking the following steps:

- Start with small emittance beams. This is achieved by using a gun plus injector intended for ERL operation [7].

- Avoid multiturn injection. By injecting directly onto the equilibrium orbit, injection-induced betatron oscillations are avoided. By this principle, the so-called "top-off" injection (to replace Touschek losses) should be avoided. To preserve this injection capability defeats the dynamic aperture scaling emphasized above (for example because of unavoidable minimum septum thickness) and restricts the peak brilliance.

- Defeat the Touschek effect by storing individual electrons for times measured only in tens of seconds, far shorter than current Touschek lifetimes. In this regime ϵ_x and ϵ_y can be reduced by big factors before the Touschek effect again becomes important.

- However, choose a storage time long enough to take advantage of the damping of ϵ_y to a small value characteristic of conventional light sources. The electron bunch equilibration time (measured in numbers of turns) is about $\mathcal{E}_e/\Delta\mathcal{E}_e \approx 5000/1.3 \approx 4000$ turns, where the energy loss per turn at $\mathcal{E}_e = 5$ GeV in the lattice is $\Delta\mathcal{E}_e \approx 1.3$ MeV. This "long storage time" condition is therefore satisfied by storing for times long compared to 10 ms.

11.3.3
Hypothetical Utilization of an Existing Large Ring

When designing a dedicated x-ray source from the ground up there is a strong cost-saving bias toward minimizing the ring circumference. When contemplating conversion of a colliding beam facility into a dedicated x-ray source, this incentive is absent, since the ring tunnel already exists.

For the sake of definiteness let us contemplate the design of a 5.7 GeV ring that will fit in a more or less circular tunnel of 750 m circumference. One knows that, to reduce ϵ_x, one will need very strong focusing and hence short cells. It is important to consider only designs in which the magnet strengths are achievable and the dynamic aperture remains acceptably large. In this sec-

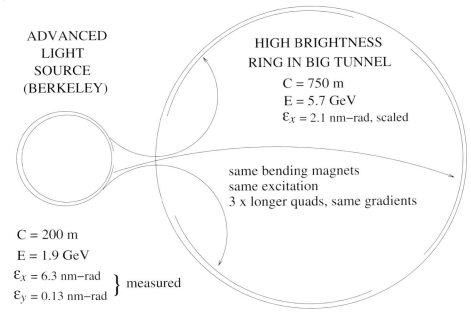

Fig. 11.4 Hypothetical triplication of the ALS 1.9 GeV storage ring to make a 5.7 GeV storage ring in a bigger tunnel.

tion a semi-quantitative (but intended to provide a persuasive introduction) hypothetical design will be described.

To start let us take the ALS lattice—magnets, vacuum chamber, power supplies and all—and string it into the existing 750 m tunnel. Because the ALS circumference is 200 m this will fill only about 1/3 of the new tunnel. Let us therefore make two more ALS copies and string them in the tunnel as well, still leaving a bit of space to preserve flexibility. This is illustrated in Figure 11.4.

The nominal ALS electron energy is 1.9 GeV and the new energy is to be 5.7 GeV. The bending magnets, even including their excitation, are more or less correct as is, since the required bend angle is reduced by roughly the same factor that the beam stiffness is increased. The sagittal offset of the magnets (already small for ALS) will be reduced, with no significant impact on the magnet design or cost. The combined-focusing field index will also not be correct, as its focusing strength will be three times weakened by the increased beam stiffness. Estimated crudely, if the pole tilts are left unchanged, the J_x partition number will change from about 1.6 to about 1.2. If nothing were done about this, the resulting 25% increase in ϵ_x would have to be tolerated. This (relatively unimportant) issue will be considered more carefully later.

The quadrupole magnets are more problematical. When powered as in the ALS these magnets will be too weak by a factor of three because of the increased beam stiffness. The simplest remedy would be to increase the magnet lengths by a factor of three, leaving the magnet transverse dimensions and fields otherwise unchanged. Looking at Figure 11.2(a) there appears to be room in the lattice to accomodate the vertically focusing quads, but not for the horizontally focusing quads. To make space, the cell length can be increased from 16 m to, say, 18 m. This will use up some of the tunnel circumference preserved earlier. The effect of this change will have been to allot circumferential arc at the rate of 6 m per bending magnet. It will be useful to keep this figure, 6 m per bending magnet, as setting a tentative scale. Of course the bending magnet length itself is closer to 2 m.

The allotment of circumference to sextupoles is harder. They will also need to be strengthened by, at least, the same factor of three for beam stiffness. Their actual strengths depend on lattice details too complicated for the hypothetical design to be applicable. In particular, their strengths have to be increased to compensate for reduced dispersion. More detailed consideration of this issue will have to be deferred. This sextupole issue, along with the potentially deleterious effects of increased energy spread, are considerations for which the hypothetical design being described is not definitively persuasive.

With these changes, how will the new storage ring perform as an accelerator? For small amplitudes the answer is easy. Since the bending magnets are *inert* as regards focal properties, the new ring will perform just like three turns around the existing ALS. But the sextupoles (present for chromaticity compensation) limit the stability of large amplitude particles. Fortunately, in triplicating the ALS, the sextupole strengths increase by roughly the same factor that the beam emittance decreases. Expressing particle amplitudes in units of the equilibrium beam sizes, the stability of a particle in the new machine will therefore be roughly the same as in the ALS. Of course the RF will have to be beefed up to make up for the increased energy loss due to synchrotron radiation. In summary, when contemplating the new ring, one need not be terrified of its tunes, even if they are some three times as large as the ALS tunes, which are $Q_x \approx 14.4$ and $Q_y \approx 8.8$—the three-times-around tunes are, by definition, three times larger than the once-around tunes.

The discussion so far has been too glib in that the practicalities of injection, when the transverse beam size greatly exceeds its eventual equilibrium size, has so far been only casually addressed by declaring, in the previous section, that single turn injection is obligatory. Already at the functioning ALS there has been the need for adequate acceptance of oversized injected beams. It will be necessary, in what follows, to ensure that injection efficiency into the new ring will be similarly high.

For the ALS the nominal horizontal emittance is $\epsilon_x = 6.3 \times 10^{-9}$ m at 1.9 GeV. Once various loose ends have been cleared up, based as before on Eq. (11.4), we expect a value, $\epsilon_x \approx 2.1 \times 10^{-9}$ m for the new ring operating at 5.7 GeV. For comparison with the (round, rather than flat) 5 GeV ERL beam this can be further scaled to predict $\epsilon_x = \epsilon_y = 0.81 \times 10^{-9}$ at 5 GeV. [4] Expressed as a normalized emittance this is $\epsilon^{(N)} = \epsilon/\gamma \approx 7.9\,\mu\text{m}$.

This normalized emittance is roughly four times the value specified for both $\epsilon_x^{(N)}$ and $\epsilon_y^{(N)}$ for the ERL x-ray source described in Chapter 10. (Recall that perfect preservation of emittance through the full acceleration cycle was assumed for that ERL.) However, at the cost of giving back the factor of two round/flat factor, the storage ring takes advantage of vertical damping to produce a vertical emittance ϵ_y some hundred times smaller than ϵ_x.[5] Altogether (except possibly for soft x-rays) a brilliance for FCSR can be expected to be an order of magnitude higher than for the ERL. The minimum emittance lattice design to be described later results in further improvement of the brilliance.

One characteristic of the lattice design just described is a very large number of magnetic elements. In spite of the fanciful nature of the "tripled ALS" design, it can form a basis for estimating the cost of producing all these elements. For example the cost of bending magnets will not exceed three times the cost (properly inflated) of the ALS bending magnets.

11.4
Acceleration Scenario

As emphasized so far, the theme of the discussion is how to retrofit an existing e+/e- colliding beam facility as an x-ray source. Facilities for which such modification is possible include PEP, PETRA, KEK, and CESR. At the same time numerical comparisons will continue to be made with an energy recovery linac (ERL), which is a competitor for designation as favored "fourth generation" x-ray source.

For definiteness here, all numerical values will apply to the CESR ring in the Wilson Laboratory tunnel. The lattice will be similar to, but not identical to the hypothetical ALS-like lattice just described. The CESR tunnel has a (very slightly) racetracked shape with straight sections symmetrically opposite, North and South. Furthermore the East and West arcs each consist of three identical sectors, separated by short straight sections. Simply stringing out 3 ALS copies would have produced superperiodicity of 36, when 6 is what the tunnel calls for. The lattice shown in Figure 11.5 is matched to the existing tunnel geometry. It is based on cells shown in Figure 11.6.

4) There is a factor of two reduction in horizontal emittance when the beam is made round. Also, for the lattice design to be explained in what follows the theoretical value of emittance will be considerably smaller than the estimate contained in this paragraph.
5) For such small values of ϵ_y the brightness becomes controversial. Some of the issues involved will be mentioned at the end.

Schematically, the acceleration sequence is essentially the same as existing gun/linac/synchrotron/storage-ring facilities such as CESR. This is shown in Figure 11.7. However the timing sequence (shown in Figure 11.8) is very different. In particular, any single electron circulates in the fast cycling storage ring (FCSR) for several seconds, instead of for an hour.

Once every several seconds (in the range from 2 to 20) the gun produces a 2 µs-long batch of electron bunches with interbunch spacing corresponding to 1.3 GHz. These electrons are accelerated to, say, 200 MeV, in the injector linac, with average current during each batch being 0.1 A. This linac is quite similar in design to the TESLA/ILC linac [8] currently under development. However, cycling at $0.05 - 0.5$ Hz, its duty factor is far less than in the TESLA design 5 Hz rate. Also, with total length not being a significant consideration, energy gain per meter in the injector can be less by a substantial factor than in a linear collider.

After injection into synchrotron SYNCH, the electrons are accelerated up to a maximum energy taken nominally to be 5 GeV. This acceleration is performed in the element labeled SYNCH-RF in the figure. This superconducting linac resembles the ERL injector linac. It has to be capable of accelerating an 0.1 A beam on a nearly-CW basis. Compared to ERL operation, though the duty factor reduction is modest—by a factor of only about two depending on the magnetic field cycle of SYNCH—there are more important relaxations. The required longitudinal gradient is small and there is no minimum gradient imposed by space charge considerations. Compared to the ERL injector linac, the gradient of SYNCH-RF can therefore be much less, perhaps by a factor of ten. Reducing the gradient incurs only the construction costs of more cryomodules and of taking up more space in the synchrotron.

At full energy the 2 µs batch of electrons is extracted from SYNCH and single-turn injected into FCSR. The same pulsed inflector into FCSR that injects the present batch also ejects the previous batch, which is conveyed to the beam dump. During smooth operation this beam transfer should be invisible[6] to x-ray users of FCSR. There will, of course, need to be a gap of, say, 200 ns, to allow for turn on and turn off of the inflector/ejector. Injection jitter into FCSR can be eliminated immediately by single turn feedback.

There are no challenging RF components needed for FCSR. Theoretical questions concerning emittance preservation, halo formation, insertion sections and so on are discussed in other sections.

---

6) The emittances of a replacement batch of electrons will be substantially greater than the emittances of the batch being replaced. Experiments relying on ultrahigh brightness will see a downward brightness transient lasting for many milliseconds after each replacement event.

340 | *11 A Fourth Generation, Fast Cycling, Conventional Light Source*

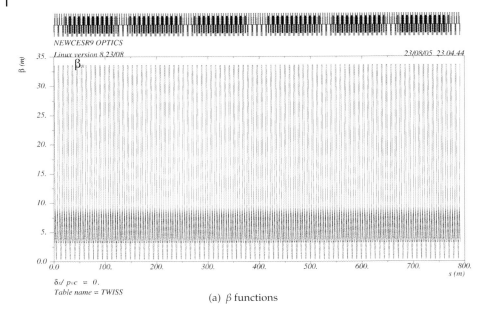

(a) β functions

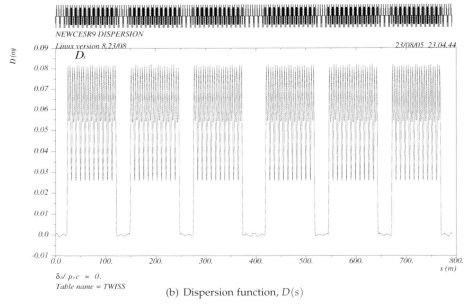

(b) Dispersion function, $D(s)$

Fig. 11.5 Lattice functions for the ultralow emittance storage ring FCSR. To support more x-ray beamlines each sextant could be configured as two 5-bend achromats instead of as the 11-bend achromat shown.

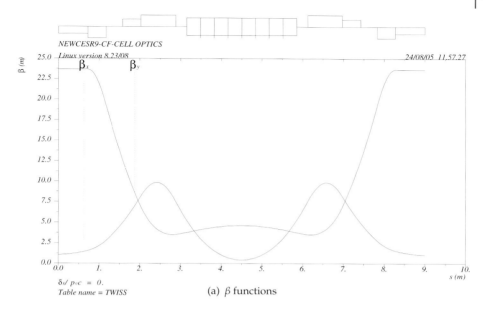

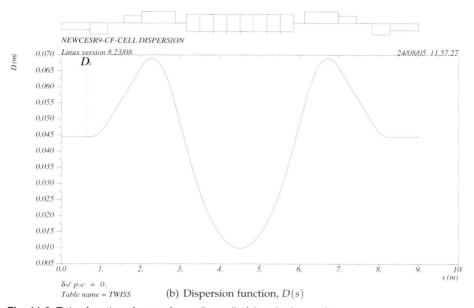

Fig. 11.6 Twiss functions for one $L_C = 9$ m cell of the ultralow emittance storage ring FCSR. This forms one section of a multibend acromat (MBA).

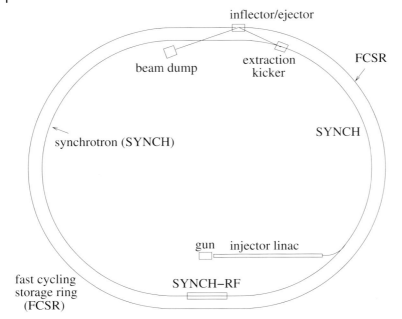

Fig. 11.7 Schematic layout of the full acceleration system.

11.5
Power Considerations

11.5.1
Average Power

In this section the rationale for the design is discussed using "ballpark" parameter values. They are not supposed to be optimal but are adopted to simplify discussion.

The original rationale for building an ERL went as follows. One wants to have a relatively high electron current such as $I \stackrel{e.g.}{=} 0.1\,\text{A}$ at an electron energy $\mathcal{E}_e \stackrel{e.g.}{=} 5\,\text{GeV}$. This corresponds to a supplied power of 500 MW, which is impractically large by a factor of at least one hundred. For present purposes let us strive for reduction of this power by a factor of at least one thousand. (Since the power radiated in undulators is expected to approach a megawatt, there is little justification for striving for a power reduction factor greater than one thousand.) In the ERL design a power reduction factor in the range from one hundred to one thousand is obtained by decelerating the electrons to recover most of their energy. In the scheme described here the reduction factor is achieved by reducing the *number* of electrons accelerated per second by a factor greater than one thousand—more than a million in fact.

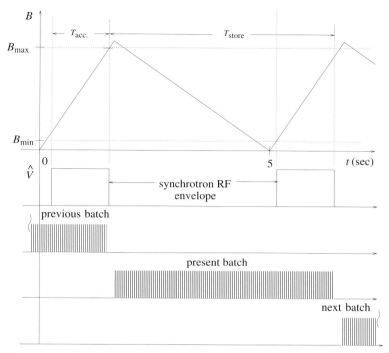

Fig. 11.8 Acceleration cycle showing the synchrotron (SYNCH) magnetic field, the SYNCH-RF modulation $\hat{V}$, and the bunch pattern in FCSR. The cycle time T_C, labelled 5 s in the figure, is expected to be in the range from 2 to 20 s.

In round numbers the revolution period of electrons in the tunnel is $T_{\text{rev.}} \approx 2\,\mu\text{s}$, which gives 500 000 turns per second. In the "fast cycling" approach the electrons are dumped from FCSR every $T_{\text{cycle}} \approx 2 - 20\,\text{s}$, after having completed $N_{\text{cycle}} = 10^6 - 10^7$ turns. This will have abundantly achieved the factor of at least one thousand power reduction called for above.

Since the electron gun and injector are required only every several seconds, their duty factors, and hence their power consumption, will have been reduced by a big factor compared to the ERL which runs CW. (As an aside, this should give a major improvement in gun cathode survival.)

Completing the argument of the first paragraph, the nominal average power delivered to the beam and dumped at full energy will be 50 − 500 W.

11.5.2
Instantaneous Power

Continuing with the parameters assigned tentatively in the previous section, for ease of comparison with the existing ERL design, let us copy parameters that are typical for ERL designs: the same beam current $I = 0.1$ A; the same bunch bunch repetition rate $f_{bunch} = 1.3$ GHz; and the same charge per bunch $Q_1 = 77$ pC.

The acceleration cycle shown in Figure 11.8 has been chosen to be so slow that energy gain per turn requirements in SYNCH are modest, both for acceleration and to replace synchrotron radiation power loss—comfortably less than what is provided by one ERL injection cryomodule, some parameters of which are: five 1300 MHz 2-cell cavities, each producing 1 MeV of acceleration, for a power transfer to the beam of 100 kW at 0.1 A; length approximately 1 m. The RF acceleration in the synchrotron can therefore be provided by a one (or, for above-5 GeV operation, a small number) of such modules, for example located at the same positions as in the original synchrotron.

11.6
Critical Components and Parameter Dependencies

An issue of interest is the relative performance of the fast cycling ring under discussion and an ERL. To simplify this comparison parameter values conventionally assumed for ERLs will continue to be used. Some elements of the fast cycling scheme are "conservative" only in the sense of trusting that the specifications for the ERL to which it is being compared, are conservative. Some of the critical components and theoretical issues are listed here:

- As with the ERL, excellent gun performance is essential for operation at the design brilliance. In detail the dependence on gun *current* and gun *emittance* specifications is different for the two schemes.

- The effect of *reduced gun current* is quite different in the ERL and the more nearly conventional fast cycling approach. One expects that reduced gun current will be accompanied by improved (smaller) emittance. Assuming the reduced emittance can be preserved throughout the entire acceleration phase, the ERL can exploit the small ϵ_x value to recover, or even exceed, the design brightness. The scheme being discussed here cannot do this; the brightness is strictly proportional to whatever charge can be captured during *one* 2 µs revolution period.

- In contrast, the fast cycling storage ring design can compensate for *too-large emittance* by increasing the synchrotron acceptance or injection energy. At the design injection current this could recover the design brightness, independent of gun emittance. Compared to the gun specifications assumed here, and for the ERL, electron sources with similar bunching, charge per bunch and/or average current roughly ten times greater, and emittances roughly ten times worse, are described, for example, by Borland et al. [9] and by Sheffield et al. [10]. Simulations by Bazarov and Sinclair [7] predict emittances more or less proportional to charge per bunch at the gun output. This is consistent with statements made in the papers of Borland and of Sheffield.

- The injector linac energy is assumed to be at least 200 MeV. The critical front end will be available upon completion of projects currently in progress at Cornell and elsewhere. The extra cost will come from acceleration from 10 MeV to 200 MeV which will, however, be at very low duty factor. Existing linacs, for example the 400 MeV linac at Cornell, appears capable of performing almost up to the specifications assumed here. This linac would certainly be satisfactory for initial commissioning, which would permit the expense of upgrading it to be deferred. Because of the high pulsed current requirement, a linac frequency lower than the 2850 MHz of this copper linac would seem to be preferable, but the linac need not be superconducting.

- To obtain the small value of ϵ_x in the present design requires an extremely high horizontal tune Q_x as has already been explained. By the argument already given, and according to a tracking simulation to be described shortly, the resulting aperture is adequate, but further investigation is clearly required.

- Because of their short focal lengths, the synchrotron quadrupoles are challenging, and their high inductances may limit the ramp rate. Eddy currents in the aluminum vacuum chamber also limit the cycling rate. These are reasons the total cycle time may have to be as long as $T_{\text{cycle}} = 20$ s or higher. A substantial further increase of T_{cycle} is likely to bring the Touschek effect back into play.

- Accepting $L_C = 9$ m as the nominal design, from Table 11.1 the strongest quadrupole has focusing coefficient `k1[q2]=1.196` m^{-2}, in MAD-convention units (which are simply MKS). This requires a pole tip field $B_{\text{pole}}[\text{T}] = 0.20\, R_Q[\text{cm}]$ where R_Q is the pole radius. How big R_Q *must be* depends on beam impedance considerations. A conservative rule of thumb gives $B_{\text{pole}}^{\text{max}} = 0.8$ T as the maximum pole tip field in an iron core quadrupole. This permits $R_Q \leq 4$ cm which is comfortably greater, for example, than the present CESR vacuum chamber. The same calculation applied to sextupole strength `k2[S1]=51.5`m^{-3} in Table 11.1 permits $R_S \leq 4.3$ cm as the pole tip radius in sextupoles for the pole field to remain less than 0.8 T. To reduce magnet power and to simplify their coils, combined electomagnet/permanent magnet designs may be appropriate for both quadrupoles and sextupoles.

- Beam current limitation due to multibunch instability will have the same effect as reduced gun current. If the current is limited to a value less than the 100 mA assumed here, the brilliance will be reduced proportionally. Any such beam current limitation is likely to be set just after injection into the synchrotron. This would be ameliorated by increased injection energy or increased chamber size. The ALS comparison stressed earlier lends confidence on this score in that the ALS's current limit is 400 mA, which is four times greater than assumed here. This may not be as conservative as it sounds though, as the instability threshold could be governed by total stored charge which will be comparable in the two cases.

- As already stated, the focusing in SYNCH will be made weaker than in FCSR. This causes no degradation of overall performance. In fact it improves injection into SYNCH and may permit the injection energy to be reduced. This is investigated in Table 11.2. Doubling the cell length from $L_C = 9$ m to $L_C = 18$ m does however increase the transverse beam size, during beam transfer from SYNCH to FCSR. Table 11.3 shows that this transfer efficiency will be satisfactory.

11.7
Trbojevic-Courant Minimum Emittance Cells

11.7.1
Basic Formulas

Trbocevic and Courant [11] describe a lattice design which minimizes the emittance (under their assumed hypotheses.) Their cell configuration is shown in Figure 11.9. There is a single dipole magnet in each cell, but it is convenient to regard the cell ends as occurring at the dipole center, as shown in the lower part of the figure.

The main results of their paper can be distilled into the following few formulas. The horizontal emittance is given by

$$\epsilon_x = \frac{C_q \gamma^2}{J_x \rho} < \mathcal{H} >, \tag{11.7}$$

where $C_q = 3.84 \; 10^{-13}$ m, $\gamma = \mathcal{E}_0/(mc^2) = 0.978 \times 10^4$ for $\mathcal{E}_0 = 5$ Gev, J_x, the partition number, can, for present purposes, be taken to be 1, and ρ is the bend radius in bending magnets. The angle brackets indicate averaging over longitudinal coordinate s over the full ring and, at any value of s, $1/\rho$ is assumed either to vanish or to have the same "isomagnetic" non-zero magnitude. The factor $\mathcal{H}$, known as Sands's "curly H", is given in terms of Twiss functions and horizontal dispersion function D by

$$\mathcal{H} = \gamma_x D^2 + 2\alpha_x DD' + \beta_x D'^2. \tag{11.8}$$

Starting from their assumed values β_0 and D_0 at the origin, the s-dependences of β and D within the dipole are

$$\beta = \beta_0 + \frac{s^2}{\beta_0}, \quad \alpha \equiv -\frac{\beta'}{2} = -\frac{s}{\beta_0}, \quad D = D_0 + \frac{s^2}{2\rho}, \tag{11.9}$$

and the mean value of curly-H is given by

$$<\mathcal{H}> = \frac{1}{L}\int_{-L/2}^{L/2} \frac{D^2 + (D\alpha + D'\beta)^2}{\beta}\,ds$$

$$= \frac{D_0^2}{\beta_0}\left(1 + \frac{1}{12}\left(\frac{\theta\beta_0}{D_0}\right)^2 - \frac{1}{12}\frac{L\theta}{D_0} + \frac{1}{320}\left(\frac{L\theta}{D_0}\right)^2\right), \tag{11.10}$$

where L is the length of the bending magnet and $\theta = L/\rho$ is the total bend angle in the magnet. This formula does not rely on the minimum emittance condition—it can be used to calculate $<\mathcal{H}>$ from the values of β_x and D at the magnet center whenever both functions are symmetric about that point. As such the lattice functions determined by a lattice fitting program can be used to calculate the emittance, even if the lattice functions do not meet the optimization conditions to be determined next. The T-C mimimization procedure is to select values of β_0 and D_0 that minimize $<\mathcal{H}>$ with L and θ fixed. The result is

$$\beta_0 = \frac{L}{2\sqrt{15}}, \quad D_0 = \frac{\theta L}{24}, \quad <\mathcal{H}> = \frac{\sqrt{15}}{180} L\theta^2 = \frac{\sqrt{15}}{180}\rho\theta^3. \tag{11.11}$$

All that remains is to adjust lattice parameters to match these values of β_0 and D_0. These conditions fix the lattice design, modulo a certain amount of freedom in choosing element lengths and locations. Without providing details, Trbojevic and Courant exhibit one cell design that meets the requirements. In the next section I give a more detailed design prescription for design of individual cells and for the lattice as a whole.

The parameter of greatest interest, the horizontal emittance ϵ_x, can be obtained using Eqs. (11.7) and (11.11);

$$\epsilon_x = \frac{C_q\gamma^2}{J_x}\frac{\sqrt{15}}{180}\left(\frac{2\pi}{N_C}\right)^3 \left(\stackrel{e.g.}{=} 1.96\times 10^{-10}\left(\frac{100}{N_C}\right)^3 \text{m rad at 5 GeV}\right), \tag{11.12}$$

where N_C is the number of full cells in the lattice. The numerical estimate given here is based on parameter values given in Table 11.1. The length and strength parameters in this table are calculated in the next section.

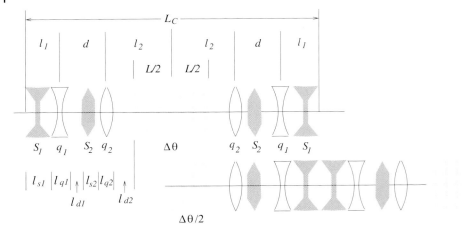

Fig. 11.9 Thick and thin lens lengths and strengths for Trbojevic-Courant minimum emittance cell. The lower element ordering is convenient for relating design values of β_0 and D_0 to the Twiss functions of sectors consisting of repeated cells. The rectangular shaded region represents a bend magnet and the angular figures represent sextupoles. As the lenses are drawn, it is horizontal optics being illustrated. The parameter $ld3$, though tentatively set to zero, is available for flexibility.

The calculation described so far has concentrated on horizontal emittance ϵ_x, as is appropriate for electron storage rings. In such rings, over times large compared to 10 ms, ϵ_x evolves toward the value given in Eq. (11.12). For multiturn injection this will normally constitute *shrinkage*, but in single turn injection from a low emittance source it is more likely to be *growth*. Meanwhile the vertical emittance ϵ_y damps down to a (typically) much smaller value, which depends on imperfections such as coupling and vertical dispersion.

11.7.2
Thin Lens Treatment

Though the T-C prescription certainly minimizes $<\mathcal{H}>$ under the assumed hypotheses, this is not quite equivalent to minimizing ϵ_x in a practical lattice intended to fit in an existing circular tunnel, having given circumference $\mathcal{C}$. To avoid worrying about sections needed for RF accelerating structures, wigglers or undulators, instrumentation, etc., let us introduce an "arc circumference" $\mathcal{C}_a \stackrel{e.g.}{=} 660$ m, that leaves length $\mathcal{C} - \mathcal{C}_a \approx 100$ m available for miscellaneous elements. By far the most important design choice is that of cell length L_C or, equivalently, the number of cells N_C. These are related by

$$\theta = \frac{2\pi}{N_C}, \quad \text{and} \quad L_C = \frac{\mathcal{C}_a}{N_C}. \tag{11.13}$$

From this and the last of Eqs. (11.11) the emittance apparently varies as N_C^{-3}, but considerations not included in the discussion so far will set a set a practical upper limit on N_C or, equivalently, a lower limit on L_C.

Another effect worth mentioning is that, because of the factor ρ in the denominator of Eq. (11.7), minimizing $\mathcal{H}$ is not equivalent to minimizing ϵ_x. In terms of quantities defined so far, ρ is given by[7]

$$\rho = \frac{C_a}{2\pi} \frac{L}{L_C}, \qquad (11.14)$$

To the extent that L fills most of the cell length L_C, minimizing $\mathcal{H}$ is equivalent to minimizing ϵ_x but, especially as L_C is reduced, an increasingly large fraction of the cell is required for the quadrupoles and sextupoles that are needed to achieve the optimal values of β_0 and D_0. This leads to difficult dependences, even in designing the linear optics, and the ultimate limit is set by nonlinear effects that can only be studied by simulation using particle tracking. (This should not be surprising when the length required for sextupoles exceeds the quadrupole length allotment.)[8] Obviously a final design will have to be the result of iterating these steps. Here I make only a single, tentative, choice of element locations, and study the performance as a function of L_C. I do not attempt to iterate beyond this first cut design, and I ignore complications such as fringe fields or the circumference allotment required for magnet ends.

Especially for conventional (multi-hour) storage ring design, other, more important, phenomena limit the maximum practical value of N_C. With increasing N_C the dynamic aperture decreases. Though the equilibrium beam size decreases more or less proportionally, the effective beam size during multiturn injection is independent of N_C and injection efficiency suffers with increasing N_C. Yet another similar limitation results from the Touschek effect in which an individual electron occasionally receives a kick sending it into a halo where it may be lost. For the scheme described here the occasional particle will be lost because of the Touschek effect, but this will have little effect on operations.

For linear lattice design, following Trbojevic and Courant, though the bend region is crucial to evolution of the dispersion function, its "optical" effect on $\beta_x(s)$ is ignored. Also the quadrupoles are initially treated as thin lenses—this assumption is retracted in a final design step using the TEAPOT program (and later, UAL) to study the dynamic acceptance of the lattice.

---

7) Following Sands originally, as do T-C, here we assume an "isomagnetic" lattice in which the magnetic field has the same value everywhere except where it vanishes.

8) In some accelerators sextupoles are built into bending magnets by shaping their poles. This would not make sense for the present ring because of the intentionally small values of D at bending magnet locations.

Trbojevic and Courant are correct in asserting that there are enough free parameters to allow the β_x and D_x-function values to be achieved. They do not, however, analyse the restriction that follows from the need for the vertical motion to be stable. In fact, some solutions have both q_1 and q_2 horizontally focusing, which is obviously unsatisfactory. Hence, in seeking solutions, one should certainly impose the condition $q_1 q_2 < 0$. After a certain amount of trial and error, it appeared that the best place for the vertical focusing quad is as far from the dipole as possible. This suggested the virtue of pushing it to the very end of the cell, as shown in Figure 11.9. This has the further virtue of halving the number of vertical focusing quads, by permitting elements from adjacent cells to be combined. The T-C conditions can be met in this case, while retaining vertical stability.

Ignoring any optical effect due to the bending magnet, transfer matrix analysis of the thin lens doublet configuration shown in the left half of the upper lattice cell shown in Figure 11.9 yields horizontal and vertical transfer matrices

$$\begin{pmatrix} X_x & X_{x'} \\ F_x & F_{x'} \end{pmatrix} = \begin{pmatrix} 1 & l_2 \\ 0 & 1 \end{pmatrix} \begin{pmatrix} 1 & 0 \\ -q_2 & 1 \end{pmatrix} \begin{pmatrix} 1 & d \\ 0 & 1 \end{pmatrix} \begin{pmatrix} 1 & 0 \\ -q_1 & 1 \end{pmatrix} \begin{pmatrix} 1 & l_1 \\ 0 & 1 \end{pmatrix}, \quad (11.15)$$

$$\begin{pmatrix} Y_y & Y_{y'} \\ G_y & G_{y'} \end{pmatrix} = \begin{pmatrix} 1 & l_2 \\ 0 & 1 \end{pmatrix} \begin{pmatrix} 1 & 0 \\ q_2 & 1 \end{pmatrix} \begin{pmatrix} 1 & d \\ 0 & 1 \end{pmatrix} \begin{pmatrix} 1 & 0 \\ q_1 & 1 \end{pmatrix} \begin{pmatrix} 1 & l_1 \\ 0 & 1 \end{pmatrix}.$$

Notice that *positive q* values correspond to *horizontal focusing*, so $q_1 < 0$ and $q_2 > 0$. Completing the multiplications yields, in the horizontal plane,

$$\begin{aligned} X_x &= 1 - (d + l_2)q_1 - l_2 q_2 + dl_2 q_1 q_2, \\ X_{x'} &= l_1 + d + l_2 - l_1(d + l_2)q_1 - (l_1 + d)l_2 q_2 + l_1 dl_2 q_1 q_2, \\ F_x &= -q_1 - q_2 + dq_1 q_2, \\ F_{x'} &= 1 - l_1 q_1 - (l_1 + d)q_2 + l_1 dq_1 q_2. \end{aligned} \quad (11.16)$$

In the vertical plane all negative signs are replaced by positive signs;

$$\begin{aligned} Y_y &= 1 + (d + l_2)q_1 + l_2 q_2 + dl_2 q_1 q_2, \\ Y_{y'} &= l_1 + d + l_2 + l_1(d + l_2)q_1 + (l_1 + d)l_2 q_2 + l_1 dl_2 q_1 q_2, \\ G_y &= q_1 + q_2 + dq_1 q_2, \\ G_{y'} &= 1 + l_1 q_1 + (l_1 + d)q_2 + l_1 dq_1 q_2. \end{aligned} \quad (11.17)$$

Letting **X** be the 4 × 4 transfer matrix of the left half cell in the upper part of Figure 11.9, the input and output coordinates of a general trajectory are related by the usual relations,

$$\mathbf{x}_{\text{out}} = \mathbf{X} \mathbf{x}_{\text{in}}, \text{ and } \mathbf{x}_{\text{in}} = \mathbf{X}^{-1} \mathbf{x}_{\text{out}}. \quad (11.18)$$

The second half of the cell consists of the same elements, but encountered in the opposite order. Let $\check{\mathbf{X}}$ stand for the transfer matrix of this sequence. Symmetry under this reflection is closely related to time reversal invariance that relates forward and backward evolving trajectories. To express this mathematically, introduce a matrix σ_3 that reverses the slopes,

$$\begin{pmatrix} x \\ -x' \\ y \\ -y' \end{pmatrix} = \begin{pmatrix} 1 & 0 & 0 & 0 \\ 0 & -1 & 0 & 0 \\ 0 & 0 & 1 & 0 \\ 0 & 0 & 0 & -1 \end{pmatrix} \begin{pmatrix} x \\ x' \\ y \\ y' \end{pmatrix} \equiv \sigma_3 \mathbf{x}, \qquad (11.19)$$

From the vector $\mathbf{x}_{\text{out}}$ appearing in Eq. (11.18), form the vector $\sigma_3 \mathbf{x}_{\text{out}}$ and take it as input to the reversed-order half-cell. By symmetry, after propagating through the half-cell, one should recover $\sigma_3 \mathbf{x}_{\text{in}}$, where $\mathbf{x}_{\text{in}}$ appears in Eq. (11.18). That is

$$\sigma_3 \mathbf{X}^{-1} \mathbf{x}_{\text{out}} = \sigma_3 \mathbf{x}_{\text{in}} = \check{\mathbf{X}} \sigma_3 \mathbf{x}_{\text{out}}, \qquad (11.20)$$

and it follows from this that

$$\check{\mathbf{X}} = \sigma_3 \mathbf{X}^{-1} \sigma_3. \qquad (11.21)$$

Because the determinants are equal to 1, the inverse matrices are given by

$$\begin{pmatrix} X_x & X_{x'} \\ F_x & F_{x'} \end{pmatrix}^{-1} = \begin{pmatrix} F_{x'} & -X_{x'} \\ -F_x & X_x \end{pmatrix}, \quad \begin{pmatrix} Y_y & Y_{y'} \\ G_y & G_{y'} \end{pmatrix}^{-1} = \begin{pmatrix} G_{y'} & -Y_{y'} \\ -G_y & Y_y \end{pmatrix}. \qquad (11.22)$$

As mentioned before, it is most convenient to select the center of the dipole magnet as origin. The transfer matrix $\mathbf{M}$, through the full cell from this point, is given by

$$\mathbf{M} = \mathbf{X}\check{\mathbf{X}} = \begin{pmatrix} X_x F_{x'} + X_{x'} F_x & 2X_x X_{x'} & 0 & 0 \\ 2F_x F_{x'} & X_x F_{x'} + X_{x'} F_x & 0 & 0 \\ 0 & 0 & Y_y G_{y'} + Y_{y'} G_y & 2Y_y Y_{y'} \\ 0 & 0 & 2G_y G_{y'} & Y_y G_{y'} + Y_{y'} G_y \end{pmatrix}. \qquad (11.23)$$

If the cell is part of a periodic lattice, its transfer matrix can be written in the form

$$\mathbf{M} = \begin{pmatrix} \cos \mu_x & \beta_x \sin \mu_x & 0 & 0 \\ -\sin \mu_x / \beta_x & \cos \mu_x & 0 & 0 \\ 0 & 0 & \cos \mu_y & \beta_y \sin \mu_y \\ 0 & 0 & -\sin \mu_y / \beta_y & \cos \mu_y \end{pmatrix}. \qquad (11.24)$$

Note that the vanishing of the αs is consistent with Eq. (11.23) because the 11,22 and 33,44 element are pairwise equal. The reason for this is that the full cell is invariant to element reversal, so Eq. (11.21) implies that $\mathbf{M} = \sigma_3 \mathbf{M}^{-1} \sigma_3$. We have therefore obtained formulas for the Twiss parameters at the cell ends;

$$\cos \mu_x = X_x F_{x'} + X_{x'} F_x, \quad \sin^2 \mu_x = -4 X_x X_{x'} F_x F_{x'}, \beta_x^2 = -\frac{X_x X_{x'}}{F_x F_{x'}},$$

$$\cos \mu_y = Y_y G_{y'} + Y_{y'} G_y, \quad \sin^2 \mu_y = -4 Y_y Y_{y'} G_y G_{y'}, \beta_y^2 = -\frac{Y_y Y_{y'}}{G_y G_{y'}}.$$

(11.25)

Because of the multiple-valued nature of inverse trigonometric functions, a certain amount of care is required in extracting the Twiss parameters from these relation. For the numerical values given in Table 11.1:

$$Q_x = 1 - \frac{\cos^{-1}(X_x F_{x'} + X_{x'} F_x)}{2\pi} = 1 - \frac{\sin^{-1}(\sqrt{-4 X_x X_{x'} F_x F_{x'}})}{2\pi},$$

$$Q_y = \frac{\cos^{-1}(Y_y G_{y'} + Y_{y'} G_y)}{2\pi} = 0.5 - \frac{\sin^{-1}(-\sqrt{-4 Y_y Y_{y'} G_y G_{y'}})}{2\pi}. \quad (11.26)$$

The T-C requirement on β_0 can be expressed by equating the ratio of 12 and 21 elements of Eq. (11.23) with the first of Eqs. (11.11) to yield the significant design formula;

$$\sqrt{-\frac{X_x X_{x'}}{F_x F_{x'}}} = \frac{L}{2\sqrt{15}}. \quad (11.27)$$

If we regard all lengths as fixed, then this equation provides one of the two conditions needed to fix the adjustable parameters q_1 and q_2.

Except within the dipole region, the dispersion function has the same s-dependence as does a particle trajectory. At the dipole end, using Eq. (11.9), we have $D(L/2) = D_0 + L\theta/4$ and $D'(L/2) = \theta$. Since $D(s)$ must have even symmetry about the cell center, and $D'(s)$ is odd, we must have

$$\begin{pmatrix} D_0 + L\theta/8 \\ -\theta/2 \end{pmatrix} = \mathbf{M}_{l_2 \to l_2 - L/2} \begin{pmatrix} D_0 + L\theta/8 \\ \theta/2 \end{pmatrix}, \quad (11.28)$$

because the transfer matrix through the bend-less section differs from the full transfer matrix only by the reduced entrance and exit distances indicated by the subscript on $\mathbf{M}$. This equation, along with Eq. (11.27), fix both q_1 and q_2. (The redundancy resulting because Eq. (11.28) contains two equations provides a consistency check.)

After having fixed all length parameters, Eqs. (11.27) and (11.28), constrained by the requirement $q_1 q_2 > 0$, were solved using Maple. Because they are polynomial equations, Maple is able to solve the equations analytically, without using numerical methods. This is more than apple-polishing since parameter dependences are analytic rather than numerical and the desired solution can be picked automatically, circumventing the need to manually scrutinize all of the numerous solutions, some of which are complex, others unstable vertically.

11.7.3
Thick Lens Treatment

The thin lens treatment just completed led to polynomial equations that could be solved in closed form. For a more accurate lattice representation it is essential to use thick lens formulas. In this representation the constraint equations needed to fix q_1 and q_2 are transcendental. These equations have to be solved numerically. This is only practical if the sought for solution is quite close to a known *stable* solution. From that point a numerical solver, for example MAD, can, by making only small stability-preserving excursions, satisfy the constraints exactly while faithfully treating the magnets as thick elements.

Fortunately, for the minimum emittance lattices under study, after replacing thin elements by thick elements of the same integrated strengths it is found that the lattice is still either stable or nearly so. Then, as just stated, it is relatively straightforward to adjust strengths so that the exact conditions are satisfied. All the numbers entering into this procedure, for various choices of cell length, are recorded in Table 11.1. For this table the fractional length assignments to bends, quadrupoles and sextupoles has been held independent of cell length (even though, in practice, one would tend to increase the fractional assignment to bends as the cell gets longer.)

This procedure has been followed to produce the thick lens lattice functions shown in Figure 11.5 and Figure 11.6, as well as in the form of numbers in Table 11.1. For the lattice design described by one column in this table, the cell length L_C has been fixed, along with the configuration of magnets in the cell, and the target values of β_0 and D_0 known to minimize the emittance. These values, along with element strengths obtained following the procedure just described, are listed in the second-from-bottom block of the table. Especially for the $L_C = 9\,\text{m}$ case favored for the new storage ring, the achieved values are fairly close to the optimal values.

Though not necessarily welcome, a certain amount of *craft* sometimes creeps into the design of accelerator lattices. The row in the table labeled `k1[b], init.` is a case in point. Simply replacing thin elements with thick elements of the same strength, in some cases, gave an unstable lattice. Empirically assigning the vertical combined-focusing strengths shown in this row of the table led to good stability in all cases. *Ex post facto* it was also found that this permitted the maximum vertical beta function to be controlled with little effect on the horizontal optics. This can be used to prevent the vertical beta function from "getting out of hand". The bending magnet gradient is also implicated in adjusting the partition number J_x. The fact that the final lattices do not quite fit the optimization conditions is due to this *ad hoc* inclusion of combined function bend magnet optics. A possible red flag concerning the series of lattices shown in the table is that the vertical tune ν_x varies more erratically than one would have expected. This should to be looked into.

In any case, as at CLS and ALS, squeezing out a final reduction in emittance will require optimizing the cell design, including adjusting the J_x partition number. All this assumes that compensating for inevitable field errors and tuning up such a delicately balanced lattice can be successfully accomplished.

The sextupole strengths shown in the table have been determined (using UAL, Unified Accelerated Libraries [13]) to set both horizontal and vertical chromaticities to zero. For the shorter cell lengths these sextupole are very strong. The values of k2 are as low as they are only because the sextupoles have been allowed to be so (surprisingly) long. The vertical sextupole coefficients are weaker than the horizontal partly because the horizontal sextupoles can be no longer than the length of the region over which ν is appreciably large. For the longer cell lengths it would not be necessary to have such long sextupoles. Shortening them will have no effect on the linear optics.

The scaling behavior of the lattice data of Table 11.1 is shown in Figure 11.10. As expected, ϵ_x decreases with decreasing L_C. What may be surprising is the nearly exact power law scaling of all parameters as functions of L_C. It has been stated earlier that the dynamic aperture and the transverse beam size scale proportionally. This can be expressed equivalently by the statement that the dynamic aperture, when quoted in units of transverse beam sigmas, is independent of L_C. The beam sigma is given by $\sqrt{\beta\epsilon}$, which, from the scaling exhibited in the figure, scales as L_C^2. The actual numerical value of the dynamic aperture can, of course, only be obtained by numerical tracking. But its scaling behavior can be expressed as the transverse amplitude at which the sextupole bend is some given fraction of the quadrupole bend. In other words the dynamic aperture scales as q/S which scales as L_C^2, just like the emittance. This confirms the proportionality of equilibrium beam size and dynamic aperture.

In the absence of all other considerations the emittance could therefore be made arbitrarily small by reducing L_C arbitrarily. Of course one cannot ignore other considerations. It is probably the sextupole length limitation mentioned earlier that excludes the possibility of choosing an ultra-short L_C-value. It will be tentatively assumed in what follows, that $L_C = 9$ m has been chosen as the FCSR cell length. A longer value, in particular $L_C = 18$ m, is the tentative choice for the SYNCH lattice.

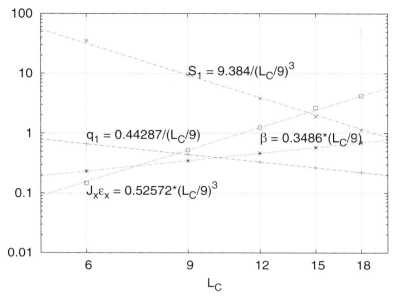

Fig. 11.10 Scaling dependence on cell length L_C for some parameters of the minimum emittance lattices shown in Table 11.1. Quad strength q, sextupole strength S, beta function β, and emittance ϵ scale, respectively, as $1/L_C$, $1/L_C^3$, L_C, and L_C^3.

Various radiation parameters are common to all entries in Table 11.1, since E, B and ρ values are approximately the same for all cases. Assuming 5 GeV operation, their values are:

$$U_0 = \text{energy radiated per turn} = 0.885 \times 10^{-4} \,\text{m GeV}^{-3} \frac{5^4}{30.9} = 1.79 \,\text{MeV},$$

$$u_c = \text{critical energy} = 9.0 \,\text{keV}, \tag{11.29}$$

$$\sigma_\delta = \text{fractional energy spread} = 0.00076,$$

$$P_{\text{tot}} = \text{total radiated power} = 1.79 \,(\text{MW A}^{-1})\, I_{\text{tot}}.$$

11.7.4
Zero Dispersion Straight Sections

It is appropriate to install most insertion devices at positions of zero dispersion. In this section such a zero dispersion region is designed. A more practical design would include the vertical focusing due to the undulator (presumed to be) located in the straight section.

11 A Fourth Generation, Fast Cycling, Conventional Light Source

Tab. 11.1 Numerical values for minimal emittance lattices of various cell lengths for $\mathcal{E}_0 = 5\,\text{GeV}$. The entries in `typewriter font` are in MAD units. Some 100 m (out of a $\mathcal{C} = 750\,\text{m}$ tunnel circumference) have been reserved for straight sections and their contributions are not included in this table. In the rows labeled $J_x \epsilon_x$ one has $1.0 < J_x < 1.6$.

Parameter	Unit	$L_C = 6\,\text{m}$	$L_C = 9\,\text{m}$	$L_C = 12\,\text{m}$	$L_C = 15\,\text{m}$	$L_C = 18\,\text{m}$
			FCSR			SYNCH
N_C		110	72	54	42	36
$\mathcal{C}_a$	m	660	648	648	630	648
θ	rad	0.057120	0.087266	0.11636	0.14960	0.17453
ρ	m	31.513	30.940	30.940	30.080	30.940
B	T	0.52925	0.53906	0.53906	0.55446	0.53905
L	m	1.8	2.7	3.6	4.5	5.4
l_{q1}	m	0.300	0.450	0.600	0.750	0.900
l_{q2}	m	0.570	0.855	1.140	1.425	1.710
l_{s1}	m	0.480	0.720	0.960	1.20	1.440
l_{s2}	m	0.300	0.450	0.600	0.750	0.900
$ld2$	m	0.180	0.270	0.360	0.450	0.540
$ld1$	m	0.270	0.405	0.540	0.675	0.810
l_2	m	1.365	2.0475	2.730	3.4125	4.095
d	m	1.005	1.5075	2.010	2.5125	3.015
l_1	m	0.630	0.9450	1.260	1.575	1.890
`k1[b], init.`	1/m	−0.20	−0.16	−0.12	−0.09	−0.07
$q_{1,thin}$	1/m	−0.66430	−0.44287	−0.33215	−0.26572	−0.22143
$q_{1,thin}/l_{q1}$	1/m	−2.2143	−0.98416	−0.55358	−0.35429	−0.24603
`k1[q1]`, thick	m^{-2}	−2.4246	−0.98578	−0.55448	−0.35480	−0.24513
$q_{2,thin}$	1/m	1.33732	0.89155	0.66866	0.53493	0.44577
$q_{2,thin}/l_{q2}$	1/m	2.3462	1.03232	0.58654	0.37539	0.26068
`k1[q2]`, thick	m^{-2}	2.7734	1.19642	0.67293	0.43048	0.29551
tune, ν_x		87.12	53.93	38.77	29.61	24.52
tune, ν_y		37.95	17.71	16.04	13.73	12.71
S_1	m^{-2}	−35.27	−9.384	−3.800	−1.900	−1.120
`k2[S1]`, thick	m^{-3}	−147.0	−26.07	−7.917	−3.167	−1.555
S_2	m^{-2}	43.19	11.59	4.466	2.157	1.211
`k2[S2]`, thick	m^{-3}	287.9	51.5	14.89	5.752	2.691
β_0	m	0.2324	0.3486	0.4648	0.5810	0.6971
$\beta_{0,achieved}$	m	0.2302	0.4244	0.6645	0.8811	1.151
D_0	m	0.00428	0.00982	0.01745	0.02805	0.03927
$D_{0,achieved}$	m	0.00418	0.00995	0.02307	0.03363	0.05271
$<\mathcal{H}>_{theory}$	mm	0.12636	0.44242	1.0487	2.1670	3.5393
$J_x \epsilon_{x,theory,flat}$	nm	0.14742	0.52572	1.2461	2.6485	4.2057
$J_x \epsilon_{x,theory,round}^{(N)}$	μm	0.77	2.57	6.1	13.0	20.6

Within a pure dipole magnet the equation satisfied by the dispersion function is

$$\frac{d^2 D}{ds^2} = \frac{1}{\rho} \quad \text{and hence} \quad \Delta D' = \frac{\Delta s}{\rho}. \tag{11.30}$$

In the T-C cell D' increases from $-\theta/2$ to $\theta/2$ in a magnet of length L. Correspondingly, D' would increase from 0 to $\theta/2$ in a magnet of length $L/2$. One therefore envisages a dispersion matching cell that is identical to the T-C cell, with the exception that the magnet is only half length, and is also appropriately located, with its edge position shifted by an amount $s = \tilde{s}$ so that D merges smoothly to zero at the other edge. The condition for this match is

$$\frac{(L/2)^2}{2\rho} - \tilde{s}\frac{\theta}{2} = D_0 + \frac{L\theta}{8}, \tag{11.31}$$

which reduces to

$$\tilde{s} = -\frac{D_0}{\theta/2}. \tag{11.32}$$

One could substitute for D_0 using Eq. (11.11) but, in case the thick and thin lens optics do not agree very well, we anticipate fixing $\tilde{s}$ empirically to give perfect dispersion suppression. The effectiveness of this dispersion suppression is exhibited in Figure 11.5(b).

11.7.5
Nonlinearity and Dynamic Aperture

Because of sensitivity to resonances, the stability of large amplitude particles in accelerators can only be investigated numerically. This is done for on-momentum particles in Figure 11.11. What is plotted in this graph are boundaries, in the x-y plane, inside which the motion is stable (for at least 256 turns) and outside which it is unstable.

The somewhat complicated axes used in this plot makes it universally applicable at any point in the same lattice. An example may be helpful to interpret this: consider the barely stable $\Delta p/p = 0$, $L_C = 6$ m point, with coordinates $x = 0$, $y/\sqrt{\beta_y} = 0.00077 \sqrt{m}$. For the proposed high brightness lattice one has $\beta_y = 5$ m at the center of the bending magnet, giving stability out to $y = 1.72$ mm. This point sets a vertical limit, at the dipole center, for a beam with large vertical motion (present, for example, because of coupling.) However, one may be more concerned with the aperture at the vertical focusing quads where $\beta_y \approx 23$ m, and maximum $y \approx 3.69$ mm.

These contours are roughly elliptical and, to simplify the discussion without making it less conservative for beams having height much less than width, the

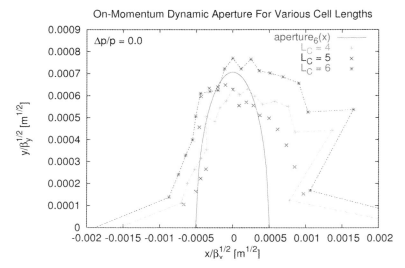

Fig. 11.11 On-momentum transverse acceptance for $L_C = 4, 5$ and 6 m. Particles are launched from a dipole center with $x' = y' = 0$, with increasingly large amplitudes, for rays emanating from the origin at various azimuthal angles. The maximum points for 256 turn stable motion are plotted at $x/\sqrt{\beta_x}, y/\sqrt{\beta_y}$. The smooth curve is an *ad hoc*, more or less maximal ellipse inscribed within the $L_C = 6$ m data in this and the off-momentum plots.

maximum amplitude contour can be replaced by an ellipse passing through the worst-case combination of x and y. Motion is stable through *all* the interior of this ellipse and unstable through *most* of its exterior. The $L_C = 6$ m ellipse is shown in the figures. The way that this ellipse simplifies the discussion is that it relieves the need for discussing horizontal and vertical motion separately. The size of this ellipse is expressed by a value ϵ_{accept}. This quantity has dimensions of length and is independent of the instantaneous beam energy of a beam being accelerated.

For historical reasons, as yet unremedied, these aperture calculations were performed for $L_C = 4, 5, 6$ m, rather than for the $L_C = 9$ m and $L_C = 18$ m cell lengths now considered to be appropriate for FCSR and SYNCH. In the figures the $L_C = 6$ m case is singled out by representing its boundary by the ellipse shown. It is the largest ellipse that is everywere inside the stability boundary for $L_C = 6$ m. Referring to purely horizontal motion, this boundary can be expressed as

$$\frac{x}{\sqrt{\beta_x}} < 0.0005 \left(\frac{L_C[\text{m}]}{6\,\text{m}}\right)^{3/2}, \quad \text{or as} \quad \epsilon_x < \left(\frac{L_C[\text{m}]}{6\,\text{m}}\right)^3 0.25 \times 10^{-6}\,\text{m}.$$

(11.33)

Two assumptions have gone into this formula. The first is that stability is governed by horizontal motion. This requires (roughly) that ϵ_y not exceed ϵ_x. This will be amply true except, perhaps, just after injection into SYNCH or, if the FCSR beam is made intentionally round to reduce ϵ_x.[9] Secondly, the scaling of dynamic aperture with L_C has been assumed to be given by the dimensional arguments discussed previously. With fractional lengths and strengths being held constant, as they are, the only quantity involved that has dimensions of length is L_C.[10] Though the extrapolation is only from $L_C = 6$ m to $L_C = 9$ m, this weak dimensional argument needs to be replaced by tracking of the actual lattices with their correct parameters.

Figure 11.11, especially as parametrized by Eq. (11.33), is applicable to both SYNCH and FCSR. With $x/\sqrt{\beta_x}$ and $y/\sqrt{\beta_y}$ for axes, the same plots are applicable everywhere in individual rings. Treating these plots as "phase diagrams", the contours separate stable phase (inside) from unstable phase (outside), for a particular choice of parameters.

When quoted in millimeters the apertures seem very small. But when they are quoted in units of beam σs, they can be quite satisfactory because of the exceedingly low values of ϵ_x. This issue is pursued in the next section.

Especially during initial operation of FCSR it will not be surprising to find closed orbit deviations of one, or a few, millimeters. As such these deviations will be comparable to, or even larger than, the dynamic apertures just discussed. *This is not paradoxical.* One knows, for example from experience with hadron accelerators, that the nonlinearity-imposed aperture is centered on the closed orbit. When the closed orbit moves, the stable region moves with it, little reduced by the orbit's deviation from design.

11.8
Emittance Evolution During Acceleration

The physical dimensions of the electron beam vary as the beam accelerates. To determine whether the accelerator aperture is sufficiently large, after establishing the normalized emittance $\epsilon^{(N)}(\gamma)$, it is necessary to determine the "unnormalized" or "geometric" emittance $\epsilon(\gamma)$. These quantities are related

---

9) By intentionally making the electron beam "round", ϵ_x can be reduced from its nominal 0.5 nm value at 5 GeV. One would then have $\epsilon_x = \epsilon_y = 0.25$ nm.

10) The phenomenon of resonance can cause erratic variation of dynamic aperture on scales short compared to the scales on which the parameters are being held constant. This certainly invalidates the dimensional analysis argument applied to particular lattices. However, general trends should follow from the dimensions.

by

$$\epsilon(\gamma) = \frac{\epsilon^{(N)}}{\gamma}. \tag{11.34}$$

Constancy of $\epsilon^{(N)}$, implicitly masks the beam size variation due to adiabatic damping. But there can also be non-Hamiltonian shrinkage due to synchrotron radiation damping and growth due to quantum fluctuation. During acceleration in SYNCH there will be transition from input-dominated emittance to radiation-dominated emittance. This transition occurs in the vicinity of the horizontal line shown in a table to be introduced next.

Detailed data describing the evolution of beam parameters throughout an acceleration cycle is given in Table 11.2. This table assumes a normalized injection emittance $\epsilon^{(N)} = 2\,\mu m$, which is the design specification from the injection linac, both for the fast cycling approach and for the competing ERL option. The lattice acceptance given by Eq. (11.33) is assumed. Based on the numbers in this table the injection energy would be chosen to be at least 100 MeV, where the beam stay clear in SYNCH is 8.65σ, including momentum spread. More conservative injection would be at 200 MeV where the stay clear is 12.2σ. As mentioned earlier, out-of-spec injection emittance could be handled by increased injection energy, but sub-par injection current would give proportionally sub-par x-ray flux.

Entries well above the horizontal line in Table 11.2 are determined by the injection emittance and adiabatic damping. In this region the entries in the three columns on the right are only academic—they apply to an equilibrium that would apply after times long compared to the time spent at that energy. The line is drawn at the point where the radiation equilibration time is one second, which is roughly equal to the time that electrons spend ramping up to full energy in SYNCH.

Entries well below the horizontal line in Table 11.2 are determined by the radiation equilibrium at that energy. In this region the adiabatically-damped entries on the left have become academic. There is fairly conservative beam headroom up to 7 or 8 GeV, but this does not necessarily imply that 8 GeV operation is practical, as the maximum magnet strengths and RF voltage may be insufficient, either in SYNCH or in FCSR.

Near the horizontal line in the table the valid entries switch from the left part of the table to the right. Technically speaking, none of the entries in these rows are valid, but the stay clear is comfortable through this region.

Beam transfer from SYNCH to FCSR is analysed in Table 11.3. The beam is assumed to have come into equilibrium in SYNCH. As such its emittance is given by the $\epsilon_{\text{equi.}}$ column in Table 11.2. This entry is copied to the ϵ_{in} column of Table 11.3 for each of the potential transfer energies. Since the energy of

FCSR will be fixed this transfer has to occur at the full FCSR energy. Nominally this will occur at energy $\mathcal{E}_e = 5\,\text{GeV}$. One sees that the acceptance at this energy is 10.1σ, where σ is the r.m.s. transverse beam size at this energy. This is comfortably large and would, presumably, permit 100% transfer efficiency. After some 10s of milliseconds the beam will have damped to the value of $\epsilon_{\text{equi.}}$ given in the second last column. From then on the aperture will be 28.7σ which is large enough to avoid all but Touschek and residual vacuum scattering losses.

Tab. 11.2 Emittance evolution in synchrotron SYNCH, with its cell length taken to be $L_C = 18\,\text{m}$, twice as great as in FCSR. The invariant emittance at injection is taken to be $\epsilon_x^{(N)} = 2\,\mu\text{m}$. $\Delta\mathcal{E}_e$ is the energy loss per turn and τ is a characteristic damping time. Transition from adiabatic damping dominance to radiation dominance occurs near the horizontal line part way down the table. Column 5 gives the aperture in units of the adiabatic damped beam size, which is applicable well above the line. The last column gives the aperture in units of equilibrium beam size, which is applicable well below the horizontal line.

$\mathcal{E}_e$	$\Delta\mathcal{E}_e$	τ	ϵ_{in}	$\sqrt{\epsilon_{\text{accept}}/\epsilon_{\text{in}}}$	$\epsilon_{\text{equi.}}^{(N)}$	$\epsilon_{\text{equi.}}$	$\sqrt{\epsilon_{\text{accept}}/\epsilon_{\text{equi.}}}$
GeV	MeV	s	m		m	m	
0.1	2.1E-7	1042	1.02E-8	8.65			
0.2	3.4E-6	130	5.11E-9	12.2			
0.3	1.7E-5	39	3.41E-9	15.0			
0.4	5.4E-5	16.2	2.56E-9	17.3	1.86E-8	2.37E-11	180
0.5	1.3E-4	8.3	2.04E-9	19.3	3.63E-8	3.71E-11	144
0.7	5.1E-4	3.0	1.46E-9	22.9	9.95E-8	7.26E-11	103
1.0	2.1E-3	1.04	1.02E-9	27.3	2.90E-7	1.48E-10	72
1.5	1.1E-2	0.31	6.81E-10	33.5	9.79E-7	3.34E-10	48
2.0	3.4E-2	0.130	5.11E-10	38.7	2.32E-6	5.93E-10	36
3.0	0.171	0.039	3.41E-10	47.4	7.83E-6	1.33E-9	23.9
4.0	0.541	0.0163	2.56E-10	54.7	1.86E-5	2.37E-9	18.0
5.0	1.32	0.0083	2.04E-10	61	3.63E-5	3.71E-9	14.4
6.0	2.74				6.27E-5	5.34E-9	12.0
7.0	5.07				9.95E-5	7.26E-9	10.3
8.0	8.65				1.49E-5	9.49E-9	9.0

Entries in Table 11.3 continue on up to 8 GeV transfer energy. At that point the aperture is only 6.3σ. The damping time being 2 ms translates to a 1000 turn betatron exponential decay time. This is the number of turns required with aperture at 6σ of a Gaussian distribution. On paper this could lead to satisfactory injection efficiency since the area outside 6σ in a Gaussian distribution is very small and a typical growth time from, say, 3σ to 6σ is several damping times. Nevertheless, these conditions could not be considered conservative enough to guarantee successful transfer at 8 GeV. Errors of one sort or another are sure to reduce the aperture somewhat and to increase the beam size in SYNCH.

Tab. 11.3 Emittance evolution and aperture requirements in FCSR. The emittance ϵ_{in} column four of this table is copied from the $\epsilon_{equi.}$ entry in the second last column of Table 11.2.

$\mathcal{E}_e$	$\Delta\mathcal{E}_e$	τ	ϵ_{in}	$\sqrt{\epsilon_{accept}/\epsilon_{in}}$	$\epsilon_{equi.}^{(N)}$	$\epsilon_{equi.}$	$\sqrt{\epsilon_{accept}/\epsilon_{equi.}}$
GeV	MeV	s	m		m	m	
1.0	2.1E-3	1.04	1.48E-10	50.8	3.62E-8	1.85E-11	144
1.5	1.1E-2	0.31	3.34E-10	33.8	1.22E-7	4.17E-11	96
2.0	3.4E-2	0.13	5.93E-10	25.4	2.90E-7	7.41E-11	72
3.0	0.171	0.039	1.33E-9	17.0	9.79E-7	1.67E-10	47.9
4.0	0.541	0.0163	2.37E-9	12.7	2.32E-6	2.97E-10	35.9
5.0	1.32	0.0083	3.71E-9	10.1	4.53E-6	4.63E-10	28.7
6.0	2.73	0.0048	5.34E-9	8.46	7.83E-6	6.67E-10	23.9
7.0	5.07	3.0E-3	7.26E-9	7.26	1.24E-5	9.08E-10	20.5
8.0	8.65	2.0E-3	9.49E-9	6.34	1.86E-5	1.19E-9	18.0

The design changes this suggests can be considered by contemplating first the top row of Table 11.2, which implies almost but not quite acceptable injection into SYNCH. To make this more conservative one has previously chosen the injection energy into SYNCH above, say, 200 MeV. It is harder to improve injection efficiency into FCSR at a given energy. Increasing the acceptance of FCSR would harm the equilibrium emittance, which is unattractive. Better would be to reduce the beam emittance from SYNCH. This would further increase the needed injection energy into SYNCH which might, however, be affordable.

To guarantee successful operation, say at 7 GeV, one would have, therefore, to change the parameters somewhat. This is the basis for the statement made earlier that maintaining the option of increasing the energy above 5 GeV will compromise the brilliance at 5 GeV to some extent. Accepting the numbers in the tables as truly valid (when they are really just estimates) the trade-off seems fairly mild. That is, the injection acceptance could be jiggered to, say, 10σ at 7 GeV without harming $\epsilon_x(5\,\text{GeV})$ very much. This is not the whole story however, as the magnets and RF would also have to function satisfactorily at the higher energy.

11.9
Touschek Lifetime Estimate

Results of some rudimentary Touschek effect calculations are shown in Figure 11.12. The formula labeled "Bruck" [14] comes from a chapter describing a theory due to Haissinski, in a book by Bruck. The formula labeled "LeDuff" comes from an article by Bocchetta [15]. These theories apparently differ primarily in their treatments of dispersion. Since the LeDuff theory is more re-

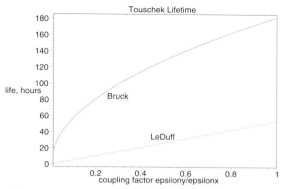

Fig. 11.12 Beam lifetime caused by the Touschek effect. $L_C = 6\,\text{m}$, $\epsilon_x = 4.0 \times 10^{-10}\,\text{m}$, $\mathcal{E}_0 = 5.0\,\text{GeV}$, $\sigma_z = 0.1\,\text{m}$, $N_B = 1280$.

cent, it may be more reliable. The lifetime increases proportional to γ^4, or to an even higher power of γ, depending on assumptions concerning the beam σs [16].

Experts in the field can undoubtedly produce far more reliable data than those in Figure 11.12. But, to support the present design, all that is needed is the observation that Touschek lifetimes for existing storage rings are measured in *hours*. The several *second* storage time in the fast cycling approach ensures the possibility of a huge reduction in beam emittances before the Touschek effect again limits operations. An uncertainty factor of two or more in the Touschek lifetime is somewhat irrelevant, since other factors are likely to set less optimistic lower limits on the emittances.

11.10
Performance as X-ray Source

11.10.1
Brilliance from Short Undulator

Using Eq. (11.6), a first, coarse, comparison between the brilliance achievable from the fast cycling x-ray source described here, and the proposed Cornell ERL source, can be made easily under the assumption of equal electron energy ($\mathcal{E}_e = 5\,\text{GeV}$) and equal beam currents ($I = 0.1\,\text{A}$) in both cases. Using subscripts FCSR (for fast cycling) and ERL,

$$\frac{\mathcal{B}_{\text{FCSR,flat}}}{\mathcal{B}_{\text{ERL,round}}} = \frac{\epsilon_{x,\text{ERL}}}{\epsilon_{x,\text{FCSR}}} \frac{\epsilon_{y,\text{ERL}}}{\epsilon_{y,\text{FCSR}}}$$
$$\approx \frac{2.04 \times 10^{-10}\,\text{m}}{4.63 \times 10^{-10}} \frac{2.04 \times 10^{-10}\,\text{m}}{4.63 \times 10^{-10}/100} \approx 20. \tag{11.35}$$

The ERL beam is necessarily round and this formula has assumed a round 2 μm-emittance beam from the gun, with no further emittance growth throughout the ERL. For FCSR the same beam current from the gun has been assumed, as well as a (more or less arbitrary, but *commonly* achievable) emittance ratio $\epsilon_y/\epsilon_x \approx 1/100$.

This comparison has been between a round ERL beam and a flat FCSR beam. Probably the main motivation for striving for high brilliance is to achieve high coherence. One tries therefore to operate in a *diffraction limited* regime which is characterized by the emittance goal $\epsilon < \lambda/\pi$ where λ is the wavelength of the x-rays being produced. Reduction of either ϵ_x or ϵ_y very far below this value provides little benefit in coherency. For example, for 12 KeV x-rays, which have one Ångstrom wavelength, the vertical emittance assumed in Eq. (11.35) is unnecessarily small, which makes the factor of 20 favoring FCSR somewhat too high. This issue is discussed in greater detail in the next section.

On the other hand, one of the attractive applications made poossible by high brightness beams, namely phase contrast imaging in thick samples, relies on hard x-rays, for which the full reduction in ϵ_y assumed in Eq. (11.35) pays off. There is a far more important consideration than this, however. It is the strong dependence of hard x-ray flux on electron energy. From the same undulator, the energies of undulator harmonics are proportional to γ^2. This would argue for increasing the beam energy to, say, 6 GeV as at ESRF, or 7 GeV, as at APS, or to 8 GeV as at SPRING-8, to quote the energies of the the presently leading centers for this work.

Another comparison can be based on the fact than some x-ray beamlines favor round beams and cannot profit from ultralow vertical emittance. Roughly round beams can be produced by collimation but it is cleaner to start with a round electron beam. A round beam in FCSR is easily achieved by introducing coupling. When this is done the FCSR emittances at 5 GeV are $\epsilon_x = \epsilon_y = 2.3 \times 10^{-10}$ m and the brightness comparison is.

$$\frac{\mathcal{B}_{\text{FCSR,round}}}{\mathcal{B}_{\text{ERL,round}}} = \frac{\epsilon_{x,\text{ERL}}}{\epsilon_{x,\text{FCSR}}} \frac{\epsilon_{y,\text{ERL}}}{\epsilon_{y,\text{FCSR}}} \approx \frac{2.04}{2.3} \frac{2.04}{2.3} \approx 0.79. \tag{11.36}$$

These and other brilliance comparisons are shown in Table 11.4. For typical undulators and typical x-ray energies, the brilliance achievable with the fast cycling scheme is an order of magnitude greater than the so-called "third generation" light sources, ESRF and SPRING-8. (It is, of course, presumptuous to compare futuristic paper-projections, to actually achieved performance, with possibly quite different parameters and conditions.)

Tab. 11.4 Comparison of brilliance $\mathcal{B}$ for various storage-ring-plus-undulator x-ray sources. "FCSR" stands for the fast cycling design. For this and for the Cornell-ERL case, a 4 meter long, 117 pole, $K = 1$, undulator, with $\lambda_w = 3.4$ cm, is assumed, and Eq. (11.6) s used to calculate $\mathcal{B}$. All entries are in (standard) units for brilliance, as given in that equation. The ESRF and SPRING-8 undulators are similar, but different, and the entries are published values for $\mathcal{B}$. Cases with $\sim$ symbols are obtained by crude scaling. Some possible sources of overestimate in the FCSR, 5 GeV-flat column are discussed in the next section. For these, and all other estimates, the horizontal partition number has been taken to be $J_x = 1$. It may be possible to increase the FCSR brightnesses by as much as a factor of two by tuning to $J_x \approx 1.5$, as is done at ALS and CLS.

$\mathcal{E}_\gamma$ keV	n	FCSR 5 GeV 4 m flat	FCSR 8.1 GeV 4 m flat	FCSR 5 GeV 4 m round	Cornell-ERL 5 GeV 4 m round	ESRF, U35 6 GeV flat	SPRING-8 8 GeV 5 m flat
4.7	1	7.3E21		2.9E20	3.7E20	6E20	
14	3	$\sim$6E21		$\sim$3E20	$\sim$3E20	5E20	
23	5	$\sim$5E21					
12.4	1		1.1E22				6E20
37.2	3		$\sim$1E22				5E20

11.10.2
Refinement of Brilliance Calculation

This section includes qualifications suggested by (but not necessarily endorsed by) Ivan Bazarov. The central question is whether or not ϵ_y is unnecessarily small in the FCSR design.

The entries in Table 11.4 for potential Cornell x-ray sources are based on Eq. (11.6). This formula has the benefit of simplicity and definiteness, but it also hides certain physical properties of undulator radiation that enter into assessing the relative effectiveness of different configurations. Some physical considerations not incorporated in Eq. (11.6) are:

- As was explained in an introductory section, the main way a storage ring maximizes brilliance $\mathcal{B}$ is by minimizing a six-dimensional phase space volume $\mathcal{L}_3 \mathcal{P}_3$ which contains most of the produced photons. As $\mathcal{L}_3 \mathcal{P}_3$ gets smaller and smaller, the fraction of experiments matched to such a small phase space volume decreases. This means, for example, that many experiments cannnot really exploit some part of the two orders of magnitude decrease in ϵ_y that gives the FCSR its main brightness advantage over the ERL.

- For ultralow emittance beams the radiation may be "diffraction limited". Many "hypermodern" applications of x-rays rely on the ccherence properties of beams in this limit. Near a beam waist, either transverse dimension of the electron beam envelope evolves longitudinally as

$$y(s) = \sqrt{\epsilon_y \beta_0} \sqrt{1 + \left(\frac{s}{\beta_0}\right)^2}. \tag{11.37}$$

As such, the parameter β_0 can be referred to as "waist length" or "doubling length". This variation resembles the s-variation of the envelope function w of an electromagnetic wave having wave number $k = 2\pi/\lambda$, in its lowest round, open cavity mode, near a point focus [17];

$$w(s) = w_0 \sqrt{1 + \left(\frac{s}{w_0^2 k/2}\right)^2}. \tag{11.38}$$

For electron and photon beams having the same (round) aspect ratio there is a wave number k_{dl} for which these shapes superimpose. Its corresponding "diffraction limited" wavelength λ_{dl} is given by

$$\lambda_{dl} = \pi \epsilon_0. \tag{11.39}$$

Another way of obtaining this result is to express the transverse phase space volume of the photon beam (matched to the electron beam) as

$$\mathcal{L}_2^T \mathcal{P}_2^T = \left(\pi \epsilon_0 \frac{h}{\lambda}\right)^2. \tag{11.40}$$

Condition (11.39) makes this equal to h^2 which is the phase space volume of one state in phase space. This gives, for example for $\lambda_{dl} = 1.0 \times 10^{-10}$ m, a diffraction limited emittance $\epsilon_0 = 0.3 \times 10^{-10}$ m. This means that, for a (typical) x-ray energy 12 keV, the vertical emittance yielding the flat beam FCSR brightness in Table 11.4 is unnecessarily small by a factor of five or so. The flat beam 5 GeV FCSR brightness values are too high by this factor. For hard x-rays and for operation above $\mathcal{E}_e = 5$ GeV the estimated brightnesses becomes more nearly valid.

The degree of transverse coherence is related to λ_{dl}. For wavelengths much longer than λ_{dl} all electrons radiate with transverse coherence. For wavelengths much shorter than λ_{dl} the electrons radiate with negligible transverse coherence. For the FCSR vertical emittance the vertical coherence remains excellent for all x-ray energies and the horizontal coherence is good up to a few keV.

- Brilliance $\mathcal{B}$ has been taken to be proportional to the number of undulator periods N_w. This may be surprising when the electric field amplitude radiated at a fixed angle is proportional to N_w, making the intensity (to which the photon flux is proportional) proportional to N_w^2 at the center of an undulator resonance line. It is therefore possible to design a beamline giving flux proportional to N_w^2, though this is not typical. For apparatus with energy acceptance broader than the undulator line, the N_w^2 peak intensity and the $1/N_w$ line width produce flux proportional to N_w, as given by Eq. (11.6).

- Most x-ray experiments require x-rays as nearly monochromatic as possible, and the feature making undulators invaluable is that the line width (in energy) is proportional to $1/N_w$. Commonly there is a monochromator near the front of an x-ray beamline. Even for very large, but realistic, values of N_w, the monochromator fractional energy acceptance is typically far less than $1/N_w$. Even in this case the x-ray flux down the beam line is usually proportional to N_w. This will be true when the angular spread of the electron beam is larger than the x-ray beam collimation that is matched to the (post-monochromator) radiation core of the undulator radiation;

$$\sqrt{\frac{\epsilon}{\beta}} > \frac{1}{\gamma}\sqrt{\frac{1}{nN_w}}, \qquad (11.41)$$

where n is the undulator resonance order. Consider a collimation system centered on the undulator axis and narrowly matched to the central undulator resonance "spike". The probability that a particular radiating electron lines up with this collimation is proportional to $1/N_w$. In this case Eq. (11.6) continues to be approximately valid. For $\epsilon_x \approx 10^{-10}$ and $\beta_x = 1$ m at the undulator described by Table 11.4, this inequality is typically satisfied. For $\epsilon_y \approx 10^{-12}$ the electron vertical angular divergence is smaller than the undulator collimation, again making the spread of vertical electron angles unnecessarily small and making the brilliance estimate in Table 11.4 unduly optimistic.

But undulator harmonic numbers n large compared to 1 are frequently used, primarily to obtain a harder x-ray spectrum. In this case lowering ϵ_y is advantageous. Furthermore undulator radiation is subject to other complications, for example giving undulator flux density proportional to N_w^2. In short, the performance of any particular physical setup depends on details far too complicated to be accurately described by a simple formula like Eq. (11.6); this formula may give either an underestimate or overestimate, depending on conditions.

- The fractional energy width of the FCSR electron beam at 5 GeV is $\sigma_\delta = 0.00076$. The energy spread of the ERL is expected to be 0.0002. Formula (11.6) does not take into account the improvement in brilliance that could result in the ERL by exploiting this advantage by allowing longer undulators. Bazarov points out that, as an example, this could permit full brightness for $N_w = 1000$ in the ERL, but only $N_w = 200$ in FCSR. In any case, at most two such undulators could be incorporated in FCSR, while more than two could be incorporated in ERL.

- Practical accelerator considerations make it difficult to use very long undulators. In particular both β-functions have to be comparable with or larger than the undulator length. The excellent match of electron beam to photon cone (see Eqs. (11.37) and (11.38)) is applicable at the beam waist, but "detunes" elsewhere. The angular spread of electron beam angles is constant through a drift section, but the effective spot size increases away from the waist, and with it the effective emittance. In the FCSR design, the extremely small value of ϵ_y makes it possible to preserve the diffraction limited condition even for undulator length appreciably longer than β. The ERL will not have this advantage but, because of its less-constrained geometry, it may not need it.

- When the more nearly conventional, fast cycling ring is regarded as R&D for the ILC, R&D for the extremely small ϵ_y performance is very appropriate.

References

1 Turner, S. (1998), (Ed.), *1996 CERN Accelerator School: Synchrotron Radiation and Free Electron Lasers*, CERN Yellow Report 98-04.

2 Ropert, A. (1998), *Lattices and Emittances*, in CERN Yellow Report 98-04, Turner, S., (Ed.).

3 Chasman, R., Green, K. (1975), *Preliminary Design of a Dedicated Synchtotron Radiation Facility*, IEEE Trans. Nucl. Sci.,**22**, 1765.

4 Grote, H. and Iselin, F. (1996), *The MAD Program (Methodical Accelerator Design) Version 8.19, User's Reference Manual*, CERN/SL/90-13 (AP).

5 Wiedemann, H. (1980), *Brightness of Synchrotron Radiation from Electron Storage Rings*, Nucl. Instrum. Methods, **172**.

6 Krinsky, S. (2002), *Fundamentals of Hard X-Ray Synchrotron Radiation Sources*, in Third Generation Hard X-Ray Synchrotron Radiation Sources, D. Mills (Ed.), John Wiley, New York.

7 Bazarov, I., Sinclair, C. (2005), *PRST-AB*, **8**, 034202.

8 Brinkmann, R. et al., (2001), TESLA Technical Design Report, Part II, DESY 2001-011, ECFA 2001-209.

9 Borland, M. (1991), et al., SLAC Report-402.

10 Sheffield, R. et al., PAC 95, 882.

11 Trbojevic, D. and Courant, E. (1994), *Low emittance lattices for electron storage rings, revisited*, in Proc. EPAC94, 1000.

12 Steffen, K. (1995), *High Energy Beam Optics*, Interscience Publishers, New York, 17.

13 Malitsky, N., Talman, R. (2005), *Accelerator Simulation Using the Unified Accelerator Libraries (UAL)*, U.S. Particle Accelerator School, Ithaca. Available at http://www.ual.bnl.gov.

14 Bruck, H., *Accélérateurs Circulaires de Particules*.

15 Bocchetta, C. (1998), *Lifetime and Beam Quality* in CERN 98-04.

16 Miyahara, Y. (1985), *IEEE Trans. Nucl. Sci.*, **32**, 3821.

17 Mandel, L., Wolf, E. (1995), *Optical Coherence and Quantum Optics*, Cambridge University Press, Cambridge, 269.

12
Compton Scattered Beams And "Laser Wire" Diagnostics

12.1
Preview

Compton scattering is the scattering of a photon by an electron. Though the topics "Compton Scattering" and "Undulator Radiation" may seem to be distinct, they are, in fact closely related. Undulator radiation can, for example, be treated as Compton scattering (in the Thomson scattering limit). Conversely, Compton scattering of a photon from an electromagnetic wave by an electron beam, can be treated as undulator radiation. Both of these approaches are investigated in this chapter, but with emphasis on the Compton scattering perspective.

Also to be covered is the use of laser beams for measuring the phase space distributions of electron beams. This material is especially relevant for ERLs and linear collider damping rings, which require the accurate measurement of the beam parameters of short, low emittance beams, at low and high energies. For this task the representation of Compton scattering as undulator radiation is quite satisfactory. The Compton scattering interpretation of undulator radiation is also appropriate for understanding undulator radiation in its fundamental mode, but understanding higher undulator harmonics in the quantum picture would require more sophisticated treatment than is attempted here. On the other hand, the theory of undulator and wiggler radiation in full generality, based on classical electromagnetic theory is relatively straightforward. This was the subject of Chapter 7.

In the first few sections of this chapter a "modern" approach, based on relativistic invariants, will be taken. In this approach all quantities are evaluated in the same frame of reference. (Parameters of incident particles in this frame have no primes and scattered particles have primes.) This greatly reduces the number of steps required compared to more elementary approaches (the subjects of later sections) which involve transformation to an intermediate frame of reference and then back to the original frame. For those approaches single primed parameters relate to the intermediate frame and double primed parameters are applied to parameters transformed back to the original frame

Accelerator X-Ray Sources. Richard Talman
Copyright © 2006 WILEY-VCH Verlag GmbH & Co. KGaA, Weinheim
ISBN: 3-527-40590-9

(which are therefore directly comparable to singly-primed quantities in the invariant approach).

To constrain the geometry a bit in this chapter, though the photon beam or wave can have arbitrary direction, the electron will always be assumed to be traveling along the negative x-axis toward a collision at the origin.

12.2
Compton Scattering Kinematics

Before specializing to particular experimental configurations, the Compton scattering process—photon scatters from electron—will be formulated in modern, manifestly relativistic, terms. This initial discussion follows Berestetskii, Lifshitz, and Pitaevskii (BLP) [1] quite closely. This includes taking $\hbar = c = 1$. In these units the velocity is $v \equiv \beta = p/\mathcal{E}$. Also the sign of e, the electron charge, will be ignored; i.e. taken as positive.

The four-momentum of any particle, $\underline{p} = (\mathcal{E}, \mathbf{p})$, incorporates energy and momenta, which are related by

$$\mathcal{E} = \sqrt{p^2 + m^2}, \tag{12.1}$$

Energies, momenta, and masses will all be expressed in eV units. The invariant scalar product of four-vectors a and b is defined by

$$(\underline{a}, \underline{b}) \equiv a_0 b_0 - \mathbf{a} \cdot \mathbf{b}, \quad \text{so} \quad (\underline{p}, \underline{p}) = m^2. \tag{12.2}$$

The Compton scattering momentum vectors are illustrated in Figure 12.1. Discussions of Compton scattering sometimes employ either the electron rest system (sometimes called the "laboratory system") or a "center of mass" system in which the electron and photon momenta are equal and opposite. We will work in a different "laboratory system"; in it the incident electron travels in the positive x-direction along the negative x-axis, toward a collision at the origin, and the incident photons travel at angle α relative to the electron. The incident electron has four-momentum

$$\underline{p} \equiv (\mathcal{E}, \mathbf{p}) \equiv (\mathcal{E}, p, 0, 0), \tag{12.3}$$

and the incident photon four-momentum is

$$\underline{k} \equiv \omega(1, \hat{\mathbf{k}}) \equiv \omega(1, \cos\alpha, \sin\alpha, 0), \tag{12.4}$$

where ω is the (radian) frequency of the photon. The incident photon and electron therefore define the (x, y) plane in the laboratory. The final state four-momenta are

$$\begin{aligned}
\underline{k}' &= (\omega', k'_x, k'_y, k'_z) = \omega'(1, \hat{\mathbf{k}}') \\
&= \omega'(1, \cos\vartheta \cos\alpha + \sin\vartheta \cos\varphi \sin\alpha, \cos\vartheta \sin\alpha \\
&\quad - \sin\vartheta \cos\varphi \cos\alpha, \sin\vartheta \sin\varphi), \\
\underline{p}' &= (\mathcal{E}', p'_x, p'_y, p'_z) = (\mathcal{E}', \mathbf{p}').
\end{aligned} \qquad (12.5)$$

With incident momenta $\underline{k}$ and $\underline{p}$ given, there are two degrees of freedom in the outgoing configuration. The outgoing photon direction $\hat{\mathbf{k}}'$ can, for example, be chosen arbitrarily. Then the scattered photon energy ω' can be determined and, finally, the scattered electron four-momentum can then be determined. The strategy to be employed for fixing all kinematic variables is to first eliminate $\underline{p}'$. Conservation of momentum and energy yields

$$\underline{p}' = \underline{k} + \underline{p} - \underline{k}'. \qquad (12.6)$$

Standard kinematic variables s, t, and u are defined by

$$\begin{aligned}
s &= (\underline{p} + \underline{k}, \underline{p} + \underline{k}) = m^2 + 2(\underline{p}, \underline{k}) = m^2 + 2\omega(\mathcal{E} - p\cos\alpha), \\
t &= (\underline{p} - \underline{p}', \underline{p} - \underline{p}') = -2(\underline{k}', \underline{k}) = -2\omega\omega'(1 - \cos\vartheta), \\
u &= (\underline{p} - \underline{k}', \underline{p} - \underline{k}') = m^2 - 2(\underline{p}, \underline{k}') \\
&= m^2 - 2\omega'(\mathcal{E} - p\cos\vartheta \cos\alpha - p\sin\vartheta \cos\varphi \sin\alpha).
\end{aligned} \qquad (12.7)$$

Using Eq. (12.6) the sum of these variables can be worked out to be

$$s + t + u = 2m^2. \qquad (12.8)$$

Substituting from Eq. (12.7), this yields $(\underline{p}, \underline{k}) = (\underline{p}, \underline{k}') + (\underline{k}, \underline{k}')$, which can be expressed as

$$\frac{\omega}{\omega'} = \frac{(\underline{p}, \hat{\underline{k}}') + (\underline{k}, \hat{\underline{k}}')}{(\underline{p}, \hat{\underline{k}})}. \qquad (12.9)$$

This determines the energy of the scattered photon. More explicitly,

$$\frac{\omega}{\omega'} = \frac{\mathcal{E} - p\cos\vartheta \cos\alpha - p\sin\vartheta \cos\varphi \sin\alpha}{\mathcal{E} - p\cos\alpha} + \frac{\omega(1 - \cos\vartheta)}{\mathcal{E} - p\cos\alpha}. \qquad (12.10)$$

A simple special case of this (needed later for laser wire treatment) is obtained for $\varphi = 0$, $\vartheta = \alpha$, (which yields scattered photon parallel to incident electron), and with $\omega \ll m$ and $v \approx 1$;

$$\frac{\omega}{\omega'} \approx \frac{1}{\gamma^2} \frac{1}{(1 - v\cos\alpha)(1 + v)}, \qquad (12.11)$$

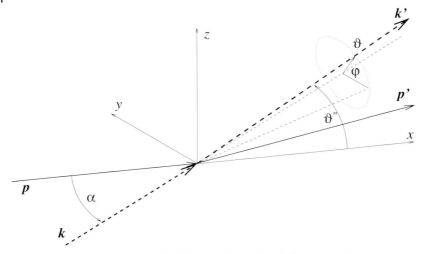

Fig. 12.1 Momentum vectors of incident and outgoing electrons **p** and **p'** and incident and Compton scattered photons **k** and **k'**. **p** defines the x-axis and **k** lies in the x, y plane.

where $\gamma = \mathcal{E}_e/m$ is the usual relativistic factor.

Equations (12.7) and (12.8) are also the bases for Figure 12.2 in which the variable s, t, and u are marked off along lines orthogonal to the three sides of the "Mandelstam" equilateral triangle shown. At each of the corners two of the three quantities vanish, so all three triangle altitudes, such as the one shown, have value $2m^2$. As explained in the caption, all points in this plot satisfy Eq. (12.8).

The "physical region", in which real photons scatter from real electrons, is restricted to the shaded region of this plot. From Eqs (12.7) some inequalities satisfied in the physical region are obvious: $s \geq m^2$, $t \leq 0$, and $u \leq m^2$. In the limit of zero scattering angle we have $t = 0$; this defines the $t = 0$ axis as the "glancing collision" boundary of the physical region. In the center of mass system, with incident photon and electron collinear, the four-vectors for straight backscattering are $\underline{p} = (\mathcal{E}, \omega, 0, 0)$, $\underline{k} = (\omega, -\omega, 0, 0)$, and $\underline{k}' = (\omega, \omega, 0, 0)$. Substituting these into Eq. (12.7) one finds

$$su = m^4, \tag{12.12}$$

as the equation of the "backscattering" boundary of the physical region. Since both sides of this equation are invariants, this equation continues to be valid in any frame. This boundary is shown in Figure 12.2.

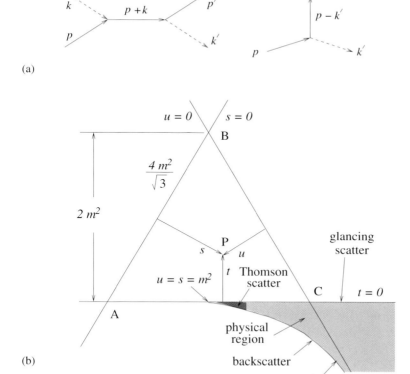

Fig. 12.2 (a) Feynman diagrams for Compton scattering. (b) The Mandelstam kinematic variables and triangle. Triangles APB, APC, and BPC have areas $(2/\sqrt{3})m^2 s$, $(2/\sqrt{3})m^2 t$, and $(2/\sqrt{3})m^2 s$, and triangle ABC has area $(4/\sqrt{3})m^4$ which is equal to their sum. This ensures that Eq. (12.8) is satisfied at every point in the plane.

12.3
Some Specialized Laser, Electron Beam Configurations

12.3.1
Back-scattered Photons

Compton back-scattering can be used to produce (weak) high energy photon beams from intense low energy photon beams, supplied, for example, by a visible light laser. Also, undulator radiation can be understood as the back scattering of a beam of (virtual) long wavelength photons "produced" by an undulator magnet. Both of these applications are characterized by relations

$$\alpha = \pi, \quad \vartheta = \pi - \theta, \quad 0 < \theta \ll 1. \tag{12.13}$$

in which case Eq. (12.10) reduces to

$$\frac{\omega}{\omega'} = \frac{\mathcal{E} - p\cos\theta}{\mathcal{E} + p} + \frac{\omega(1 + \cos\theta)}{\mathcal{E} + p} \approx \frac{1 - (p/\mathcal{E})\cos\theta}{1 + p/\mathcal{E}}, \tag{12.14}$$

where dropping the second term remains to be justified. The kinematic invariants are

$$\begin{aligned} s &= m^2 + 2\omega(\mathcal{E} + p) \\ t &= -2\omega\omega'(1 + \cos\theta) \\ u &= m^2 - 2\omega'(\mathcal{E} - p\cos\theta). \end{aligned} \tag{12.15}$$

For relativistic electrons, and $\theta \ll 1$, one can use

$$\frac{p}{\mathcal{E}} \approx 1 - \frac{1}{2\gamma^2}, \quad \cos\theta \approx 1 - \frac{\theta^2}{2} \tag{12.16}$$

Dropping the second term in Eq. (12.14) is valid, even at $\theta = 0$, provided that

$$\mathcal{E}\omega \ll m^2, \tag{12.17}$$

which will normally be true for ω corresponding to visible light. In this case

$$\omega'(\theta) \approx \frac{2\omega}{\frac{1}{2\gamma^2} + \frac{\theta^2}{2}}. \tag{12.18}$$

A few back-scattering examples are given in Table 12.1.

Tab. 12.1 Some exactly back-scattered Compton examples. All energies and momenta are expressed in eV units.

| ω | $\mathcal{E}$ | α | ϑ | ω' | photon type |
eV	eV			eV	after scatter
1	10^8	π	π	1.530×10^5	back-scattered gamma
1	10^7	π	π	1.530×10^3	back-scattered x-ray
1	10^6	π	π	13.24	back-scattered UV

12.3.2
Orthogonal Photon Incidence in the Laboratory

A diagnostic technique developed recently has been to illuminate an electron beam from the side by an intense laser beam and detect the scattered photons.

With the laser beam being capable of being focused to a spot as small as a few microns, there is the possibility of high resolution, nondestructive profile measurement of the beam profile. In this configuration the light beam is sometimes referred to as a "laser wire". Another possibility is to use a broad, parallel, laser beam and to image the scattered photons with a camera.

The formulas needed were derived in Section 12.2. It may be advantageous to shine the laser beam at an arbitrary angle but, for simplicity, let us take the laser beam exactly transverse to the electron beam (in the laboratory). Substituting $\alpha = \pi/2$ into Eq. (12.7) we obtain

$$\begin{aligned} s &= = m^2 + 2\omega\mathcal{E}, \\ t &= = -2\omega\omega'(1 - \cos\vartheta), \\ u &= m^2 - 2\omega'(\mathcal{E} - p\sin\vartheta\cos\varphi), \end{aligned} \quad (12.19)$$

and into Eq. (12.10) to get

$$\frac{\omega}{\omega'} = 1 - \frac{p}{\mathcal{E}}\sin\vartheta\cos\varphi + \frac{\omega}{\mathcal{E}}(1 - \cos\vartheta). \quad (12.20)$$

12.3.3
Orthogonal Electron Frame Incidence

A general Lorentz transformation from a frame with no subscript to a frame with subscript "0" moving with velocity $\mathbf{v}$, can be expressed as [2]

$$\begin{aligned} \mathbf{p}_0 &= \mathbf{p} + \mathbf{v}\gamma\left(\frac{\gamma}{\gamma+1}\mathbf{v}\cdot\mathbf{p} - \mathcal{E}\right) \\ \mathcal{E}_0 &= \gamma(\mathcal{E} - \mathbf{v}\cdot\mathbf{p}). \end{aligned} \quad (12.21)$$

By choosing $\mathbf{v}$ to be the laboratory electron velocity, and identifying quantities in the electron rest system by subscript "0", these equations serve to translate four-vectors in the lab system to the electron rest system. When applied to the electron's four-momentum this transformation yields, $\mathbf{p}_0 = 0, \mathcal{E}_0 = m$, as it should. The incident photon four-momentum in the electron rest system is given by

$$\begin{aligned} \mathbf{k}_0 &= \mathbf{k} + \frac{p\omega}{m}\left(\frac{\mathbf{p}\cdot\hat{\mathbf{k}}}{\mathcal{E}+m} - 1\right), \\ \omega_0 &= \frac{\omega}{m}(\mathcal{E} - \mathbf{p}\cdot\hat{\mathbf{k}}) = \omega\gamma(1 - v\cos\alpha). \end{aligned} \quad (12.22)$$

and the scattered photon four-momentum in the electron rest system is given by

$$\mathbf{k}'_0 = \mathbf{k}' + \frac{\mathbf{p}\omega'}{m}\left(\frac{\mathbf{p}\cdot\hat{\mathbf{k}}'}{\mathcal{E}+m} - 1\right),$$

$$\omega'_0 = \frac{\omega'}{m}(\mathcal{E} - \mathbf{p}\cdot\hat{\mathbf{k}}') = \omega'\gamma(1 - v\cos\vartheta''), \quad (12.23)$$

where ϑ'' is the angle shown in Figure 12.1.

Problem 12.3.1 *These formulas can be used to find, for example, the incident lab photon angle α that yields an incident photon angle (in the electron's rest frame) normal to $\hat{\mathbf{x}}$ (the common axis of the transformation). Show that*

$$\cos\alpha = v = \sqrt{1 - \frac{m^2}{(m+K)^2}}, \quad (12.24)$$

where K is the electron kinetic energy.

Parameters corresponding to orthogonal electron frame operation are shown in Table 12.2 and plotted in Figure 12.3.

A possible diagnostic application, using a visible laser, of the configuration of the previous problem, would exploit the polarization dependence of Compton scattering. The laser electric field component in the x, y plane would be directed along the electron's direction of motion, and would therefore give no radiation along the electron's direction. For electron orbits lying not quite in the x, y plane there *would* be radiation along the x-axis. This could be the basis of a determination of the angular spread of an electron beam. For fully relativistic motion Eq. (12.24) yields $\alpha \approx 1/\gamma$, but the exact relation is almost as simple.

12.3.3.1 Orthogonal Electron Frame in, Parallel to Electron out

With incident photon satisfying the orthogonal electron frame condition of the first section we can find the energy of a photon radiated parallel to the electron direction. Conditions specifying this configuration are

$$\cos\alpha = v, \quad \vartheta = \alpha, \quad \varphi = 0. \quad (12.25)$$

Problem 12.3.2 *Using Eq. (12.10), define laboratory conditions such that the photon's approach angle in the electron's rest frame is at right angle to flight path and the photon scatters parallel to the electron. Assuming $\omega \ll m$ show that*

$$\frac{\omega}{\omega'} = 4 - 2\sqrt{3} = 0.5359. \quad (12.26)$$

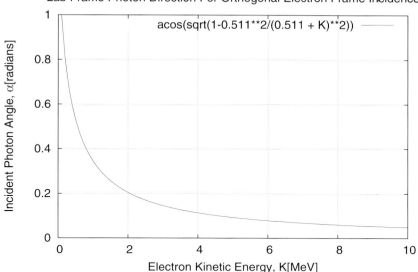

Fig. 12.3 With the horizontal cordinate being the incident electron kinetic energy K in MeV, the vertical coordinate is the incident photon angle α in the laboratory that yields orthogonal incidence in the electron's rest frame, as given by Eq. (12.24).

Tab. 12.2 Laboratory photon direction giving orthogonal electron frame photon incidence for a few values of electron kinetic energy K, where $\mathcal{E} = m + K$. Entries in the columns labeled x contain values of the kinematic variable x appearing in Eqs. (12.35) and (12.38).

Kinetic energy	γ	Incident angle, α	$\omega = 2\,\mathrm{eV}$	$\omega = 4\,\mathrm{eV}$
MeV		degrees	x	x
0	1	90	7.8×10^{-6}	1.56×10^{-5}
0.1	1.196	56.75	6.5×10^{-6}	1.31×10^{-5}
0.2	1.391	45.95	5.6×10^{-6}	1.13×10^{-5}
0.5	1.978	30.36	3.9×10^{-6}	0.79×10^{-5}
1.0	2.957	19.77	2.6×10^{-6}	0.53×10^{-5}
2.0	4.914	11.74	1.59×10^{-6}	0.318×10^{-5}
5.0	10.78	5.32	0.726×10^{-6}	0.145×10^{-5}
10.0	20.57	3.277	0.381×10^{-6}	0.076×10^{-5}

In visual terms, a red incident photon will convert to a soft ultraviolet photon scattering parallel to the electron beam axis. A schematic of a diagnostic apparatus based on this Compton scattering configuration has been shown in

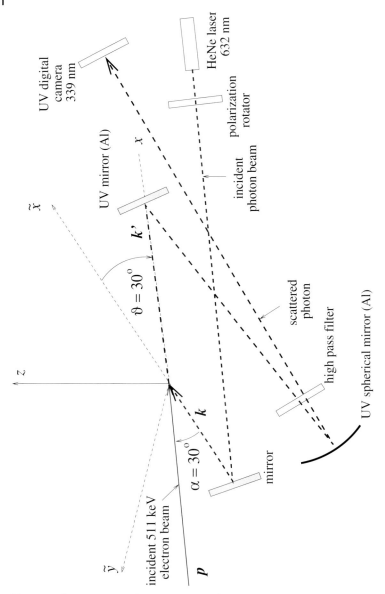

Fig. 12.4 Schematic of a (nondestructive) Compton scattering or "laser wire camera" to be used to measure the distributions of a low energy electron beam. The configuration has been arranged to give 90° incidence angle in the electron's rest frame. A steering element that prevents the electron beam from striking the detection apparatus is not shown.

Figure 12.4. The spherical mirror produces an image of the transverse electron beam profile by focusing the Compton-produced ultraviolet. By symmetry the color of the scattered radiation is insensitive to the vertical angle distribution of the electron beam. But the ultraviolet wavelength depends on horizontal electron angle, making it possible, in principle, to measure the electron angular divergence by measuring the wavelength spread of the scattered UV.

Problem 12.3.3 *For the configuration shown in Figure 12.1, show that the electron frame angle of incidence α_0 is related to the laboratory frame angle α by the relation*

$$\tan \alpha_0 = \frac{\sin \alpha}{\gamma(\cos \alpha - v)}. \qquad (12.27)$$

and that ω_0, the photon frequency in the electron rest frame, is given by

$$\omega_0 = \gamma \omega (1 - v \cos \alpha). \qquad (12.28)$$

Write formulas, inverse to these, that give angle and energy in the lab frame in terms of angle and energy in the electron rest frame.

Problem 12.3.4 *Because of the double valued nature of inverse trigonometric functions, it is useful (especially in the vicinity of $v = \cos \alpha$) to corroborate angle transformation results with more than one formula. Show that the angles are also related by*

$$\cos \alpha_0 = \frac{\cos \alpha - v}{1 - v \cos \alpha}. \qquad (12.29)$$

Problem 12.3.5 *Conservation of energy and momentum in the electron's rest system can be expressed by the four-vector equation*

$$\underline{k_0} + \underline{p_0} = \underline{k_0}' + \underline{p_0}'. \qquad (12.30)$$

In this frame $\underline{p_0} = (m, 0)$ is especially simple. Starting by forming the invariant product of left and right sides of Eq. (12.30), and then equating the results, show that the frequency of the scattered photon (in the electron's rest frame) is given by

$$\omega_0' = \frac{\omega_0}{1 + (1 - \hat{\mathbf{k}}_0' \cdot \hat{\mathbf{k}}_0)\omega_0/m}. \qquad (12.31)$$

This, followed by use of the inverse transformations of the previous problem, enables determination of laboratory frame energy and direction of the scattered photon. For Thomson scattering the second term in the denominator of Eq. (12.31) can be neglected. [This and the previous two problems are due to Ken Finkelstein.]

Problem 12.3.6 *The electron beam diagnostic mentioned in the text and illustrated in Figure 12.4 can, potentially, be used to measure the angular width of the electron beam by measuring the wavelength spread of the scattered light. The parameters are* adjusted for perpendicular incidence in the electron's rest frame and emission parallel to the electron beam. As a start at investigating the sensitivity that can be expected, show that

$$\frac{\partial}{\partial \alpha}\left(\frac{\omega}{\omega'}\right) = \frac{\gamma^2}{2}\frac{\omega}{\omega'}\sin 2\alpha. \qquad (12.32)$$

Also evaluate $\partial/\partial\vartheta(\omega/\omega')$ and show (numerically, for visible incident laser) that the ratio ω/ω' is far less sensitive to ϑ than to α.

12.4
Total Compton Cross Section

The total laboratory Compton scattering cross section for photons incident on an electron at rest is well known, especially in the limit in which $\omega \ll m$, where the cross section is known simply as the "Thomson Cross Section";

$$\sigma_T = \frac{8\pi r_e^2}{3}, \quad \text{where} \quad r_e = \frac{e^2}{4\pi\epsilon_0 mc^2} \qquad (12.33)$$

is the classical electron radius. To calculate scattering rates in a diagnostic set-up in which a laser beam illuminates a relativistic electron beam from the side, one needs the effective cross section orthogonal to the laser beam's direction. In this frame, which has axes differentiated by tildes, and is shown in Figure 12.4, the relativistic invariants are the same as in Eqs. (12.7).

BLP give a formula for the differential Compton scattering cross section, expressed in terms of dimensionless invariant quantities;

$$d\sigma = 8\pi r_e^2 \frac{dt}{x^2 m^2}\left(\left(\frac{1}{x}-\frac{1}{y}\right)^2 + \left(\frac{1}{x}-\frac{1}{y}\right) + \frac{1}{4}\left(\frac{x}{y}+\frac{y}{x}\right)\right), \qquad (12.34)$$

where

$$x = \frac{s-m^2}{m^2} = \frac{2\omega(\mathcal{E}-p\cos\alpha)}{m^2},$$
$$y = \frac{m^2-u}{m^2} = \frac{2\omega'(\mathcal{E}-p\cos\vartheta\cos\alpha - p\sin\vartheta\cos\varphi\sin\alpha)}{m^2}. \qquad (12.35)$$

For a detection apparatus that accepts only a limited range of azimuth one can replace Eq. (12.34) by

$$d^2\sigma = 8\pi r_e^2 \frac{1}{x^2 m^2}\left(\left(\frac{1}{x}-\frac{1}{y}\right)^2 + \left(\frac{1}{x}-\frac{1}{y}\right) + \frac{1}{4}\left(\frac{x}{y}+\frac{y}{x}\right)\right) dt \frac{d\varphi_\parallel}{2\pi}. \qquad (12.36)$$

The extra factor $d\varphi_\parallel$ restricts the azimuthal range under the hypothesis that the scattering is azimuthally symmetric about the common axis (which will be true for unpolarized beams, but may not be in the presence of polarized beams or polarization sensitive detectors). Since $\varphi_\|$ is preserved in transforming between collinear frames, Eq. (12.36), like Eq. (12.34), is relativistically invariant.

This equation is still not completely general as regards frame of reference. It describes Compton scattering only in frames in which the photon and electron paths are collinear. The most important applications of the theory, such as gamma ray scattering from electrons at rest, back-scattered laser radiation, and undulator radiation, satisfy this condition. There is, however, a configuration of recent interest, namely the "laser wire", in which the photon beam approaches the electron beam at some arbitrary skew angle. This configuration will be discussed in the concluding section of this chapter.

Returning to Eq. (12.34), for visible photons, except at extraordinarily high electron energy, one will have $x \ll 1$. To find the total cross section it is convenient to change the variable of integration from t to u, by using Eq. (12.8) and exploiting the constancy of s. The integration boundaries have earlier been shown to be $t = 0$ and $su = m^4$. The total cross section is therefore given by

$$\sigma = \frac{8\pi r_e^2}{3} \frac{3}{x^2} \int_{x/(x+1)}^{x} \left(\left(\frac{1}{x} - \frac{1}{y}\right)^2 + \left(\frac{1}{x} - \frac{1}{y}\right) + \frac{1}{4}\left(\frac{x}{y} + \frac{y}{x}\right) \right) dy. \quad (12.37)$$

After integration, this yields

$$\sigma = \frac{8\pi r_e^2}{3} \frac{3}{4x} \left(\left(1 - \frac{4}{x} - \frac{8}{x^2}\right) \log(1+x) + \frac{1}{2} + \frac{8}{x} - \frac{1}{2(1+x)^2} \right) \approx \sigma_T (1-x), \quad (12.38)$$

where the final approximation is valid for $x \ll 1$. Normally the $1-x$ factor can even be set to 1 for visible laser radiation and practical electron energies.[1] The leading factor in Eq. (12.38), $8\pi r_e^2/3$, has been recognized to be the Thompson cross section σ_T. This formula is applicable when the photon energy in the electron rest frame is negligible relative to electron mass m. Values of x are given for visible incident photons $\omega = 2\,\text{eV}$, and $\omega = 4\,\text{eV}$ for the various electron energies given in Table 12.2.

12.5
The Photon Beam Treated as an Electromagnetic Wave

Especially if it is produced by a laser, an incident photon beam can be represented as an electromagnetic wave rather than as a collection of photons.

---

[1] As it happens, because of cancellation of leading terms, the exact formula (12.38) needs to be evaluated to extended precision to produce the correct result for the extremely small values of x typical for photons that are visible in the lab.

Consistent with this classical treatment, the velocity of light will be expressed as c rather than as 1 in this section. All quantities will be evaluated in the laboratory system, in which the electrons travel along the x-axis and the photons along the $\tilde{x}$-axis, an axis that is also designated by the wave (unit) vector $\hat{\mathbf{k}}$.

An electron in the electron beam, as it passes through the electromagnetic wave, feels an oscillating force not unlike the force on an electron passing through an undulator magnet. The wavelength of a light beam is very much shorter than the wavelength of an undulator but, otherwise, the orbits are similar. In this section the motion of the electron is derived. Radiation due to this motion can be calculated using the conventional undulator formalism described in Chapter 7.

There is one inescapable difference between the force caused by an an undulator magnet and that caused by an electromagetic wave: unlike the undulator, there is no frame of reference in which the electric field of a traveling wave vanishes. This correlates with the fact that energy transfer between particle and field is possible for a traveling wave, but not for a pure magnetic field. This energy transfer is fundamental to the operation of linear accelerators and free electron lasers, but it will be seen to be inessential for the small electron angular excursions that are typical for a classical visualization of Compton scattering.

12.5.1
Determination of the Electron's Velocity Modulation

Since the electron is highly relativistic, it is essential for relativistically valid formulas to be used. Fortunatately, exact equations of motion are known for motion of a charged particle in an electromagnetic wave [3].

For an electromagnetic plane wave traveling in direction $\hat{\mathbf{k}}$, the electric and magnetic fields are related by

$$\mathbf{B} = \frac{1}{c}\hat{\mathbf{k}} \times \mathbf{E}, \quad \text{and} \quad \mathbf{B} \cdot \hat{\mathbf{k}} = \mathbf{E} \cdot \hat{\mathbf{k}} = 0. \tag{12.39}$$

For a monochromatic wave of frequency ω, dependences on both position $\mathbf{r}$ and time t can be expressed in terms of a single independent (phase) variable

$$\Phi = \omega\left(t - \frac{\hat{\mathbf{k}} \cdot \mathbf{r}}{c}\right). \tag{12.40}$$

The position vector $\mathbf{r}$ will be taken to be the position of a particle moving with velocity $\mathbf{v} = d\mathbf{r}/dt$, in which case

$$\frac{d\Phi}{dt} = \omega\left(1 - \frac{\hat{\mathbf{k}} \cdot \mathbf{v}}{c}\right). \tag{12.41}$$

In effect, the variable Φ locates the particle relative to the wave by giving the instantaneous longitudinal projection of the particle position onto the wave axis. One configuration of interest has the wave traveling almost anti-parallel to a relativistic electron, $\hat{\mathbf{k}} = -\hat{\mathbf{v}}$, in which case $\Phi \approx 2\omega t$.

The mechanical energy γmc^2 of an electron of velocity $\mathbf{v}$ is governed by

$$\frac{d\gamma}{dt} = \frac{e}{mc^2} \mathbf{E} \cdot \mathbf{v}, \tag{12.42}$$

and Newton's equation is

$$\frac{d}{dt}\left(\gamma \frac{\mathbf{v}}{c}\right) = \frac{e}{mc}(\mathbf{E} + \mathbf{v} \times \mathbf{B}) = \frac{e}{mc}\left(1 - \frac{\hat{\mathbf{k}} \cdot \mathbf{v}}{c}\right)\mathbf{E} + \frac{e}{mc^2}(\mathbf{E} \cdot \mathbf{v})\,\hat{\mathbf{k}}. \tag{12.43}$$

Problem 12.5.1 *Use Eqs. (12.42) and (12.43) to evaluate $d\mathcal{L}/dt$, where*

$$\mathcal{L} = \gamma\left(1 - \frac{\hat{\mathbf{k}} \cdot \mathbf{v}}{c}\right), \tag{12.44}$$

and use the result to show that $\mathcal{L}$ is a constant of the motion.

In what follows the constancy demonstrated in the previous problem will be exploited by replacing the right-hand side, when it appears, by the constant $\mathcal{L}$. For exactly anti-parallel, fully relativistic motion $\mathcal{L} \approx 2\gamma$.

We now set about changing the independent variable from t to Φ in the electron's equation of motion, using primes to indicate $d/d\Phi$. Then, using Eq. (12.41),

$$\mathbf{r}' = \frac{d\mathbf{r}/dt}{d\Phi/dt} = \frac{\mathbf{v}}{\omega(1 - \hat{\mathbf{k}} \cdot \mathbf{v}/c)}. \tag{12.45}$$

Differentiating again yields

$$\mathbf{r}'' = \frac{1}{\omega^2(1 - \hat{\mathbf{k}} \cdot \mathbf{v}/c)} \frac{d}{dt}\left(\frac{\mathbf{v}}{1 - \hat{\mathbf{k}} \cdot \mathbf{v}/c}\right). \tag{12.46}$$

Problem 12.5.2 *Using Eq. (12.46), after first substituting from Eq. (12.44), show that the left-hand side of Eq. (12.43) can be rewritten as*

$$\frac{d}{dt}\left(\gamma \frac{\mathbf{v}}{c}\right) = \mathcal{L} \frac{d}{dt}\left(\frac{\mathbf{v}/c}{1 - \hat{\mathbf{k}} \cdot \mathbf{v}/c}\right) = \omega^2 \mathcal{L}\left(1 - \hat{\mathbf{k}} \cdot \frac{\mathbf{v}}{c}\right)\mathbf{r}''. \tag{12.47}$$

and, again using Eq. (12.45), that the right-hand side of the same equation is

$$\left(1 - \hat{\mathbf{k}} \cdot \frac{\mathbf{v}}{c}\right)\frac{e}{m}\left(\mathbf{E} + \frac{\omega}{c}\mathbf{E} \cdot \mathbf{r}'\,\hat{\mathbf{k}}\right). \tag{12.48}$$

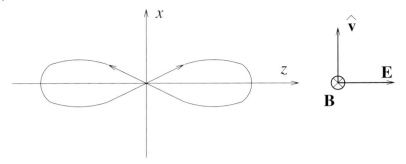

Fig. 12.5 Motion of electron, relative to its average motion, in a plane polarized, electromagnetic plane wave. The figure is distorted, in the sense that, typically, the longitudinal motion (along x) is far less than the transverse motion (along z).

As a result of the previous problem the common factor $1 - \hat{\mathbf{k}} \cdot \mathbf{v}/c$ can be cancelled, and the equation of motion becomes

$$\mathbf{r}'' = \frac{e}{\mathcal{L} m \omega^2}\left(\mathbf{E} + \frac{\omega}{c} \mathbf{E} \cdot \mathbf{r}' \hat{\mathbf{k}}\right). \tag{12.49}$$

With Φ as independent variable, this equation gives a kind of "acceleration". The equation is exact relativistically, except for not including the radiation reaction force. We know that the only important effect of this force of reaction is the "slowing down" required by energy conservation, which is unimportant in the present context.

Clemmow and Dougherty [3] solve Eq. (12.49) in the electron's rest frame and show that the electron's motion follows the "figure 8" pattern shown in Figure 12.5. This pattern results because the longitudinal oscillation frequency is twice the transverse frequency, which is ω. As the figure caption states, this figure exaggerates the longitudinal amplitude which, in practical cases, is much less than the transverse amplitude.

Problem 12.5.3 *Perform the calculation just mentioned, describing the electron's motion in its own (average) rest frame, and assuming the electron's velocity remains small compared to c, to obtain the result illustrated in Figure 12.5. In particular, calculate the ratio of longitudinal amplitude divided by the transverse amplitude.*

We intend to solve Eq. (12.49) in the laboratory system, while intentionally neglecting the longitudinal velocity modulation caused by the interchange of energy between wave and electron.

For the special case in which the wave is incident at angle α (see Figure 12.4), with electric field in the x, y plane, the (x, y, z) components of $\hat{\mathbf{k}}$ and $\mathbf{E}$, and

$\mathbf{E} \cdot \mathbf{r}'$ are

$$\hat{\mathbf{k}} = (\cos\alpha, \sin\alpha, 0),$$
$$\mathbf{E} = E_{\tilde{y}}(-\sin\alpha, \cos\alpha, 0)\cos\Phi, \qquad (12.50)$$
$$\mathbf{E} \cdot \mathbf{r}' = E_{\tilde{y}}(-x'\sin\alpha + y'\cos\alpha)\cos\Phi.$$

The electron's equations of motion are

$$\frac{dx'}{d\Phi} = \frac{eE_{\tilde{y}}}{\mathcal{L}m\omega^2}\left(-\sin\alpha + \frac{\omega}{c}(-x'\sin\alpha + y'\cos\alpha)\cos\alpha\right)\cos\Phi,$$
$$\frac{dy'}{d\Phi} = \frac{eE_{\tilde{y}}}{\mathcal{L}m\omega^2}\left(\cos\alpha + \frac{\omega}{c}(-x'\sin\alpha + y'\cos\alpha)\sin\alpha\right)\cos\Phi, \qquad (12.51)$$
$$\frac{dz'}{d\Phi} = 0.$$

Problem 12.5.4 *Treating the velocity-like quantities x' and y' on the right-hand sides of these equations as constant (which amounts to averaging the electron's motion and will be tantamount to retaining only electron response at first harmonic of $d\Phi/dt$) and using Eq. (12.45), show that the solutions to these equations are*

$$x' = \frac{v}{\omega(1 - \cos\alpha\, v/c)}$$
$$+ \frac{eE_{\tilde{y}}}{\mathcal{L}m\omega^2}\left(-\sin\alpha + \frac{\omega}{c}(-<x'>\sin\alpha + <y'>\cos\alpha)\cos\alpha\right)\sin\Phi,$$
$$y' = \frac{eE_{\tilde{y}}}{\mathcal{L}m\omega^2}\left(\cos\alpha + \frac{\omega}{c}(-<x'>\sin\alpha + <y'>\cos\alpha)\sin\alpha\right)\sin\Phi. \qquad (12.52)$$

The z equation has been dropped.

The x' component just found, being parallel to the electron's average velocity, causes negligible radiation. But the y-motion gives velocity and acceleration components transverse to the electron's motion. This is the only important source of radiation that is caused by the $E_{\tilde{y}}$ laser electric field component.

The component E_z of the incident laser field also leads to radiation because it causes transverse acceleration of the electron;

$$\frac{dz'}{d\Phi} = \frac{eE_z}{\mathcal{L}m\omega^2}\cos\Phi, \qquad (12.53)$$

Problem 12.5.5 *After solving Eq. (12.53), again using Eq. (12.44) and (12.45), and retaining only transverse velocity components, obtain the following formulas for the transverse velocity of the electron;*

$$v_y = \frac{eE_{\tilde{y}}}{\gamma m\omega}\left(\cos\alpha - \frac{\sin\alpha\, v/c}{1 - \cos\alpha\, v/c}\right)\sin\Phi, \quad v_z = \frac{eE_z}{\gamma m\omega}\sin\Phi. \qquad (12.54)$$

In Chapter 7, which analyses undulator radiation, these formulas are seen to be directly applicable to the determination of the frequency spectrum of undulator radiation.

Problem 12.5.6 Use the result of the previous problem, along with Eq. (7.27), to obtain the Fourier transform $\tilde{E}_y(\omega)$ of the radiation emitted from the electron because of its transverse motion. Because the wavelength is so short it is surely valid to neglect the t_r^2 term in the exponent.

Co-traveling electron and photon beams. When an electron beam is traveling parallel to, or almost parallel to, an intense electromagnetic wave, it may be necessary to account for energy transfer between electrons and wave. As an electron acquires a velocity component parallel to the electric field of the wave, the electron's energy varies according to

$$\frac{d\gamma}{d\Phi} = \frac{e}{mc^2} \cdot \mathbf{v'} = \frac{1}{\mathcal{L}} \left(\frac{eE}{mc\omega}\right)^2 \sin\Phi, \quad \gamma = -\frac{1}{\mathcal{L}} \left(\frac{eE}{mc\omega}\right)^2 \cos\Phi + \gamma_0.$$
(12.55)

Any decrease in electron energy must correspondingly increase the electromagnetic beam energy. This is known as *free electron laser* radiation. If the wave increases the electron energy (at the expense of its own intensity) it is known as an *inverse free electron laser*. Whether the electron gains or loses energy depends on where it rides on the wave (i.e. its phase). The net acceleration of the electron, though it can be appreciable, is invariably limited by the fact that the electron eventually falls out of phase with the wave. Then acceleration converts to deceleration and vice versa. This limits the useful length that such a wave (regarded as an particle accelerator) can have. For a wave longer than this the longitudinal acceleration averages to zero. In a linear accelerator these considerations are overcome by using a mode having longitudinal electric field, with its phase velocity "slowed" to match the electron velocity.

This coherent energy transfer between wave and particle is insignificant when the electron beam and electromagnetic wave are approximately antiparallel, as applies to the production of high energy photons by the backscattering of photons in a laser beam. In this case the phase averaging occurs almost instantaneously. The energy acquired by the scattered photon is incoherently extracted from one electron.

12.5.2
Undulator Parametrization of Electron Motion in a Wave

The undulator constant K of an ordinary magnetic undulator is defined by the formula

$$\Theta = \frac{K}{\gamma},$$
(12.56)

where Θ is the maximum deviation angle of the electron relative to its average orbit. (For mnemonic one can remember K as being the maximum angle of the orbit expressed in units of $1/\gamma$.) This same formula can be used to re-express Eqs. (12.54) as "effective K values" for the electromagnetic wave;

$$(K_{y,\text{eff.}}, K_{z,\text{eff.}}) = \frac{\lambda}{2\pi} \frac{c}{\langle v \rangle} \left(\frac{E_{\tilde{y}}}{mc^2/e} \left(\cos \alpha - \frac{\sin \alpha \, v/c}{1 - \cos \alpha \, v/c} \right), \frac{E_z}{mc^2/e} \right), \quad (12.57)$$

where λ is the wavelength of the laser radiation.

With a_γ being the width of the laser beam, one can also define an effective number of undulator periods $N_{w,\text{eff.}}$ by keeping track of the number of phase reversals as an electron passes through the laser beam.

Problem 12.5.7 *Find the time t_{ib} an electron of speed v and angle α spends traveling through an electromagnetic wave of width a_γ. Defining "wavefronts" as planes orthogonal to the wave direction and separated by λ, find the number of wavefronts the electron passes (or rather is passed by) while it is in the beam. Show that this number is given by*

$$N_{w,\text{eff.}} = \frac{a_\gamma}{\sin \alpha} \frac{c/v - \cos \alpha}{\lambda}, \quad (12.58)$$

and the effective undulator period length is

$$\lambda_{w,\text{eff.}} = \frac{\lambda}{c/v - \cos \alpha}, \quad (12.59)$$

Laser field in terms of laser power. It is usually the power of a laser beam, rather than its electric field amplitude that is known. For a waveguide mode having wavelength λ in a waveguide of width a_γ and height b_γ, the maximum electric field is related to the beam power and other guide parameters by

$$\mathcal{P} = \frac{|E_{\max}|^2}{Z_0} \sqrt{1 - \left(\frac{\lambda}{2a_\gamma}\right)^2} \frac{a_\gamma b_\gamma}{2}, \quad (12.60)$$

where $Z_0 = \sqrt{\mu_0/\epsilon_0} = 1/(\epsilon_0 c)$ is the impedance of free space. For a highly "overmoded" wave one has $\lambda \ll 2a_\gamma$ and the square root term in Eq. (12.60) is well approximated by 1. Since a laser beam is very much like an overmoded waveguide wave, Eq. (12.60) can be applied to the laser beam in our Compton scattering configuration. Using this result in the z-component of Eq. (12.57), for a laser having only a z-directed electric field, the laser's effective undulator constant $K_{\text{eff.},z}$ can be expressed as

$$K_{z,\text{eff.}}^2 = \frac{\lambda^2}{4\pi^2} \frac{c^2}{v^2} \frac{1}{(mc^2/e)^2} \frac{2Z_0}{a_\gamma b_\gamma} \mathcal{P}_{\text{laser}}. \quad (12.61)$$

Having worked out $N_{w,\text{eff.}}$ and $K_{\text{eff.}}$, one can treat the laser beam as an undulator and (later) use standard undulator formulas to calculate the radiation from an electron as it passes through a laser beam. The factors $N_{w,\text{eff.}}$, $\lambda_{w,\text{eff.}}$, and (for example) $K_{z,\text{eff.}}$ enter scattering rate formulas as the product

$$\frac{N_{w,\text{eff.}} K_{z,\text{eff.}}^2}{\lambda_{w,\text{eff.}}} = \frac{1}{2\pi^2} \frac{c^2}{v^2} \frac{1}{b_\gamma \sin\alpha} \left(\frac{c}{v} - \cos\alpha\right)^2 \frac{Z_0 \mathcal{P}_{\text{laser}}}{(mc^2/e)^2}. \tag{12.62}$$

Except for a numerical example at the end of this chapter, the derivation of the radiation from an undulator, using classical electromagnetism, is discussed in Chapter 7.

Microwave Undulators, Laser Back-Scattering and Photon Colliders. In a *microwave undulator* electrons travel parallel to the axis of a waveguide in the presence of a waveguide mode. The peak electric field can be calculated from Eq. (12.60) and the other undulator parameters can be obtained similarly. Then the observed radiation (except for different parameters, such as shorter undulator period λ_w) is just like that coming from a magnetic undulator. The same comment can be made about beams produced by *laser back-scattering*, typically using a visible light laser. But, in this case, the value of λ_w can be very short, with the consequence that the fundamental undulator resonance energy can be hundreds of MeV, or even more. Taken at face value, the kinematic formulas seem to show, in this case, that the scattered photon energy can exceed the energy of the incident electron. This is, of course, not possible; it would just indicates that the formulation is being employed outside its region of validity. A correct quantum calculation is then required. This regime is of interest for the so-called *photon colliders* that are associated with high-energy electron–positron linear colliders such as the ILC. There is a detailed analysis of this regime in the TESLA proposal [4].

12.6
Undulator Fields in Electron Rest Frame

Just as Compton scattering can be understood as undulator radiation, undulator radiation can be understood in terms of Compton scattering. It must first be demonstrated that the undulator fields in the electron rest frame resemble the fields in an electromagnetic wave.

An electron propagates at positive velocity v toward the origin along the negative x-axis. Unlike usage up to this point, *single primes will indicate coordinates in the electron rest frame* In this frame an undulator magnet fixed in the laboratory appears to be approaching with velocity $-v$ along the positive x-axis. As the magnet passes the electron, the electron radiates photons. After working out the distributions of these photons in the electron's frame, the distributions will be transferred back into the laboratory frame. Double primes will be used to indicate coordinates of the scattered photon in the laboratory frame.

12.6.1
Some Formulas from Special Relativity [7]

12.6.1.1 Lorentz Transformation

For motion along the common x-axis, the Lorentz transformation from the (unprimed) laboratory frame to the electron's frame (primed), and its inverse, are:

$$\begin{pmatrix} x' \\ y' \\ z' \\ ct' \end{pmatrix} = \begin{pmatrix} \gamma(x - \beta ct) \\ y \\ z \\ \gamma(-\beta x + ct) \end{pmatrix}, \quad \begin{pmatrix} x \\ y \\ z \\ ct \end{pmatrix} = \begin{pmatrix} \gamma(x' + \beta ct') \\ y' \\ z' \\ \gamma(\beta x' + ct') \end{pmatrix}, \quad (12.63)$$

where $\beta = |v|/c$ and $\gamma = 1/\sqrt{1-\beta^2}$.

12.6.1.2 Energy and Momentum

Energy and momentum components transform as in Eq. (12.63) after the replacement $x \to p_x, y \to p_y, z \to p_z, ct \to \mathcal{E}/c$,

$$\begin{pmatrix} p'_x \\ p'_y \\ p'_z \\ \mathcal{E}'/c \end{pmatrix} = \begin{pmatrix} \gamma(p_x - \beta\mathcal{E}/c) \\ p_y \\ p_z \\ \gamma(-\beta p_x + \mathcal{E}/c) \end{pmatrix}, \quad \begin{pmatrix} p_x \\ p_y \\ p_z \\ \mathcal{E}/c \end{pmatrix} = \begin{pmatrix} \gamma(p'_x + \beta\mathcal{E}'/c) \\ p'_y \\ p'_z \\ \gamma(\beta p'_x + \mathcal{E}/c') \end{pmatrix}. \quad (12.64)$$

12.6.1.3 Transformation of E and B

Between the same frames the electric and magnetic field transformations are

$$E'_x = E_x, \quad E'_y = \gamma(E_y - vB_z), \quad E'_z = \gamma(E_z + vB_y),$$
$$B'_x = B_x, \quad B'_y = \gamma(B_y + \frac{v}{c^2}E_z), \quad B'_z = \gamma(B_z - \frac{v}{c^2}E_y). \quad (12.65)$$

If B_z is the only nonvanishing field component in the laboratory frame then, in the frame of an electron moving with velocity $\mathbf{v} = v\hat{x}$ the fields are

$$E'_y = -\gamma v B_z, \quad B'_z = \gamma B_z. \quad (12.66)$$

12.6.2
Treatment of an Undulator Magnet as an Electromagnetic Wave

Undulator radiation can be treated as Compton back-scattering of the "photons" that are "provided" by the wiggler magnet. Both $\hbar$ and c will be exhibited explicitly in the following formulas. Single primes will continue to indicate coordinates in the electron rest frame and double primes will indicate coordinates used when describing scattered photons in the laboratory frame. The usual Planck/deBroglie associations of frequency and wavelength with energy and momentum will be assumed. The frequency of a "photon" associated with an undulator at rest in the laboratory is zero (because the field is constant in time) and the photon wavelength is $\lambda_w \equiv 2\pi/k_w$, which is the period length of the periodic magnetic field of the wiggler/undulator. These quantities do not satisfy relation (12.1) between energy and momentum that is required for *real* particles. Such a "particle" is therefore said to be *virtual*.

For a horizontal-bending undulator aligned with the x-axis, the only non-vanishing field component is

$$\mathbf{B} = B_0 \cos(k_w x)\, \hat{\mathbf{z}}. \tag{12.67}$$

From the electron's point of view the undulator propagates at velocity $-v$ back toward the origin along the positive x-axis. Taking its arrival time at the origin to be 0, its laboratory equation of motion is $x = -vt$. Continuing to treat $\beta = v/c$ as a positive quantity and using Eq. (12.63), the Lorentz transformation equations between electron frame and lab frame coordinates are

$$x = \gamma(x' + \beta ct'), \quad ct = \gamma(\beta x' + ct'). \tag{12.68}$$

For the field given by Eq. (12.67) to be useful to an observer at rest in the electron's rest frame it is necessary to derive from it (using Eq. (12.65)) the electic and magnetic field components in the local frame. Furthermore it is necessary to express these components as functions of the (x', t') coordinates. The result is

$$\mathbf{E}' = -\gamma v B_0 \cos\left(k_w \gamma(x' + vt')\right) \hat{\mathbf{y}}', \quad \text{and} \quad \mathbf{B}' = -\frac{1}{v} \hat{\mathbf{x}}' \times \mathbf{E}'. \tag{12.69}$$

These are very nearly the same as the relations describing the electric and magnetic fields belonging to a plane-polarized, free space plane wave. In fact, in the limit $v \to c$ the correspondance becomes exact. Making the replacement $v = c$ and ascribing the force (which in this case is due to a magnet) to a traveling electromagnetic wave, is known, for example in elementary particle physics, as the Weiszäcker-Williams approximation.

The wave just derived is said to be made up of "virtual" photons, and these photons can Compton scatter off the electrons in an electron beam. The rest energy of one of these virtual photons is calculated most easily in the laboratory frame because its frequency vanishes there. Using Eq. (12.1), $|m_\gamma|$ the (magnitude of) the photons "rest mass" can be obtained from its energy and momentum;

$$|m_\gamma c^2| = \left|\sqrt{(\hbar \omega)^2 - (\hbar k_w)^2 c^2}\right| = \hbar k_w c. \tag{12.70}$$

Next consider this "photon" in the electron rest frame. One sees from the cosine's argument in Eq. (12.69) that the photon's frequency is $\omega' = k_w \gamma v$, so its energy is $\hbar k_w \gamma v$. The condition for this energy to be small compared to the electron rest energy is

$$\gamma (\hbar k_w c) \frac{v}{c} \ll mc^2. \tag{12.71}$$

The parenthesized expression here is the energy of a photon having wavelength equal to the undulator period—typically at least several centimeters. Even though γ can be a large number such as 10^4, one sees that inequality (12.71) will be amply satisfied for any feasible undulator magnet.

We should still investigate the potential influence of the virtual photon mass. For this mass (calculated in Eq. (12.70)) to be negligible would require its corresponding rest energy to be small compared to the photon energy (in the electron rest frame);

$$\hbar k_w c \ll \gamma \hbar k_w c \frac{v}{c}. \tag{12.72}$$

Since this reduces to $\gamma \gg 1$, it will be abundantly true in practice.

This discussion has validated treating the undulator as an electromagnetic wave and the further assumption, in the electron rest frame, that a scattered photon has the same energy as the incident photon. This is the condition assumed in the Thomson scattering calculation.

12.7
Classical Derivation of Thomson Scattering

12.7.1
Introduction

In this and the next section calculations are to be performed in the electron's rest frame, which has been indicated by primed symbols up to this point. Even in this frame two coordinate systems are "natural". The one having polar axis aligned with the undulator axis will continue to be represented by primes. The other has the polar axis aligned with the axis of the dipole moment that is induced by the undulator acting on the electron. This frame will be represented by overhead bars. Only angles will be differentiated in this way by primes and bars. The same Cartesian axes $\bar{x}$, $\bar{y}$, and $\bar{z}$, will be used for both orientations. The relation between area elements in the two frames,

$$d\bar{\vartheta} \sin \bar{\vartheta} \, d\bar{\varphi} = d\vartheta' \sin \vartheta' \, d\varphi'. \tag{12.73}$$

will be required shortly. Note the following notational inconsistency concerning the angles ϑ and φ—unlike the angles denoted by the same symbols in Figure 12.1, the present angles are referenced to the electron's line of travel.

Many of the following calculations have already been exhibited in fully relativistic fashion in earlier sections. Elementary, nonrelativistic calculations are adequate for the present purpose of calculating Thomson scattering.

12.7.2
Free Electron Oscillating in Electromagnetic Wave [5]

Consider an electron at rest except for electric field $\bar{E}_0 \cos \bar{\omega} \bar{t} \, \hat{\bar{z}}$. The electron oscillates along the $\bar{z}$-axis,

$$\bar{z} = \frac{-e/m}{\bar{\omega}^2} \bar{E}_0 \cos \bar{\omega} \bar{t}, \tag{12.74}$$

effectively making a time-varying dipole moment

$$\bar{p} = \frac{-e^2/m}{\bar{\omega}^2} \bar{E}_0 \cos \bar{\omega} \bar{t}. \tag{12.75}$$

Supposing that $\bar{E}_0$ here is due to an electromagnetic wave, incident from positive $\bar{x}$ toward negative $\bar{x}$, the Poynting vector of the wave is

$$\mathbf{S} = -c\epsilon_0 \bar{E}_0^2 \cos^2(\bar{k}\bar{x} + \bar{\omega}\bar{t}) \, \hat{\bar{x}}, \tag{12.76}$$

and the intensity (average power per unit area per unit time) is

$$\bar{\mathcal{I}} = \frac{1}{2} c\epsilon_0 \bar{E}_0^2. \tag{12.77}$$

Problem 12.7.1 *Show that the force due to the magnetic component of an electromagnetic wave acting on an electron that is at rest on the average can be neglected as long as the electron's oscillatory motion remains nonrelativistic.*

12.7.3
Electric Dipole Radiation [6]

Consider two time varying charges $q(t) = \pm q_0 \cos \bar{\omega} t$, placed at $\bar{z} = \pm d/2$. (They could be little conductors joined by a wire with current flowing back and forth between them, but it will be valid to neglect any magnetic field due to the current.) This makes a time-varying electric dipole moment $\bar{p}(t) = \bar{p}_0 \cos \bar{\omega} t$ where $\bar{p}_0 = q_0 d$.

The retarded potential, evaluated at field point $\mathbf{r}$, is easy to write;

$$V(\mathbf{r}, t) = \frac{q_0}{4\pi\epsilon_0} \left(\frac{\cos(\bar{\omega}(t - \mathcal{R}_+/c))}{\mathcal{R}_+} - \frac{\cos(\bar{\omega}(t - \mathcal{R}_-/c))}{\mathcal{R}_-} \right), \tag{12.78}$$

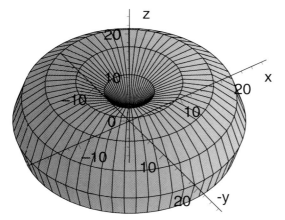

Fig. 12.6 Electric dipole intensity pattern of radiation from electric charge oscillating along the z-axis.

where $\mathcal{R}_+$ and $\mathcal{R}_+$ are the distances from the point charges to the observation point. In the radiation zone, $d \ll r, r \gg c/\bar{\omega}$, the radiated electric and magnetic fields, expressed in terms of polar unit vector $\hat{\vartheta}$ and azimuthal unit vector $\hat{\varphi}$, are

$$\bar{\mathbf{E}} = -\frac{\mu_0 \bar{p}_0 \bar{\omega}^2}{4\pi} \frac{\sin \bar{\vartheta}}{r} \cos\left(\bar{\omega}\left(t - \frac{r}{c}\right)\right) \hat{\vartheta}, \quad \bar{\mathbf{B}} = \frac{E}{c} \hat{\varphi}. \tag{12.79}$$

Integrating the Poynting vector over angles gives the total radiated power to be

$$\bar{\mathcal{P}} = \frac{\mu_0 \bar{p}_0^2 \bar{\omega}^4}{12\pi c}, \tag{12.80}$$

and the intensity pattern of the radiation is

$$\frac{d\bar{\mathcal{P}}}{d\Omega} = \frac{3\bar{\mathcal{P}}}{8\pi} \sin^2 \bar{\vartheta}, \quad \frac{d\bar{\mathcal{P}}}{d\bar{\vartheta} d\bar{\varphi}} = \frac{3\bar{\mathcal{P}}}{8\pi} \sin^3 \bar{\vartheta}. \tag{12.81}$$

This familiar "donut-shaped" radiation pattern is shown in Figure 12.6. There is no radiation along the $\bar{z}$-axis and the ring of maximum intensity lies in $\bar{x}, \bar{y}$ plane.

Using area relation (12.73) and the result of the next problem, the intensity pattern referred to a polar axis lying along the $\bar{x}$ (undulator) axis is

$$\frac{d^2 \mathcal{P}'}{d\vartheta' d\varphi'} = \frac{3\mathcal{P}'}{8\pi} \sin^2 \bar{\vartheta} \sin \vartheta' = \frac{3\mathcal{P}'}{8\pi} (1 - \sin^2 \vartheta' \sin^2 \varphi') \sin \vartheta' \tag{12.82}$$

Problem 12.7.2 *The derivation of Eq. (12.82) relies on the orthogonality of spherical coordinates. Using the fact that the same unit vector $\hat{\mathbf{n}}$ can be expanded using either* set of coordinates

$$\hat{\mathbf{n}} = \cos\vartheta'\,\hat{\mathbf{x}} + \sin\vartheta'\sin\varphi'\,\hat{\mathbf{y}} + \sin\vartheta'\cos\varphi'\,\hat{\mathbf{z}}$$
$$= \sin\bar{\vartheta}\sin\bar{\varphi}\,\hat{\mathbf{x}} + \sin\bar{\vartheta}\cos\bar{\varphi}\,\hat{\mathbf{y}} + \cos\bar{\vartheta}\,\hat{\mathbf{z}}, \qquad (12.83)$$

confirm the substitution for $\sin^2\bar{\vartheta}$ that was made in Eq. (12.82),

$$\sin\bar{\vartheta} = \sqrt{1 - \sin^2\vartheta'\sin^2\varphi'}, \qquad (12.84)$$

and calculate the Jacobian of the transformation explicitly to complete the derivation of Eq. (12.82) from Eq. (12.81).

To make Eq. (12.82) appear more homogeneous the replacement $\bar{\mathcal{P}} \to \mathcal{P}'$ has been made, even though these symbols refer to the same quantity. Especially when applying this formula to undulator radiation it will be useful, by integrating over φ', to find $d\mathcal{P}'/d\vartheta'$, which describes how the power is distributed in ϑ', or equivalently, in $\cos\vartheta'$. Performing the integration, the result is

$$\frac{d\mathcal{P}'}{\mathcal{P}'} = -\frac{3}{4}\left(1 - \frac{1}{2}\sin^2\vartheta'\right)d\cos\vartheta'. \qquad (12.85)$$

The expression on the left is "normalized" by definition, meaning that its integral over the complete range is equal to 1. Letting the radiated power be regarded as being made up of photons, since all the photons have the same energy, the total number of photons N' is given by $N' = \mathcal{P}'/\mathcal{E}'_\gamma$. Defining $N'_{\cos\vartheta'}(\cos\vartheta')\,d\cos\vartheta'$ to be the number of photons in range $d\cos\vartheta'$, we have

$$N'_{\cos\vartheta'}(\cos\vartheta') = N'\frac{3}{8}(1 + \cos^2\vartheta'), \qquad (12.86)$$

and

$$\frac{dN'}{d\cos\vartheta'} = \frac{\mathcal{P}'}{\mathcal{E}'_\gamma}\frac{3}{8}(1 + \cos^2\vartheta'), \qquad (12.87)$$

In this relation the minus sign has been suppressed, corresponding to the interval $d\cos\vartheta'$ being taken to be implicitly positive. Corresponding to the donut-shaped radiation pattern this distribution is greater near $\vartheta' = 0$ and $\vartheta' = \pi$ than near $\vartheta' = \pi/2$. But the distribution is not very strongly peaked and can be thought of qualitatively as more or less isotropic. The differential cross section in the forward $\theta' = 0$ direction is proportional to $3/(8\pi)$ which can be compared to the $1/(4\pi)$ factor that would apply in the case of isotropy.

12.7.4
Scattering Rate Expressed as a Total Cross Section

By conservation of energy, whatever power has been radiated by the oscillating electron must have been removed from the electromagnetic wave. This amount of power can be expressed as an area σ—the transverse area through which that amount of power flows through in the wave. Expressed in terms of intensity $\bar{\mathcal{I}}$,

$$\bar{\mathcal{I}}\sigma = \bar{\mathcal{P}}. \tag{12.88}$$

Substituting from Eqs. (12.80), (12.75) and (12.77) yields

$$\sigma_T = \frac{8\pi}{3}\left(\frac{e^2}{4\pi\epsilon_0 mc^2}\right)^2 = \frac{8\pi}{3}r_e^2 = 0.665 \times 10^{-28}\,\text{m}^2. \tag{12.89}$$

This is known as the "Thomson cross section" and the quantity in large parentheses is known as the "classical electron radius", $r_e = 2.81784 \times 10^{-15}$ m. If the incident wave is visualized as made up of photons, then the loss of forward-directed energy from the incident beam and the appearance of the more-or-less isotropically directed energy can be interpreted as "scattering" of the incident photons. The reason σ_T deserves the name "cross section" is that photons within that area can be regarded as the ones that have scattered. For low energy photons the Thomson cross section gives a good account of the interaction of light with free electrons. The polarization of the scattered light is also well described. The total cross section σ_T has been calculated in the electron rest frame, but in the lab frame (which might, for example, be the rest frame of an undulator magnet) the total cross section is the same.

From Eq. (12.85) the fractional emission into solid angle $\Delta\Omega'$ in the forward direction is

$$\frac{\Delta\sigma'}{\sigma_T} = \frac{3}{8}(1+\cos^2\vartheta')_{\vartheta'=0}\Delta\cos\vartheta'\,\frac{\Delta\varphi'}{2\pi}. \tag{12.90}$$

From this and Eq. (12.89) one obtains

$$\left.\frac{d\sigma'}{d\Omega'}\right|_{\vartheta'=0} = r_e^2. \tag{12.91}$$

12.8
Transformation of Photon Distributions to the Laboratory

12.8.1
Solid Angle Transformation

The energy and momentum components of the radiated photon in the electron rest system (single primes) and laboratory system (double primes) are related by the Lorentz transformation equations:

$$\mathcal{E}' \sin \vartheta' = \mathcal{E}'' \sin \vartheta''$$
$$\mathcal{E}' \cos \vartheta' = \gamma \mathcal{E}'' \left(\cos \vartheta'' - \frac{v}{c} \right) \tag{12.92}$$
$$\mathcal{E}' = \gamma \mathcal{E}'' \left(1 - \frac{v}{c} \cos \vartheta'' \right).$$

As in Eq. (12.29), the resulting angular transformation can be expressed as

$$\cos \vartheta' = \frac{\cos \vartheta'' - v/c}{1 - \cos \vartheta'' v/c}. \tag{12.93}$$

Differentiating this relation, and using the equality $d\varphi' = d\varphi''$, one obtains the solid angle transformation one obtains

$$\frac{d\Omega'}{d\Omega''} = \frac{d \cos \vartheta'}{d \cos \vartheta''} = \left(\frac{1/\gamma}{1 - \cos \vartheta'' v/c} \right)^2 = \left(\frac{(1 + \cos \vartheta'' v/c)/\gamma}{1 - \cos^2 \vartheta'' v^2/c^2} \right)^2. \tag{12.94}$$

The last manipulation seems to be an unsimplification. Its purpose is to show that, in the forward, $\vartheta'' = 0$, direction, the solid angle transformation is

$$\left. \frac{d\Omega'}{d\Omega''} \right|_{\vartheta''=0} = \gamma^2 \left(1 + \frac{v}{c} \right)^2 \tag{12.95}$$

12.8.2
Photon Energy Distribution in the Laboratory

Continuing the Thomson scattering representation of undulator radiation, to calculate the radiation pattern in the laboratory one starts from the angular distribution in the rest system of the electron and Lorentz transforms it. Though the scattered photons are monoenergetic in the electron rest system, the photon energy depends on direction in the laboratory system. We can write down the *maximum* lab energy directly, since it clearly results when the photon is exactly back-scattered. After back-scattering the photon energy/momentum four-vector is $\hbar(\gamma k_w v/c, \gamma k_w v/c, 0, 0)$. Transforming this to the lab frame, the frequency is

$$\omega_0'' = \left(1 + \frac{v}{c} \right) \gamma^2 k_w v \approx 2\gamma^2 \, 2\pi \frac{c}{\lambda_w}, \tag{12.96}$$

since $v \approx c$. This is just one of the several ways in which the wavelength of the fundamental, pseudo-monochromatic, undulator radiation peak can be shown to be equal to the undulator wavelength divided by $2\gamma^2$.

Since $\mathcal{E}'$ is independent of ϑ', the last of these equations yields $\mathcal{E}''$ as a function of ϑ''. For the small angles that are typically of interest, the result, expressed as frequencies, is

$$\omega''(\vartheta'') = \frac{\omega'/\gamma}{1 - \frac{v}{c}\cos\vartheta''}$$
$$\approx \frac{\omega'/\gamma}{1 - \left(1 - \frac{1}{2\gamma^2}\right)\left(1 - \frac{\vartheta''^2}{2}\right)} \approx \frac{2\gamma^2}{1 + \gamma^2\vartheta''^2}\frac{2\pi c}{\lambda_w}, \quad (12.97)$$

where Eq. (12.96) has been used, and is now extended to all (small) angles.

This dependence of energy on angle is identical to that given by calculation of undulator radiation by classical electromagnetic theory. This corroborates the essential equivalence of undulator radiation and Compton scattered radiation.

The discussion to this point has implicitly assumed $K \ll 1$, where K is the so-called "undulator parameter". The main effect of finite K is to cause the electron's average longitudinal velocity to be

$$\frac{<v_x>}{c} \approx 1 - \frac{1}{2\gamma^2}\left(1 + \frac{K^2}{2}\right), \quad <\gamma> \approx \frac{\gamma}{\sqrt{1 + K^2/2}}. \quad (12.98)$$

After this alteration, Eq. (12.97) becomes

$$\omega''(\vartheta'') = \frac{\omega'/\gamma}{1 - \left(1 - \frac{1}{2\gamma^2}\right)\left(1 - \frac{\vartheta''^2}{2}\right)} \approx \frac{\omega''(0)}{1 + K^2/2 + \gamma^2\vartheta''^2}. \quad (12.99)$$

This formula also agrees with standard undulator formulas (especially when the double primes are suppressed to indicate that transformation back to the laboratory frame has been completed). It shows that increasing K causes a reduction in peak x-ray energy by factor $1 + K^2/2$, and an increase in angular width by factor $\sqrt{1 + K^2/2}$.

It is important to recognize that the treatment of undulator radiation as Compton scattering is valid only for the fundamental undulator peak. As K is increased toward 1 or higher, higher harmonics become progressively more important and these can only be understood quantum mechanically as the coherent scattering of more than one photon. By contrast, in classical electromagnetic theory, higher undulator radiation modes can be calculated by straightforward Fourier analysis of the electron's motion, followed by application of classical radiation formalism applied to each Fourier amplitude of the electron's motion.

There is another refinement needed to obtain a good understanding of undulator radiation from the present point of view. Since any practical undulator has a finite number $2N_w$ of poles, the representation of the undulator field by a pure sinusoid in the electron's rest frame is not quite valid. The true field has to be represented as a Fourier transform having finite width in frequency.

As a result the energy spectrum of scattered photons in the electron rest frame is not quite monoenergetic. Their fractional energy width is of order $1/N_w$. As a result the photons observed at fixed angle in the laboratory, rather than being monochromatic, also have fractional width of order $1/N_w$. These issues are explored in detail in the undulator chapter.

12.8.3
A Theorem Applicable to Isotropic Distributions

The result to be obtained in this section applies, though only approximately, to both Thomson scattering and undulator radiation. Though this result will be superceded by an exact formula in the next section, it is of more than academic interest. The value of the theorem is its extreme simplicity for providing qualitatively correct results with no need for analytical calculation. For our purposes the theorem will be applied only to photons, which are massless, but it is proved for massive particles with no extra effort.

Consider a particle having momentum p' in a primed frame (which will shortly be taken to be the electron rest frame) and traveling at angle ϑ' relative to the x' axis. The particle's momentum can therefore be written

$$\mathbf{p}' = p' \sin \vartheta' \, \hat{\mathbf{y}}' + p' \cos \vartheta' \, \hat{\mathbf{x}}'. \tag{12.100}$$

Using Eq. (12.64), its energy in the unprimed system (which will shortly be taken to be the laboratory system) is given by

$$\mathcal{E} = \gamma(vp' \cos \vartheta' + \mathcal{E}'). \tag{12.101}$$

Suppose that all such particles, like the photons in a dipole radiation pattern, have energy independent of direction in the primed system. When viewed in the unprimed system, the particle energies depend on the angle ϑ. Let us further assume that the particles are distributed isotropically in the primed system. In terms of solid angle $d\Omega'$ their angular distribution is therefore given by

$$\frac{dN}{d\Omega'} = \frac{1}{4\pi}, \quad \text{or} \quad \frac{dN}{d(\cos \vartheta')} = -\frac{1}{2}, \tag{12.102}$$

since $d\Omega' = 2\pi \sin \vartheta' \, d\vartheta' = -2\pi d(\cos \vartheta')$.

Continuing to assume the particles are isotropic and monoenergetic in the primed frame, Eqs. (12.101) and (12.102) can be used to work out the distribution of energies as viewed in the unprimed frame. The result is

$$\frac{dN}{d\mathcal{E}} = \frac{dN}{d(\cos \vartheta')} \frac{d(\cos \vartheta')}{d\mathcal{E}} = -\frac{1}{2\gamma v p'}, \tag{12.103}$$

which is constant, independent of ϑ. So we have proved the striking theorem: *radiation that is isotropic and monochromatic in one frame of reference is uniformly distributed in energy in any other frame of reference.*

Given that the energy distribution is uniform, to completely specify the energy distribution, we need only find the minimum and maximum values of $\mathcal{E}$ and adjust the constant probability distribution between these limits to fix the normalization. Minimum and maximum energies are determined by $\cos\vartheta' = \pm 1$ which correspond to particles parallel to and anti-parallel to **v**. So Eq. (12.101) yields

$$\mathcal{E}_\pm = \gamma(\pm vp' + \mathcal{E}'). \tag{12.104}$$

Specializing to our (massless) photons, and assuming $v \approx c$, the laboratory energy ranges from $\gamma(-cp' + \mathcal{E}') \approx 0$ up to $\gamma(cp' + \mathcal{E}') \approx 2\gamma\mathcal{E}'$. Of course the dipole radiation pattern in the electron rest frame is not precisely isotropic so its laboratory energy distribution is not precisely uniform. But the donut-shaped dipole pattern is *more-or-less* isotropic in the electron's rest frame, so the laboratory energy distribution is *more-or-less* uniform.

From the point of view of users of the x-rays being produced, their energy distribution is far from optimal. Most x-ray experiments make use of only a very narrow band of energies, typically centered near $\mathcal{E}_+$, the top end of the energy range. With a typical fractional energy acceptance being 10^{-4}, the fraction of x-rays with no chance of being counted is 0.9999. In this sense light sources are very inefficient, producing vast numbers of x-rays that have no chance of contributing to any experiment.

12.8.4
Energy Distribution of Undulator Radiation

At this point we are primarily interested in the fundamental resonance radiation from an undulator. When viewed in the rest frame of an electron passing through an undulator, this is electric dipole radiation. To the extent we are willing to describe this pattern as being "more or less isotropic", the energy distribution of x-rays in the laboratory will be "more or less uniform", as has just been shown.

It is not difficult to do better than this though, since the directional dependence of the dipole radiation in the electron's frame is well known, and the transformation needed from primed to unprimed frames is Eq. (12.101), which can be re-expressed as

$$\cos\vartheta' = \frac{\mathcal{E} - \gamma\mathcal{E}'}{\gamma v p'} \approx \frac{\mathcal{E}}{\gamma\mathcal{E}'} - 1. \tag{12.105}$$

The final approximation has assumed $v \approx c$. The distribution in $\mathcal{E}$ can be obtained from the distribution in $\cos\vartheta'$, derived earlier as Eq. (12.87), by equating probabilities;

$$N_{\cos\vartheta'}(\cos\vartheta')\,d\cos\vartheta' = N_{\mathcal{E}}(\mathcal{E})\,d\mathcal{E}. \tag{12.106}$$

Using Eq. (12.87), the result is

$$N_{\mathcal{E}}(\mathcal{E}) = \frac{3}{8}(1+\cos^2\vartheta')\frac{d\cos\vartheta'}{d\mathcal{E}'} = \frac{3}{8\gamma\mathcal{E}'}\left(2 + \frac{\mathcal{E}^2}{\gamma^2\mathcal{E}'^2} - 2\frac{\mathcal{E}}{\gamma\mathcal{E}'}\right). \tag{12.107}$$

This can be simplified somewhat by introducing a fractional (laboratory) energy variable ν_J defined by[2]

$$\nu_J = \frac{\mathcal{E}}{\mathcal{E}_{\max.}}, \tag{12.108}$$

where

$$\mathcal{E}_{\max.} \approx 2\gamma\mathcal{E}' \tag{12.109}$$

is the maximum x-ray energy. The distribution in ν_J is then given by

$$N_{\nu_J}(\nu_J) = \frac{3}{2}(1 - 2\nu_J + 2\nu_J^2). \tag{12.110}$$

Being normalized to 1, this is a probability distribution. Multiplied by some total number N_0 of photons it would give a photon number distribution. This distribution is plotted in Figure 12.7. As expected the distribution is more-or-less uniform, with a deficit in the middle corresponding to the reduced average intensity near 90° in the electron's frame.

Since the energy carried by one photon is proportional to ν_J the same distribution can be converted to a power distribution by multiplying by ν_J. After renormalization, the distribution of power (also referred to as the power spectrum) within a beam of total power $\mathcal{P}$ is given by

$$\mathcal{P}(\nu_J) = 3\mathcal{P}\nu_J(1 - 2\nu_J + 2\nu_J^2), \tag{12.111}$$

Here $\mathcal{P}(\nu_J)\,d\nu_J$ is the power in fractional range $d\nu_J$. This equation agrees with Jackson's Eq. (14.118). The spectrum is plotted in Figure 12.8.

Applying Eq. (12.111) at $\nu_J = 1$, the fractional power $\Delta\mathcal{P}/\mathcal{P}$ in the (differential) range $1 - \Delta\nu_J < \nu_J < 1$ is given by

$$\frac{\Delta\mathcal{P}}{\mathcal{P}} = 3\Delta\nu_J. \tag{12.112}$$

[2] The "J" subscript on ν_J stands for "Jackson", who defines this quantity in his Eq. (14.118). This subscript will serve to differentiate ν_J from a similar, but differently defined, fractional energy to be encountered later.

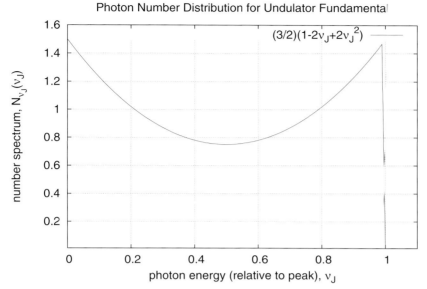

Fig. 12.7 The number distribution of photons emitted from the first undulator harmonic, as given by Eq. (12.110).

According to Eq. (12.99) (with $K \approx 0$) there is a functional dependence of frequency on scattering angle ϑ. For $\vartheta \ll 1$ this relation reduces to

$$\Delta \nu_J = \gamma^2 \vartheta^2 = \gamma^2 \Delta \Omega, \tag{12.113}$$

where $\Delta \Omega$ is the solid angle corresponding to a square cone with sides ϑ centered on the forward direction. Combining these formulas, the power radiated into solid angle $\Delta \Omega$, centered on the forward direction, is

$$\Delta \mathcal{P} = 3 \gamma^2 \Delta \Omega \, \mathcal{P}. \tag{12.114}$$

Problem 12.8.1 *(a) Integrate the first of Eqs. (12.81) over the full solid angle $\Omega = 4\pi$ range and confirm that the result is consistent—meaning that the power actually integrates up to $\bar{\mathcal{P}}$. (b) Confirm the normalizations of Eqs. (12.107), (12.110), and (12.111) similarly.*

12.9
Rate Estimates for a Laser Wire Diagnostic Apparatus

In this section we apply some of the formalism that has been developed to estimate the counting rates to be expected in the diagnostic apparatus shown

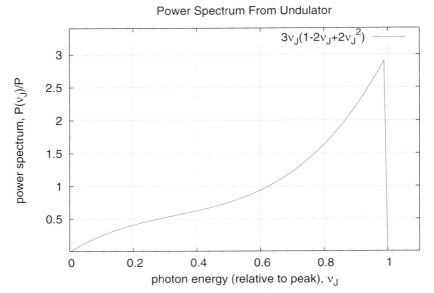

Fig. 12.8 The undulator power spectrum as given by Eq. (12.111).

in Figure 12.4. After working out the rate in the nonrelativistic case, the laser beam will be treated as an effective undulator and then as a beam of photons being Compton scattered. Finally the invariant formalism derived at the beginning of this chapter will be used to obtain the counting rate. For simplicity we assume the laser is perfectly polarized with E_z its only electric field component. When treating the laser as an undulator, this permits us to use only the second component in Eq. (12.57). For simplicity, some calculations will be specialized to the precise configuration shown in Figure 12.4. As a practical matter, for example to increase the energy of the scattered photon, it may be necessary to increase the angle α from the 30 degree value.

We are interested in the counting rate of photons scattered from a laser beam into a detector by an electron beam. In practice, either by necessity, or to increase the counting rate, either or both of the beams will be bunched. But, to simplify the following calculations, both beams will be treated as continuous. The electron beam current I will be assumed to be uniformly distributed over a rectangular area of width a_e, height b_e. The laboratory line charge density is therefore given by $\lambda = I/v$. The laser beam power $\mathcal{P}_{\text{laser}}$ is uniformly distributed over width a_γ and height b_γ. With both incident beams lying in a horizontal plane, and perfectly matched and aligned, dependence on beam heights is straightforward, and we can assume $b_e = b_\gamma = b$.

The incident laser angle α is a free variable, as is the direction at which the detector solid angle $\Delta\Omega''$ is located.[3]

Exact agreement cannot be expected, since the undulator treatment has been only approximate. But, to simplify the treatment, and to facilitate comparison with the undulator formalism, the detector will be assumed to count photons scattered more or less along the electron beam direction. (This pre-supposes the existence of beam steering that prevents the electron beam itself from running into the detector.)

12.9.1
The Nonrelativistic Limit

To calculate photon scattering rates from slow, speed v electrons the Thomson scattering formula can be used directly. For beam current I, the number of electrons illuminated by a laser beam of width a_γ is

$$N_e = \frac{I/e}{v} \frac{a_\gamma}{\sin\alpha}. \tag{12.115}$$

The incident photon flux (photons/sec-m^2) is given by

$$\mathcal{F} = \frac{\mathcal{P}_{\text{laser}}}{\mathcal{E}_\gamma} \frac{1}{a_\gamma b_\gamma}. \tag{12.116}$$

Copying the forward differential cross section from Eq. (12.91), the number of counts per second in forward (i.e. parallel to the electron beam) solid angle $\Delta\Omega''$ is

$$N_{\gamma,\text{scat.}} = \mathcal{F} N_e \frac{d\sigma'}{d\Omega'} \Delta\Omega'' = \frac{\mathcal{P}_{\text{laser}}}{\mathcal{E}_\gamma} \frac{I}{ev} \frac{r_e^2}{b_\gamma \sin\alpha} \Delta\Omega''. \tag{12.117}$$

In this nonrelativistic regime, rates and polarization dependences in all other directions can be calculated using Thomson scattering formulas derived earlier.

12.9.2
Laser Wire Treated as Undulator

In this section we assume $v \approx c$. The energy U_0 radiated from one electron of energy $\mathcal{E}_e$ completing one full revolution in a circle of radius R_0, is given by

$$U_0 = \frac{4\pi}{3} \frac{r_e}{(mc^2)^3} \frac{\mathcal{E}_e^4}{R_0}, \quad \text{where} \quad r_e = \frac{e^2}{4\pi\epsilon_0 mc^2}. \tag{12.118}$$

[3] The detector solid angle is symbolized by $\Delta\Omega''$ in the following sections only for ease of comparison with the method of calculation that transforms to the electron rest system and then back to the laboratory.

In analysing the effects of synchrotron radiation in accelerators it is customary to express this using energy units more convenient for high energy particles;

$$U_0[\text{GeV}] = C_\gamma \frac{\mathcal{E}_e^4[\text{GeV}^4]}{R_0[\text{m}]}, \quad \text{where} \quad C_\gamma = 0.885 \times 10^{-4}\,\text{m}\,\text{GeV}^{-3}. \quad (12.119)$$

This formula (first version) can be applied to approximate the total energy radiated from our incident laser beam treated as if it were an undulator. Recall that $\Theta = K/\gamma$ is the maximum angle in an undulator. The energy radiated per pole $U_{1\,\text{pole}}$ can be approximated by U_0 multiplied by the fraction of a full circle corresponding to one pole of the undulator, giving $U_{1\,\text{pole}} \approx U_0 \Theta/\pi$. The bend radius R_0 in the undulator can also be estimated in terms of Θ; $R_0 \approx (\lambda_w/4)/\Theta$. From an undulator with $2N_w$ poles, these estimates can be combined to give the total energy radiated by one electron passing through the laser "undulator";

$$
\begin{aligned}
U_{1\,\text{electron}} &\approx \frac{4}{\pi}\left(\frac{4\pi}{3}\frac{r_e}{\lambda_w}\frac{\mathcal{E}_e^4}{(mc^2)^3}\right)\frac{2N_w K^2}{\gamma^2} \\
&= \frac{8}{\pi\gamma^2}\left(\frac{4\pi}{3}r_e\frac{\mathcal{E}_e^4}{(mc^2)^3}\right)\left(\frac{1}{2\pi^2}\frac{c^2}{v^2}\frac{1}{b_\gamma \sin\alpha}\left(\frac{c}{v}-\cos\alpha\right)^2 \frac{Z_0 \mathcal{P}_{\text{laser}}}{(mc^2/e)^2}\right) \\
&= \frac{64\gamma^2}{3\pi}\frac{r_e^2}{b_\gamma \sin\alpha}\frac{1}{c}\left(\frac{c}{v}-\cos\alpha\right)^2 \mathcal{P}_{\text{laser}}. \quad (12.120)
\end{aligned}
$$

where the effective values of K, N_w, and λ_w calculated in Section 12.5 have been substituted in the combination given in Eq. (12.62). λ is the laser wavelength and b_γ is the laser beam height.

The way this energy is distributed is given, in terms of numbers of photons by Eq. (12.87) and, in terms of radiant energy carried by the photons by Eq. (12.111). There is a one-to-one correspondance between observation angle ϑ'' and frequency ω'' of the scattered radiation. The relation is given by Eq. (12.99) (with $K \approx 0$). Applying Eq. (12.114), the energy radiated by one electron into (small) forward solid angle range $\Delta\Omega''$ is

$$\Delta U_{1\,\text{electron}} = 3\gamma^2 \Delta\Omega'' U_{1\,\text{electron}}. \quad (12.121)$$

This energy comes in the form of photons of energy $\mathcal{E}_\gamma''$. When counted in a detector subtending solid angle $\Delta\Omega''$, the number of scattered photon counts $N_{\gamma,\text{scat.}}$ coming from $N_e = I/e$ electrons per second is

$$
\begin{aligned}
N_{\gamma,\text{scat.}} &= \frac{I}{e} 3\gamma^2 \frac{U_{1\,\text{electron}}}{\mathcal{E}_\gamma''}\Delta\Omega'' \\
&= \frac{\mathcal{P}_{\text{laser}}}{\mathcal{E}_\gamma}\frac{I}{ec}\frac{32}{\pi}\frac{r_e^2}{b_\gamma}\frac{1-\cos\alpha}{\sin\alpha}\gamma^2\Delta\Omega''. \quad (12.122)
\end{aligned}
$$

where $\mathcal{E}''_\gamma$, with $\beta = 1$, has been obtained from Eq. (12.11). (The divergence at $\alpha = \pi$ can be understood as applying to a physically unrealizable configuration in which, because the photon and electron beams are counter-traveling, every photon scatters eventually. For a physically meaningful result in this configuration it is necessary to take proper account of beam bunching.)

12.9.3
Laser Wire Treated via Electron Rest Frame

We concentrate on an apparatus that counts photons radiated parallel to the electron flight path. The Thomson differential cross section for scattering in the forward direction in the electron's rest frame was obtained in Eq. (12.91);

$$\left.\frac{d\sigma'}{d\Omega'}\right|_{\vartheta'=0} = r_e^2 = \left(\frac{e}{4\pi\epsilon_0(mc^2/e)}\right)^2 = (2.81784 \times 10^{-15}\,\text{m})^2. \tag{12.123}$$

For the electron frame differential cross section, which is given by

$$\frac{d\sigma'}{d\Omega''} = \frac{d\sigma'}{d\Omega'}\frac{d\Omega'}{d\Omega''}, \tag{12.124}$$

we know the first factor, and the second factor has been derived previously in Eq. (12.95). We therefore obtain

$$\frac{d\sigma'}{d\Omega''} = r_e^2 \gamma^2 \left(1 + \frac{v}{c}\right)^2. \tag{12.125}$$

To employ the electron frame cross section we must work out the electron frame flux. Fortunately, since the electron frame cross section is independent of photon energy, we need only work out the photon number flux and the number of electrons illuminated.

Let the length of the laser-illuminated region in the electron frame be $\Delta x'_{\text{illum.}}$ and let the laser beam width in the electron frame be a'_γ. The ratio of these quantities can be obtained from the incidence angle;

$$\frac{a'_\gamma}{\Delta x'_{\text{illum.}}} = \sin\alpha' = \frac{\sin\alpha}{\gamma(1 - \cos\alpha\, v/c)}, \tag{12.126}$$

where the final equation comes from the product of Eqs. (12.27) and (12.29). The total number of photons in the laser beam passing a fixed point in the laboratory in one second is $N_\gamma = \mathcal{P}_{\text{laser}}/\mathcal{E}_\gamma$. Making allowance for time dilation, since photons preserve their identity in passing from the lab frame to the electron rest frame, the total number of photons per second of electron frame time is

$$N'_\gamma = \frac{\mathcal{P}_{\text{laser}}}{\gamma \mathcal{E}_\gamma}. \tag{12.127}$$

The flux of laser beam photons crossing unit area transverse to the laser beam in unit time in the electron frame is therefore

$$\mathcal{F}' = \frac{\mathcal{P}_{\text{laser}}}{\gamma \mathcal{E}_\gamma} \frac{1}{a'_\gamma b_\gamma}. \tag{12.128}$$

Taking account of the Lorentz length contraction, the number N'_e of electrons being illuminated is equal to lab distance corresponding to $\Delta x'_{\text{illum.}}$ multiplied by the laboratory number of charges per unit length;

$$N'_e = \frac{I}{ev} \frac{\Delta x'_{\text{illum.}}}{\gamma}, \tag{12.129}$$

The scattering rate per second of electron frame time into forward rest frame solid angle $\Delta \Omega'$ is

$$r_e^2 \Delta \Omega' \, \mathcal{F}' \, N'_e = r_e^2 \Delta \Omega' \frac{\mathcal{P}_{\text{laser}}}{\gamma \mathcal{E}_\gamma} \frac{1}{b_\gamma} \frac{I}{ev} \frac{\Delta x'_{\text{illum.}}}{\gamma a'_\gamma}$$

$$= \frac{r_e^2}{b_\gamma} \frac{\mathcal{P}_{\text{laser}}}{\mathcal{E}_\gamma} \frac{I}{ev} \frac{1 - \cos \alpha \, v/c}{\gamma \sin \alpha} \Delta \Omega'. \tag{12.130}$$

The solid angle conversion factor from electron rest frame to lab was given by Eq. (12.95). Converting from electron frame time back to laboratory frame time cancels the $1/\gamma$ factor introduced in Eq. (12.127) and the expected laboratory counting rate into forward laboratory solid angle $\Delta \Omega''$ per second of laboratory time is

$$N_{\gamma,\text{scat}} = \frac{r_e^2}{b_\gamma} \frac{\mathcal{P}_{\text{laser}}}{\mathcal{E}_\gamma} \frac{I}{ev} \frac{1 - \cos \alpha \, v/c}{\sin \alpha} \left(1 + \frac{v}{c}\right)^2 \gamma^2 \Delta \Omega''. \tag{12.131}$$

This reduces to Eq. (12.117) in the nonrelativistic region, as it should. Except for a dimensionless numerical factor it also agrees with Eq. (12.122). The numerical $(8/\pi)$ discrepancy may be due to the crudeness of the undulator power estimate.

12.9.4
Invariant Cross Section Applied to Laser Wire

To calculate the rates expected under arbitrary conditions the most straightforward approach seems to be to use Eq. (12.36);[4]

$$d^2 \sigma = 8 \pi r_e^2 \frac{1}{x^2 m^2} \left(\left(\frac{1}{x} - \frac{1}{y} \right)^2 + \left(\frac{1}{x} - \frac{1}{y} \right) + \frac{1}{4} \left(\frac{x}{y} + \frac{y}{x} \right) \right) \frac{dt \, d\varphi_\parallel}{2\pi}. \tag{12.132}$$

[4] To confirm that Eq. (12.132) reproduces result (12.91), namely $d\sigma'/d\Omega'|_0 = r_e^2$, it is necessary to go to the glancing $\vartheta \approx 0$ region. Otherwise, $\omega' \neq \omega$, even for $v = 0$. This causes $1/x - 1/y$ not to cancel, as it should in the Thomson limit. This difference (for visible light) is then not small compared to the last term in large parentheses, which is equal to 1/2.

It is important to remember that invariants t, x, and y, as defined by Eqs. (12.7) and (12.35), were expressed in terms of the angles ϑ and φ, which are measured relative to the *incident* photon *direction in the laboratory*. At this point all notations revert back to the usage in section 12.2, including $c = 1$. However, to avoid its misinterpretation, the scattered photon frequency in the lab frame will be expressed as ω''.

By its very nature a laser wire will typically shine toward the electron beam from some arbitrary skew angle α. For reasons to be made clear shortly, let us assume that α is a "big" angle, which will be taken to mean $\alpha \gg 1/\gamma$. This relation will be satisfied in typical practice, especially for high energy electron beams. In this case we know the scattered radiation in the laboratory frame is mainly contained in a cone of angle $1/\gamma$ centered on the electron direction, so we are assuming the incident beam is well outside this cone.

The apparatus in which scattered photons are detected has laboratory angular detector acceptances $\Delta\vartheta$ and $\Delta\varphi$. To apply Eq. (12.132) we therefore need the Jacobian determinant for the transformation $(\vartheta, \varphi) \to (t, \varphi_{\parallel})$. This will be complicated and difficult to calculate in general. This complication reflects the fact that the angular distribution, including all directions, is complicated. But we are primarily interested in scattered radiation within the previously-mentioned cone.

One frame of reference in which the incident photon and incident electron are collinear is the electron rest frame. Coordinate transformations to that frame have been obtained previously in Eq. (12.22). We know that the scattered photon is *not* more or less parallel to the electron in this frame; in fact, the radiation is more or less isotropic. Furthermore, after transformation back to the laboratory, the strong peaking along the electron direction is *not* due to the Jacobian. It is due to the near vanishing of one or both of the x and y denominator factors in Eq. 12.132). It will therefore be an adequate approximation to evaluate the Jacobian precisely in the direction that will eventually transform back to be parallel to the electron beam in the laboratory. This will produce a seriously incorrect Jacobian factor only in regions where the cross section is negligible. Throughout the $1/\gamma$ cone containing most of the scattered photons the Jacobian will be taken to be constant.

The incident photon's angle α_0 in the electron rest frame has been calculated previously in Eq. (12.29). The incident frequency ω_0 is given by Eq. (12.22) and, because the Thomson approximation is valid, ω_0' has the same value;

$$\omega_0 = \omega_0' = \omega\gamma(1 - v\cos\alpha). \tag{12.133}$$

Evaluated in this frame the invariant t is given by

$$t = -2\omega_0^2(1 - \cos\vartheta_0), \tag{12.134}$$

where ϑ_0 is the photon scattering angle in the electron rest frame. The condition for emission to be along the electron direction is

$$\vartheta_0 = \alpha_0. \tag{12.135}$$

Based on the discussion of the previous paragraph, we wish to calculate the Jacobian at this point. Let $d\theta_{z0}$ be an angular, out-of-plane, deviation from the horizontal plane in the electron rest frame. It and $d\vartheta_0$ are related to t and $d\varphi_\parallel$ by

$$dt = -2\omega_0^2 \sin\vartheta_0 \, d\vartheta_0, \quad d\varphi_\parallel = \frac{d\theta_{z0}}{\sin\vartheta_0}. \tag{12.136}$$

(Since $\vartheta_0 > 1/\gamma$ there is no difficulty with singularity in the second of these relations.) The differential factor needed in Eq. (12.132) can therefore be written as

$$|dt d\varphi_\parallel|_0 = 2\omega_0^2 \, d\vartheta_0 d\theta_{z0} = 2\omega_0^2 \, d\Omega_0, \tag{12.137}$$

where $d\Omega_0 \equiv d\Omega'$ is a forward solid angle element in the electron rest frame. We have seen previously that forward solid angles in electron rest frame and lab are related by Eq. (12.94). Combining factors we obtain

$$dt d\varphi_\parallel = 2\omega_0^2 \gamma^2 (1+v)^2 d\Omega'' \tag{12.138}$$

as the differential factor needed for Eq. (12.132). Using Eq. (12.133) to express ω_0 in terms of ω and combining factors, we get

$$\frac{d^2\sigma}{d\Omega''} = 8r_e^2 \frac{\omega^2 \gamma^4}{x^2 m^2} (1+v)^2 (1-v\cos\alpha)^2 \left(\left(\frac{1}{x} - \frac{1}{y}\right)^2 + \left(\frac{1}{x} - \frac{1}{y}\right) + \frac{1}{4}\left(\frac{x}{y} + \frac{y}{x}\right) \right). \tag{12.139}$$

12.9.5
Bunched Beam Rates

A configuration in which a laser bunch collides with an electron bunch is shown in Figure 12.9. For simplicity both bunches are assumed to form uniformly-filled parallelopipeds They are perfectly matched vertically and the electron bunch is assumed to be short enough that all electrons cross in the interior of the laser pulse. Scatters occur only in the shaded region, for which the volume is

$$V = a_e b \frac{a_\gamma}{\sin\alpha}. \tag{12.140}$$

More general bunch distributions can be represented as integrals over such interaction parallelopipeds. The time taken for the electron bunch to cross the photon pulse is

$$T = \frac{l_e}{v}. \tag{12.141}$$

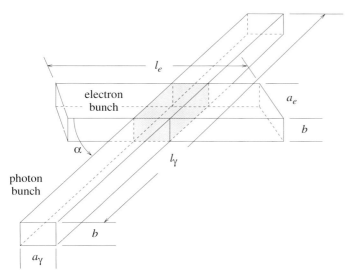

Fig. 12.9 Electron and photon bunches are vertically matched and the photon bunch is long compared to the electron bunch and their centers are time-synchronized. Scatters occur only within the shaded region.

The cross section σ for some type of scattering in the electron rest frame can, by definition, be said to be "invariant" because the frame in which it is determined has been unambiguously specified. In this frame the number of events N occurring in time T is

$$N = \sigma v_{\text{rel.}} n_e n_\gamma V T, \tag{12.142}$$

where n_e and n_γ are number densities in the two beams. The quantity $v_{\text{rel.}}$ is (invariantly) defined to be the velocity of photon beam relative to electron beam *in the electron rest frame*. It is shown in Landau and Lifshitz, *The Classical Theory of Fields*, that $v_{\text{rel.}}$ can be evaluated from laboratory frame velocities by the formula

$$v_{\text{rel.}} = \frac{\sqrt{(\mathbf{v}_\gamma - \mathbf{v_e})^2 - (\mathbf{v}_\gamma \times \mathbf{v_e})^2}}{1 - \mathbf{v}_\gamma \cdot \mathbf{v_e}}, \tag{12.143}$$

but this is superfluous in our case since, with the incident particle being a photon, $v_{\text{rel.}} = 1$.

It is convenient to re-write Eq. (12.142) as

$$N = \sigma v_{\text{rel.}} (\underline{p}_e, \underline{p}_\gamma) \frac{n_e}{\mathcal{E}_e} \frac{n_\gamma}{\mathcal{E}_\gamma} (VT), \tag{12.144}$$

where the four-scalar product notation introduced in Section 12.2 is used. What makes this form useful is that all its factors are relativistic invariants.

The $n/\mathcal{E}$ factors are invariant because both numerator and denominator transform like the time component of a four-vector. One must, of course, check that the newly inserted factors, when evaluated in the electron rest frame, cause Eq. (12.144) to reduce to Eq. (12.142). Once this is done one can use Eq. (12.144) in any arbitrary frame of reference. In our case we intend to apply it in the laboratory. For example

$$\frac{(\underline{p}_e, \underline{p}_\gamma)}{\mathcal{E}_e \mathcal{E}_\gamma} = 1 - v \cos \alpha. \tag{12.145}$$

From the laser power $\mathcal{P}$ and the electron beam current I, and the bunch structures of the beams, one can find Q_1, which is the total charge in one electron bunch and U_1 which is the total energy in one photon bunch. From these we can find the number densities

$$n_e = \frac{Q_1/e}{a_e b l_e} \equiv \frac{N_{1,e}}{a_e b l_e}, \quad n_\gamma = \frac{U_1/\mathcal{E}_\gamma}{a_\gamma b l_\gamma} \equiv \frac{N_{1,\gamma}}{a_\gamma b l_\gamma}. \tag{12.146}$$

Using Eq. (12.89) for the total cross section, and combining the various factors, we obtain, for $N_{1,\text{scat.}}$, the total number of scattered photons per bunch crossing, in terms of numbers $N_{1,e}$ and $N_{1,\gamma}$ of particles per bunch of each type

$$N_{1,\text{scat.}} = \frac{8\pi}{3} \frac{r_e^2}{b l_\gamma} \frac{1 - \beta \cos \alpha}{\sin \alpha} \frac{N_{1,e}}{\beta} N_{1,\gamma}. \tag{12.147}$$

In this case the seeming divergence for, $\alpha = 0$ or $\alpha = \pi$, is explainable by the fact that the geometric assumptions, as illustrated in Figure 12.9, cannot be satisfied for too-glancing incidence. In fact we have to enforce a geometric inequality

$$\sin \alpha > \frac{a_\gamma}{l_e}, \tag{12.148}$$

or rather a somewhat more complicated version of this formula that depends also on the electron bunch width.

A natural, slightly more general, situation than that discussed so far would have the electron beam height b_e not equal to the photon beam height b_γ, but all other conditions unchanged. Defining b to be the larger of b_e and b_γ, the rate would still be given by Eq. (12.147). This is because *lowering* either beam height from the matched condition, leaves the rate unchanged.

The way the scattered photons are distributed in energy and direction can be obtained to excellent accuracy using Eq. (12.139). The energy distribution can be approximated as uniform from peak energy down to zero, or, better for head-on collisions, by Eq. (12.111). In either case, the approximate distribution in angles (relative to the electron beam) can be obtained from Eq. (12.97), using the one-to-one relationship between angle and energy.

As mentioned before, for more realistic particle distributions, the interaction volume has to be subdivided into small parallelopipeds for which Eq. (12.147) can be applied, and the results summed.

References

1 V.Berestetskii, E.Lifshitz, L.Pitaevskii (1982), *Quantum Electrodynamics*, Pergamon Press, New York.

2 Hagedorn, R. (1964), *Relativistic Kinematics*, W.A. Benjamin, New York, 33.

3 Clemmow, P., Dougherty, J. (1990), *Electrodynamics of Particles and Plasmas*, Addison-Wesley, Redwood City, CA, ascribe this theory to Kolomenskii A.A. and Lebedev A.N., *Sov. Phys. Doklady*, **7**, 745 (1963) and *Sov. Phys. JETP*, **17**, 179 (1963).

4 *TESLA Design Report, Part VI Appendices*, (2001), DESY 2001-011.

5 Griffiths, D. (1999),*Introduction to Electrodynamics*, Prentice Hall, Englewood Cliffs, NJ, Ch. 9.

6 Griffiths, D. (1999), *Introduction to Electrodynamics*, Prentice Hall, Englewood Cliffs, NJ, Ch. 11.

7 Griffiths, D. (1999), *Introduction to Electrodynamics*, Prentice Hall, Englewood Cliffs, NJ, Ch. 12.

13
Space Charge Effects and Coherent Radiation

13.1
Acknowledgement and Preview

Much of the material in this chapter has been generated in collaboration with Nikolay Malitsky. He is the architect of the UAL simulation environment as well as most of the computer code implementing the *string space charge model* that is the basic theoretical formulation of the problem described in this chapter.

Incoherent synchrotron radiation, characterized by its proportionality to the number N of particles in one bunch, has been analysed at great length in earlier chapters. Coherent synchrotron radiation (CSR), though proportional to N^2, has, until recently, been unimportant in the operation of electron accelerators. This lack of importance has been due to the comparatively long wavelengths of the radiation, much of which is suppressed by the vacuum tube acting as a high pass waveguide, and because the total power radiated would, in any case, be small. But recent applications, in particular ERLs and FELs, generate ultrashort, ultra-intense bunches for which coherent radiation is important.

Incoherent radiation has been seen to make its presence known in accelerators through its fluctuations and through the long-time-averaged deceleration it causes. The coherent space charge force influences a bunch of particles in a storage ring through the spatial-dependence of its accompanying longitudinal self-force, as well as by its accompanying spatially-dependent transverse force, called the centrifugal space charge force (CSCF). See Talman [15].

In historical practice it has been customary to treat "space charge", Touschek effect, and CSR as disjoint topics, subject to disconnected theoretical treatment. The separation has been based on domains of importance: "space charge" limits beam intensities at low energy without reference to external fields, the Touschek effect causes particle loss due to intrabeam scattering in dense bunches, and CSR influences ultrashort bunches in bending magnets. This separation is somewhat unnatural since all these effects are due to the same Coulomb interaction and all theories have the same goal—understanding the bunch disruption caused by this force.

Accelerator X-Ray Sources. Richard Talman
Copyright © 2006 WILEY-VCH Verlag GmbH & Co. KGaA, Weinheim
ISBN: 3-527-40590-9

My article *String Formulation of Space Charge Forces in a Deflecting Bunch* [1] departed from this conventional approach by integrating all space charge effects into a single formalism, irrespective of whether the beam is in a magnet or in a field free region. Most of this chapter derives from this article. Parts too technical for inclusion here can be obtained from the original. Some work done subsequent to that article is also described here. In particular the UAL computer code STRINGSC, mainly written by Nikolay Malitsky [2], and results obtained using that code, are described.

That low energy space charge effects and Touschek effect have been subsumed into a common (string space charge) program is somewhat academic here, since this chapter contains little discussion of these other effects. The issue to be emphasized is the emittance growth of intense short bunches in magnetic fields.

The force between two moving point charges, because of its inverse square law singularity, cannot be applied directly to the numerical simulation of particle dynamics; radiative effects make this especially true for short bunches being deflected by magnets. The formalism overcoming this problem has, as a basic ingredient, the *total*, electric plus magnetic, force on a point charge co-moving with a longitudinally-aligned, uniformly-charged string. With the bunch represented as a distribution of strings, one per particle, bunch evolution can then be treated using direct particle-to-particle, intrabeam scattering, with no need for an intermediate, particle-in-cell, field calculation step. Since the basic formulas are both exact (in paraxial approximation) and fully relativistic, they are applicable to beams of all particle types and all energies. But the theory emphasizes the calculation of emittance growth of the ultra-short electron bunches of current interest for ERLs (energy recovery linacs) and FELs (free electron lasers). The CSR and CSCF effects important in this context are subsumed into the theory. Regularized, on-axis, longitudinal field components are in excellent agreement with values from Saldin et al. [3].

13.2
Introduction to the String Space Charge Formalism

Though CSR and CSCF are known to influence intense short bunches, there have been few reliable measurements of the properties or this radiation or the effects of these forces. This chapter therefore emphasizes the theory, especially computational, of the process.

The motivation for developing a string (or line) approach to space charge modeling has been the requirement to estimate the growth of transverse emittance due to space charge forces as a slender bunch passes through a magnetic field region. Over and above space charge effects present in straight line motion, the main longitudinal curvature effect goes by the name "coherent synchrotron radiation" (CSR) and the main transverse curvature effect is called the "centrifugal space charge force" (CSCF). Since both effects can lead to growth of transverse emittance it is necessary to treat them together consistently. In spite of the fact that the problem is thoroughly relativistic, and therefore thoroughly non-static, much of the calculation can be recast as electro- and magnetostatics. The strategy will be, as far as possible, to transform the problem into an electro/magneto-static calculation, along the lines of Bassetti and Brandt [6]. Their formulas have to be altered by retarded time calculations leading to changed integration limits of "effective" charge distributions, and by the inclusion of "string end" fields.

The electric and magnetic forces between two point charges are proportional to the inverse square of their separation distance. For real electrons in an intense bunch the occasional near miss results in a large deflection and possible loss through a mechanism known as the Touschek effect. Because of the vastly greater charge per macroparticle in a numerical calculation of the evolution of a bunch of N particles, this singular behavior results in intolerably erratic behavior during close encounters unless some sort of smoothing or averaging procedure, such as the PIC (particle-in-cell) code, has been adopted to overcome this problem. Here a different workaround is described in which the singularity is made less severe by representing point charges as longitudinally-aligned, line charges or needles or "strings". These strings have zero transverse extent and length $2L$. Until a brief discussion in a later section, it is left open whether L is matched to the actual bunch length or is purely artificial (though short compared to the actual bunch length.)

The basic ingredient of the string space charge formalism is a closed-form expression for the force on a point charge due to a co-moving charged string. By "softening" near miss forces between particles this expression forms the basis of treatments of space charge effects as direct "intrabeam scattering"—there is no need for charge distribution tallying followed by field solving. Though the formalism is applicable to particles of arbitrary type and energy, the emphasis here will be on short electron bunches for which radiative effects are important. The longitudinal force component is used to calculate CSR, the transverse component for calculating CSCF. Some comments concerning the history of CSCF are contained in a paper by Li and Derbenev [4] and there is further discussion in a paper by Geloni et al. [5]. Other early papers were written by Piwinski [13], Decker [14], Lee [16], Laslett [17], and Derbenev and Shiltsev [18].

The early discussion will tend to emphasize longitudinal forces. This is *not* because they are more important—both longitudinal and transverse forces cause emittance growth, and both require regularization. But the longitudinal component is directly related to coherent synchrotron radiation, which can be calculated by an alternative method and can therefore provide an important test of the validity of the string formalism.

In a numerical simulation of bunch dynamics, each of the N particles is to be treated as a point particle as far as its own dynamics is concerned, but as a line charge for the purpose of calculating the electromagnetic fields it generates. The $N-1$ forces on each particle due to the other particles are added for each of the N particles. These N^2 scaling and computation time considerations limit the number of macroparticles per bunch that can be modeled. To permit larger values of N the forces could be calculated on a three-dimensional grid, from which forces on individual particles are then calculated by interpolation. For the single pass problems the theory has been applied to so far this grid strategy has not been needed. In any case the basic ingredient of the theory is the force on a point charge due to a charged string.

Since the (vector) force is the result of an integration along a line charge distribution, it can be expressed as a closed form, indefinite integral, to be evaluated at the string ends. Once formulas have been made available for these integrals, as they will be, all that remains is to determine the limits of these integrals, which is to say, to locate the "effective" string ends, using the retarded time formalism; Though this is simple in principle, the near cancelation of terms, encountered earlier during incoherent synchrotron radiation calculations, makes this, perhaps, the hardest part of the calculation, especially in the case of bunches entering and leaving magnets. Robust numerical procedures will be derived for locating bunch ends of bunches either wholly inside or wholly outside magnets. For strings of millimeter length and magnets of meter length it might be thought that the legitimacy of ignoring the occasional overlap would be obvious, but the apparent or retarded length of a short string can be much longer than the length observed in the laboratory. A prescription for treating bunches entering and leaving magnets is discussed here, but explicit formulas are not given.

A serious conceptual problem that arises, especially when calculating longitudinal effects, is that Maxwell theory is incomplete when describing the self-force of a moving point charge. To recover self-consistency for point charges it is necessary to acknowledge the presence of *ad hoc*, internal, non-electromagnetic stress forces that oppose the electromagnetic forces to leave the total self-force finite.

There are at least two ways to calculate CSR. Most elementary is to integrate the Poynting-vector-calculated energy flow in the far field radiation. Formulas are given below. Alternatively, by energy conservation, the radiated energy should equal the work done by the self-force. The aforementioned divergence complicates the calculation of this self-force. There is a trick, called "renormalization" by Saldin et al. [3].[1] (It may be due to earlier authors such as Iogansen and Rabinovich [7], or Tamm [8].) Their trick is to subtract, from the force calculated in curved motion, the longitudinal force that would be present in straight line motion. This cancels the infinite self-force mentioned in the previous paragraph; the residual force is due entirely to the curvature. This may leave residual internal longitudinal forces but they (a) will be already present in linear motion, and hence calculable (and subtractable) as if the bunch is in free space, and (b) will cause no net acceleration of the bunch as a whole.

Calculating the force due to the presence of a charge in a magnet by subtracting the force it would feel in free space, as in the regularized approach just described, has a disadvantage. Especially at low energies, the space charge force even in free space is not at all negligible and should not simply be subtracted. If the free space force is subtracted when a particle is in a magnet it will have to be put back by some other method, such as PIC.

In a later section (13.8) the regularized, on-axis, longitudinal force is calculated by formulas from this chapter and by formulas of Saldin et al., and the results are found to be in excellent agreement. This chapter can therefore be regarded as extending the formulas of Saldin et al. both to off-axis locations and to include the transverse force components. For on-axis particles the transverse force has also been calculated by Geloni et al. [9] in a paper that also discusses transitions from outside to inside magnets. A later paper by Geloni et al. [10] discusses the off-axis problem, for vertical, though not radial, offsets. As stated already, the present treatment avoids the need for regularization during numerical simulation of bunch evolution.

The strategy is to reduce divergence problems by working only with line charges. The calculation starts gently, by calculating the self-force of a longitudinally-aligned, uniformly-charged, string moving in free space. Though leading-end and trailing-end forces will be found, they are only logarithmically divergent, and furthermore they are equal in magnitude but opposite in sign, thereby producing no net force on the string as a whole. This means that the regularization mentioned in the previous paragraph, though not necessarily incorrect, is an artifact of the mathematics of intermediate stages of the calculation rather than being an inherent feature of electromagnetic theory.

Though the word "string" is used, the treatment being described is not "string theory" as that term is currently understood. But the use of the recently-fancy noun "string" is not entirely inappropriate because, as in el-

---

1) The term "regularization" seems be more appropriate than "renormalization" for a process in which the theoretical formula for a measurable physical effect is the difference of two terms that are individually divergent.

ementary particle theory, the spurious (or at least non-electromagnetic) self-force of a string is less divergent than is the self-force of a point charge. Though this may still leave a regularization process necessary, the sensitivity of the procedure is greatly reduced by the gentler, only logarithmic, divergence. The chapter begins with the easiest part of the calculation—the self-force of a straight charged string in field free space.

13.3
Self-force of Moving Straight Charged String

A preliminary calculation will be to find the force exerted on itself by a uniformly traveling charged string. One purpose for this is to illustrate the style of calculation to be performed. The other is to prepare to handle a complication that will arise when calculating the space charge forces on a particle within a bunch traveling in an external magnetic field.

A uniformly charged string of length $2L$, charge q, travels with velocity $v = \beta c \hat{z}$ along the z-axis. Its line charge density is therefore $\lambda_0 = q/(2L)$. Within the charge distribution there are electric forces parallel to the axis but no magnetic forces. At position z on the z-axis, using a formula that Jackson calls a "preliminary form" (of equations that both Griffiths and Jackson refer to as "Jefimenko's equations"), the electric field $\mathbf{E}(z,t)$ is given by

$$\mathbf{E}(z,t) = -\frac{1}{4\pi\epsilon_0} \int \frac{dz'}{|z-z'|} \left[\nabla'\lambda + \frac{1}{c^2}\frac{\partial \mathbf{I}}{\partial t'} \right]_{\text{ret}}, \tag{13.1}$$

where z' is the source point and z is the field point, and where the "ret" subscript implies that the quantity in square brackets is to be evaluated at the "retarded time" t_r appropriate for the particular value of z', i.e. at $t_r = t - |z-z'|/c$. To facilitate calculation later on it is useful to multiply the integrand by a factor $\delta(\tau - (t - |z-z'|/c))$. Integration over τ then undoes the damage and restores the value of the expression. Then, since both integration ranges are infinite, the order of integration can be reversed. After these operations the integral (without numerical factor) becomes

$$\int d\tau \int \frac{dz'}{R} \left(\nabla'\lambda(\tau) + \frac{1}{c^2}\frac{\partial \mathbf{I}}{\partial \tau}(\tau) \right) \delta(\tau - (t - R/c)), \tag{13.2}$$

where $R = |z - z'|$. In terms of step function U the charge and current densities are

$$\begin{aligned} \lambda(z',\tau) &= \lambda_0 \left(U(z' - \underline{z}'(\tau)) - U(z' - \overline{z}'(\tau)) \right) \\ I(z',\tau) &= \beta c \lambda_0 \left(U(z' - \underline{z}'(\tau)) - U(z' - \overline{z}'(\tau)) \right). \end{aligned} \tag{13.3}$$

Here $\underline{z}'(\tau)$ and $\overline{z}'(\tau)$ are, respectively, tail and head positions of an "effective" charge distribution to be specified more explicitly below. The only dependences of the charge and current distributions are through these locations; otherwise the charge density is constant, either zero or λ_0. The derivatives needed for Eq. (13.2) are

$$\nabla'\lambda|_z = \lambda_0 \left(\delta(z' - \underline{z}'(\tau)) - \delta(z' - \overline{z}'(\tau)) \right)$$
$$\frac{\partial I}{\partial \tau} = -\beta^2 c^2 \lambda_0 \left(\delta(z' - \underline{z}'(\tau)) - \delta(z' - \overline{z}'(\tau)) \right). \tag{13.4}$$

The purpose of introducing the artificial variable τ is so that these derivatives can be evaluated without reference to any subsequent specialization of the meaning of τ. The time derivative is constant everywhere except at the string ends, which are always moving at speed βc. The gradient $\nabla'\lambda$ also has transverse components; they will not be required. For an on-axis point z the only component of electric field is longitudinal. Substitution into Eq. (13.2), using $1 - \beta^2 = 1/\gamma^2$, and (without loss of generality) setting $t = 0$ yields

$$E_z(z,0)$$
$$= -\frac{\lambda_0}{4\pi\epsilon_0 \gamma^2} \int d\tau \int \frac{dz'}{|z-z'|} \left(\delta(z' - \underline{z}'(\tau)) - \delta(z' - \overline{z}'(\tau)) \right) \delta(\tau + R/c)$$
$$= -\frac{\lambda_0}{4\pi\epsilon_0 \gamma^2} \int \frac{d\tau}{R(\tau)} \left(\delta(\tau + \underline{R}(\tau)/c) - \delta(\tau + \overline{R}(\tau)/c) \right). \tag{13.5}$$

Even for this preliminary one-dimensional problem it is convenient to use unit vectors $\hat{\mathbf{z}}$ and (source point to field point unit vector) $\hat{\mathbf{r}}$ which, though equal in direction and magnitude, may be opposite in sign. The δ-functions can be transformed to handle the implicit τ-dependence. For example, introducing "retardation factor" $\underline{\kappa}$ defined by

$$\underline{\kappa} \equiv \frac{\partial}{\partial \tau} \left(\tau + \frac{|z - \underline{z}'(\tau)|}{c} \right) = 1 - \beta \hat{\mathbf{v}} \cdot \hat{\mathbf{r}}, \tag{13.6}$$

the first δ-function becomes

$$\delta \left(\tau + \frac{|z - \underline{z}'(\tau)|}{c} \right) = \frac{1}{1 - \beta \hat{\mathbf{v}} \cdot \hat{\mathbf{r}}} \delta(\tau - \underline{t}'), \tag{13.7}$$

where $\underline{t}'$ is the retarded time from which the tail particle has influence at the observation point at $t = 0$. For trailing and leading ends the on-axis retardation factors are

$$\underline{\kappa} = 1 - \beta, \quad \overline{\kappa} = 1 + \beta. \tag{13.8}$$

Combining factors, the on-axis electric field is

$$E_z(z,0) = \frac{\lambda_0}{4\pi\epsilon_0 \gamma^2} \left(\frac{1}{1+\beta} \frac{1}{|z - \overline{z}'|} - \frac{1}{1-\beta} \frac{1}{|z - \underline{z}'|} \right). \tag{13.9}$$

The electric field at $z = 0$ (now assumed to be internal to the string) is given by

$$E_z(0,0) = \frac{\lambda_0}{4\pi\epsilon_0 \gamma^2} \left(\frac{1}{1+\beta} \frac{1}{\overline{z}'} + \frac{1}{1-\beta} \frac{1}{\underline{z}'} \right). \tag{13.10}$$

(In simplifying this formula the "obvious" relations $\underline{z}' < 0$ and $0 < \overline{z}'$ have been assumed. For points exactly on the line of charge these are correct. But, anticipating later discussion, for a point displaced transversely, say by a finite height y, no matter how small, one or the other of these inequalities may be incorrect.) Next the retarded times and positions will be determined.

Referring to Figure 13.1, a test point within the string, displaced by z_t from the string center, arrives at point P ($z' = 0$) at time t. The head particle emits a signal at time $\overline{t}_r$ as it passes the point $z' = \overline{z}'$. For this signal's arrival to coincide with the test point's arrival at P requires

$$\overline{t}_r + \frac{\overline{z}'}{c} = t, \quad \underline{t}_r - \frac{\underline{z}'}{c} = t, \tag{13.11}$$

where the tail equation has also been written. (The times t and t_r in Eqs. (13.11) are "laboratory time", as measured by clocks stationary in the laboratory. Since the events are at different locations it is necessary to use two previously synchronized clocks, one at each event. This same comment applies to all times appearing in this chapter and there will never be any introduction of moving coordinate frames, nor Lorentz transformation.) Note that both $\underline{t}_r$ and $\overline{t}_r$ precede t.

Let us set $t = 0$; that is, the test point arrives at point P at $t = 0$. At that instant, relative to the test point, the head particle is at $z = L - z_t$. The equations of motion of head (and similarly for tail) are therefore

$$\overline{z}' = (L - z_t) + \beta c \overline{t}_r, \quad \underline{z}' = (-L - z_t) + \beta c \underline{t}_r. \tag{13.12}$$

Eliminating $\overline{z}'$ and $\underline{z}'$ from these equations yields

$$c\overline{t}_r = \frac{-L + z_t}{1+\beta}, \quad c\underline{t}_r = -\frac{L + z_t}{1-\beta} \tag{13.13}$$

(Since both values of t_r are necessarily negative it would be more apt to refer to t_r as "earlier time" rather than as "retarded time" which is the customary terminology.) Solving for $\overline{z}'$ and $\underline{z}'$ yields

$$\overline{z}' = \frac{L - z_t}{1+\beta}, \quad \underline{z}' = \frac{-L - z_t}{1-\beta}. \tag{13.14}$$

These points delimit the "effective" charge distribution.

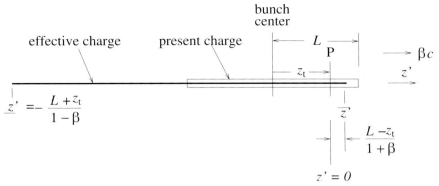

Fig. 13.1 Definition of coordinates. Snapshot of moving string (the open rectangle) at the "present" time $t = 0$. The solid line indicates the electrostatic configuration temporarily "equivalent" to the "true" electrodynamic system of a moving charged string; there are charges in just those locations that, at some time $t < 0$ contribute to the electric field at test point P at $t = 0$. In this figure that point is labeled P and is instantaneously located at $z = 0$.

Returning to the self-force calculation, the end coordinates can be substituted into Eq. (13.10) to produce

$$E_z(z_t) = \frac{\lambda_0}{4\pi\epsilon_0 \gamma^2} \left(\frac{1}{L - z_t} - \frac{1}{L + z_t} \right). \tag{13.15}$$

Another way of calculating this electric field is to start from the Heaviside, Poincaré, Schott formula for the electric field accompanying a uniformly moving point charge $\lambda_0|d\zeta|$ as viewed at transverse position y and longitudinal position ζ;

$$d\mathbf{E}(\mathbf{r}, t) = \frac{\lambda_0|d\zeta|}{4\pi\epsilon_0 \gamma^2} \frac{1}{(1 - \beta^2 \sin^2\theta)^{3/2}} \frac{\hat{\mathbf{r}}}{\zeta^2 + y^2}, \tag{13.16}$$

where θ is the angle between the vector from the source point to the field point and the trajectory direction. When expressed in this way, in terms of θ, the formula is uniformly valid both in front of and behind the moving charge, and the longitudinal integration yields for the longitudinal electric field

$$E_z(\mathbf{r}, t) = \frac{\lambda_0}{4\pi\epsilon_0 \gamma^2 y} \int_{\underline{\theta}}^{\overline{\theta}} \frac{d(\sin\theta)}{(1 - \beta^2 \sin^2\theta)^{3/2}}$$

$$= \frac{\lambda_0}{4\pi\epsilon_0 \gamma^2} \left[\frac{\sin\theta / y}{(1 - \beta^2 \sin^2\theta)^{1/2}} \right]_{\underline{\theta}}^{\overline{\theta}}. \tag{13.17}$$

On-axis, this agrees with Eq. (13.15). Another check is contained in a paper by Jefimenko [11], which yields the same result.

Later we will take yet another (only slightly different) approach, starting from the final Jefimenko equation for $\mathbf{E}(\mathbf{r},t)$;

$$\frac{1}{4\pi\epsilon_0}\int\left[\frac{\lambda(z',t')\hat{\mathbf{r}}}{(z-z')^2}+\frac{(\partial\lambda/\partial t')\hat{\mathbf{r}}/c-(\partial I/\partial t')\hat{\mathbf{z}}/c^2}{|z-z'|}\right]dz'. \tag{13.18}$$

Though this formula has great heuristic virtue, in order for it to be valid, as Jackson emphasizes, extreme care is needed in interpreting the partial derivative symbols. In particular, as well as the square bracket being evaluated at retarded time t', the observation time t is to be held constant as the partials are evaluated. This is harder than it sounds.

The first two terms of Eq. (13.18) result from a perhaps subtle interpretation and a differentiation by parts, but the third term has simply been copied from Eq. (13.1); indicating it by superscript (3), its value has already been calculated to be

$$E_z^{(3)}(z_t)=-\frac{\lambda_0\beta^2}{4\pi\epsilon_0}\left(\frac{1}{1+\beta}\frac{1}{\overline{z}'}+\frac{1}{1-\beta}\frac{1}{\underline{z}'}\right) \tag{13.19}$$

The first term of Eq. (13.18) can be interpreted as Coulomb's law applied to the effective linear charge distribution running from $\underline{z}'$ to $\overline{z}'$—in fact this is the motivation underlying the Jefimenko equation. Again referring to Figure 13.1, the electric field at a point z_t closer to the front than to the back of the line charge, can be calculated at time $t=0$. The result, after setting $z=0$, is

$$E_z^{(1)}(z_t)=\frac{\lambda_0}{4\pi\epsilon_0}\int_{\underline{z}'}^{-\overline{z}'}\frac{dz'}{z'^2}=\frac{\lambda_0}{4\pi\epsilon_0}\left(\frac{1}{\overline{z}'}+\frac{1}{\underline{z}'}\right). \tag{13.20}$$

In this integral, coming from the first term of Eq. (13.18), z has been set to zero in order to find the electric field at z_t. The (divergent) contributions from intervals adjacent and symmetric relative to the test point have cancelled by symmetry. This comment is is only valid for $-\overline{z}>\underline{z}$, but repeating the calculation for points closer to the tail gives the same result.

All that remains is to evaluate the second term of Eq. (13.18). For a particular feature, say the trailing end for definiteness, consider the evaluation of $\partial U(z'-\underline{z}')/\partial t'$. The implicit dependence of argument on t' and the need to hold t fixed strains the partial derivative notation enough to make the derivative perhaps ambiguous. Proceeding cautiously, the derivative is

$$\frac{\partial}{\partial t'}U(z'-\underline{z}'(t'))=\delta(z'-\underline{z}'(t'))(-\beta c), \tag{13.21}$$

which makes the trailing edge term be

$$-\frac{\lambda_0}{4\pi\epsilon_0}\beta\int dz'\frac{\hat{\mathbf{r}}}{|z-z'|}\delta(z'-\underline{z}'(t')). \tag{13.22}$$

Performing the integral as before and combining factors, we obtain, for the second term of Eq. (13.18),

$$\overline{E}_z^{(2)}(z_t) = \frac{\lambda_0}{4\pi\epsilon_0} \frac{\beta}{1+\beta} \frac{-1}{|z-\overline{z}'|_{z=0}} = \frac{\lambda_0}{4\pi\epsilon_0} \frac{-\beta}{1+\beta} \frac{1}{\overline{z}'},$$

$$E_z^{(2)}(z_t) = -\frac{\lambda_0}{4\pi\epsilon_0} \frac{\beta}{1-\beta} \frac{1}{|z-\underline{z}'|_{z=0}} = \frac{\lambda_0}{4\pi\epsilon_0} \frac{\beta}{1-\beta} \frac{1}{\underline{z}'}. \tag{13.23}$$

Summing the fields given by Eqs. (13.19), (13.20) and (13.23), the result agrees with Eq. (13.10). This confirms the validity of the recent, possibly dubious, differentiation.

The total force acting on the string is obtained by integrating Eq. (13.15) over z_t;

$$\int_{-L}^{L} E_z(z_t) \lambda_0 \, dz_t = \frac{q^2/(4L^2)}{4\pi\epsilon_0 \gamma^2} \int_{-L}^{L} dz_t \left(\frac{1}{L-z_t} - \frac{1}{L+z_t} \right). \tag{13.24}$$

In spite of the logarithmic divergence at each end, their sum vanishes—a very satisfactory result. But the string tension is infinite and when L is allowed to approach zero, the total force becomes ambiguous. This is presumably a manifestation of the fact that charges of zero extent cannot be consistently incorporated into electromagnetism without introducing further effects such as non-electromagnetic internal stresses. Unlike point charges, where the divergence is worse, the present end divergences are only logarithmic (and suppressed by factor $1/\gamma^2$ at that) and for straight line motion they cancel. It is the only-weak divergence of the self-force that makes me feel justified in using the term "string", since it is presumably very similar considerations that simplify the renormalizability of string theory. For motion of the string along a curved path in a magnetic field the end cancellation may no longer be perfect. What to do about this possibility will be discussed next.

13.4
Self-force of Moving Straight Charged Ribbon

A way of regularizing the self-force is to give the string of charge some transverse size, say a height d, as shown in Figure 13.2, and call it a moving ribbon. To calculate the self-force let us first find the force of a sub-ribbon at y_2 due to a sub-ribbon at y_1. These ribbons have transverse separation $y = y_2 - y_1$.

A new feature that has to be appreciated is that the transverse displacement y has a *big* effect on the longitudinal force. One aspect of this is indicated qualitatively in the caption to the figure. Another aspect (for relativistic motion) concerns signal propagation between two initially side-by-side particles, one on sub-ribbon (1) the other on (2). The longitudinal distance side-by-side

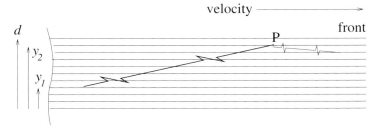

Fig. 13.2 Configuration used to calculate the self-force of a moving ribbon of charge. For sub-ribbons transversely close, the point P is largely influenced by points further forward in the bunch that take advantage of their earlier arrival to "send signals back". Sub-ribbons transversely farther away from the point P can be influential only by sending a signal from far enough back for the slight $c > v$ signal speed advantage to make up for the extra signal path length.

particles have traveled when the signal from one arrives at the other is given by

$$\delta_{12} \approx \gamma |y|. \tag{13.25}$$

Even for small y this can be large, even far larger than the length of a slim bunch.

The effect of non-zero y on the retarded time calculation is illustrated by Figure 13.3. The lower branch of hyperbola

$$c^2 t_r^2 - z'^2 = y^2 \tag{13.26}$$

contains points with time and position coordinates (t_r, z') on the string axis from which emitted signals arrive at observation point at $t = 0$. (The upper branch would apply to an irrelevant "advanced time" calculation.) Consider the charge at position z_s relative to string center; relative to this charge the test point z_t is displaced by $\Delta z = z_t - z_s$. With time and space origins adjusted to vanish instantaneously at the test point, the world line of the source point is

$$ct_r = \frac{z' + \Delta z}{\beta}. \tag{13.27}$$

The intersection of line and hyperbola gives the (ct_r, z') coordinates of the particular source charge influencing the particular test charge at $t = 0$. In particular the straight lines passing through the string extremes define the effective electrostatic line charge. Contrary to a valid only for $y = 0$ assumption mentioned just below Eq. (13.11), one sees that the most forward source point can be behind (i.e. at less positive longitudinal position) than the test point. The dividing condition, for the effective charge front to coincide longitudinally with the test point, is

$$\bar{z}' = 0, \quad \text{or} \quad \Delta z = -\beta y, \quad \text{or} \quad z_s = z_t + \beta y. \tag{13.28}$$

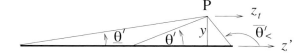

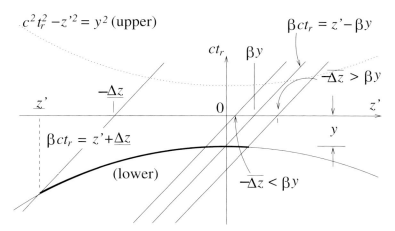

Fig. 13.3 Figure illustrating the $y \neq 0$ retarded time calculation. The upper figure shows the effective electrostatic configuration. In the lower figure the lower hyperbolic branch contains points from which emitted signals arrive at observation point P at $t = 0$. Straight lines are world lines of point charges in the line charge distribution. Heavy lines in both figures correspond to the "effective" charge distribution. Depending on y, the head may be longitudinally in front of, or behind, the test point.

These conditions are illustrated in Figure 13.4. For charges in the interval of length βy at the front of the actual bunch, even apart from any dependence of electric field on y, this invalidates some relative-position assumptions made while obtaining the $y = 0$ field.

A work-around for this problem is suggested by the upper part of Figure 13.3. It is to express the range of the effective charge distribution by angular coordinate θ' instead of by rectilinear coordinate z'. For $y > 0$ the entire range of θ' is well behaved. Setting up the integral analogous to Eq. (13.20) and performing the integration, one finds that the longitudinal electric field at P due to the effective line charge (end contributions temporarily neglected) is given by

$$dE_z(z_t, y) = \frac{\lambda_0 (dy_1/d)}{4\pi\epsilon_0 y} \left(\sin \overline{\theta}' - \sin \underline{\theta}' \right). \tag{13.29}$$

For consistency, having admitted finite y, it is necessary to consider magnetic fields as well. With **r** being the vector from the source point to the field

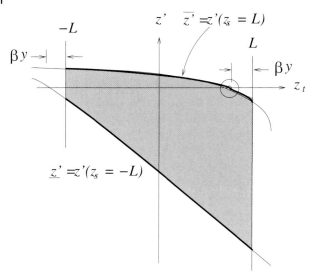

Fig. 13.4 The region of effective charge contributing to the total bunch self-force. At given test point z_t the uniform charge distribution runs from $\underline{z'} = z'(z_s = -L, y)$ to $\overline{z'} = z'(z_s = -L, y)$. The open circle indicates a troublesome point at which there is a "one-sided" singularity in the integral over z_t for $y \to 0$.

point, so $r^2 = z'^2 + y^2$, the complete Jefimenko equations are

$$\mathbf{E}(P, t) = \frac{1}{4\pi\epsilon_0} \int \left(\frac{\lambda(z', t_r)}{r^2} \hat{\mathbf{r}} + \frac{\dot{\lambda}(z', t_r)}{cr} \hat{\mathbf{r}} - \frac{\dot{\mathbf{I}}(z', t_r)}{c^2 r} \right) dz', \tag{13.30}$$

$$\mathbf{B}(P, t) = \frac{\mu_0}{4\pi} \int \left(\frac{\mathbf{I}(z', t_r)}{r^2} + \frac{\dot{\mathbf{I}}(z', t_r)}{cr} \right) \times \hat{\mathbf{r}} \, dz', \tag{13.31}$$

Here, following Griffiths's somewhat casual notation, partial derivatives with respect to "time" are indicated by overhead dots. These derivatives assume z' is held fixed, and the resulting quantity evaluated at the retarded time appropriate for that value of z'. In other words the overhead dots have the same meaning as the partial derivatives in Eq. (13.18). (This is not at all the same as the convention invented by Newton and customary in mechanics.) We are mainly interested in the total force, say on charge e, which is given by

$$\frac{\mathbf{F}(P)}{e} = \mathbf{E}(P) + \beta c \hat{\mathbf{z}} \times \mathbf{B}(P). \tag{13.32}$$

Substituting from Eqs. (13.30) and (13.31) and reducing the result produces

13.4 Self-force of Moving Straight Charged Ribbon

body contribution

$$d\mathbf{F}^{body}/e = \frac{\lambda_0 dy_1/d}{4\pi\epsilon_0 y}\left(\frac{\sin\theta'}{\gamma^2}\hat{\mathbf{y}} + \cos\theta'\hat{\mathbf{z}}\right)dz'$$

$$= \frac{\lambda_0 dy_1/d}{4\pi\epsilon_0 y}\left(-\frac{\cos\theta'}{\gamma^2}\hat{\mathbf{y}} + \sin\theta'\hat{\mathbf{z}}\right)\Bigg|_{\theta'}^{\overline{\theta'}}, \quad (13.33)$$

and end contribution

$$d\mathbf{F}^{ends}/e = \frac{dy_1/d}{4\pi\epsilon_0}\int\frac{\dot{\lambda}}{cr}\left(\frac{\sin\theta'}{\gamma^2}\hat{\mathbf{y}} + (\cos\theta' - \beta)\hat{\mathbf{z}}\right)dz'$$

$$= \frac{\lambda_0 dy_1/d}{4\pi\epsilon_0}\left[\frac{\beta}{1 - \beta\cos\theta'}\left(\frac{\sin\theta'}{\gamma^2 r'}\hat{\mathbf{y}} + \frac{\cos\theta' - \beta}{r'}\hat{\mathbf{z}}\right)\right]_{\theta', r'}^{\overline{\theta'}, \overline{r'}}. \quad (13.34)$$

The end corrections have been calculated as in Eq. (13.22), using $\kappa = 1 - \beta\cos\theta'$. Combining body and end contributions, for $z = 0$ the longitudinal force is

$$\frac{dF_z}{e} = \frac{\lambda_0 dy_1/d}{4\pi\epsilon_0\gamma^2}\left[\frac{1}{r' - \beta r'\cos\theta'}\right]_{\theta', r'}^{\overline{\theta'}, \overline{r'}}. \quad (13.35)$$

Since no approximations have been made, nor assumptions about the relative position of source points and field point, this formula is valid everywhere in space. But it is inconveniently expressed in terms of retarded coordinates. It will be convenient to convert to coordinates related to the actual, i. e. present, position of the string. For this calculation Figure 13.5 is useful. The master relationships, as shown in the figure, are

$$-\underline{z}' - (L + z_t) = \beta\underline{r}' \quad \text{and} \quad (L - z_t) - \overline{z}' = \beta\overline{r}'. \quad (13.36)$$

The first of these equations shows that the trailing edge travels from its retarded position to its present position at speed v while a signal travels from the retarded position to the observation point at speed c. Also one has

$$\underline{r}'\cos\underline{\theta}' = -\underline{z}' \quad \text{and} \quad \overline{r}'\cos\overline{\theta}' = -\overline{z}'. \quad (13.37)$$

Equations (13.36) and (13.37) remain valid even if the test point lies longitudinally outside the true charge distribution. Using the figure, and following a manipulation suggested by Jefimenko, the needed, trailing-edge, denominator factor $\underline{r}' - \beta\underline{r}'\cos\underline{\theta}'$ can be expressed as

$$\sqrt{\underline{r}'^2 + 2\beta\underline{r}'\underline{z}' + \beta^2\underline{z}'^2 + \underline{z}'^2 - \underline{z}'^2 + \beta^2\underline{r}'^2 - \beta^2\underline{r}'^2}, \quad (13.38)$$

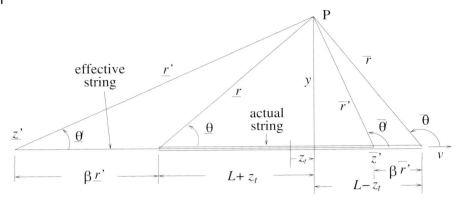

Fig. 13.5 Geometric relationships between present string coordinates and retarded string coordinates. As drawn $\bar{z}'$ is positive (i. e. to the right of P) but, for y as large as shown, $\bar{z}'$ would be negative for highly relativistic motion.

with terms having been judiciously added and subtracted. The terms can be regrouped using $\underline{r}'^2 - \underline{z}'^2 = y^2$ and (from Eq. (13.36)) $(L+z_t)^2 = \underline{z}'^2 + 2\beta \underline{z}' \underline{r}' + \beta^2 \underline{r}'^2$. The result is

$$\underline{r}' - \beta \underline{r}' \cos \underline{\theta}' = \sqrt{(L+z_t)^2 + y^2/\gamma^2}. \tag{13.39}$$

This result can be generalized to

$$r' - \beta r' \cos \theta' = \sqrt{\Delta z^2 + y^2/\gamma^2}, \tag{13.40}$$

where $\Delta z = z_t - z_s$ is the relative longitudinal coordinate introduced previously. Using this result, the differential longitudinal force is

$$\frac{\mathrm{d}F_z}{e} = \frac{\lambda_0 \mathrm{d}y_1/d}{4\pi\epsilon_0 \gamma^2} \left(\frac{1}{\sqrt{(L-z_t)^2 + y^2/\gamma^2}} - \frac{1}{\sqrt{(L+z_t)^2 + y^2/\gamma^2}} \right). \tag{13.41}$$

The total longitudinal force on sub-ribbon (2) due to sub-ribbon (1) is given by

$$\mathrm{d}^2 F_z(y) = \frac{\lambda_0 \mathrm{d}y_2}{d} \int_{-L}^{L} \mathrm{d}z_t \frac{\mathrm{d}F_z}{e}. \tag{13.42}$$

This integral vanishes because the integrand is an odd function of z_t. If one insists on calculating a fractional force, say on the front half of the ribbon, after integration over z_t, y_1, and y_2, the resulting force will be convergent; the divergent force has been moderated to produce only a term proportional to $\ln d$ which is only weakly divergent as $d \to 0$.

13.4 Self-force of Moving Straight Charged Ribbon

Since the magnetic field contributes no longitudinal force, Eq. (13.42) should also be equivalent to Eq. (13.17). To confirm this observe that

$$\frac{\sin\theta/y}{\sqrt{1-\beta^2\sin^2\theta}} = \frac{1}{\sqrt{y^2/\tan^2\theta + y^2/\gamma^2}} = \frac{1}{\sqrt{\Delta z^2 + y^2/\gamma^2}}. \quad (13.43)$$

where θ is the angle from present source point to present test point.

The transverse force can be calculated similarly. Summing body and end contributions, for $z = 0$, analogous to Eq. (13.35), the y-component of force is

$$\frac{dF_y}{e} = \frac{\lambda_0 dy_1/d}{4\pi\epsilon_0\gamma^2}\left[-\frac{\cos\theta'}{y} + \frac{\beta\sin\theta'}{r' - \beta r'\cos\theta'}\right]_{\theta',r'}^{\theta',\bar{r}'}. \quad (13.44)$$

Manipulations like those used for the longitudinal force lead to the identity

$$-\frac{\cos\theta'}{y} + \frac{\beta\sin\theta'}{r' - \beta r'\cos\theta'} = \frac{-\Delta z/y}{\sqrt{\Delta z^2 + y^2/\gamma^2}}, \quad (13.45)$$

The square bracket expression in Eq. (13.44) therefore reduces to

$$-\frac{1}{y}\left[\frac{L-z_t}{\sqrt{(L-z_t)^2 + y^2/\gamma^2}} - \frac{L+z_t}{\sqrt{(L+z_t)^2 + y^2/\gamma^2}}\right]. \quad (13.46)$$

The structures of Eq. (13.41) and (13.46) are similar and noteworthy. Naturally, being the result of integrations, they are differences of indefinite integrals evaluated at string head and tail. But the indefinite integral, especially for dF_z, is a function only of a "generalized" present distance $\sqrt{(L\pm z_t)^2 + y^2/\gamma^2}$ from field point P to the string end. There is a kind of "inverse distance force law", attractive toward the leading end (for test particles within the string), repulsion from the trailing end. This force is far from being directed along the line from source charge to test charge however and dF_z does not necessarily reverse sign as the test point is moved longitudinally past the head of the string. One expects the transverse force to depend primarily on y. Equation (13.46) confirms this by showing a leading $1/y$ dependence. But this factor is "modulated" by a difference of terms depending only on distances to the string ends, or rather on the ratios of actual distance divided by generalized distance. Except for z_t very close to either end, cancellation of the terms within the square brackets suppresses the $1/y$ singularity at $y = 0$.

It is consistent to pretend that the total force is due entirely to two "sources", one at the head, one at the tail, with source "strength" given by the value of the indefinite integral. In this picture there are *no* force contributions from the charge in the interior of the string—not even proportional to $1/\gamma^2$. This suggests a simple numerical procedure for calculating the effects of space charge forces on a beam traveling in a field free region. With the bunch represented as a superposition of longitudinal strings, the force on any single particle depends only on its relation to the ends of the other strings. Because of the inverse dependence on y in Eq. (13.46) the end forces are not, however, directed radially along the line joining the end points and the field point. These forces are intended to form the basis for treating space charge effects as direct intrabeam scattering forces.

Accelerator beams are typically considerably longer than they are broad or high. This large aspect ratio gets effectively increased by the further factor γ by which y is multiplied in the "generalized distance" appearing in the denominators. This provides further heuristic explanation of the surprisingly small space charge effects in accelerator beams.

13.5
Curve End Point Determination

Finally we start on the real problem, which is to calculate the space charge force from a longitudinally-aligned, curved (due to its presence in a magnetic field) string of charge of length $2L$, uniform charge density λ_0, acting on a co-moving "test charge" as it passes some nominal lattice point P. The plan is, to, as far as possible, recast this electrodynamic problem into an "equivalent" electro/magneto-static problem, following the pattern of the previous sections.

As far as possible notation will be carried over from the preceeding discussion. In particular retarded locations will be indicated by symbols with primes. (With retarded locations being regarded as source points, this is quite intuitive. The only important exception is that $\mathbf{r}'$ is to be the vector directed from the retarded source point to the test point P, as in Figure 13.5.)

Consider the distribution of charges illustrated in Figure 13.6. An arc of charge, linear charge density λ_0, flows along the arc of a circle of radius R with speed βc. The single test charge (charge = e, longitudinal position s_t, transverse coordinates x, y, all relative to string center) is traveling with velocity $\beta c \hat{\mathbf{s}}$ parallel to the charged string. After finding the force, due to incremental length $R d\alpha'$ at source point P', we will find the total force on the test charge at P by integrating along the arc.

Choosing the origin of time at the instant the test charge passes point P in the storage ring, the test charge's equation of motion is

$$s_t = \beta c t. \tag{13.47}$$

Let the longitudinal coordinate of the test particle relative to some particular source particle be

$$\Delta s \equiv s_t - s_s. \tag{13.48}$$

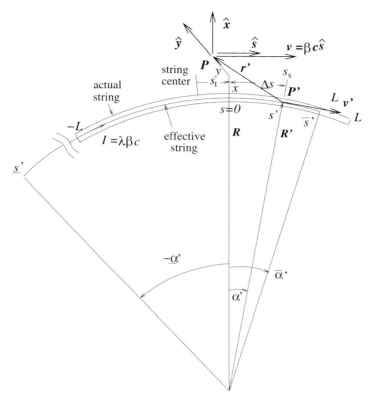

Fig. 13.6 Definition of coordinates. A snapshot of the present, actual, charge distribution is shown and, superimposed on it is an electro/magneto-static configuration "equivalent" to the "true" electrodynamic system of a moving charged string. There are moving charges in just those locations of charges in the true system that, at some earlier time $t < 0$ contribute to fields at P at $t = 0$.

(For example, for the source particle being at the head of the string, $\Delta s = s_t - L \equiv \overline{\Delta s}$, which is negative (or zero).) The equation of motion of the source particle is therefore

$$s' = -\Delta s + \beta c t_r. \tag{13.49}$$

This corresponds to Eqs. (13.12) but differs in that the equation describes a general source particle rather than the head or tail. Since it will be convenient to use angular coordinates rather than arc length, define

$$\alpha' = \frac{s'}{R}. \tag{13.50}$$

The equation of motion of the source particle can therefore also be expressed as

$$R\alpha' = \beta ct_r - \Delta s. \tag{13.51}$$

Charge and current densities are to be evaluated at time t_r, which is related to observation time t by

$$t = t_r + \frac{r'}{c}, \tag{13.52}$$

where **r**$'$ is the vector from source point to field point. The length r' of this vector is given by

$$r' = \sqrt{R^2 + (R+x)^2 - 2R(R+x)\cos\alpha' + y^2}. \tag{13.53}$$

Setting $t = 0$ in Eq. (13.52), and combining it with Eqs. (13.51) and (13.53), the condition for time t_r at which charge at P' influences fields at P at $t = 0$, is

$$\frac{R\alpha' + \Delta s}{\beta} = -\sqrt{2R^2\left(1 + \frac{x}{R}\right)(1 - \cos\alpha') + x^2 + y^2}. \tag{13.54}$$

With the minus sign explicitly included, the r.h.s. of this equation is necessarily negative, which forces t_r to be negative, as it must be, and corresponds to picking the intersection with the lower branch in Figure 13.3. This equation (implicitly) determines α' as a function of Δs. Equivalently, in view of Eq. (13.51), the equation determines t_r as a function of Δs. The solution of Eq. (13.54) for a typical choice of parameter values is shown in Figure 13.7.

Two values are special; they are $s_s = \pm L$, at the head and tail of the moving string. These values determine the extreme angles in Figure 13.6. The position defined by $\bar{\alpha}'$ is the location at which the particle at $s_s = L$ has an influence at P at $t = 0$. Because the head particle is spatially ahead, as it arrives at $\bar{\alpha}'$ there is still time to get a signal back to P coincident with the test charge's arrival there. On the other hand a signal launched from the tail particle at $\underline{\alpha}'$ can catch up with the test charge at P by taking the path "as the crow flies". For the head particle Eq. (13.54) yields

$$\frac{R\overline{\alpha'} + s_t - L}{\beta} = -\sqrt{2R^2\left(1 + \frac{x}{R}\right)(1 - \cos\overline{\alpha'}) + x^2 + y^2}, \tag{13.55}$$

and the tail equation is obtained by the replacement $-L \rightarrow L$. At least in the range $-L < s_s < L$, when Maple is instructed to find a root of this equation, it finds the correct root $\overline{\alpha'}$. The angles found in this way are plotted in Figure 13.8.

For $x = y = 0$, for the end points, Eq. (13.54) simplifies to

$$\begin{aligned} R\overline{\alpha}' + s_t - L &= -2R\beta\sin(\overline{\alpha}'/2), \\ R\underline{\alpha}' + s_t + L &= 2R\beta\sin(\underline{\alpha}'/2). \end{aligned} \tag{13.56}$$

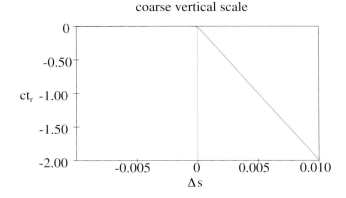

Fig. 13.7 Plot of ct_r given by Eq. (13.54) versus $\Delta s = s_t - s_s$ with numerical values $R = 80$ m, $x = y = 1$ mm. For $2L = 0.02$ m, $s_t = 0$, $\underline{\alpha}$ can be obtained from the right edge of the upper graph, $\bar{\alpha}$ from the left edge of the lower one.

For the highly relativistic condition of interest, e.g. relativistic factor $\gamma = 10^4$, Figure 13.6 is distorted in various ways. For physical systems of interest (except in tiny intervals near the ends) one will have the following inequalities

$$x, y \ll s_s \sim s_t \sim \Delta s \sim L \ll R. \tag{13.57}$$

For example, one system of interest has values $R = 80$ m, $2L = 2$ cm, $x \sim y \sim 1$ mm. In this case $\bar{\alpha}'$ is greatly exaggerated in the figure (and could even be negative if the head particle is only slightly longitudinally ahead of the test particle.) On the other hand it is normal for the condition $|\underline{\alpha}'|R \gg L$ to hold, as shown, because, with particle velocities close to the signal speed c, a signal from the tail particle can only catch up by "cutting across" a substantial arc.

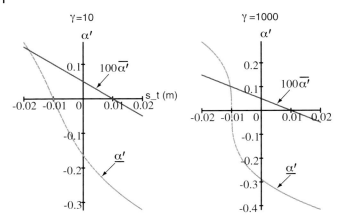

Fig. 13.8 The effective bunch end point angles $\underline{\alpha'}$ and $100\overline{\alpha'}$, for $x = y = 0$, as determined by Eqs. (13.56), are plotted against the test point longitudinal coordinate s_t. $R = 10$ m, $L = 0.01$ m.

In realistic storage ring configurations it is even possible for the shortcut just mentioned to be blocked by some obstacle, such as the inner wall of the vacuum system. This indicates that "screening effects" due to nearby conductors are likely to have significant effects on the space charge forces. Nevertheless I will neglect such things by assuming all charges are in free space. The rationale for this neglect is that the "image fields" by which such screening effects could be represented will vary but little over the extremely short transverse distances to be emphasized in the present chapter. The validity of this assumption can be investigated numerically later while applying the formulas to emittance growth. I will also neglect the longitudinal "slippage" between string charge and test charge, having identical speeds, that occurs for $x \neq 0$.

13.6
Field Calculation

One way of proceeding with the electromagnetic calculation would be to use the Liénard Wiechert expressions for the electric and magnetic fields of a moving point charge. Another would be to integrate over arc length the retarded time formulas for scalar and vector potentials and then to differentiate these potentials to find **E** and **B**—a notoriously complicated enterprise. As already stated, neither of these approaches is to be taken here. Rather **E** and **B** are to be obtained using the laws of Coulomb and Biot-Savart. In using these laws with moving charges, as well as evaluating the integrands at retarded times, it is necessary to include the extra "Jefimenko" terms in Eqs. (13.30) and (13.31).

Referring to Fig 13.6, with components spelled out in the order s, x, y,

$$\mathbf{r}' = \begin{pmatrix} -R\sin\alpha' \\ R(1-\cos\alpha') + x \\ y \end{pmatrix}, \hat{\mathbf{v}}' = \begin{pmatrix} \cos\alpha' \\ -\sin\alpha' \\ 0 \end{pmatrix}. \tag{13.58}$$

Note that $\mathbf{r}'$ is the vector from retarded source point to (present) field point. The charge and current densities are to be evaluated at time t_r calculated using Eqs. (13.52) through (13.54). The time derivative terms will be referred to as "string end" terms, consistent with the assumption that, within the string, the charge and current densities are independent of longitudinal position and hence also of time.

For a longitudinally uniform charge distribution the charge and current densities can be expressed in terms of step function U;

$$\begin{aligned}\lambda(\alpha') &= \lambda_0 (U(\alpha' - \underline{\alpha}') - U(\alpha' - \overline{\alpha}')) \\ \mathbf{I}(\alpha') &= \beta c \lambda_0 \hat{\mathbf{v}}'(\alpha')(U(\alpha' - \underline{\alpha}') - U(\alpha' - \overline{\alpha}')).\end{aligned} \tag{13.59}$$

The only dependence of λ on α' and t is through the string end locations—otherwise its value is either λ_0 or zero. The *direction* of $\mathbf{I}$ depends on α', but not on time.

It is well known that there is a substantial cancellation of magnetic and electric forces between parallel-traveling, relativistic particles. It is therefore appropriate to express explicitly the force $\mathbf{F}$ acting on the test charge as it passes point P;

$$\frac{\mathbf{F}(P,0)}{e} = \mathbf{E}(P,0) + \beta c \hat{\mathbf{s}} \times \mathbf{B}(P,0). \tag{13.60}$$

Temporarily setting aside the string end terms in Eqs. (13.30) and Eqs. (13.31), and copying a manipulation due to Bassetti and Brandt [6], and using the relations $\epsilon_0 \mu_0 = 1/c^2$ and $\beta^2 = 1 - 1/\gamma^2$, $\hat{\mathbf{s}} \cdot \hat{\mathbf{v}}' = \cos\alpha'$, and $\hat{\mathbf{s}} \cdot \mathbf{r}' = -R\sin\alpha'$, the force is given by

$$\begin{aligned}\frac{\mathbf{F}^{(\text{body})}}{e} &= \frac{R\lambda_0}{4\pi\epsilon_0} \int_{\underline{\alpha}'}^{\overline{\alpha}'} \frac{d\alpha'}{r'^3} \left(\mathbf{r}' + \beta^2 \hat{\mathbf{s}} \times (\hat{\mathbf{v}}' \times \mathbf{r}') \right) \\ &= \frac{R\lambda_0}{4\pi\epsilon_0} \int_{\underline{\alpha}'}^{\overline{\alpha}'} \frac{d\alpha'}{r'^3} \left((1-\cos\alpha')\mathbf{r}' - R\sin\alpha'\,\hat{\mathbf{v}}' + \frac{1}{\gamma^2}(\cos\alpha'\,\mathbf{r}' + R\sin\alpha'\,\hat{\mathbf{v}}') \right) \\ &\equiv \mathbf{F}_0^{(\text{body})}/e + \mathbf{F}_{1/\gamma^2}^{(\text{body})}/e.\end{aligned} \tag{13.61}$$

In the highly relativistic $\gamma \gg 1$ domain, the force $\mathbf{F}_{1/\gamma^2}^{(\text{body})}$, because it is multiplied by the small factor $1/\gamma^2$, is likely to be negligible. (However, this ceases to hold in the $R \to \infty$ limit.) The individual components of $\mathbf{F}_0^{(body)}$ are[2]

$$\frac{\mathbf{F}_0^{(body)}}{e} =$$

$$= \frac{R\lambda_0}{4\pi\epsilon_0} \int_{\underline{\alpha}'}^{\overline{\alpha}'} \frac{d\alpha'}{r'^3} \left((1-\cos\alpha') \begin{pmatrix} -R\sin\alpha' \\ R(1-\cos\alpha') + x \\ y \end{pmatrix} - R\sin\alpha' \begin{pmatrix} \cos\alpha' \\ -\sin\alpha' \\ 0 \end{pmatrix} \right)$$

$$= \frac{\lambda_0}{4\pi\epsilon_0} \int_{\underline{s}'}^{\overline{s}'} ds' \begin{pmatrix} -R\sin(s'R_i)/r'^3 \\ (2+xR_i)R(1-\cos(s'R_i))/r'^3 \\ (1-\cos(s'R_i))\, y/r'^3 \end{pmatrix} \quad (13.62)$$

where, to make evaluation easier in drift regions, the integration variable has been changed and abbreviation $R_i = 1/R$ has been introduced, The components of $\mathbf{F}_{1/\gamma^2}^{(body)}$ are

$$\frac{\mathbf{F}_{1/\gamma^2}^{(body)}}{e} =$$

$$= \frac{R\lambda_0}{4\pi\epsilon_0\gamma^2} \int_{\underline{\alpha}'}^{\overline{\alpha}'} \frac{d\alpha'}{r'^3} \left(\cos\alpha' \begin{pmatrix} -R\sin\alpha' \\ R(1-\cos\alpha') + x \\ y \end{pmatrix} + R\sin\alpha' \begin{pmatrix} \cos\alpha' \\ -\sin\alpha' \\ 0 \end{pmatrix} \right)$$

$$= \frac{\lambda_0}{4\pi\epsilon_0\gamma^2} \int_{\underline{s}'}^{\overline{s}'} ds' \begin{pmatrix} 0 \\ (-R(1-\cos(s'R_i)) + x\cos(s'R_i))/r'^3 \\ y\cos(s'R_i)/r'^3 \end{pmatrix} \quad (13.63)$$

The present electrodynamic calculation is closely connected (in fact patterned after) the static calculation of Bassetti and Brandt [6]. (Incidentally these authors show that $\mathbf{F}_{1/\gamma^2}^{(body)}$ *always* becomes essential as $R \to \infty$.) Formula (13.61) differs from the Bassetti-Brandt formula mainly because, working with a complete ring of current, their integration range is $-\pi < \alpha' < \pi$. For the same reason their force is purely transverse. Another consequence of the newly-introduced forward/back asymmetry is a non-vanishing longitudinal force. Of course, another deviation from the static calculation will result from the not-yet-included string end forces. It has been our assumption of uniform longitudinal charge distribution that has permitted the force exhibited so far to be represented as a straightforward sum of an electrostatic, Coulomb's law force and a magnetostatic, Biot-Savart force.

In the static calculation there are charges and currents present everywhere on a circle and at all times. In the dynamic calculation these distributions

---

[2] The final formulas in Eqs. (13.62), and similar equations below, have been manipulated slightly, from the versions given in the original paper, to facilitate the treatment of space charge forces in field free regions, where $R \to \infty$. This is explained further in Section 13.13. Mainly it amounts to replacing $1/R$ by R_i.

vanish over most of this path, most of the time; they are non-vanishing only at times and places capable of producing fields at point P at $t = 0$. In either case the time-dependent charge density is everywhere constant, with value either 0 or λ_0, except at the ends of the string.

Turning to string end terms, the step function expression in Eq. (13.59), with arguments fully spelled out, is

$$U\left(\alpha' - \frac{\beta c \underline{t_r} - \Delta \underline{s}}{R}\right) - U\left(\alpha' - \frac{\beta c \overline{t_r} - \overline{\Delta s}}{R}\right). \tag{13.64}$$

Time differentiation of these densities yields

$$\frac{\dot{\lambda}\hat{\mathbf{r}}'}{cr'} = -\frac{\beta \lambda_0}{R r'^2} \mathbf{r}' [\delta(\alpha' - \underline{\alpha}') - \delta(\alpha' - \overline{\alpha}')],$$

$$\frac{\dot{\mathbf{I}}}{c^2 r'} = -\frac{\beta^2 \lambda_0}{R r'} \hat{\mathbf{v}}'(\alpha') [\delta(\alpha' - \underline{\alpha}') - \delta(\alpha' - \overline{\alpha}')]. \tag{13.65}$$

(As mentioned before, overhead dots indicate *partial* derivatives with respect of time.) There is no term proportional to $\dot{\hat{\mathbf{v}}}'$ since $\hat{\mathbf{v}}'$ is a time-independent vector. The combination appearing in Eq. (13.30) is

$$\frac{\dot{\lambda}\mathbf{r}'}{cr'^2} - \frac{\dot{\mathbf{I}}}{c^2 r'} = -\frac{\beta \lambda_0}{R r'^2} (\mathbf{r}' - \beta r' \hat{\mathbf{v}}') [\delta(\alpha' - \underline{\alpha}') - \delta(\alpha' - \overline{\alpha}')]. \tag{13.66}$$

The string end forces are then given by

$$\frac{\mathbf{F}^{(\text{ends})}}{e} = \frac{R}{4\pi\epsilon_0} \int \frac{d\alpha'}{\kappa r'^2} \left(\frac{\dot{\lambda}\mathbf{r}'}{c} - \frac{r'\dot{\mathbf{I}}}{c^2} + \beta \hat{\mathbf{s}} \times \left(\dot{\mathbf{I}} \times \mathbf{r}'\right)\right)$$

$$= -\frac{\beta \lambda_0}{4\pi\epsilon_0} \left[\frac{1}{\kappa r'^2} \left(\mathbf{r}' - \beta r' \hat{\mathbf{v}}' + \beta^2 (\hat{\mathbf{s}} \cdot \mathbf{r}') \hat{\mathbf{v}}' - \beta^2 (\hat{\mathbf{s}} \cdot \hat{\mathbf{v}}') \mathbf{r}'\right)\right]_{\underline{\alpha}'}^{\overline{\alpha}'}$$

$$= \frac{\beta \lambda_0}{4\pi\epsilon_0} \left[\frac{1}{\kappa r'^2} \left((1 - \cos\alpha') \mathbf{r}' - (\beta r' + R \sin\alpha') \hat{\mathbf{v}}'\right)\right.$$

$$\left. + \frac{1}{\gamma^2} \left(\cos\alpha' \mathbf{r}' + R \sin\alpha' \hat{\mathbf{v}}'\right)\right]_{\underline{\alpha}'}^{\overline{\alpha}'} \tag{13.67}$$

The denominator factor κ is the "retardation factor" first introduced below Eq. (13.7). Here its value is

$$\kappa = 1 - \beta \hat{\mathbf{v}}' \cdot \hat{\mathbf{r}}' = 1 + \beta \frac{1 + x R_i}{r'} R \sin(s' R_i), \tag{13.68}$$

where inverse radius R_i has again been used. Spelling out the components, the γ-independent part is

$$\frac{\mathbf{F}_0^{(ends)}}{e} = \frac{\beta\lambda_0}{4\pi\epsilon_0}\left[\frac{1-\cos\alpha'}{\kappa r'^2}\begin{pmatrix} -R\sin\alpha' \\ R(1-\cos\alpha')+x \\ y \end{pmatrix} - \frac{R\sin\alpha'+\beta r'}{\kappa r'^2}\begin{pmatrix} \cos\alpha' \\ -\sin\alpha' \\ 0 \end{pmatrix}\right]$$

$$= \frac{\beta\lambda_0}{4\pi\epsilon_0}\left[\begin{array}{c} -R\sin(s'R_i)/\kappa r'^2 - \beta\cos(s'R_i)/\kappa r' \\ (2+xR_i)R(1-\cos(s'R_i))/\kappa r'^2 + \beta\sin(s'R_i)/\kappa r' \\ (1-\cos(s'R_i))\,y/\kappa r'^2 \end{array}\right]_{\underline{s}'}^{\overline{s}'}. \qquad (13.69)$$

The $1/\gamma^2$ term in Eq. (13.67) will typically be negligible but, if needed, it is

$$\frac{\mathbf{F}_{1/\gamma^2}^{(ends)}}{e} = \frac{\beta\lambda_0}{4\pi\epsilon_0\gamma^2}\begin{pmatrix} 0 \\ -R(1-\cos(s'R_i)) + x\cos(s'R_i) \\ y\cos(s'R_i) \end{pmatrix}\frac{1}{\kappa r'^2}. \qquad (13.70)$$

Note the absence of longitudinal component in the $1/\gamma^2$ term in both body and end contributions.

13.7
"Regularization" of the Longitudinal Force

It was established earlier that there is no net self-force for a straight string in a field free region. This largely obviates the need for any regularization. But for a curved string, even if it is subject to zero acceleration, since the ends are not quite parallel, a "small" logarithmically divergent self-force seems likely to survive. Furthermore, the presence of divergent terms at string ends, even if they sum to zero in the total force, prevent the internal forces from being calculated reliably. For calculating effects due entirely to acceleration, forces present even with acceleration absent need to be subtracted or otherwise accounted for.

The straight string field calculated in Eq. (13.35) should be derivable from the formulas just derived as a special case. To do this let us change the variable of integration from α' to $s' = \alpha' R$ using $\cos\alpha' \to \pm 1$ and $R\sin\alpha' \to -r'\cos\theta'$, where θ' is an angle shown in Figure 13.5. As a result

$$\frac{F_s^{(\infty)}(s_t, 0, y)}{e} = \lim_{R\to\infty}\frac{\lambda_0}{4\pi\epsilon_0}\left(\frac{1-\beta^2\cos\overline{\alpha'}}{\overline{r'}+\beta R\sin\overline{\alpha'}} - \frac{1-\beta^2\cos\underline{\alpha'}}{\underline{r'}+\beta R\sin\underline{\alpha'}}\right)$$

$$= \frac{\lambda_0}{4\pi\epsilon_0}\left(\frac{1/\gamma^2}{\overline{r'}(1-\beta\cos\overline{\theta'})} - \frac{1/\gamma^2}{\underline{r'}(1-\beta\cos\underline{\theta'})}\right). \qquad (13.71)$$

This formula agrees with Eq. (13.35). For $y=0$ the end points are given by Eq. (13.14).

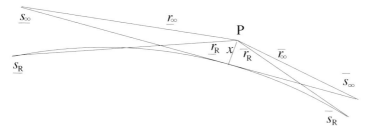

Fig. 13.9 Superposition of straight line and curved charge distribution for the purpose of regularizing the self-force due to the curvature of the path. At each point P the regularized force **G** is obtained by subtracting the tangentially-matched straight line force $\mathbf{F}^{(\infty)}$ from the curved line force $\mathbf{F}^{(R)}$.

We are now in a position to perform the regularization process mentioned in the introduction. Symbolizing the curvature-present force by $\mathbf{F}^{(R)}$, the curvature-absent force by $\mathbf{F}^{(\infty)}$, and their difference by **G** one has

$$\mathbf{F}^{(R)} = \mathbf{F}^{(\infty)} + \mathbf{G}, \text{ where } \mathbf{G} = \mathbf{F}^{(R)} - \mathbf{F}^{(\infty)}. \tag{13.72}$$

The definitions of these terms is made more explicit by referring to Figure 13.9. Force $\mathbf{F}^{(R)}(x, y, s)$ at position P is to be calculated based on the curved charge distribution running from $\underline{s}_R$ to $\bar{s}_R$ and force $\mathbf{F}(\infty)(x, y, s)$ is to be calculated based on the straight charge distribution running from $\underline{s}_\infty$ to $\bar{s}_\infty$. For transverse positions x, y very close to either end of the charge distribution $\mathbf{F}^{(R)}$ and $\mathbf{F}^{(\infty)}$ will exhibit the same logarithmic divergence. As a result the divergence will be absent from **G**.

The "regularized" force **G** can be used to calculate the emittance growth due purely to orbit curvature. But there may also be emittance growth caused by space charge forces, even in field free regions. That growth will have to be estimated using **F**, in which case the divergence has to be handled differently. The intention of treating point particles as strings is to obviate the need for regularization. By starting all particles in a simulation with non-zero interparticle separation the divergence can only occur after some bunch evolution has occurred, and then the probability of exact spatial coincidence later on is small, especially for high precision computation. The occasional close encounter will be insignificant since the singularity is only logarithmic. Of course these comments have to be confirmed quantitatively by numerical investigation.

Sample evaluation of the longitudinal force are shown in Figure 13.10. For purposes of comparison values due to Saldin et al. [3] are also plotted. The lower, $\gamma = 10$, value is close to, and the higher, $\gamma = 1000$, much greater than, the energy above which Eq. (13.77) becomes a good approximation. It

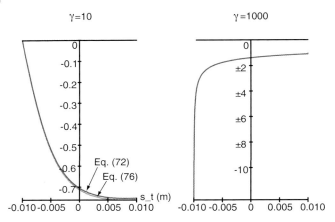

Fig. 13.10 Plot of $G_s(s_t, 0, 0)$, for $R = 10$ m, $L = 0.01$ m, for two γ-values. In both cases the result from this chapter, Eq. (13.72), is plotted as a solid line. The factor $(e\lambda_0)/(4\pi\epsilon_0)$ is suppressed in this and the next plot. For $\gamma = 10$ the exact Saldin formula (13.76) is plotted as a broken line. For $\gamma = 1000$ the high-energy approximate formula (13.77), $G_s^{(\gamma \gg 1/u_s)}$, is plotted as a broken line. Especially for the $\gamma = 1000$ graph, the curves superimpose well enough to be scarcely distinguishable.

can be seen that the agreement with Saldin is excellent and that the $G_s^{(\gamma \gg 1/u_s)}$ approximation is excellent at large γ. But this approximation, independent of γ as it is, greatly overestimates the self-force (and hence the CSR) at $\gamma = 10$. For the same two γ values the unregularized longitudinal force F_s is exhibited for both curving and straight strings in Figure 13.11. Here the longitudinal range is extended both before and after the actual string. The vertical axis has been artificially distorted (cube root) in order to expand the dynamic range while leaving the internal force visible.

13.8
Coherent Synchrotron Radiation

The power radiated by N electrons in the form of synchrotron radiation has been calculated by Schwinger [19], both in the form of coherent and incoherent radiation:

$$P_{\text{coh.}}^{(N)} = \beta N^2 \frac{ce^2}{4\pi\epsilon_0 R^2} \left(\frac{\sqrt{3}R}{2L}\right)^{4/3}, \quad P_{\text{incoh.}}^{(N)} = \beta N \frac{2}{3} \frac{ce^2}{4\pi\epsilon_0 R^2} \gamma^4. \tag{13.73}$$

These contributions are equal for

$$\gamma_{\text{crit.}} = \left(\frac{3N}{2}\right)^{1/4} \left(\frac{\sqrt{3}R}{2L}\right)^{1/3}. \tag{13.74}$$

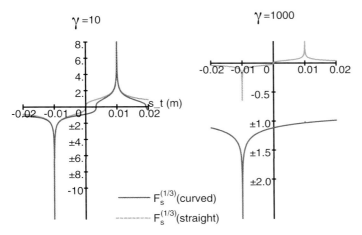

Fig. 13.11 The curved string and straight string longitudinal forces (after taking their cube roots) are shown over a range extended in front of and behind the string. (There is nothing fundamental about the cube root; this function has been chosen only to expand the scale at low amplitude relative to that at large amplitude while otherwise preserving the general shape.) As γ increases the straight string force becomes negligible compared to the curving string case. $R = 10\,\text{m}$, $L = 0.01\,\text{m}$.

The coherent power radiated should be equal to the rate at which the bunch does work on itself, which is given by

$$P = \beta c \int_{-L}^{L} \frac{G_s(s_t, 0, 0)}{e} \lambda_0 ds_t, \tag{13.75}$$

where $\lambda_0 = Ne/(2L)$.

The on-axis regularized longitudinal field G_s can be compared with results of Saldin et al. [3] For this comparison, our $s_t = s_{\text{Saldin}} + L$, and their symbol u_s, in our notation, is

$$u_s \equiv \underline{\alpha}'.$$

They give the formula

$$\frac{G_s(s_t, 0, 0)}{e} = -\frac{\lambda_0}{4\pi\epsilon_0} \frac{4\gamma}{R} \frac{(\gamma\underline{\alpha}')(8 + \gamma^2\underline{\alpha}'^2)}{(4 + \gamma^2\underline{\alpha}'^2)(12 + \gamma^2\underline{\alpha}'^2)}. \tag{13.76}$$

This reduces to an approximate form, valid for $\gamma \gg 1/u_s$ which, expressed in our notation, is

$$\frac{G_s^{(\gamma \gg 1/u_s)}}{e} = \frac{\lambda_0}{4\pi\epsilon_0} \frac{2}{3^{1/3}} \frac{1}{R^{2/3}} \frac{1}{(L+s_t)^{1/3}}. \tag{13.77}$$

This approximation, which, after integration from $-L$ to L, gives perfect agreement with Eq. (13.73) for $P_{\text{coh.}}^{(N)}$, is given by Saldin et al. [3] only in the range $-L < s_t < L$.

Tab. 13.1 Coherent power, as calculated by self-force, P, and as by Schwinger from far fields, $P_{\text{coh.}}^{(N)}$. $L = 0.01\,\text{m}$, $R = 10\,\text{m}$, $\gamma = 1000$, $N = 10^{10}$.

| L | R | γ | P | $P_{\text{coh.}}^{(N)}$ | $\gamma_{\text{crit.}}$ |
m	m		W	W	
0.01	10	10	207	568	3336
		100	555	571	3336
		1000	563	571	3336
		10000	563	571	3336
0.001	10	1000	12179	12298	7187
0.01	100	1000	121.8	123.0	7187

Some numerical comparisons of the two methods of determining coherent power are given in Table 13.1. High energy approximation (13.77) becomes an excellent approximation only for surprisingly large values of γ, such as $\gamma = 100$. As explained by Saldin et al. [3], roughly speaking, the approximation is valid only for $\gamma \gg (R/L)^{1/3}$. (The roughly one percent disagreement of P and $P_{\text{coh.}}^{(N)}$ for an ultra-large value such as $\gamma = 10000$ is presumably ascribable to the only rudimentary evaluation of the integral by the parallelogram rule—a more accurate integration prescription was unable to handle the singular behavior of the integrand at the end points.)

In summary, as stated in the introduction, agreement with on-axis results of Saldin et al. is excellent.

13.9
Evaluation of Integrals

The string end forces have been given in closed form and the longitudinal body force integral evaluated for $x = 0$. For $x \neq 0$ the integals can also be evaluated in closed form, but they are more complicated. Depending, as they do, on r' as given by Eq. (13.53), the integrals appearing in Eqs. (13.62) and (13.63) depend on factors of the form $\sqrt{A + B \cos \alpha'}$ where, satisfying $A > 0$ and $A + B > 0$,

$$A \equiv 2R(R+x) + x^2 + y^2, \quad B \equiv -2R(R+x). \tag{13.78}$$

(Using, for example, Maple) the required integrals can be expressed in terms of incomplete elliptic integrals [12] (The Maple argument k is related to the Abramowitz and Stegun argument m by $m = k^2$.)

$$E(z,k) = \int_0^z \sqrt{1-k^2t^2}/\sqrt{1-t^2}\,dt,$$
$$F(z,k) = \int_0^z 1/\sqrt{(1-k^2t^2)(1-t^2)}\,dt, \tag{13.79}$$
$$\text{Pi}(z,\nu,k) = \int_0^z 1/\sqrt{(1-\nu t^2)(1-t^2)(1-k^2t^2)}\,dt$$

$$I_s = \int \frac{\sin\alpha'}{(A+B\cos\alpha')^{3/2}}\,d\alpha' = \frac{2}{B\sqrt{A+B\cos\alpha'}},$$
$$I_0 = \int \frac{\cos\alpha'}{(A+B\cos\alpha')^{3/2}}\,d\alpha', \tag{13.80}$$
$$I_{x,y} = \int \frac{1-\cos\alpha'}{(A+B\cos\alpha')^{3/2}}\,d\alpha'.$$

Explicit formulas for these and some other integrals, plus hints and cautions concerning the numerical treatment of the formulas are given in the original paper.

13.10
Calculational Practicalities

The length parameter L has been left undetermined so far. There has to be a trade-off between numerical stability of the calculation (which favors large L) and faithful representation (which favors small L). For the present discussion we assume the true longitudinal beam distribution is Gaussian, with standard deviation σ_s. Three possible prescriptions for the specification of L, in order of decreasing L, and hence increasingly faithful representation are:

- A coarse estimate of space charge effects can be obtained by selecting $L \approx \sqrt{3}\sigma_s$ (to match the standard deviation) and simulating only the evolution of transverse coordinates.

- A model for which L is as large as possible without distorting the longitudinal distribution excessively is to represent the longitudinal beam distribution as the convolution of a distribution of reduced (relative to σ_s) standard deviation $\sigma_s' = \sqrt{\sigma_s^2 - L^2/3}$. When convoluted with the uniform distribution corresponding to string length $2L$, this produces an approximately correct longitudinal effective source distribution having the correct standard deviation. However the effective source distribution will agree with the actual distribution only for sufficiently small L value.

- To most faithfully represent the longitudinal distribution, the parameter L can be chosen increasingly small compared to σ_s until all results are sufficiently unaffected by further reduction.

Another practicality, probably more important, concerns emittance dilution due to CSR radiation. The present chapter has shown that the self-force caused energy lost to CSR agrees with the various far field calculations of Schiff, Nodvick and Saxon, Schwinger, Tamm, and many more recent authors. Furthermore, in agreement with Saldin et al., at least in the large γ region, this energy has been shown to come primarily at the expense of the energy of particles near the tail of the bunch. As the bunch is conveyed toward the region where its small emittance is needed it passes through optical elements tuned to focus the beam achromatically. But (barring esoteric RF cavities) there can be nothing in the optics of arcs, chicanes, etc. that treats front particles differently than back particles. So the optics that perfectly focuses, say, the front particles, will not focus the back particles perfectly. The result will be an effective emittance growth of the bunch treated as a whole, at the location where the small emittance is critical to its intended application.

13.11
Suppression of CSR by Wall Shielding

Another serious issue is the possible reduction of coherent synchrotron radiation due to the "shielding" effect of the conductive beam tube. This effect, first accurately calculated by Schwinger [19], is known to suppress the long wavelength components of CSR.

Whether or not shielding is important depends on the length of bunch $2L$ (assumed uniform) relative to the chamber height a (assumed to be smaller than the width). For long bunches, $2L > a$, the radiation is strongly suppressed. For short bunches, $a \ll 2L$, the shielding becomes negligible. For the ultrashort bunches of x-ray ERLs the latter condition holds and the suppression of CSR effects by wall shielding can probably be neglected.

Numerical values are given in Table 13.2 for a configuration in which coherent synchrotron radiation *is* strongly suppressed by shielding. The entries in this table apply to a planned ERL electron cooler (planned for cooling the heavy ions circulating in their Relativistic Heavy Ion Collider (RHIC)). The formulas in the table come from Schwinger's paper. A useful parameter is $\gamma_{\text{crit.}}$ which is the relativistic gamma factor for which, with no shielding, coherent and incoherent radiated power would be equal. CSR is primarily significant for $\gamma < \gamma_{\text{crit.}}$. One sees from the entries in the table that they apply to a relatively low energy case in which the coherent radiation dominates. Furthermore, because the bunches are long compared to the chamber height, the coherent radiation is strongly suppressed.

Tab. 13.2 Shielded and unshielded coherent synchrotron radiation in the beam stretcher section of a proposed electron ERL at Brookhaven.

symbol	quantity	formula	numerical value	unit
γ	relativistic factor		100	
Q	charge/bunch		10	nC
N	number of electrons		0.62×10^{11}	
R	radius of curvature		1	m
$2L$	bunch length (uniform)		10	cm
$P_{coh.}$	coherent (peak) power	$\beta N^2 \frac{ce^2}{4\pi\epsilon_0 R^2} \left(\frac{\sqrt{3}R}{2L}\right)^{4/3}$	12.1	kW
$P_{inc.}$	incoherent power	$\beta N \frac{2}{3} \frac{ce^2}{4\pi\epsilon_0 R^2} \gamma^4$	negligible	
$\gamma_{crit.}$	value for $P_{coh.} = P_{inc.}$	$\left(\frac{3N}{2}\right)^{1/4} \left(\frac{\sqrt{3}R}{2L}\right)^{1/3}$	1430	
F_s/e	longitudinal force per charge	$\frac{P_{coh.}}{Nec}$	4030	V/m
$\frac{P_{coh.,shielded}}{P_{coh.}}$		$\left(\frac{2L}{\sqrt{3}R}\right)^{1/3} \frac{a}{4L}$	0.039	
$\frac{F_{s,shielded}}{e}$		$\frac{P_{coh.}}{Nec} \left(\frac{2L}{\sqrt{3}R}\right)^{1/3} \frac{a}{4L}$	158	V/m

In the string space charge model a crude estimate of the importance of the inner wall of the vacuum chamber could start by first ignoring the wall and finding the tail angle $\underline{\alpha'}$. Then the line joining source point and field point can be checked to see whether it misses the inner chamber wall. If the line misses the chamber wall then the formulas derived so far apply. Otherwise the effective charge distribution is "cut off" at a point determined by a tangency condition, effectively bringing $\underline{\alpha'}$ closer to the bunch. From this picture the outer wall of the chamber would seem to have no effect on CSR.

In the string space charge model the upper and lower walls can be accounted for by the method of images. This calculation requires the retarded head and tail locations to be determined for each image in the series of images in the upper and lower walls (assumed to be plane). Though the series of images in parallel planes is poorly convergent, the rapid dependence on angle should cause the series of forces caused by the images to converge much more quickly. This calculation would be lengthy, but would use formulas given earlier in straighforward fashion. For a round vacuum chamber there would be a single image, reducing the computation time greatly.

13.12
Effects of Entering and Leaving Magnets

Any practical accelerator consists of alternating bend and drift regions. As a result the space charge force due to a moving string needs special treatment at entrances and exits of bending regions. In the procedure being described this presents little conceptual difficulty, but it does lead to substantial calculational complication. For calculating the force on a particular particle the first thing to be calculated is the location of the ends of the effective (i.e. retarded) charge distribution. If the field point is inside a magnet one or the other of the effective ends may be outside, or vice versa. We have seen that the effective *head* is very close to the field point, so it seems to be a good approximation to declare this always to be the case. This reduces the problem to locating the effective *tail*. If this location is in the same magnetic element as the field point, be it drift or bend, the force is given by the formulas already derived.

Suppose the effective charge distribution crosses a field boundary. (In principle it could cross more than one but, for simplicity, let us ignore that possibility.) By calculation one can therefore locate the point along the effective charge distribution that coincides with the boundary. It is natural then to break the string at that point into a head segment and a tail segment. Within the head segment, the analytic retarded time formula being used is valid, so the force due to the front segment can be calculated directly, using curved or straight string as appropriate. The only substantial new calculation is to locate the tail of the tail segment. Though straightforward this calculation will be tedious and will require formulas not derived so far. Once the ends of the tail segment have been located, the tail segment's contribution to the force can be obtained using formulas from this chapter. This contribution is likely to be quite different from what would be given by treating the entire effective charge distribution as if in the same field as the field point.

Though the procedure just described is straightforward in principle it certainly adds unwanted complexity to the problem. Any justification for simply ignoring the problems at bend boundaries would be welcome. Since we are primarily interested in *short* multi-micrometer bunches, let us assume that even the length of a retarded time bunch is short compared to the lengths of the magnet the bunch passes through.

It is difficult, even in principle, to segregate direct space charge forces from other current dependent beam-wall forces.

It is common, and valid, to associate radiation with accelerated charges. But, when contemplating self-forces, it may be helpful to visualize the important effect of transverse acceleration as moving a charge into an electromagnetic field to which it would not otherwise be subjected. As the bunch exits from a magnet the interval over which it is subject to self-force is comparable to the interval contributing to its instantaneous force while it is in the magnet. Since this interval has been assumed to be small compared to the magnet length the effect of exiting the magnet can be legitimately regarded to be a fractionally small end effect.

The situation on entry is more complicated. The fields accompanying a charged bunch traveling in a straight line inside a conducting chamber are the superposition of fields generated at all previous times. Contributions can come from an arbitrarily large number of sufficiently glancing reflections from the walls of the vacuum chamber. In other words the effects of nearby vacuum chamber walls cannot be ignored. For a perfectly conducting, perfectly cylindrical, chamber the fields caused by the walls add up to zero—the self-field is the same as if the bunch were in free space.

Realistic vacuum chambers have discontinuities which cause "wake fields", some of which are responsible for the "shielding" discussed earlier. These wake fields contribute to the self-force felt by a bunch of particles as it enters a magnetic field. In particular its transverse acceleration moves it from a region where the wall fields may approximately vanish to a region where they may not. Contributions to the force may come from anywhere in the preceeding drift region. It seems impossible, therefore, to separate the problem of magnet entry from the problem of wake fields. Since calculations of such wake fields goes well beyond the string space charge model capabilities, we are forced to assume that, like the exit forces, the CSR and CSCF forces on entry are fractionally small.

Fortunately the difficulties mentioned in the previous paragraph do not prevent the estimation of emittance growth for some configurations of current interest. The beam line required to bend a short and intense electron beam through an angle that is some substantial fraction of 2π is made up mainly of bending magnets. For beam lines like this the overestimate of forces within magnets may tend to be compensated by neglected forces in drift regions. In any case, *high precision* in the estimation of a coarse parameter (emittance) is rarely what is being sought.

13.13
Space Charge Calculations Using Unified Accelerator Libraries

The space charge formalism derived in this chapter has been incorporated into the UAL (Unified Accelerator Libraries) [20] accelerator simulation framework. The Physics User's Guide to the UAL code is by Malitsky and Talman [21]. This code can be used to calculate, among many other accelerator physics effects, space charge effects in realistic lattices. UAL provides a modularized framework, based on C++, for merging bunch processing methods encapsulated in various libraries. Following [22], this section discusses numerical issues that arise during the application of the string space charge model.

All formulas needed have already been derived in earlier sections. These formulas give the forces on particles in a bunch in a magnetic field of strength such that the orbit radius is R. Of course, in a real ring, the bunches will be in bending magnets only some of the time. The rest of the time they will be in drift spaces, quadrupoles, sextupoles, RF cavities, etc. (all of which are treated as drifts for present, space charge, purposes). In a drift space, where the magnetic field vanishes, the bend radius is $R = \infty$.

Because the string space charge formulation avoids the need for regularization, it can and does subsume all space charge forces, irrespective of the local magnetic field—this is a major advantage of the formalism. But, because of the frequently infinite value of R, there are serious programming complications in writing computer code that uses formulas expressed in terms of R. How this problem is handled, without introducing a drift/bend flag to steer the calculation to separate (and hence bug-prone) code blocks, is the main topic of this section.

The formulas for the components of forces need to gracefully provide finite results in the drift space limit. A useful first step in this direction is to introduce a new variable

$$R_i = \frac{1}{R}, \qquad (13.81)$$

which vanishes in field free regions. Since R_i is proportional to B it is necessarily finite everywhere. A next useful step is to re-express all formulas in terms of longitudinal coordinate s' instead of angular coordinate α'. (This step is needed because α' is not useful in drift regions.) Repeating Eq. (13.50), these quantities are related by

$$\alpha' = R_i s'. \qquad (13.82)$$

Confirmation of the limiting behavior of the analytical formulas as $R_i \to 0$ is a straightforward procedure that will not be exhibited here. The final formulas in equations such as Eq. (13.61) have been organized to facilitate this check.

The UAL environment is far too complex to be explained fully here. Documentation can be found at the UAL website, `http://www.ual.bnl.gov`. Documentation written at the same pedagogical level as this text is in *Text for UAL Simulation Course* by N. Malitsky and R. Talman. As well as serving as text for a U.S. Particle Accelerator School course, this source serves as the *UAL Physics Users Guide*. This guide also contains problems having roughly the same level of difficulty as problems in this text.

13.13.1
Numerical Procedures Used by UAL

While in a magnetic field a uniformly-charged longitudinally-aligned needle, or string, forms a circular arc of length $2L$ traveling along a radius R circular path, as shown in Figure 13.6. The line charge density is $\lambda = q/(2L)$, the speed is $\beta = v/c$, and the corresponding current is $\beta c \lambda$, within the string, and zero otherwise. The length $2L$ can initially be regarded as arbitrary (small compared to the bunch length of the charged bunch being analysed but large enough to reduce erratic variation of results).

There are two major tasks in a fully-relativistic calculation of the force acting on a co-moving point charge q at point P, due to the charged string.

The first task is to locate string end points. The electromagnetic field components at point P at, say, time $t = 0$, reflect not the instantaneous charge density at that time (which is indicated by an open curved box in Figure 13.6) but rather the retarded time or "effective" charge distribution, shown as a curved arc in the figure. It is necessary to find the positions of the effective head and tail of the string. The equation determining the head is Eq. (13.55), which can be re-expressed, using Eq. (13.82), as

$$\frac{\overline{s'} + s_t - L}{\beta} = -\sqrt{(1 + xR_i)\,\overline{s'}^2 \left(\frac{\sin(\overline{s'}R_i/2)}{\overline{s'}R_i/2}\right)^2 + x^2 + y^2}, \qquad (13.83)$$

where the symbols are defined in the figure. In particular, s_t is the londitudinal displacement of the "test point" P from the bunch center. The tail equation is similar. This formula exhibits the numerical problem in the $R_i \to 0$ limit. If it were valid to keep only the leading term in a Taylor series expansion of the sine function, the cancellation $RR_i = 1$ would be immediate, irrespective of the value of R. If higher order terms were retained, they would explicitly give zero in drift regions.

This procedure can be made more robust by introducing the function

$$j_0(x) = \frac{\sin x}{x} \equiv \text{sinc}(x), \qquad (13.84)$$

which is known in some quarters as a "spherical Bessel function" and in other as a "sinc" function. In the GNU Scientific Library this function is known as `sf_bessel_j0(x)`. In terms of this function Eq. (13.83) becomes

$$\frac{\overline{s'} + s_t - L}{\beta} = -\sqrt{(1 + xR_i)\,\overline{s'}^2\, j_0^2(\overline{s'}R_i/2) + x^2 + y^2}, \qquad (13.85)$$

which is numerically well behaved for all physical values of R.

The second major task is to evaluate the various force components. In the string formalism both electric and magnetic fields are subsumed into a fundamental charged-string/point-charge force using the Lorentz force formula. The magnetic field, for example, is given by the "Jefimenko equation":

$$\mathbf{B}(P,t) = \frac{\mu_0}{4\pi} \int \left(\frac{\mathbf{I}(z',t_r)}{r^2} + \frac{\dot{\mathbf{I}}(z',t_r)}{cr}\right) \times \hat{\mathbf{r}}\, dz'. \qquad (13.86)$$

The first term is just the Biot-Savart law, though applied to the retarded distribution. This is referred to as the "body" term. The second term vanishes everwhere except at the string ends, where it gives δ-function contributions that are referred to as "end" terms. There is a similar equation for the electric field and a similar division into Coulomb's law body terms and string end terms. After finding the effective string ends, the second computational task is to evaluate all these terms. The extreme near-cancellation of electric and magnetic forces makes it essential to handle the cancellation explicitly, which is why the formalism is restricted to give just the total force vector.

The components of force on the test particle due to a source particles are known analytic functions of the head and tail angles. For example, from Eq. (13.62), the (fully-relativistic) horizontal force component F_{0x} is given by

$$F_{0x} = \frac{\lambda}{4\pi\epsilon_0} \int_{\underline{s}'}^{\overline{s}'} ds' \frac{(2+xR_i)}{r'^3} R(1 - \cos(s'R_i))$$
$$= \frac{\lambda}{4\pi\epsilon_0} \frac{(2+xR_i)R_i}{2} \int_{\underline{s}'}^{\overline{s}'} ds' \frac{s'^2 j_0^2(s'R_i/2)}{r'^3}. \tag{13.87}$$

where r' is the distance from source point to test point, and where manipulations like those in Eq. (13.85) have been performed. In this form F_{0x} can be evaluated numerically without drift spaces being exceptional. The Bessel function make the integral look formidable but, since it is the same integral as before, just expressed differently, it can be evaluated in closed form.

In Section 13.9 all components were expressed in terms of parameters A and B, which can now be replaced by $A' = AR_i^2$, $B' = BR_i^2$, where

$$A' \equiv 2(1 + xR_i) + R_i^2(x^2 + y^2), \quad B' \equiv -2(1 + xR_i), \tag{13.88}$$

to be substituted into indefinite integrals such as

$$I_s(\alpha'; A, B) = \int \frac{\sin \alpha'}{(A + B \cos \alpha')^{3/2}} d\alpha'$$
$$= R_i^3 \int \frac{\sin(s'R_i)}{(A' + B' \cos(s'R_i))^{3/2}} d(s'R_i)$$
$$= R_i^3 I_s(s'R_i; A', B'). \tag{13.89}$$

A subroutine written to evaluate the function $I_s(\alpha'; A, B)$ can obtain the same answer by evaluating $R_i^3 I_s(s'R_i; A', B')$. In this form drift sections are not exceptional. The other integrals needed can be evaluated in the same way.

13.13.2
Program Architecture

Discussion will be limited to calculation of the emittance growth of a dense electron bunch making a single pass through a short accelerator sector. To simplify the discussion space charge forces will be neglected everywhere except within bending magnets.

In a single particle simulation code a particle is deflected at each lattice element. In the intrabeam scattering code being described, a particle suffers a further deflection due to every other charge in the bunch, when, and only when, it is in a bending magnet. For a bunch containing N particles this requires N^2 calculations at each bending magnet. Computation time imposes a practical upper limit, perhaps $N < 10^4$. As well as depending on the relative particle positions, the deflections depend on the bend radius and are proportional to the magnet length. With the sum of all space charge forces on a particle ascribed to the bending magnet, particle evolution then proceeds as in a single particle tracking code. For improved precision the magnet can be subdivided as finely as needed.

The general code architecture can be inferred from the APDF (Accelerator Propagator Description Format) file shown next:

```
<apdf>
 <propagator id="stringsc" accelerator="ring">
  <create>
   <link algorithm="TEAPOT::DriftTracker"
                   types="Default" />
   <link algorithm="TEAPOT::DriftTracker"
                  types="Marker|Drift" />
   <link algorithm="TEAPOT::DipoleTracker"
                   types="Sbend" />
   <link algorithm="TEAPOT::MltTracker"
    types="Quadrupole|Sextupole|[VH]kicker"/>
   <link algorithm="TIBETAN::RfCavityTracker"
      types="RfCavity"/>
   <link algorithm="TEAPOT::StringSCKick"
      types="Kicker"/>
  </create>
 </propagator>
</apdf>
```

This XML file associates propagation algorithms with accelerator elements. Since this file is largely self-explanatory it should need only partial explanation. Most lattice elements are treated as they would be in the absence of space charge. Space charge kicks are associated with thin *kicker* elements located at the centers of every bending magnet. These artificial elements have to have been inserted artificially into the original lattice description file. For elements of this type the APDF line containing `TEAPOT::StringSCKick` delegates all string calculations to `StringSCSolver` methods.

13.13.3
Numerical Procedures

Equation solving routines and special function evaluation is performed using the the GNU Scientific Library [23]. The only new class necessitated by space charge is `StringSCSolver` whose methods include (numerical) retarded time equation solving and (analytical) definite integral evaluation.

To find the effective (i. e. retarded) angle of the head of a source string from the position of a test particle, it is necessary to solve Eq. (13.55) for $\overline{\alpha'}$. Though this equation looks fairly simple, its coefficients are so close to irregular points (i. e. $\beta \approx 1$, $\alpha' \approx 0$, and $x, y \ll R$) that solution is not simple. A procedure that has been found to be robust in all cases tried is to start by finding a coarse solution using the root-bracketing procedure `gsl_root_fsolver_brent` which uses the so-called *Brent-Decker* method. From the physics there is certain to be just one solution of the equation and this method, though slow, is guaranteed to find it (approximately). It has not been investigated whether it is economical to iterate this method to a sufficiently accurate result. Rather, iteration is stopped when the relative change in $\overline{\alpha'}$ is less than 10^{-4}. Then the method `gsl_root_fdfsolver_steffenson`, employs the *Steffenson* method to "polish" the result to arbitrarily high accuracy. This method, which uses derivatives, is not guaranteed to converge to the correct root, but we have observed no anomalous behavior. The same sequence of methods works for finding the effective tail angle $\underline{\alpha'}$.

The various terms of integral (13.86) are expressible in terms of *elliptic integral* special functions, given in the GNU scientific library as `gsl_sf_ellint_E` and `gsl_sf_ellint_F`. All other integrals, for both transverse and longitudinal forces, are similarly expressible, even in the non-relativistic regime where γ is of order 1.

It is the near vanishing of denominators in integrals like (13.86) that make their evaluation delicate. In fact, if the range of integration in (13.86) includes the origin, especially for the longitudinal force component, it is necessary to split the range and to exclude an "infinitesimal" range centered on the origin. Though the integrand is singular and discontinuous at the origin, the contribution from the excluded range cancels by symmetry. The end contributions are relatively simple (though also nearly singular) trigonometric functions of the end angles.

The formalism that has been described has various limitations. The restriction to $\gamma \gg 1$ is superficially not very serious, since fully relativistic versions of all formulas have been given. But non-relativistic conditions bring in other complications that are not being handled consistently except under relativistic conditions. The assumptions concerning entry to and exit from magnets have been discussed earlier. Also quadrupoles and sextupoles are treated as drifts, as far as space charge effects are concerned.

13.13.4
Comparison with TRAFIC4 [24]

There is a computer code TRAFIC4 [24], due to Dohlus et al., that describes very much the same physics as does the string space charge model described in this chapter. Their code also (and for the same reason) subdivides the charge distributions into longitudinally elongated elements. Unlike the present model, however, these elements are not associated with individual particles. Rather, they constitute a purely numerical subdivision of a continuous charge distribution. TRAFIC4 proceeds using numerical methods to find the space charge forces by solving the partial differential equations with

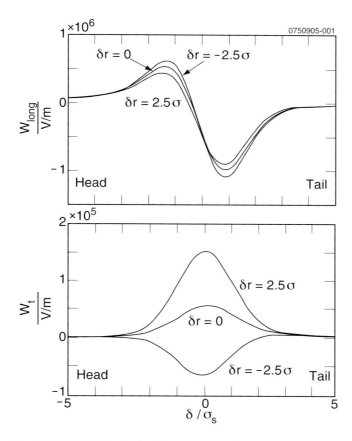

Fig. 13.12 TRAFIC4-calculated space charge forces on a 1 nC, $\gamma = 100$, Gaussian, spherically shaped, bunch of electrons of r.m.s. dimension $\sigma = 50\,\mu$m, traveling on a circular orbit of radius $R = 10$ m. The abscissa δ/σ_s is the longitudinal coordinate in units of the r.m.s. bunch length. These graphs have been copied from Ref. [24].

these source terms. This seems to me like an eminently correct approach. It is, however, extremely computer intensive, and seems to require a farm of computers for reconstructing the space charge forces in realistic situations. Some results copied from their publications are shown in Figure 13.12. These graphs show the dependence on position within a bunch of the longitudinal and transverse forces in an electron bunch traveling on a circular orbit. The bunch parameters are listed in the caption to the figure. Their radial coordinate δr corresponds to our x, their longitudinal position δ/σ_s corresponds to our s/σ_s. Their longitudinal force W_{long} is our F_s/e. Their transverse force W_t is our F_x/e. In each case the dependence of the force component on s at each of three radially displaced positions is shown.

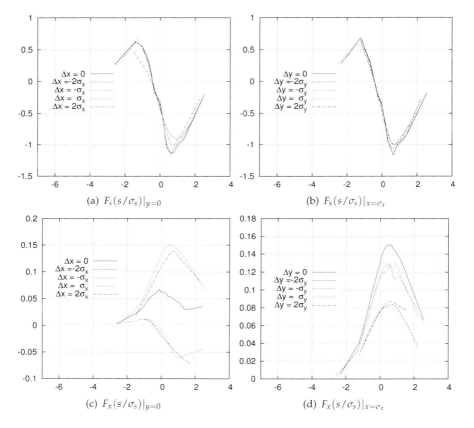

Fig. 13.13 UAL, `stringsc` calculation of space charge forces on a 1 nC, $\gamma = 100$, spherically-shaped, Gaussian longitudinal, uniform transverse, bunch of electrons of r.m.s. dimension $\sigma = 50\,\mu\text{m}$, traveling on a circular orbit of radius $R = 10\,\text{m}$. These conditions are (almost) identical to those used in the TRAFIC4 calculation shown in Fig 13.12. As in that figure, the *head* of the bunch is on the left. The units of the vertical scales are MV/m. The horizontal scales are the same as in Fig 13.12 except for a possible small translation.

Corresponding results obtained using the UAL `stringsc` code are shown in Figure 13.13. All parameters are the same as in Figure 13.12. (In the case of transverse bunch distribution, though the r.m.s. sizes are identical, the UAL distributions are uniform while the TRAFIC4 distributions are Gaussian.) Longitudinal bunches are Gaussian in both cases, and γ, bunch charge, radius of curvature, are identical. The curves plotted in Figure 13.13 are based on plotting smooth curves obtained from the actual forces in a simulation in which 4000 particles were propagated through a single magnet. That is why the curves in the graphs are plotted only at points where the particle distributions differ appreciably from zero. Any disagreements beween UAL and TRAFIC4 rarely exceed 10%. This provides further confirmation of the string space charge formulation.

References

1. Talman, R. (2004), *String Formulation of Space Charge Forces in a Deflecting Bunch*, PRST-AB, **7**, 100701.
2. Malitsky, N., Talman, R. (2004) *UAL Implementation of String Space Charge Formalism*, WEPLT154, EPAC Conference, Lucerne, Switzerland.
3. Saldin, E., Schneidmiller, E., Yurkov, M. (1995), *On the Coherent Radiation of an Electron Bunch Moving in an Arc of a Circle*, DESY-TESLA-FEL-96-14, and *Nucl. Instrum. Methods A*, **417**, 158 (1998).
4. Li, R., Derbenev, Y. (2003), *Canonical Formulations and Cancellation Effect in Electrodynamics of relativistic Beams on a Curved Trajectory*, Jefferson lab preprint, LJAB-TN-02-054.
5. Geloni, G. et al., (2003), *Misconceptions regarding the cancellation of self forces in the transverse equation of motion for an electron in bunch*, DESY 03-165, October.
6. Bassetti, D., Brandt, D. (1896), *Transverse Electro-Magnetic Forces in Circular Trajectory*, CERN/LEP-TH/86-04.
7. Iogansen, L., Rabinovich, M. (1958), *Coherent Electron Radiation in a Synchrotron, I, II, and III*, JETP (U.S.S.R) **35**, 1013; **37**, 118 (1959), **38**, 1183 (1960).
8. A personal communication from Vladimar Shiltsev reports that early work by Tamm, little known even in Russia, preceeded most of the known work on coherent synchrotron radiation.
9. Geloni, G. et al. (2002), *Transverse self-fields within an electron bunch moving in an arc of a circle*, DESY Report 02-048
10. Geloni, G. et al. (2003), *Transverse self-fields within an electron bunch moving in an arc of a circle (generalized)*, DESY Report 03-044
11. Jefimenko, O. (1995), *Retardation and relativity: The case of a moving line charge, Am. J. Phys.* **63** (5).
12. Abramowitz, M., Stegun, I. (1965), *Handbook of Mathematical Functions*, Dover Publications Inc, New York.
13. Piwinski, A. (1985), *On the Transverse Forces Caused by the Curvature*, CERN 85-43.
14. Decker, G. (1986), Cornell PhD thesis.
15. Talman, R. (1986), *Novel Relativistic Effect Important in Accelerators*. PRL **56**, 1429.
16. Lee, E. (1990), LBID-1376 (1988), and *Particle Accelerators*, **25**, 241.
17. ERAN-30 Note (1969) included in *Selected Works of L. Jackson Laslett*, LBL PUB-616, p. 13 (1987).
18. Derbenev, Y., Shiltsev, V. (1996), SLAC-Pub-7181.
19. Schwinger, J. (1945), *On Radiation by Electrons in a Betatron*, transcribed by M. Furman and reprinted as Lawrence Berekley Lab Report, LBNL-39088, (1999).
20. Malitsky, N., Talman, R. (1998), *The Framework of Unified Accelerator Libraries*, ICAP98, Monterey, and *UAL User Guide*, BNL-71010-2003, and http://www.ual.bnl.gov.
21. Malitsky, N., Talman, R. (2005), *Text for UAL Accelerator Simulation Course*, U.S. Particle Accelerator School, Ithaca, N.Y. Available at http://www.ual.bnl.gov.
22. Malitsky, N., Talman, R. (2004), *UAL Implementation of String Space Charge Formalism*, European Particle Accelerator Conference, Lucerne, Switzerland.
23. Galassi, M. et al., (2005), *GNU Scientific Library Reference Manual*, Network Theory Limited, 15 Royal Park, Clifton, Bristol BS8 3AL, England.
24. Dohlus, M., Kabel, A., and Limberg, T. (2000), *Efficient field calculation of 3D bunches on general trajectories*, Nucl. Instrum. Methods A **445**, p. 338.

14
The X-ray FEL

14.1
Absorption and Spontaneous and Stimulated Emission

It was Einstein who, using statistical mechanics, first established the important relations governing emission and absorption of radiation. Let the rate of spontaneous emission of a photon by an electron be given by A_{21} and assume that the rates of absorption and stimulated emission are both proportional to $\rho(\nu)$, the photon energy density per unit frequency range at the position of the electron. Transition rates for emission of a photon, W_{21}, or absorption of a photon, W_{12}, can then be expressed as

$$W_{21} = B_{21}\rho(\nu) + A_{21},$$
$$W_{12} = B_{12}\rho(\nu), \qquad (14.1)$$

where B_{21} and B_{12} are the coefficients of stimulated emission and absorption.

There is no reason to suppose the electrons and photons we are dealing with are anywhere near thermal equilibrium at any temperature T, but if they were, $\rho(\nu)$ would be given by

$$\rho(\nu) = \frac{8\pi h \nu^3}{c^3} \frac{1}{e^{h\nu/kT} - 1}, \qquad (14.2)$$

and the number densities of electrons, N_2 and N_1, with and without the energy difference corresponding to that of one photon, would be in the ratio

$$\frac{N_2}{N_1} = e^{-h\nu/kT}, \qquad (14.3)$$

The reason these relations are germane is that the coefficients A_{21}, B_{21} and B_{12}, themselves independent of the local distribution of radiation, must be consistent with this equilibrium. This implies

$$N_2 \left(B_{21}\rho(\nu) + A_{21} \right) = N_1 B_{12}\rho(\nu). \qquad (14.4)$$

Solving Eqs. (14.4) for N_2/N_1 and substituting the result into Eq. (14.3) yields

$$\rho(\nu) = \frac{A_{21}}{B_{12}e^{h\nu/kT} - B_{21}}. \qquad (14.5)$$

Accelerator X-Ray Sources. Richard Talman
Copyright © 2006 WILEY-VCH Verlag GmbH & Co. KGaA, Weinheim
ISBN: 3-527-40590-9

For this equation to be consistent with Eq. (14.2) requires

$$B_{12} = B_{21} = \frac{c^3}{8\pi h \nu^3} A_{21}. \tag{14.6}$$

The first equation expresses the equality of absorption and induced emission. The second expresses the stimulated enhancement over and above spontaneous emission.

As stated before, FEL operation depends on this stimulated emission. But, as with any laser, to get more emission than absorption, it is necessary for the ratio N_2/N_1 to be altered artificially from its thermal equilibrium value.

14.2
Closed and Open FELs

The general operation of the closed FEL can be inferred from Figure 14.1. A circulating bunch of electrons or positrons is arranged to pass through the electromagnetic wave in an optical resonator in such a way that energy is extracted from the electron beam in the form of radiation. Some of the radiation passes through the partial mirror for its intended use, the rest replenishes the resonator energy. For the device to be useful the energy radiated per passage has to be greater than the energy lost in the resonator between passages.

The radiation can be analysed (semi-classically using conservation of energy) as the radiation that accompanies deceleration of the electrons by the electromagnetic wave, or (quantum mechanically) as Compton scattering of the virtual photons of the undulator, as stimulated by the pre-existing photons trapped in the optical resonator. Historically the quantum picture came first, but the classical analysis is more elementary and, as far as I know, is always valid in practical devices.

The so-called "optical klystron" uses two undulators, separated by a dispersive region. The radiation from the first wiggler introduces energy oscillation along the electron bunch, which the dispersive section converts to a kind of "population inversion" in the form of particle density bunching. This bunching can be synchronized into optimal phase relationship with the pre-existing radiation, to yield greater radiation gain. In the context of the present book the closed FEL is only of academic interest since mirrors giving appreciable near-normal reflection of the short wavelengths of interest do not exist.

However the physics is very much the same for an open FEL, in which the radiation from an electron bunch making a single pass through an undulator is self-amplifying. The approximate wavelength of FEL radiation is governed by the undulator period and the electron energy, but these parameters cannot determine the precise frequency and phase of the radiation. In the so-called SASE (self-amplified spontaneous emission) process the precise param-

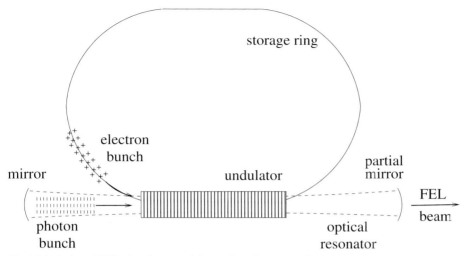

Fig. 14.1 A closed FEL. One long straight section of a racetrack-shaped storage ring is shared with an optical resonator and a magnetic wiggler to form a free electron laser.

eters are established by the particular random density fluctuation which first gives substantial coherent synchrotron radiation. An alternative (but as yet unachieved approach) would be to "seed" the process from an external x-ray source. Until some mechanism has been found for stabilizing the frequency, phase, and, especially, of amplitude of the radiation the FEL is unlikely to surpass conventional storage rings for most x-ray experiments.

14.3
Interpretation of Undulator Radiation as Compton Scattering

The interpretation of undulator radiation as Compton scattering has been exhaustively analysed in Chapter 12. The undulator field can be said to be made up of "virtual" photons which can Compton scatter from the electrons. The magnitude of the mass-squared of these virtual photons was determined in Eq. (12.70) to be $|m_\gamma c^2| = \hbar k_w c$, which was shown to be negligibly small in every relevant frame except the frame in which the undulator is at rest. Furthermore, in the electron rest frame the photon energy is small compared to the electron rest mass. This is the condition needed to justify treating the interaction as Thompson scattering, for which the incident and scattered photon energies are the same. (In spite of the smallness of the photon energy, the virtual photon mass is still negligible.)

We have reviewed enough to be able to introduce the fundamental FEL motivation. When the Thomson cross section just calculated is calculated quan-

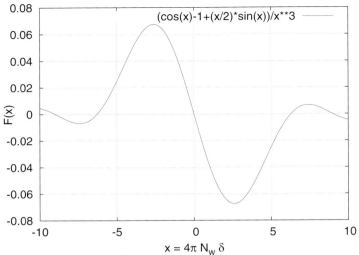

Fig. 14.2 Dependence of intensity gain of an FEL as a function of fractional momentum offset δ, relative to the central momentum, in an undulator with N_w periods. The vertical scale is arbitrary. The analytical formula is shown on the figure.

tum mechanically the probabilites of emission to, and absorption from, each state are enhanced proportionally to the pre-existing population of that state. This is the role of the photons trapped in the optical resonator. The stimulated emission enhances the intensity of this radiation. Unfortunately the absorption reduces the intensity. On the average, and under ordinary circumstances these intensity contributions are equal and opposite, and there is no significant intensity enhancement.

The trick to enhancing the emission relative to the absorption is to make all the electrons have energy slightly less than the "resonant energy" (the energy for which the laser wavelength is $\lambda_w/(2\gamma^2)$) since the absorption is then weakened relative to the emission. Too great an offset, however, leads to reduced radiation as the electron loses synch with the wiggler—an effect that becomes increasingly important as N_w is increased. The combination of these factors leads to the "gain" dependence shown (arbitrary units) in Figure 14.2 [1]. The fact that this curve is antisymmetric about $\delta = 0$ corresponds to the previously stated tendency for absorption and emission to cancel.

Figure 14.2 also can be used to estimate the maximum spread of momenta such that all particles stay close enough to the optimal phase through the whole undulator length:

$$\delta \lesssim \frac{1}{\pi N_w}. \tag{14.7}$$

14.4
Applicability Condition for Semi-Classical Treatment

Since elementary particles and photons are involved in the radiation process one can inquire whether quantum mechanical considerations are important. For a start, one can observe that it is inconsistent to start with an electron at rest. It is not possible to localize the electron better than by describing it by a Gaussian wave packet, with spreads Δx_0 and Δp_x that satisfy the Heisenberg uncertainty condition,

$$\Delta x_0 \Delta p_x \geq \hbar. \tag{14.8}$$

As the packet spreads with time, let its spatial spread be $\Delta x(t)$. Accepting the equality condition as determining Δp_x, and treating the electron motion non-relativistically, the spatial width evolves with time according to

$$\Delta x^2(t) = \Delta x_0^2 + \left(\frac{\hbar t}{m \Delta x_0}\right)^2. \tag{14.9}$$

The time it takes for half the wiggler to pass the electron is $N_w \lambda_w/(2\gamma c)$. To minimize the importance of spreading, we can choose the value for Δx_0 that minimizes $\Delta x(t)$ after this time, namely

$$\Delta x_0^2(t) = \frac{\hbar t}{m} = \frac{1}{4\pi} \frac{N_w \lambda_w}{\gamma} \frac{hc}{mc^2}. \tag{14.10}$$

The final factor, $\lambda_C = hc/(mc^2) = 2.43 \times 10^{-12}$ m, is the electron Compton wavelength. To maintain coherence over all wiggler radiation, one requires the electron's Heisenberg motion to be small compared to the wavelength of its radiation. For $\Delta x_0(t)$ to be much less than λ_w/γ, requires

$$N_w \ll \frac{4\pi}{\gamma} \frac{\lambda_w}{\lambda_C}. \tag{14.11}$$

For undulator period of order one centimeter and realistically small values of γ, the number of poles can be many tens of thousands with quantum uncertainty concerning initial conditions still negligible.

In striving for ever more brilliant beams one reduces the emittance of the electron beam, perhaps even to $\epsilon_e = \sigma_x^2/\beta_w = 10^{-12}$ m. It would be inconsistent to demand the dimensions of the packets describing individual electrons to exceed the bunch dimensions. The normalized beam emittance, given by

$$\epsilon^{(N)} = \pi \sigma_x \sigma_{x'} \gamma \approx \pi \sigma_x \sigma_{x'} \frac{\mathcal{E}_e/c}{mc}, \tag{14.12}$$

can be compared with the transverse Heisenberg condition, which requires $\Delta x \Delta x' \mathcal{E}_e/c > \hbar$. Ignoring the distinction between Δs and σs, one requires

$$\epsilon^{(N)} > \frac{hc}{2mc^2} \approx 10^{-12} \text{ m}. \tag{14.13}$$

This is comparable with the assumed bunch emittance. It seems therefore, that one can come close to quantum mechanical limitations.

14.5
Comparison of Storage Ring, ERL, and FEL

A representative set of parameters for the TESLA x-ray source is given in Table 14.1.

Tab. 14.1 Some possible x-ray FEL parameters [2].

Parameter	Symbol	Unit	Value
electron energy	$\mathcal{E}_e$	GeV	30
charge per bunch	Q_B	nC	1
rms bunch length	σ_s	μm	30
bunch duration	$L_B = 2\sigma_s/c$	fs	200
normalized emittance	$\epsilon^{(N)}$	μm	2
rms energy spread	σ_δ	MeV	2.5
bunch train duration		ms	1
number of pulses per train	N_B		10^4
repetition rate	$f_{\text{rep.}}$	Hz	5
undulator length		m	100
undulator periods	N_w		
undulator wavelength	λ_w	m	0.045

To quantify (by example) the impact of superradiance let us start by comparing a current I beam of hypothetical particles having charge $2e$ with a beam of actual electrons having the same current I. Because the basic radiation amplitude is proportional to the particle charge and the intensity to the amplitude-squared, the flux of incoherently radiated photons would be twice as great in the hypothetical case. As the discussion in Chapter 13 showed, at sufficiently long wavelengths, essentially the same quadratic dependence of photon intensity on total charge per bunch, and linear dependence on beam current, is obtained in any bunched beam. By reducing the bunch length these dependences can be pushed to ever shorter wavelengths.

The point of free electron lasers (FEL) is to exploit this "superradiance" to produce more brilliant beams than can be produced with conventional light sources. If the bunch length L_B could be made as short as $\lambda_\gamma/2$, where λ_γ is the desired wavelength, then the desired intensity increase equal to the number of particles per bunch would be achieved. (The relevant formulas are given in Chapter 13.) But wavelengths of interest are typically in the

Ångstrom range, which is far shorter than any practically achievable bunch length..

What *is* possible, however, is "microbunching", in which the longitudinal bunch profile is modulated with wavelength λ_γ. The effect of this modulation is to increase the radiated intensity by roughly the number of electrons n_1 per wavelength, which is related to the total number of electrons per bunch n_B by

$$n_1 = n_B \frac{\lambda_\gamma}{L_B} \left(\stackrel{\text{e.g.}}{=} \frac{10^{-9}\,\text{C}}{1.6 \times 10^{-19}\,\text{C}} \frac{10^{-10}\,\text{m}}{30\,\mu\text{m}} \approx 2 \times 10^4 \right). \tag{14.14}$$

This is the maximum factor by which the total power of the radiation from the bunch can be increased by the superradiance. It is the possibility of making this factor large that provides one of the motivations for FELs. The other motivation (shared with ERLs) is the possibility of producing the ultrashort bunches needed for studying molecular dynamics on an appropriately short time scale.

Though much is known about the microbunching process, the process is too poorly understood for inclusion, at this time, in this text. This is regrettable, since microbunching is what makes the open FEL a promising x-ray source. Here we accept the existence of *perfect* microbunching and proceed to calculate the properties of the resulting radiation.

It has been seen previously, especially Chapter 7, how photon intensities can also be increased by using long undulators in which the magnetic field varies periodically, preferably with quite short wavelength λ_w. All x-ray sources take advantage of this enhancement but, especially after the lattice has been fixed, the length of undulator that can be inserted into a conventional storage ring is typically more limited than for ERLs or FELs.

In practice, again because short wavelength x-rays are sought, and λ_w cannot be made arbitrarily short, FELs require electrons of very high energy $\mathcal{E}_e$. Oversimplifying shamelessly, and using round numbers, a factor of ten increase in $\mathcal{E}_e$, compared to a conventional ring, (say from 3 GeV to 30 GeV) is required for an FEL to produce a comparable, Ångstrom scale x-ray beam. The costs of all types of light source are more or less proportional to $\mathcal{E}_e$. Of course, with the cost of RF accelerating structure ranging from only significant to dominant, the constant of proportionality is different for the different categories of x-ray sources. Just for the sake of the present argument let us guess, taking account of both factors, that the costs are in the ratios 1 : 2 : 10 for FCSR:ERL:FEL. It is very possible, when compared on a cost per photon basis, for the superradiance enhancement factor n_1 introduced above for the FEL, to overwhelm this differential in accelerator construction cost.

Even accepting the numbers assumed, it is not certain, however, that this cost comparison is appropriate. The cost of operating power must not be neglected. The whole point of an ERL is to "recover" the energy required to accelerate the electron beam. For the FCSR this issue is academic because, even thought it is "fast cycling" the average power spent in acceleration is negligible.

But for the FEL there is a serious power constraint. Using parameters from Table 14.1, the charge accelerated each second is the product of charge per bunch Q_B, number of bunches per train N_B and repetition rate $f_{\text{rep.}}$. The total power is therefore given by

$$P[\text{W}] = \frac{\mathcal{E}_e}{e} Q_B N_B f_{\text{rep.}} \quad \left(\stackrel{\text{e.g.}}{=} 30\,[\text{GV}] \times 1\,[\text{nC}] \times 10^4 \times 5[\text{Hz}] = 1.5\,\text{MW} \right). \tag{14.15}$$

This amount of power is comparable with the power expended in FCSR, and it is *less*, by a factor of, perhaps, 5 than figures mentioned for an ERL. The FEL power is as low as it is only because the average current is so low; it is given by

$$I_{\text{ave.}} = Q_B N_B f_{\text{rep.}} \quad \left(\stackrel{\text{e.g.}}{=} 1[\text{nC}] \times 10^4 \times 5[\text{Hz}] = 5 \times !0^{-5}[A] \right). \tag{14.16}$$

The nominal current for the ERL and for the FCSR is 0.1 A which is 2000 times higher. Folding together the estimated total cost, the superradiance factor, and the average current, the costs per photon produced end up therefore in the same ballpark for the three categories of x-ray source.

As mentioned earlier, both ERL and FEL tend to have longer undulators than in conventional rings, perhaps by an order of magnitude, and they therefore provide proportionally higher brilliance. But conventional rings usually can have far more beamlines, which more or less compensates, because numerous beamlines can be serviced simultaneously.

There *is* a class of experiments for which the FEL has no rival. These are experiments in which the radiation dose needed to measure some microscopic feature is so high as to do serious radiation damage to the sample. For experiments like this it is appropriate to maximize the photons per bunch in order to illuminate the sample with an intensity sufficient to complete the measurement before the sample is obliterated. One class of experiments of this sort are the so-called "pump–probe" experiments. In these experiments the sample is first sent into an excited state by an ultrashort, intense pulse of visible laser light. An accurately synchronized, ultrashort, x-ray pulse follows shortly (of the order of tens of femtoseconds) later, to measure some dynamic feature of the sample. Experiments like this are well matched to the low repetition rates ν_{FEL} that are typical of FELs. It is not possible to accelerate arbitrarily large bunches of charge but, if it were, the allowable current $I_{\text{ave.}}$ could be distributed into giant bunches for which

$$n_1 = \frac{I_{\text{ave.}}}{\nu_{\text{FEL}}\, e} \frac{\lambda_\gamma/2}{L_B}$$
$$\left(\stackrel{\text{e.g.}}{=} \frac{3 \times 10^{-4}\,\text{A}}{10^2 \times 1.6 \times 10^{-19}\,\text{C}} \frac{0.5 \times 10^{-10}\,\text{m}}{3 \times 10^{-5}\,\text{m}} \approx 0.3 \times 10^8\right). \tag{14.17}$$

No such huge enhancement factor can actually be achieved, but the FEL is the only serious choice for this class of experiment.

References

1 Krinsky, S. (1987), *Introduction to the Theory of Free Electron Lasers,* in *Physics of Accelerators,* M. Month, (Ed.), American Institute of Physics, vol. **153**, p. 1026.

2 *TESLA Design Report, Part VI Appendices,* (2001), DESY 2001-011.

Index

$<\xi>_\gamma, <\xi^2>_\gamma$, energy moments 102
C_q 328
C_γ 64, 262
K, wiggler 184
K, eff., undulator 387
N_0, number of photons 276
$N_{w,\text{eff.}}$, undulator 387
Q, cavity 144
Q, unloaded 164
Q'_x, Q'_y, chromaticity 91
Q_L, loaded 167
Q_x, Q_y
– Trbojevic-Courant 352
– weak focusing 89
R/Q, cavity 145
R_i, inverse radius 448
$S(\xi)$, normalized energy distribution 99
U_0, energy loss per turn 64, 97
Z_s, resonant circuit impedance 166
α definition 16
α_j, damping decrement 107
α_y, damping rate 110
β, see beta function
β_x, β_y, x-ray line 253
$\delta(\rho, \lambda)$, index of refraction, x-ray 273
δ_{Be}, index, beryllium 273
ϵ, emittance 79
$\epsilon(\gamma)$, evolving emittance 360
ϵ_x, horizontal emittance 134, 328
ϵ_x, ϵ_y, x-ray line 253
ϵ_x^B, beam emittance 127
$\epsilon_y^{(G)}$, geometric emittance 304
$\epsilon_y^{(I)}$, invariant emittance 304
$\epsilon_y^{(N)}$, normalized emittance 305
γ_t, transition gamma 117
Ω_s, synchrotron frequency 112
Θ, maximum angle in undulator 386
κ, retardation factor 419, 437
$\lambda_{1,\text{edge}}$, undulator wavelength 182
$\lambda_{w,\text{eff.}}$, undulator 387

$\mathcal{B}_1$, undulator brilliance 262
$\mathcal{B}_1$, undulator fundamental brilliance 334
$\mathcal{D}$, curly-D 116, 118
$\mathcal{H}$, curly-H 123, 134, 328
$\mathcal{N}_0$, number of photons 78
$\mathcal{N}_{2N_w}$, number of photons 293
$dN/d\nu_I$, undulator energy distribution 257
ω_0, revolution frequency 57
ω_c, critical frequency 57
σ-mode, π-mode, synchrotron radiation 61
σ_x, σ_y, x-ray line 253
σ_δ, r.m.s. energy 130
σ_θ, r.m.s. cone angle 253
$\sin\theta_B$, Bragg condition 254
$\widetilde{\mathbf{E}}(\omega)$, undulator 209
$\widetilde{\mathbf{E}}(\omega)$, undulator field 190
$\widetilde{\mathcal{B}}_1$, undulator brilliance 260
$\widetilde{\mathcal{F}}_1$, undulator flux 260
$n(\lambda)$, index of refraction 273
n_u, photon number distribution 69, 99, 103
r_e 64
s, t, u, relativistic invariants 371
u_c, critical energy 98, 235
x, y
– Compton parameters 380
$\mathbf{F}^{(\text{body})}$, string body force 435
$\mathbf{F}^{(\text{ends})}$, string end force 437
$\check{\mathbf{X}}$, element reversed matrix 351

a
ABCD
– coefficients 41
– law 40, 42
achromat
– DFA 328
– MBA 341
– TBA 330
adiabatic
– condition 30

468 | *Index*

– coupling change   138
– damping   145, 360
– damping in ERL   300
– invariance   1
– invariant   150
alternate gradient focusing   93
Alvarez linac   143
amplitude
– cylindrical wave   46
– position-dependent   29
– single period   210
– total undulator, Fourier   209
– traveling wave   28
– undulator electric field   190
– undulator field, figure   242
analogy
– diffraction grating   212
– false, with antenna array   229
– headlight   54, 283
– photon and electron optics   12
– synchrotron and simple harmonic oscillation   120
– thermodynamics   132
– trajectory, ray   33
angular distribution   62, 65
APDF, accelerator desciption   451
approximation
– adiabatic   30, 148
– by special functions   220
– Fraunhofer   197
– isomagnetic   96
– linearized   148
– radial field   60
– short wavelength   30
– Taylor   220
– thin lens   8
– Weizsäcker-Williams   390

b

back-scattering   374
beam
– -based Twiss parameters   15
– as wave   27
– camera   69
– covariance   17
– diffraction limited   46
– distribution
– – Gaussian   17
– elliptical in phase space   13
– envelope   15
– perspective   3
– power   71
– variance   17
– waist   23
– width   121
beam camera   252, 279

– Compton scattering   378
– depth of focus   285
– diffraction limit   283
– laser wire   378
– layout   282
– pin-hole   279
– source position dependence   282
– visible light   280
– x-ray   279
– x-ray cone dependence   283
beam line
– aperture-defined   77, 257
– aperture-free   286
– design   257
– perspective   3
beta function
– equation   19
– FODO   92
– in linac   314
Bragg
– scattering monochromator   251
– scattering parameters   254
brightness   71
brilliance   71, 83
– from bending magnet   79
– refinement of calculation   365
– strategy to maximize   334
Brillouin diagram   158
bunch
– length   131
– length, femtosecond   310
BW, *see* bandwidth

c

camera, *see* beam camera
capillary mirror   267
cavity
– Q 145
– R/Q 145
– decay time   145
– excitation   166
– filling time   145
central, *see* reference
central limit theorem   127
centrifugal space charge force, CSCF   321, 413
Chasman-Green lattice   328
chromaticity   91
circularized, *see* normalized
cleanliness   71
closed
– FEL   XIII
coherence   27
– longitudinal   47
– multiple deflections   196
– synchrotron radiation   70

Index | 469

– transverse 47
coherent sychrotron radiation, CSR 321
coherent synchrotron radation, CSR
– wall shielding 444
coherent synchrotron radiation, CSR 413
– table 445
combined function 93
comparison
– FEL, ERL, and conventional light source 462
complex
– radius of curvature 40
– variable, conformal plane 239
Compton scattering
– invariant cross section 380
concatenate 5
condition
– adiabatic 30
– Bragg 172
– constructive interference 208
– field periodicity 208
– geometric optics 30
– ideal undulator 198
– interference maximum 197
– long magnet 62, 180
– T-C dispersion 352
– T-C minimum emittance 352
conformal mapping 239
conservation of wave energy 35
continuity equation 35
cosine-like, see sine-like
coupling 12
Courant–Snyder invariant 15, 121, 300
covariance 17
critical frequency, ω_c 57
critical Fresnel number, N_0 277
CSCF, see centrifugal space charge force
CSR, see coherent synchrotron radiation
curly-D 116
curly-H 123
cut-off frequency 158
cylindrical wave 40

d
damping
– adiabatic 145
– decrement 107
– longitudinal 111
– partition number 119
– rate 119
– – sum rule 105
– synchrotron 111
– vertical 110
Darwin
– energy spread 254
– phase 255
– width 254
– – number of, N_D 291
DC electron gun 301
deBroglie relation 44
defocusing
– transverse in accelerating gap 151
demagnetization curves 246
design, see reference
– electromagnet 239
– high brilliance, FCSR
– – actual 338
– – hypothetical 335
– minimum emittance lattice 350
– of lattice triplet 318
– permanent magnet 244
DFA, see double focusing achromat
diffraction
– -limited x-ray beam 365
– grating 212
– Laue 274
– limited beam 46
dimensionless coordinates 302
directional derivative 35
dispersion
– $D(s)$ 311
– function
– – ALS 331
– – CLS 329
– – FCSR 340
– relation 161
– suppression 312
– weak focusing 89
distribution
– angular 62
– Gaussian 127
– horizontal 127
– longitudinal 128
– photon energy 203
– UAL-calculated 323
Doppler shift 199
dynamic aperture, FCSR 357

e
effective
– potential energy 148
eigenvalues 91, 106
eikonal 29
– equation 31
Einstein
– radiation coefficients 457
– summation convention 31
electric dipole
– intensity pattern 393
– radiation 392
electric field
– Fourier transform 65

– longitudinal, in linac   143
– radiation   58
– undulator radiation   208
electromagnet design   239
electron
– frame, orthogonal incidence   376
– orbit   184
electron gun   301
ellipse
– beam   13
– equation   15
– mirror   266
elliptic integrals   442
emittance   14, 15, 17, 79, 83
– equilibrium, SYNCH and FCSR   356
– evolution
– – during acceleration   356
– – in FCSR   362
– – in gun   301
– – in SYNCH   361
– factor, FODO   332
– geometric   304
– growth
– – CSR and CSCF   324
– – injection merge   324
– – return loop   324
– growth, ERL   322
– horizontal increment   127
– invariant   304
– longitudinal increment   129
– normalized   305
energy
– component   2
– differential   107
– distribution, undulator   257
– potential   44
– radiated   97
– recovery linac   299
– spread   128
– spread, Darwin   254
– total   44
entropy, see pseudo-entropy
equation
– continuity   35
– dispersion   357
– for dispersion   135, 311
– force   88
– Hamilton   149
– Hamilton-Jacobi   44
– Helmholz   157
– Huygens-Kirchoff   47
– hyperbolic mirror   268
– Jefimenko   426
– orbit in linac   315, 319
– Schrödinger   44
– string end point   449

– thin lens   43
– wave   28, 37
equilibrium
– damping/quantum fluctuation   120
– horizontal   121
ERL, see energy recovery linac
evolution
– ABCD parameters   42
– beam ellipse   18
– beta function   15, 18
– – drift   23
– – focusing element   24
– – in linac   314
– – pictorial   20
– complex radius of curvature   40
– intensity   36
– matrix elements in linac   6
– orbit in linac   315, 319
– paraxial wave   38
– pseudo-harmonic in linac   317
– transverse phase space   303
– Twiss functions in linac   316
– wave field   39
existing ring utilization   335

f
FCSR, see fast cycling storage ring
FEL, see fee electron laserXIII, 27
– close   458
– gain function   459, 460
FEL, free electron laser   43
field
– electric   58
– index   87
– integral   7
– magnetic   58
– point   53
fish diagram   148
fit
– empirical, to radiation distribution   100
– Gaussian   244
– polynomial   246
fluctuation, quantum   104
flux   71
– density   71
– from bending magnet   78
– magnetic   244
– through sample   252
focal length   8
– triplet quads   318
focus
– relation to beam waist   40
focusing
– alternate gradient   93
– combined function   93
– equation   4

– medium   37, 40
– strength   4
– strong   92
– weak   85, 88
FODO cell   10
formula
– Heaviside, Poincaré, Schott   421
four-
– momentum   370
– scalar product   370
Fourier transform
– bending magnet radiation   65
– undulator radiation field   209
Fraunhofer approximation   197
free electron laser, see FEL
– closed   XIII
– open   XIII
frequency
– band gap   171
– cut-off   158
– synchrotron, Ω_s 112
Fresnel
– critical number, N_0 277
– lens   249
– number   277, 291
function
– Anger   221
– grating   200
– MacDonald   99
– Weber   221

g

Gaussian
– beam
– – distribution   3, 17
– – in focusing medium   37
– – wave description   36
– – fit to pole field   244
– hyper-   224
– optics, see linearized
– single pole   188
generalized
– Hamiltonian   149
– potential function   148
generation, light source   49, 327
geometric optics   30
GNU scientific library   449
Gouy phase factor   39
grating function   200
group velocity   161
GSL, GNU Scientific Library   452
gun, electron, photoelectric   303

h

Hamilton-Jacobi equation   44
Hamiltonian   106, 149

– coordinate   5
harmonic
– odd, dominant   197
– space-, linac   174
– undulator angular pattern   213
– undulator intensity dependence on   213
Heaviside, Poincaré, Schott formula   421
Heisenberg uncertainty principle   45, 281
Helmholtz equation   157
high brightness scenario, FCSR   338
holography, x-ray   46
horizontal, see transverse
Huygens-Kirchoff equation   47

i

ideal, see reference
impedance
– complex, wavefront   42
– load   167
– resonant circuit, Z_s 166
– shunt   143
index of refraction   28
– x-ray   272
intensity   71
– evolution   36
– transport equation   34
interference maximum   197
isochronous   90
– three-cell module   312
isomagnetic assumption   96

k

kinematics   52, 92, 301, 370, 375, 389
kink   7
Kirchoff, see Huygens

l

laser
– field in terms of power   387
– wire
– – bunched beam rate   408
– – invariant treatment   406
– – kinematic sensitivity   380
– – schematic   378
– – treated as undulator   403
– – via electron frame   405
lattice
– -based   23
– – Twiss parameters   21
– Chasman-Green   323
– dog-bone ERL   307
– FODO   10
– isochronous   309
– isochronous triplet   312
– minimum emittance   350
– sprocket ERL   307

– Trbojevic-Courant   346
– wiggler-dominated   132
lattice functions, *see* Twiss parameters
Laue diffraction   274
law
– Newton's   148
– Snell   254, 275
lens
– -free x-ray imaging   46
– cylindrical   40
– Rayleigh criterion   278
– roughness criterion   278
– spherical   40
– stop, effective   291
– x-ray   270
light source generations   49, 327
linac
– Alvarez   143
– drift tube   143
– electron   153
– nonrelativistic   142
linearized
– optics   3
Liouville theorem   6, 267
load line, magnetic   245
longitudinal
– coherence   47
– coordinate   2
– coordinate evolution   2
– motion in storage ring   112
– particle motion, linac   145
– source length   282
Lorentz transformation
– along axis   389
– general   375
lowest mode, *see* mode, fundamental

m

MacDonald function   99
magnetic
– field, radiation   58
– field, undulator   233
– load line   245
Mandelstam triangle   373
Maple   64, 99, 101, 105, 204, 221, 315
matrix element, *see* transfer matrix
maximize brilliance   334
microbunching   462
minimum emittance lattice   346
mirror
– capillary   267
– elliptical   266
– hyperbolic   267
– Kirkpatrick–Baez pair   266
mode
– diffraction limited   365

– fundamental   40
– resonator   40
– synchrotron radiation   61
– waveguide   155
momentum
– angular, conservation   34
– compaction   90
– compaction factor   310
– component   2
monochromator
– post-, profile   223
– ring   225
– – numerical   227
– table   256

n

needle, *see* string
normalized
– emittance   305
– energy distribution   99
– phase space   22

o

open
– FEL   XIII
orbit
– electron in EM wave   382
– electron in undulator   184
– elliptical in phase space   13
– off-momentum, closed   113
overmoded   159

p

parameter
– A,B,C,D   3
– accelerator   252
– Bragg scattering   254
– radiation   98
– table
– – SYNCH and FCSR parameters   356
– Twiss   15
– undulator radiation ring   272
– undulator, K 184, 232
parametrization
– EM wave as undulator   386
– transfer matrix   21
– wave   41
paraxial, *see* linearized
– approximation   37
– ray   33
particle
– -wave duality   44
– acceleration by wave   154
permanent magnet design   244
perspective
– particle, beam, beam line   3

phase   29
– advance per cell   175
– betatron
– – averaging over   122
– contrast   36
– contrast imaging   46
– Darwin   255
– factor
– – Gouy   39
– from intensity   34
– observation vs. emission   207
– stability   145
– velocity   29, 157
phase space   5
– betatron
– – impulse   121
– density   6
– longitudinal   119, 148
– normalized   22
phasor
– diagram   200
– sum   210
– sum, $K \ll 1$ 199
photoelectric gun   303
photon
– beam, treated as wave   381
– energy distribution   203
– energy distribution in lab   396
– number density n_u 69, 78
– number spectrum $\mathcal{N}$ 78
– virtual   389
physical region, Compton   373
pill-box resonator   163
pin-hole camera   279
Planck relation   44
potential
– retarded   53
– scalar   53
– vector   53
power
– average   342
– instantaneous   343
– radiated   62, 104
Poynting vector   35, 62
propagation, see evolution
pseudo-
– capacity   138
– entropy   138
– harmonic   18
– harmonic orbits in linac   317
– thermodynamics   132

q
quadrupole
– lens   11
– skew, transfer matrix   12

– transfer matrix   12
quality factor, see Q
quantum fluctuation   104

r
radial field approximation   60
radiation
– bending magnet   51
– electric dipole   392
– one wiggler pole   187
– spontaneous   457
– stimulated   457
– undulator, general   222
– undulator, ideal   203
radius of curvature, complex   40
rate
– Compton, bunched beams   408
– expressed as cross section   394
– laser wire diagnostic   401
ray
– equation   32
– – linearized   33
– from wavefronts   31
Rayleigh lens quality criterion   278
reconciliation
– beam-based, lattice-based   23
– cone angles of visible and x-ray   281
– particle and wave   40
– spike through monochromator   224
reference
– particle   2, 85
– trajectory   2
reflectivity   266
– glancing   44
regularization
– photon energy distribution   99
– string force   438
relation
– deBroglie   44
– Planck   44
renomalization, see regularization
resonator   42
– lumped constant model   165
– pill-box   163
retardation factor, κ 419, 437
retarded
– potential   53
– time   53, 189, 205, 418
– time, calculation, figure   425
– time, circle geometry   431
– time, graph of end points   434
revolution frequency, ω_0 57
Robinson theorem   109
rule of thumb
– dispersion   89
– index of refraction   273

– momentum compaction   90
– number of terms retained   222
– pole tip field   345
– short wavelength approximation   30

s
Sands
– curly-D   116
– curly-H   123
SASE   458
scattering
– Bragg   254
– Compton   369
– Thompson   381
Schott formula   187
Schrödinger equation   44
self-force
– regularization   438
– ribbon   423
– straight string   418
separatrix   148
short wavelength limit   30, 45
shunt impedance   143
sine-like, cosine-like   5, 34, 42, 315
skin depth   160
Snell's law   254, 275
source point   53
space charge
– emittance growth   321
– emittance growth, figures   322
– forces, figures   453, 454
spectral, *see* spectrum
– analysis
– – arbitrary profile   188
– brightness   71, 83
– power, accurate formula   68
spectrum
– estimate from pulse durattion   65
– from single pole   192
– power density   65
– single pole, graph   194
specular reflectivity   266
stability
– condition   88
– horizontal   85
– horizontal and vertical   88
– longitudinal   148
– phase   145
– vertical   87
stay clear, beam   16
storage ring
– see light source   49
string
– end point determination   430, 449
– space charge formulation   413
strong focusing   92

superradiance   462
symplectic condition   6
SYNCH, *see* injector ring
synchronous, *see* reference
synchrotron
– radiation
– – coherent, incoherent   70, 413
– – incoherent and coherent   440
– – randomness   103
– – regularized treatment   99
– – wave or particle   45
– radiation, CSR table   442
– SYNCH   339

t
T-C, *see* Trbojevic-Courant
– dispersion suppression   355
– thick lens treatment   353
– thin lens treatment   348
table
– brilliance comparison   365
– Compton back-scattering   374
– FCSR parameters vs. energy   362
– mimimum emittance lattice parameters   356
– orthogonal incidence   377
– storage ring parameters   98
– SYNCH parameters vs. energy   361
– TESLA parameters   462
– x-ray FEL parameters   462
– x-ray intensity measures   75
tangential, *see* longitudinal
TBA, *see* triple beam achromat
TEAPOT, `StringSCSolver`   452
TESLA   462
theorem
– central limit   127
– Earnshaw   151
– Floquet   174
– isotropy→uniform in energy   398
– Liouville   6, 267
– Robinson   109
– work/energy   114
thermodynamics, *see* pseudo-thermodynamics
thick lens   9
– T-C analysis   353
thin lens   7
– equation   43
– isochronous triplet   312
– T-C analysis   348
– triplet   318
Thomson scattering, *see* Compton scattering
– classical derivation   391
Touschek effect   328, 362

TRAFIC4, comparison with UAL   453
trajectory, *see* orbit
transfer matrix   5, 34
– 4×4   12
– 6D, uncoupled   110
– determinant   106
– diagonal element   107
– differential   106
– drift   7
– eigenvalues   91
– first order differential   106
– FODO   91
– FODO cell   9
– minimum emittance lattice   350
– parameterization   21
– skew quad   12, 13
– spherical mirror   42
– T-C cell   350
– thick lens   9
– thin lens   8
– Twiss form   91
transformation
– **E** and **B**   389
– angle, lab to electron frame   379
– energy, lab to electron frame   379
– solid angle   395
transit time factor   142
transition gamma   117
transport of intensity   34
transverse
– coherence   47
– coordinate   2
– electric, TE   157
– magnetic, TM   157
Trbojevic-Courant
– lattice   346
– lens layout   348
triplet
– design   318
– in ERL straight sections   308
– isochronous   312
tune   22, 89
– Q_x 86
Twiss function   15
– ALS   331
– CLS   329
– ERL arc   312
– ERL matched as ring   313
– evolution   15, 18
– FCSR   340
– one $L_C = 9$ m cell   341
Twiss parameter, *see* Twiss function

u
UAL
– distributions   323

– Physics Users Guide   448
– Simulation Course   448
– space charge calculation   447
ultrabright light source
– time sequence   343
ultrabright light source, FCSR
– physical layout   342
undulator
– -specific   272
– angular narrowing?   227
– brilliance, $\mathcal{B}_1$ 262
– choice of K 292
– edge wavelength, λ_0 196
– electromagnet design   239
– energy distribution   399
– energy,angle distribution   259
– field near pole   239
– field plot   242
– flux estimate   293
– gap height $2g$ 233
– harmonic filter   274
– magnet
– – electro-permanent   237
– magnetic field shape   234
– microwave   388
– number distribution figure   401
– parameter, K 184
– permanent magnet design   244
– power spectrum figure   402
– radiation
– – analytic formulation   204
– radiation pattern   214, 215, 217, 228
– ring parameters   272
– spectrum
– – graph   202
– spike through monochromator   224
– treated as wave   389

v
variance   17
velocity
– average, in undulator   196
– group   161
– phase   157
vertical, *see* transverse

w
waist
– beam   23
– beam envelope near   366
– length   23, 366
– relation to focus   40
wall shielding   444
wave
– -particle duality   44
– cylindrical   40

– equation   37
– – scalar   28
– function   29
– – evolution   39
– – paraxial   46
– in waveguide   155
– near focus   39, 40
– number   28
– paraxial
– – evolution   38
– vector   3, 29
wavefront   29
– cylindrical   46
– spherical   38
waveguide mode   155
wavelength
– free space, λ_0 155
– guide, λ_g 159
weak focusing   85
wiggler, *see* undulator

– -dominated emittance   132
– -dominated lattice   132
– narrow gap   134
– wide gap   134
Wronskian   4

x

x-ray
– absorption coefficient   277
– compound lens   274
– critical Fresnel coefficient   277
– holography   46
– index of refraction   272
– interferometer   271
– lens, monochromatic   270
– lens, multichromatic   272
– lens, undulator-specific   272
– lens-free imaging   46
– phase contrast   46